Delivery Technologies for Biopharmaceuticals

Peptides, Proteins, Nucleic Acids and Vaccines

Editors

Lene Jorgensen

Faculty of Pharmaceutical Sciences
University of Copenhagen, Denmark

Hanne Mørck Nielsen

Faculty of Pharmaceutical Sciences
University of Copenhagen, Denmark

WILEY

A John Wiley & Sons, Ltd., Publication

This edition first published 2009

Registered office
John Wiley & Sons Ltd, The Atrium, Southern Gate, Chichester, West Sussex, PO19 8SQ, United Kingdom

For details of our global editorial offices, for customer services and for information about how to apply for permission to reuse the copyright material in this book please see our website at www.wiley.com.

Library of Congress Cataloging-in-Publication Data

Delivery technologies for biopharmaceuticals : peptides, proteins, nucleic acids, and vaccines / editors, Lene Jorgensen, Hanne Mørck Nielsen.
p. ; cm.
Includes bibliographical references and index.
ISBN 978-0-470-72338-8 (cloth : alk. paper) 1. Drug delivery systems. 2. Drug carriers (Pharmacy) I. Jorgensen, Lene. II. Nielsen, Hanne Mørck.
[DNLM: 1. Drug Delivery Systems–methods. 2. Biological Products. QV 785 D355 2009]
RS199.5.D45 2009
615′.6–dc22

2009025904

A catalogue record for this book is available from the British Library.

ISBN 978-0-470-72338-8 (H/B)

Set in 10 on 12 pt Times by Toppan Best-set Premedia Limited
Printed and bound in Great Britain by CPI Antony Rowe, Chippenham, Wiltshire

Contents

Preface *page* xv
List of Contributors xvii

INTRODUCTION 1

1. **Challenges in Delivery of Biopharmaceuticals; the Need for Advanced Delivery Systems** 3
Hanne Mørck Nielsen and Lene Jorgensen

1.1 Introduction 3
1.2 Overcoming Delivery Barriers 4
1.2.1 Stabilization 4
1.2.2 Enhancing Delivery 5
1.3 Drug Delivery Technologies and Excipients 5
1.4 Risks 6
1.5 Conclusion 7
References 7

DELIVERY OF BIOPHARMACEUTICALS 9

2. **Novel Formulation Approaches for Peptide and Protein Injectables** 11
Mingshi Yang and Sven Frokjaer

2.1 Introduction 11
2.2 Formulation Challenges 12
2.3 Chemical Modification of Proteins and Peptides 13
2.3.1 PEGylation 13
2.3.2 Glycoengineering 15
2.3.3 Acylation 16
2.3.4 Amino Acid Substitution 17
2.3.5 Protein Fusion 17
2.4 Depot Delivery Systems 18
2.4.1 Micro- and Nanoparticulate Systems 18
2.4.2 *In Situ* Depot-Forming Systems 21
2.4.3 Implant Systems 21

DELIVERY TECHNOLOGIES FOR BIOPHARMACEUTICALS 129

7. Lipid Nanoparticle-Based Systems for Delivery of Biomacromolecule Therapeutics 131
Susana Martins, Domingos C. Ferreira and Eliana B. Souto

7.1 Introduction 131
7.2 Definitions and Properties of Solid Lipid Nanoparticles (SLNs) 133
7.3 Definitions and Properties of Lipid-Drug Conjugates (LDCs) 135
7.4 Administration, Delivery and Targeting 136
7.5 Toxicity and Safety 138
7.6 Applications for Biopharmaceuticals 139
7.6.1 Peptides and Proteins 139
7.6.2 Nucleic Acids 139
7.6.3 Vaccines 142
7.7 Conclusions 142
References 142

8. Dendrimers in Delivery of Biopharmaceuticals 149
Chandan Thomas and Fakhrul Ahsan

8.1 Introduction 149
8.1.1 Properties of Dendrimers as a Carrier System 150
8.2 Case I – Application of Dendrimers in Delivery of Large Molecular Weight Drugs 153
8.3 Case II – Application of Dendrimers in Gene Delivery 155
8.3.1 Case IIA – Application of PAMAM-PLGA Microparticle Conjugates in Gene Delivery 156
8.3.2 Case IIB – Application of Dendrimers in siRNA Delivery 158
8.4 Case III – Application of Dendrimers in Vaccine Delivery 162
8.5 Concluding Remarks 165
References 166

PEPTIDES AND PROTEINS 169

9. Modification of Peptides and Proteins 171
Susanne Hostrup, Kasper Huus and Henrik Parshad

9.1 Introduction 171
9.2 PEGylated Peptides and Proteins 171
9.2.1 Methods 172
9.2.2 Conceptual Considerations on PEGylation 173
9.2.3 Pharmacokinetic Aspects 174
9.2.4 Pharmaceutical Aspects 175
9.2.5 PEGylated Proteins in Advanced Delivery Systems 177
9.3 Lipidization of Peptides and Proteins 177
9.3.1 Methods of Lipidization 177
9.3.2 Conceptual Considerations on Lipidization 178

9.3.3 Improved Pharmacokinetics 178
9.3.4 Improved Delivery via Non-Parenteral Delivery Routes 179
9.3.5 Pharmaceutical Aspects 180
9.3.6 Lipidization in Advanced Delivery Systems 181
9.4 Modification of the Primary Structure of Peptides and Proteins 181
9.4.1 Strategies for Designing Protein Analogues 182
9.4.2 Advantages and Disadvantages 183
9.4.3 Pharmaceutical Aspects – General Principles for Improving Protein Stability 183
9.4.4 Pharmacokinetic Aspects 185
9.4.5 Manipulating the Isoelectrical Point (pI) by Molecular Engineering 185
9.5 General Considerations on Processing and Characterization 186
9.6 Conclusions 186
References 187

10. Nanocarriers for the Delivery of Peptides and Proteins **193**
Kenneth Lundstrom

10.1 Introduction 193
10.2 Polymeric Nanoparticles 194
10.2.1 Lactic/Glycotide Polymers 194
10.2.2 Polycaprolactones 195
10.2.3 Polyphosphoesters 195
10.2.4 Polyanhydrides 196
10.2.5 Polyorthoesters 196
10.2.6 Block Copolymers with a Specific Emphasis on Medusa 197
10.2.7 Cross-Linked Dextran 198
10.3 *In Situ* Depot Forming Systems 199
10.3.1 Precipitation Systems 199
10.3.2 Thermal Gelling Systems 199
10.3.3 Cross-Linked Systems 200
10.3.4 Thermoplastic Semisolids 201
10.4 Conclusions 202
References 202

11. Polymer-Based Delivery Systems for Oral Delivery of Peptides and Proteins **207**
Bruno Sarmento, Domingos Ferreira and Teófilo Vasconcelos

11.1 Introduction 207
11.2 Advances in Oral Protein Delivery 208
11.3 Intestinal Considerations for Protein Absorption 209
11.4 Polymer-Based Delivery Systems for Oral Delivery of Proteins 211
11.4.1 Hydrogels 211
11.4.2 Patches and Tablets 212
11.4.3 Microparticles 213
11.4.4 Nanoparticles 215

11.5 Conclusions 219
References 220

12. Advanced Pulmonary Delivery of Peptides or Proteins Using Polymeric Particles **227**
Yu Seok Youn, Kang Choon Lee, You Han Bae, Kun Na and Eun Seong Lee

12.1 Introduction 227
12.2 Practical Issues in the Pulmonary Delivery of Peptides and Proteins 228
12.2.1 Physiological Features of the Lungs 228
12.2.2 Barriers to the Pulmonary Delivery of Peptides and Proteins 228
12.2.3 Factors Affecting the Pulmonary Delivery of Peptide and Protein Particles 230
12.3 Polymeric Microparticles for Delivering Peptides and Proteins 231
12.4 Porous Microparticles 232
12.5 Polymeric Nanoparticles 237
12.6 Sustained or Controlled Release Issues in Polymeric Particle Formulations 239
12.7 Stability Issues of Peptides and Proteins in Particles 240
12.8 Toxicity Issues of Inhaled Particles 240
12.9 Conclusions 240
References 241

NUCLEIC ACIDS **245**

13. Polymer Microparticles for Nucleic Acid Delivery **247**
Tim Pearce, Jared Hierman and Chun Wang

13.1 Introduction 247
13.2 Microparticles Based on Poly(Lactic-*co*-Glycolic Acids) (PLGA) 248
13.2.1 Overview 248
13.2.2 Design Parameters 248
13.2.3 General Strategies and Fabrication of Microparticles 251
13.2.4 Optimization of Microparticle Formulations 253
13.2.5 Combining Polycations with PLGA Microparticles 254
13.2.6 Current Applications 256
13.3 Other Types of Polymer Microparticles 257
13.3.1 Alginate 257
13.3.2 Chitosan 258
13.3.3 Hyaluronan (HA) 259
13.3.4 Macro-Albumin Aggregates (MAA) 260
13.3.5 Poly(D,L-Lactide-*co*-4-Hydroxy-L-Proline) (PLHP) 260
13.3.6 The Pluronics 261
13.3.7 Poly(Ortho Esters) (POEs) 262
13.3.8 Polyacrylamide 263
13.4 Conclusions 263
References 264

14. Pulmonary Delivery of Small Interfering RNA for Novel Therapeutics **269**
Qing Ge, David Evans, John J. Xu, Harry H. Yang and Patrick Y. Lu

14.1 Introduction 269
14.2 RNA Interference: Mechanism of Action 270
14.3 Lung Physiology 270
14.4 Animal Models 272
14.5 siRNA Delivery to Treat Pulmonary Diseases 273
14.5.1 Challenges 273
14.5.2 Viral Versus Non-Viral Vectors 274
14.5.3 Airway siRNA Delivery 275
14.5.4 Intravenous siRNA Delivery 278
14.5.5 Other Delivery Methods 279
14.6 siRNA-Induced Inflammatory Response 279
14.7 Case Studies 280
14.7.1 siRNA Therapeutics Against RSV and PIV Infection in Mouse 280
14.7.2 siRNA Therapeutics for Treatment of SCV Infection in Monkey 281
14.8 Faster Drug for Unknown Bugs 284
14.9 Conclusions 284
References 285

15. Lipid-Based Formulations for siRNA Delivery **291**
Camilla Foged, Pieter Vader and Raymond M. Schiffelers

15.1 Introduction 291
15.2 Cationic Lipid-Based Delivery Systems for Nucleic Acid-Based Drugs 292
15.2.1 Helper Lipids 293
15.2.2 Lipopolymers 294
15.2.3 Active Targeting 294
15.3 Neutral and Anionic Lipid-Based Drug Delivery Systems 296
15.4 Mechanisms of Internalization 296
15.5 Immune Activation 297
15.6 Conclusions 299
References 300

16. Cellular Bioavailability of Peptide Nucleic Acids (PNAs) Conjugated to Cell Penetrating Peptides **305**
Takehiko Shiraishi and Peter E. Nielsen

16.1 Introduction 305
16.2 Peptide Nucleic Acids (PNAs) 306
16.3 Cell Penetrating Peptides (CPPs) 306
16.4 Cellular Uptake Versus Bioavailability 307

16.5 PNA Conjugates 307
16.6 Efficacy Versus Efficiency 307
16.7 Auxiliary Agents 330
16.8 *In Vivo* Activity of PNA Conjugates 331
16.9 Conclusions 331
References 336

17. DEAE-dextran-MMA Graft Copolymer for Non-Viral Delivery of DNA 339
Yasuhiko Onishi, Yuki Eshita and Masaaki Mizuno

17.1 Introduction 339
17.2 Theoretical 340
17.3 Materials and Methods 341
17.3.1 Preparation of DDMC 341
17.3.2 Characterization of DDMC 341
17.3.3 Reaction Between DDMC and DNA 341
17.3.4 The Structure of the Complex Between DNA and DDMC 343
17.3.5 Transfection 345
17.3.6 Protection from DNase Degradation 350
17.3.7 DNA Delivery Pathways by DDMC 352
17.4 Conclusions 353
References 354

VACCINES 357

18. Liposomes in Adjuvant Systems for Parenteral Delivery of Vaccines 359
Dennis Christensen, Karen Korsholm, Grith Krøyer Wood, Afzal Mohammed, Vincent Bramwell, Peter Andersen, Else Marie Agger and Yvonne Perrie

18.1 The Structural Basis of Liposomes 359
18.2 Key Characteristics Which Make Liposomes Advantagous as Vaccine Delivery Systems 360
18.3 Formulation Approaches to Improve Adjuvanticity and Enhance Immunological Targeting 363
18.3.1 Immunomodulators 364
18.3.2 Physicochemical Properties 365
18.4 Instability Considerations of Liposomes 366
18.5 Techniques to Formulate Long-Term Stable Liposome Formulations 368
18.5.1 Protection During Freeze-Drying 369
18.6 Conclusion 371
References 372

19. Influenza Virosomes as Delivery Systems for Antigens 377
Mario Amacker, Stefan Moese, Andreas R. Kammer, Ari Helenius and Rinaldo Zurbriggen

19.1 Introduction 377
19.1.1 Immune Stimulating Reconstituted Influenza Virosomes (IRIVs) as Adjuvant and Delivery System 377
19.1.2 Mechanisms of Action 379
19.2 A Case Study – Antigen-Virosome Drug Delivery System 382
19.2.1 Particle Characterization 382
19.2.2 Virosome Uptake into HeLa Cells 382
19.2.3 Humoral Immune Response to Antigen Virosomes 385
19.2.4 Cellular Immune Response to Antigen Virosomes 386
19.2.5 Efficiency of the Antigen Virosome Drug Delivery System 388
19.3 Methods 389
19.3.1 Virosome Formulation 389
19.3.2 Size Determination 389
19.3.3 Freeze–Fracture Electron Microscopy 389
19.3.4 Transmission Electron Microscopy 390
19.3.5 TIRF Microscopy 390
19.3.6 Mice and Immunizations 390
19.3.7 Intracellular IFNγ Staining 390
19.3.8 ELISA 391
References 391

20. GRAZAX®: An Oromucosal Vaccine for Treating Grass Pollen Allergy with Immunotherapy 395
Annette Römmelmayer Lundegaard, Lise Lund and Jørgen Nedergaard Larsen

20.1 Introduction 395
20.2 Specific Immunotherapy 396
20.3 Description of Grass Pollen Extract 397
20.4 Technologies of Fast Dissolving Tablets 398
20.5 Formulation of GRAZAX® 399
20.6 Production of GRAZAX® 400
20.7 Specifications and Analytical Procedures 401
20.7.1 Physical and Chemical Tests 401
20.7.2 Immunochemical Tests 402
20.7.3 Microbial Tests 403
20.8 Feasibility Study: Additional Analyses 403
20.9 Conclusions 403
References 404

FINAL COMMENTS 405

21. Delivery Technologies for Biopharmaceuticals: A Critical Assessment 407
Marco van de Weert

21.1 Introduction 407

List of Contributors

Else M. Agger, Statens Serum Institute, Department of Infectious Disease Immunology, Copenhagen, Denmark

Fakhrul Ahsan, School of Pharmacy, Department of Pharmaceutical Sciences, Texas Tech University Health Sciences Center, Amarillo, TX, USA

Mario Amacker, Pevion Biotech, Ittigen/Bern, Switzerland

Peter Andersen, Statens Serum Institute, Department of Infectious Disease Immunology, Copenhagen, Denmark

You Han Bae, University of Utah, Department of Pharmaceutics and Pharmaceutical Chemistry, Salt Lake City, UT, USA

Shervin Bahrami, Aarhus University, Department of Molecular Biology, Aarhus, Denmark

Vincent Bramwell, Aston University, School of Life and Health Sciences, Birmingham, UK

Dennis Christensen, Statens Serum Institute, Department of Infectious Disease Immunology, Copenhagen, Denmark

Dongmei Cun, University of Copenhagen, Department of Pharmaceutics and Analytical Chemistry, Faculty of Pharmaceutical Sciences, Copenhagen, Denmark

Yuki Eshita, Oita University, Department of Infectious Diseases, Faculty of Medicine, Oita, Japan

David Evans, Sirnaomics, Inc., Gaithersburg, MD, USA

Domingos C. Ferreira, University of Porto, Department of Pharmaceutical Technology, Faculty of Pharmacy, Porto, Portugal

Camilla Foged, University of Copenhagen, Department of Pharmaceutics and Analytical Chemistry, Faculty of Pharmaceutical Sciences, Copenhagen, Denmark

Sven Frokjaer, University of Copenhagen, Department of Pharmaceutics and Analytical Chemistry, Faculty of Pharmaceutical Sciences, Copenhagen, Denmark

Qing Ge, Sirnaomics, Inc., Gaithersburg, MD, USA

Ari Helenius, Institute of Biochemistry, ETH Hoenggerberg, Zurich, Switzerland

Jared Hierman, University of Minnesota, Department of Biomedical Engineering, Minneapolis, MN, USA

Susanne Hostrup, Novo Nordisk A/S, Bagsværd, Denmark

Leaf Huang, University of North Carolina, Eshelman School of Pharmacy, Chapel Hill, NC, USA

Kasper Huus, Novo Nordisk A/S, Måløv, Denmark

Lene Jorgensen, University of Copenhagen, Department of Pharmaceutics and Analytical Chemistry, Faculty of Pharmaceutical Sciences, Copenhagen, Denmark

Andreas R. Kammer, Pevion Biotech Ltd., Ittigen/Bern, Switzerland

Yoshiaki Kawashima, Aichi Gakuin University, School of Pharmaceutical Science, Nagoya, Japan

Karen Korsholm, Statens Serum Institute, Department of Infectious Disease Immunology, Copenhagen, Denmark

Jørgen Nedergaard Larsen, ALK-Abelló A/S, Hørsholm, Denmark

Kang Choon Lee, SungKyunKwan University, College of Pharmacy, Jangan-ku, Suwon City, Republic of Korea

Eun Seong Lee, The Catholic University of Korea, Division of Biotechnology, Bucheon-si, Republic of Korea

Patrick Y. Lu, Sirnaomics, Inc., Gaithersburg, MD, USA

Lise Lund, ALK-Abelló A/S, Hørsholm, Denmark

Annette Römmelmayer Lundegaard, ALK-Abelló A/S, Hørsholm, Denmark

Kenneth Lundstrom, Pan Therapeutics, Lutry, Switzerland

Shirui Mao, Shenyang Pharmaceutical University, School of Pharmacy, Shenyang, China

Susana Martins, University of Porto, Department of Pharmaceutical Technology, Faculty of Pharmacy, Porto, Portugal

Masaaki Mizuno, Nagoya University, Graduate School of Medicine Program in Cell Information Medicine, Nagoya, Japan

Stefan Moese, 3-V Biosciences, Schlieren, Switzerland

S. Moein Moghimi, University of Copenhagen, Department of Pharmaceutics and Analytical Chemistry, Faculty of Pharmaceutical Sciences, Copenhagen, Denmark

Afzal Mohammed, Aston University, School of Life and Health Sciences, Birmingham, UK

Kun Na, The Catholic University of Korea, Division of Biotechnology, Bucheon-si, Republic of Korea

Hanne Mørck Nielsen, University of Copenhagen, Department of Pharmaceutics and Analytical Chemistry, Faculty of Pharmaceutical Sciences, Copenhagen, Denmark

Peter E. Nielsen, University of Copenhagen, Department of Cellular and Molecular Medicine, The Panum Institute, Faculty of Health Sciences, Copenhagen, Denmark

Yasuhiko Onishi, Ryuju Science Co. Ltd., Seto, Japan

Henrik Parshad, Novo Nordisk A/S, Bagsværd, Denmark

Tim Pearce, University of Minnesota, Department of Biomedical Engineering, Minneapolis, MN, USA

Finn Skou Pedersen, Aarhus University, Department of Molecular Biology, Aarhus, Denmark

Yvonne Perrie, Aston University, School of Life and Health Sciences, Birmingham, UK

Bruno Sarmento, University of Porto, Department of Pharmaceutical Technology, Faculty of Pharmacy, Porto, Portugal

Raymond M. Schiffelers, Utrecht University, Department of Pharmaceutical Sciences, Faculty of Science, Utrecht, The Netherlands

Takehiko Shiraishi, Utrecht University, Department of Pharmaceutical Sciences, Faculty of Science, Utrecht, The Netherlands, and University of Copenhagen, Department of Cellular and Molecular Medicine, The Panum Institute, Faculty of Health Sciences, Copenhagen, Denmark

Eliana B. Souto, Fernando Pessoa University, Department of Pharmaceutical Technology, Faculty of Health Sciences, Porto, Portugal, and Institute of Biotechnology and Bioengineering, Centre of Genetics and Biotechnology, University of Trás-os-Montes and Alto Douro, Vila Real, Portugal

Chandan Thomas, Texas Tech University Health Sciences Center, Department of Pharmaceutical Sciences, School of Pharmacy, Amarillo, TX, USA

Pieter Vader, Utrecht University, Department of Pharmaceutical Sciences, Faculty of Science, Utrecht, The Netherlands

Teófilo Vasconcelos, University of Porto, Department of Pharmaceutical Technology, Faculty of Pharmacy, Porto, Portugal

Elizabeth A. Vasievich, University of North Carolina, Eshelman School of Pharmacy, Chapel Hill, NC, USA

Chun Wang, University of Minnesota, Department of Biomedical Engineering, Minneapolis, MN, USA

Marco van de Weert, University of Copenhagen, Department of Pharmaceutics and Analytical Chemistry, Faculty of Pharmaceutical Sciences, Copenhagen, Denmark

Grith Krøyer Wood, Statens Serum Institute, Department of Infectious Disease Immunology, Copenhagen, Denmark

John J. Xu, Sirnaomics, Inc., Gaithersburg, MD, USA

Harry H. Yang, Sirnaomics, Inc., Gaithersburg, MD, USA

Mingshi Yang, University of Copenhagen, Department of Pharmaceutics and Analytical Chemistry, Faculty of Pharmaceutical Sciences, Copenhagen, Denmark

Yu Seok Youn, Pusan National University, College of Pharmacy, Geumjeong-gu, Busan, Republic of Korea

Rinaldo Zurbriggen, Pevion Biotech Ltd., Ittigen/Bern, Switzerland

Introduction

1

Challenges in Delivery of Biopharmaceuticals; the Need for Advanced Delivery Systems

Hanne Mørck Nielsen and Lene Jorgensen
University of Copenhagen, Department of Pharmaceutics and Analytical Chemistry, Faculty of Pharmaceutical Sciences, Copenhagen, Denmark

1.1 Introduction

Due to the advances in biotechnology, traditional biopharmaceuticals of biological origin are being replaced and supplemented by therapeutics developed by recombinant technologies and chemical synthesis. In addition, the rapid advances in molecular biology and genetics contribute to an in depth understanding of mechanisms involved in disease development, and generate possibilities for novel indications for biopharmaceuticals. Recent reviews list several hundred biopharmaceuticals either in development or undergoing clinical trials for treatment of a number of serious diseases [1, 2]. However, formulating biopharmaceuticals with the optimal therapeutic efficacy and the possibility of large scale production of the required formulation as well as optimal storage stability is highly challenging. Firstly, the unique structural characteristics of both amino acid-based as well as nucleic acid-based biopharmaceuticals make the formulation development more challenging than for more conventional drugs. The backbone and folding structure of the drug must be retained during processing and storage for the drug to exert its effect on for example a receptor. Secondly, upon administration to the patient the formulation must

Delivery Technologies for Biopharmaceuticals: Peptides, Proteins, Nucleic Acids and Vaccines Edited by Lene Jorgensen and Hanne Mørck Nielsen

protect the drug under degradative conditions, such as the presence of enzymes, in order to ensure efficient delivery. Optimally, the formulation should ensure targeted delivery to the site of action and controlled release of the drug from the formulation at this site, thus enabling easy administration and high efficacy of the formulated drug. Thirdly, the structure–activity complexity of the active biopharmaceutical often requires that for each drug, an individualized delivery system is designed, meaning that what may work for one biopharmaceutical may not be feasible or optimal for another biopharmaceutical drug.

1.2 Overcoming Delivery Barriers

1.2.1 Stabilization

Due to the inherent instability towards hydrolysis and enzymatic degradation and fast clearance, the half-life of biopharmaceuticals is most often relatively short after administration. Ways to overcome this are to develop, for example, stabilized analogues or prodrugs by conjugation of different chemical groups or by physical stabilization of the biopharmaceutical drug by use of excipients.

An example of a way to increase the circulation lifetime is the acylation of peptides, for example C16-acylated glucagon-like peptide 1, which possesses satisfactory pharmacokinetic properties sufficient for a once-daily administration [3]. Another approach is site-specific pegylation, for example that of coagulation factor VIII with poly(ethylene glycol) (PEG) with a molecular weight of more than 10 kDa resulting in a significant longer *in vivo* half-life as compared to the native factor VIII [4]. Likewise, it was shown that conjugation of a 20 kDa PEG chain to aptamers significantly increased aptamer residence in circulation and thus facilitate distribution to highly perfused organs [5]. Amino acid substitution in peptides and proteins [6] and modification of the degree of glycosylation [7] are other ways to enhance the therapeutic efficacy of the peptide or protein drugs. Regarding nucleic acid-based drugs, chemical modifications of the nucleic acid backbone are also used to obtain increased biological stability as well as increased target affinity [8]; one eminent example being the locked nucleic acid (LNA) modifications of RNAs intended for antisense therapy [9].

Thus, overcoming the challenges of inactivation of the biopharmaceutical by (i) physical conformational changes, (ii) complexation or association to other (naturally occurring) compounds or (iii) chemical degradation can be achieved by modification of the drug molecule.

Complexing or encapsulation of the biopharmaceutical drug in order to physically shield against protease or nuclease degradation is another generally interesting approach to prolong the half-life of amino acid-based and nucleic acid drugs, respectively. Peptides and protein drugs have thus proven to be protected by particulate encapsulation [10, 11]. Polymers capable of condensing nucleic acids (both DNA and RNA), and thereby protect them from nuclease degradation, include polycations like chitosan [12], linear and branched poly(ethylene imine) (PEI) [13] and dendrimers [14]. Lipoplex formation [15] or encapsulation in capsules such as liposomes or solid particles based on lipids or polymers is also a promising approach to shield and protect nucleic acid-based biopharmaceuticals [8, 16].

For the above mentioned formulation approaches, it is a prerequisite that the drug is not only protected during storage and administration, but also that it is released, *delivered*, at the target site irrespective of whether invasive or non-invasive administration routes are applied.

1.2.2 Enhancing Delivery

A prerequisite for the successful delivery for local as well as systemic delivery of peptides, proteins and nucleic acid derived drugs is the maximization of the delivery to the target site and a controlled and prolonged delivery time. Apart from the above mentioned strategies, one of the potential successful injectable sustained release lipid formulations for biomacromolecules is the Depofoam™ concept by SkyePharma. It differs from the conventional lipsomes in the increased surface area available, which makes the aqueous volume larger [17], and it has been shown to successfully encapsulate (60–85%) and sustain the release of insulin, luteinizing hormone-releasing factor and others [18].

The therapeutic applicability of biopharmaceuticals is hampered not only by their inherent instability, but also by their weight and polar surface characteristics. This is mainly a challenge for therapeutics that are to be administered by non-invasive techniques and intended for systemic absorption, since the drug must reach and pass the viscous mucosa lining, for example the oral and upper airway mucosa, and then permeate the tight cellular barrier in order for the drug to reach the circulation after administration by, for example, the gastrointestinal or airway routes. In theory, the drug in solution can permeate the epithelium by the para- or transcellular route. However, the size of peptides, proteins and nucleic acid-based drugs implies that the amount crossing the barrier by the paracellular route is strictly limited. The barrier properties might be affected by co-administration of chemical enhancers in the delivery system [19], but safe and reversible modulation of tight junction function might be achievable with co-excipients such as lipids [20] and cationic peptides [20, 21]. As biopharmaceuticals are usually very potent, a low absorbed amount of drug might be sufficient and the use of such membrane destabilizing excipients is promising. However, the risk of side effects by modulating the barrier properties should not be neglected. Endocytotic transport mechanisms, such as macropinocytosis, are likely to comprise the primary flux mechanisms for transcellular flux of macromolecules larger than, for example, insulin. Also, the cellular uptake of complexes, aggregates or nanoparticles is expected to occur by endocytotic mechanisms.

Even though formulation advances are evident, the above mentioned barriers still imply that most biomacromolecular drugs are currently administered by injection, since the absorption and distribution properties are better due to that the capillary lining endothelia is more permeable than the mucosal epithelium. A crucial issue in formulating novel drug delivery systems is, therefore, also to consider how to overcome the biological barrier at the site of administration.

1.3 Drug Delivery Technologies and Excipients

As indicated above, longevity of the drugs and delivery systems is important for efficient delivery. However, targetability and triggerability of the system are issues that are addressed more efficiently as materials sciences develop.

Nanocarriers such as liposomes can be prepared at a certain size and with a shielding coating to prevent both aggregation and also facilitate passive targeting by the enhanced permeation and retention effect, and thereby accumulate in fenestrated tissue like cancer or inflamed tissue. Further, the vesicles can be tailored to trigger release upon changes in, for example, pH or temperature [22, 23]. Likewise, synthetic biomimetic pH sensitive polymersomes have shown successful intracellular delivery of macromolecules [24]. Supramolecular structure synthesis on the basis of well-known chemical structures like cyclodextrins [25] is becoming increasingly applicable, also in combination with other polymers and targeting ligands [26]. Emerging materials such as carbon nanotubes can also be grafted with targeting ligands in order to obtain more efficient delivery [27]. Several promising delivery systems based on polymers that partly consist of cell penetrating peptides are reported to show high cellular transfection efficiencies and low cytotoxicity [28] and intracellular targeting [29], and thus be feasible as a non-viral delivery technology for, for example, antisense therapeutics. It has recently been shown *in vivo* that nasal [21] and oral [30] absorption of insulin in rats was markedly improved by the presence of L-penetratin or oligo-arginine, respectively, and without any detectable side effects; yet there is still a number of important factors to be addressed to elucidate the enhancing effect of co-administration of cell penetrating peptides with biopharmaceuticals [31].

Also, for vaccines, it is important not only to target the delivery system to the proper antigen presenting cells, but also to elicit an immune response. Since many excipients and drug delivery systems explored for systemic delivery of, for example, proteins used in substitution or replacement therapy in themselves show immunomodulatory effects, more and more of these advanced delivery systems are explored as adjuvants for vaccine formulations. As an example, alginate coated chitosan nanoparticels have proven effective as an adjuvant for subcutaneous administration of an antigen [32], and short peptides or nucleic acid-based immunomodulators are explored for a number of optimized vaccine formulations [33]. In addition, non-injectable formulations for, for example, airway delivery of vaccines may also introduce the use of various well-known and novel excipients [34].

1.4 Risks

All structures introduced to the patient should be biocompatible and biodegradable, so the direct effects of, for example, membrane interacting chemical enhancers or reactive degradation products of excipients must obviously be known and avoided when designing drug delivery systems. Apart from that, for many biologic pharmaceuticals, a major issue is the risk of inducing immunogenic side reactions towards the therapeutic molecule itself or towards the co-administered excipients or degradation products thereof. For all types of pharmaceutical formulations, the choice of excipients can be detrimental to the efficacy of the formulation, since minor changes in, for example, composition and purity can cause highly unwanted side effects [35]. One example is the alteration in the glycosylation pattern of erythropoietin, which is thought to cause increased immunogenicity [36]. Polymeric and lipid carrier molecules as well as their physical appearance as, for example, nanoparticles might induce significant toxicity issues that require resolution, since a number of toxicology reports have demonstrated that exposure to nanotechnology derived particles pose a serious risks to biological systems [37, 38]. Taking advantage of the technical advances in detection and prediction of unwanted side effects also helps to focus the attention on the long term effects of seemingly safe drug delivery systems.

1.5 Conclusion

Successful formulation of delivery systems for biopharmaceuticals depends on a thorough understanding of the biopharmaceutical molecular structure, stability and biological activity and the effect of processing it into a pharmaceutical dosage form. A thorough understanding of the wanted and, especially, the unwanted effects caused by the excipients and the formulation appearance in relation to the administration and target site is equally important. Developing advanced drug delivery systems thus requires interdisciplinary science, and a successful formulation of a biotechnology-based drug is very likely to be individual for each drug due to their structural complexity.

References

1. Tauzin, B. (2008) Biotechnology research continues to bolster arsenal against disease with 633 medicines in development. *Medicines in Development, Biotechnology*, http://www.phrma.org/files/Biotech%202008.pdf (accessed 7 July 2009).
2. Walsh, G. (2006) Biopharmaceutical benchmarks. *Nat. Biotech.*, **24**, 769–776.
3. Russell-Jones, D. (2009) Molecular, pharmacological and clinical aspects of liraglutide, a once-daily human GLP-1 analogue. *Mol. Cell Endocrinol*, **297**, 137–140.
4. WIPO (2007) Patent Application WO/2007/126808, Pegylated factor VIII, http://www.wipo.int/pctdb/en/wo.jsp?WO=2007126808 (accessed 7 July 2009).
5. Healy, J.M., Lewis, S.D., Kurz, M. *et al.* (2004) Pharmacokinetics and biodistribution of novel aptamer compositions. *Pharm. Res.*, **21**, 2234–2246.
6. Frokjaer, S. and Otzen, D.E. (2005) Protein drug stability: a formulation challenge. *Nat. Rev. Drug Discov.*, **4**, 298–306.
7. Egrie, J.C., Dwyer, E., Browne, J.K. *et al.* (2003) Darbepoetin alfa has a longer circulating half-life and greater *in vivo* potency than recombinant human erythropoietin. *Exp. Hematol.*, **31**, 290–299.
8. Fattal, E. and Bochot, A. (2008) State of the art and perspectives for the delivery of antisense oligonucleotides and siRNA by polymeric nanocarriers. *Int. J. Pharm.*, **364**, 237–248.
9. Elmen, J., Lindow, M., Schutz, S. *et al.* (2008) LNA-mediated microRNA silencing in non-human primates. *Nature*, **452**, 896–469.
10. Morishita, M. and Peppas, N.A. (2006) Is the oral route possible for peptide and protein drug delivery? *Drug Discov. Today*, **11**, 905–910.
11. Almeida, A.J. and Souto, E. (2007) Solid lipid nanoparticles as a drug delivery system for peptides and proteins. *Adv. Drug Del. Rev.*, **59**, 478–490.
12. Lai, W.-F. and Lin, M.C.-M. (2009) Nucleic acid delivery with chitosan and its derivatives. *J. Control. Release*, **134**, 158–168.
13. Howard, K.A., Li, X.W., Somavarapu, S. *et al.* (2004) Formulation of a microparticle carrier for oral polyplex-based DNA vaccines. *Biochim. Biophys. Acta (BBA)*, **1674**, 149–157.
14. Svenson, S. and Tomalia, D.A. (2005) Dendrimers: a Versatile Targeting Platform: dendrimers in biomedical applications – reflections on the field. *Adv. Drug Del. Rev.*, **57**, 2106–2129.
15. Wasungu, L. and Hoekstra, D. (2006) Cationic lipids, lipoplexes and intracellular delivery of genes. *J. Control. Release*, **116**, 255–264.

16. Cun, D., Boye Jensen, L., Mørck Nielsen, H. *et al.* (2008) Polymeric nanocarriers for siRNA delivery: challenges and future prospects. *J. Biomed. Nanotech.*, **4**, 1–18.
17. Ye, Q., Asherman, J., Stevenson, M. *et al.* (2000) DepoFoam(TM) technology: a vehicle for controlled delivery of protein and peptide drugs. *J. Control. Release*, **64**, 155–166.
18. Howell, S.B. (2001) Clinical application of a novel sustained-release injectable drug delivery system: Depofoam™ technology. *Cancer J.*, **7**, 219–227.
19. Khafagy, E.-S., Morishita, M., Onuki, Y. and Takayama, K. (2007) Current challenges in non-invasive insulin delivery systems: a comparative review. *Adv. Drug Del. Rev.*, **59**, 1521–1546.
20. Johnson, P.H., Frank, D. and Costantino, H.R. (2008) Discovery of tight junction modulators: significance for drug development and delivery. *Drug Discov. Today*, **13**, 261–267.
21. Khafagy, E.S., Morishita, M., Isowa, K. *et al.* (2009) Effect of cell-penetrating peptides on the nasal absorption of insulin. *J. Control. Release*, **133**, 103–108.
22. Auguste, D.T., Furman, K., Wong, A. *et al.* (2008) Triggered release of siRNA from poly(ethylene glycol)-protected, pH-dependent liposomes. *J. Control. Release*, **130**, 266–274.
23. Torchilin, V. (2008) Multifunctional and stimuli-sensitive pharmaceutical nanocarriers. *Eur. J. Pharm. Biopharm.* doi: 10.1016/j.ejpb.2008.09.026
24. Lomas, H., Canton, I., MacNeil, S. *et al.* (2007) Biomimetic pH sensitive polymersomes for efficient DNA encapsulation and delivery. *Adv. Mater.*, **19**, 4238–4243.
25. Li, J. and Loh, X.J. (2008) Cyclodextrin-based supramolecular architechtures: synthsis, structures, and applications for drug and gene delivery. *Adv. Drug Del. Rev.*, **60**, 1000–1017.
26. Hu-Lieskovan, S., Heidel, J.D., Bartlett, D.W. *et al.* (2005) Sequence-specific knockdown of EWS-FLI1 by targeted, nonviral delivery of small interfering RNA inhibits tumor growth in a murine model of metastatic Ewings sarcoma. *Cancer Res.*, **65**, 8984–8992.
27. Foldvari, M. and Bagonluri, M. (2008) Carbon nanotubes as functional excipients for nanomedicines: II. Drug delivery and biocompatibility issues. *Nanomed. Nanotechnol. Biol. Med.*, **4**, 183–200.
28. Kim, T., Ou, M., Lee, M. and Kim, S.W. (2009) Arginine-grafted bioreducible poly(disulfide amine) for gene delivery systems. *Biomaterials*, **30**, 658–664.
29. Unciti-Broceta, A., Diezmann, F., Ou-Yang, C.Y. *et al.* (2009) Synthesis, penetrability and intracellular targeting of fluorescein-tagged peptoids and peptide – peptoid hybrids. *Bioorg. Med. Chem.*, **17**, 959–966.
30. Morishita, M., Kamei, N., Ehara, J. *et al.* (2007) A novel approach using functional peptides for efficient intestinal absorption of insulin. *J. Control. Release*, **118**, 177–184.
31. Foged, C. and Nielsen, H.M. (2008) Cell-penetrating peptides for drug delivery across membrane barriers. *Expert Opin. Drug Deliv.*, **5**, 105–117.
32. Borges, O., Silva, M., de Sousa, A. *et al.* (2008) Alginate coated chitosan nanoparticles are an effective subcutaneous adjuvant for hepatitis B surface antigen. *Int. Immunopharmacol.*, **8**, 1773–1780.
33. Nordly, P., Madsen, H.B., Nielsen, H.M. and Foged, C. (2009) Status and future prospects of lipid-based particulate delivery systems and immunostimulators as vaccine adjuvants. *Expert Opin. Drug Del.*, **6**, 1–16.
34. Coucke, D., Schotsaert, M., Libert, C. *et al.* (2009) Spray-dried powders of starch and crosslinked poly(acrylic acid) as carriers for nasal delivery of inactivated influenza vaccine. *Vaccine*, **27**, 1279–1286.
35. van de Weert, M. and Moeller, E.H. (2008) *Immunogenicity of Biopharmaceuticals*, Springer, New York, ISBN: 978-0-378-75840-4.
36. Schellekens, H. and Casadevall, N. (2004) Immunogenicity of recombinant human proteins: causes and consequences. *J. Neurol.*, **251**, II/4–II/9.
37. Moghimi, S.M., Hunter, A.C. and Murray, J.C. (2005) Nanomedicine: current status and future prospects. *FASEB J.*, **19**, 311–330.
38. Wu, T.-L. and Ertl, H.C.J. (2009) Immune barriers to successful gene therapy. *Trends Mol. Med.*, **15**, 32–39.

Delivery of Biopharmaceuticals

2

Novel Formulation Approaches for Peptide and Protein Injectables

Mingshi Yang[1] *and Sven Frokjaer*[2]
[1]CMC Analytical Support, Novo Nordisk A/S, Måløv, Denmark
[2]University of Copenhagen, Department of Pharmaceutics and Analytical Chemistry, Faculty of Pharmaceutical Sciences, Copenhagen, Denmark

2.1 Introduction

Over the last two decades, protein and peptide drugs have become an important class of therapeutic agents. There are more than 100 protein and peptide therapeutic products marketed globally, and the product revenue was more than US $60 billion in 2004. Due to their intrinsic properties, that is large molecular size, physical and chemical instability, and poor permeability through biological membranes, most of the protein and peptide drugs have to be administered parenterally. This approach assures bioavailability, fast onset of action, high efficacy and reliable pharmacokinetic and pharmacodynamic properties. However, the major drawback of parenteral drug delivery is its invasive nature. Protein and peptide drugs often have short plasma half-lives, and need to be dosed frequently to achieve required therapeutic profiles. It often causes poor patient compliance. Non-invasive routes such as oral, pulmonary, nasal and transdermal delivery are ideal to improve patient compliance but the outcome of non-invasive routes is often disappointing owing to low bioavailability.

Many attempts have been made to overcome the aforementioned shortcomings of parenteral drug delivery. Modification of a therapeutic protein or peptide to modulate its pharmacological and therapeutic properties can often be very effective. In order to

Delivery Technologies for Biopharmaceuticals: Peptides, Proteins, Nucleic Acids and Vaccines Edited by Lene Jorgensen and Hanne Mørck Nielsen

improve the properties of the therapeutic protein or peptide it is often necessary either to modify it chemically at a specific position or to design a drug delivery system. The main objective is for the patient to obtain less frequent administration, greater convenience and improved efficacy.

2.2 Formulation Challenges

The inherent physicochemical properties of proteins and peptides as well as the chemical and physical instability pose a major challenge to the pharmaceutical formulation scientist. The obstacles to efficient delivery of proteins and peptides to the therapeutic target can broadly be categorized as either the enzymatic barriers that the drug encounters after administration or the physical barriers preventing an efficient transport of proteins and peptides across cell barriers (Figure 2.1).

The pharmaceutical and pharmacokinetic properties of proteins and peptides can be modulated in several ways, for example by mutating one or more amino acids using either natural or un-natural amino acids, by PEGylation, by acylation or by designing specific drug delivery systems. However, both therapeutic efficacy and potential side effects must always be taken into consideration when a protein/peptide drug molecule is chemically modified and/or formulated as a dosage form, for example a controlled delivery system.

During processing and formulating a dosage form, as well as during storage and handling, the therapeutic proteins or peptides are exposed to conditions that may be harmful to the drugs. The main factors are pH, temperature, physical interfaces and shear forces, for instance those introduced by shaking.

In contrast to low molecular weight drugs, proteins and larger peptides possess a three-dimensional structure essential to the therapeutic activity and avoidance of side effects such as immunological reactions. Hence, a major challenge is not just to secure the chemical stability but also to maintain the native structure of the protein or peptide drug. Some of the main stability issues are those stated in Table 2.1.

Therefore, a drug delivery system or a chemical modification of a protein or peptide cannot be applied fully until the effects on stability have been studied in detail during

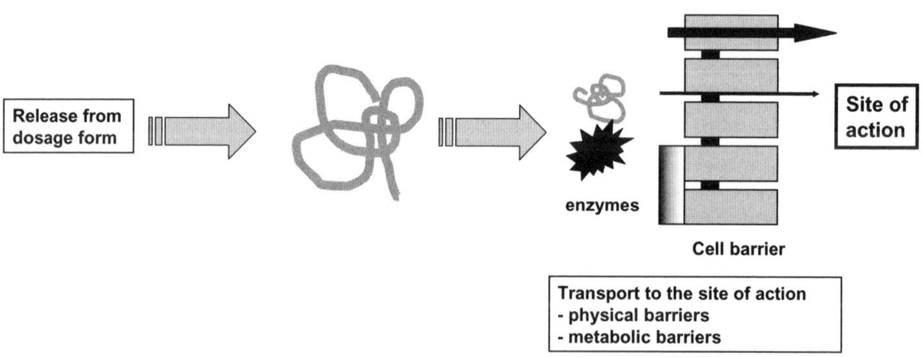

Figure 2.1 Barriers to biomacromolecular drug transport

Table 2.1 *Major degradation pathways of peptides and proteins*

Physical instability	Chemical instability
Denaturation	Hydrolysis
Adsorption	Deamidation
Aggregation	Oxidation
Precipitation	Racemization
	Isomerization
	Disulfide exchange
	B-Elimination

production, storage and handling. Already today it is a major challenge to formulate protein and peptide drugs as suspensions or aqueous solutions either in a ready-to-use form or as freeze-dried product for reconstitution. However, the future use of more advanced drug delivery systems will be an even greater challenge, as these formulations may involve the use of polymers, organic solvents, elevated temperatures and changes in pH. These are all denaturant conditions that can compromise the stability of the drug [1, 2].

2.3 Chemical Modification of Proteins and Peptides

Chemical modification is an approach that uses synthetic and/or molecular engineering techniques to modify the protein/peptide of therapeutic interest by attaching a chemical group or altering the amino acid on the backbone. The method includes PEGylation, glycoengineering, acylation, protein fusion and amino acid substitution. They all have the potential advantages of generating a new protein/peptide entity with extended plasma half-life and/or enhanced efficacy.

2.3.1 PEGylation

PEGylation refers to the attachment of a poly(ethylene glycol) (PEG) moiety either covalently or reversibly to the surface of the protein or peptide through a reactive group such as amine, thiol or carboxylic acid. PEG is generally known as a water-soluble, non-toxic, non-immunogenic and non-antigenic polymer, and it has been used in foods, cosmetics and pharmaceuticals for many years. The introduction of a high molecular weight PEG moiety to the protein effectively increases the protein's hydrodynamic radius, thus reducing renal ultrafiltration. PEGylation also generally masks the protein's surface, which leads to a reduced degradation by proteolytic enzymes and a decreased immune response. Hence, the biodistribution and solubility of the protein can be significantly altered by PEGylation [3]. It conveys to the protein drug a prolonged half-life, improved solubility, reduced side effects and increased efficiency. By PEGylating with different molecular weight, structure and/or number of PEG chains, the duration of action of proteins and peptides can be extended from minutes to hours, and even to days.

The clinical value of PEGylation has been well established. In March 1990, Enzon's Adagen® [4], a PEG-modified bovine enzyme adenosine deaminase, was the first

PEGylated product approved by the United States Food and Drug Administration (FDA). Since then, several PEGylated protein/peptide drug products have been marketed, including Oncaspar® (Enzon, 1994) [5], PEG-Intron® (Schering-Plough/Enzon, 2000) [6], Somavert® (Pfizer, 2002) [7], PEGASYS® (Roche, 2002) [8], Neulastra® (Amgen, 2002) [9], Macugen® (OSI Pharmaceutical/Pfizer UCB, 2004) [10] and Cimzia® (Celltech, 2008) [8]. Meanwhile, there are a number of PEGylated protein or peptide pharmaceuticals that are now in clinical trials or at various stages in their development [11, 12].

To couple PEG to the protein, it is necessary first to activate the PEG by converting one or both termini to some functional groups that are capable of reacting with the reactive amino acids or functional groups on the surface of the protein. The mono-methoxylated PEG (mPEG) is generally used in protein modification due to its monofunctionality, and it yields compounds with a higher purity and a less polydisperse distribution of the conjugates [13]. The typical reactive amino acids and functional groups include lysine, cysteine, histindine, arginine, aspartic acid, glutamic acid, serine, threonine, tyrosine, the N-terminal amino group and the C-terminal carboxylic acid.

The most common route for PEGylation of proteins has been to activate the PEG with functional groups suitable for reaction with either the lysine residue or the N-terminal amino group. The activated PEG, such as PEG aldehyde, PEG hydroxysuccinimide esters and PEG dichlorotriazine, can be conjugated with the amino groups of proteins by alkylation or acylation reactions [14]. However, the technique is generally plagued by PEG impurities, restriction to low molecular weight PEGs, unstable linkages and lack of selectivity in PEGylation [13]. Among the possible activated PEGs, PEG-aldehyde such as PEG-propionaldehyde and PEG-acetaldehyde may convey better selectivity in modification of the protein.

The conjugation of PEG to the thiol or hydroxyl groups and to the disulfide bridges has also been made possible by using several specific chemical or enzymatic methods [15, 16]. PEGylation of the free cysteine residue is the main approach for a site-specific modification. The activated PEGs such as PEG-maleimide, PEG-vinylsulfone, PEG-iodoacetamide and PEG-orthopyridyl disulfide can specifically react with cysteine. Even though few proteins possess cysteine residues on the surface that are suitable for PEG conjugation, one or more free cysteines may be inserted at the desired position of the sequence by genetic engineering [17]. The advantage of this approach is that it makes site-specific PEGylation possible in areas on the protein that will minimize a loss in biological activity and decrease the immunogenicity. Another method for site-directed PEGylation of proteins is oxidation of carbohydrate residues or the N-terminal serine or threonine [18, 19].

PEG-isocyanate is useful for hydroxyl group conjugation yielding a stable urethane linkage [20]. Meanwhile, it is highly reactive with amines too. There are also a few reports on linking PEG to the arginine residues by using PEG-1-3-dioxocompounds. The disadvantage of this method is related to the long reaction time and the lack of specificity, since other amino acids such as histidine and lysine may also react [21, 22].

Most PEGylation chemistry is designed to create a stable covalent link between the PEG and the protein. In most cases, the stable linkage is suitable for protein formulation in the aspect of easier purification and increased stability in long term storage. On the other hand, it is also observed that the stable linkage can reduce the activity, possibly due to the presence of the PEG chain at the active site of the protein. Hence, a prodrug concept has been applied by designing reversible PEGylated protein. The strategy involves introducing labile bonds between PEG and the conjugated moiety so that the native protein can regenerate under physiological conditions [23, 24]. These releasable PEG reagents include PEG-succinimidyl succinate, PEG-maleic anhydride, PEG-phenyl ether succinimidyl carbonates and PEG- benzamide succinimidyl carbonates.

It is always a challenge to characterize the exact composition of the PEGylated product. Problems can be encountered when conjugating the protein of interest to PEG such as variations in the degree of PEGylation, which leads to polydispersity of the newly formed molecules. Furthermore, there can also be difficulties in determining the exact sites of conjugation in the polypeptides and proteins. Edman degradation can be used to characterize the sites of conjugation by identifying missing amino acids in the step-wise degradation of a conjugate [25]. The degree of PEGylation is often determined by using colorimetric titration, GPC, MALDI mass spectrometry and capillary electrophoresis [26, 27].

In brief, as an injectable biocompatible technology, PEGylation has developed substantially over the past decades for peptide and protein delivery. The diversity of the PEG molecular weight, structure and chemistry will allow protein chemists numerous choices to generate PEGylated proteins with appropriate half-life, targeting properties, controlled pharmacokinetic/pharmacodynamic parameters and improved activity.

2.3.2 Glycoengineering

Glycoengineering of a protein is a molecular engineering approach to link saccharides to proteins of therapeutic interest or alter the protein-associated carbohydrate. It exploits the natural post-translational modification process of glycosylation to generate a new molecular entity. By using glycoengineering the activity of a protein can be increased and prolonged, thus allowing a reduced frequency of administration [28].

The saccharide chain that is attached to the target protein plays a role in molecular stability, solubility, *in vivo* activity, serum half-life and immunogenicity. For example, some proteins do not fold correctly unless they are glycosylated first. Glycosylation can maintain the molecular integrity of glycoprotein by reducing the susceptibility to proteolysis. Interferon beta (IFN-β), for example, shows that the sensitivity of glycosylated protein to thermal denaturation is decreased compared to the unglycosylated form [29].

The presence of the oligsaccharide structure on the glycoprotein often ensures a good solubility [30]. For example when fibrinogen, a N-linked glycoprotein, is deglycosylated it results in the formation of insoluble complexes [31]. Glycosylated leptin has an over 15-fold increase in solubility compared to native leptin when five additional N-linked carbohydrates are added to the native sequence of leptin via glycoengineering [32].

Glycosylation can increase the *in vivo* biological activity, which may be due to a prolonged half-life. Darbepoetin-alfa, a hyperglycosylated analogue of erythropoietin that contains two additional N-linked carbohydrates, possess a threefold increase in serum half-life and increased *in vivo* activity compared to the recombinant human erythropoietin [33]. When single injections were compared approximately 30–40-fold more recombinant human erythropoietin was required to match the activity of the lower dose of darbepoetin-alfa [34]. The ability of increasing the *in vivo* activity by glycoengineering is also seen in other proteins, such as leptin and Mpl (myeloproliferative leukaemia) ligands [28].

The immunogenicity of a protein can be affected by glycoengineering. The carbohydrates may be a part of the epitope itself or they may masque existing antigenic sites on the peptide backbone. Casadevall *et al.* reported that immunoreactivity of antibodies in pure red cell aplasia patient serums to recombinant human erythropoietin increases when carbohydrate is removed, which suggests that the carbohydrate can inhibit the antibody formation by 'shielding' the underlying protein sequences from the immune system [32].

Amgen has successfully used glycoengineering to create Darbepoetin-alfa, a synthetic form of erythropoietin under the trade name of AraNESP® [35]. In September 2001, it was approved by the FDA for treatment of anemia in patients with chronic renal failure by intravenous or subcutaneous injection.

Saccharides can be attached to protein mainly in three ways: (i) through an N-glycosidic bond between the reducing terminal N-acetylglucosamine and the side chain of Asn within the consensus sequence Asn-X-Ser/Thr, where X is any amino acid except Pro; (ii) through an O-glycosidic bond to the side chain of Ser, Thr, hydroxylysine or hydroxyproline; and (iii) as a component of the glycophosphatidylinositol (GPI) membrane anchor [36, 37]. In addition, glycosaminoglycan chains may also be used to covalently link to a protein to form proteoglycans.

The challenges presented with glycoengineering focus on the production of the desired protein, primarily owing to the heterogeneity of glycosylation in mammalian systems including glycosylation site occupancy and under sialylation [35]. Sialylation will lead to shortening plasma exposure as a result of clearance by the asialogycoprotein receptor in the liver or a host of other glycan specific receptors on endothelial or Kupffer cells.

2.3.3 Acylation

Acylation is the process of linking the carboxylic group of fatty acids to the amine group of the N-terminal residue of a peptide through a stable amide bond [38]. It can also be done at cysteine residue to obtain reversible acylated protein [39]. Acylation is another chemical modification approach that impairs therapeutic proteins or peptides an extended serum half-lives and increased stability.

One of the ideas behind this modification was that the circulation time of acylated protein could be increased to some extent by attaching to human serum albumin through its fatty acid binding site. For example, insulin detemir from Novo Nordisk A/S can extend the action time of insulin and reduce the injection frequency from three times a day to only twice or once daily. Detemir is an acylated insulin analogue in which the amino acid threonine at B30 is removed, and a C-14, myristoyl fatty acid is acylated to the lysine residue at B29 [38]. The attached fatty acid also confers an increased hydrophobic interaction of the insulin hexamer complex, thus prolonging the release of insulin monomer from the site of injection. Other proteins and peptides, such as glucagon-like peptide 1 [40], interferon-α [41], desmopressin [42] and octreotide [39], have also been reported to have increased their pharmacological effect and half-lives by the acylation modification of the molecules. The effect of acylation is naturally greater for smaller proteins and peptides because of the greater relative increase in hydrophobicity [43]. As other chemical modification technologies, acylation may also adversely affect the affinity of proteins or peptides to the receptors and thus compromises the bioactivity.

2.3.4 Amino Acid Substitution

Different from the previously described approaches, amino acid substitution is an approach where one or more amino acids in the primary structure of the protein is substituted in order to gain an altered pharmacokinetic profile *in vivo*. Amino acid substitution has been successfully applied to develop novel insulin analogues without compromising the biological function.

Some examples are insulin analogues such as Lantus, Humalog and NovoRapid. Lantus® is a 24-h long-acting recombinant insulin analogue produced by Aventis [44]. It differs from human insulin as the asparagine at A21 is replaced with a glycine and two arginine amino acids are further added to the C-terminus of the B-chain. With these alterations Lantus® has a low solubility at physiological pH and it forms a slow-dissolving depot of hexameric insulin at the subcutaneous injection site. Studies have shown that it can mimic physiological basal insulin release [45, 46]. Not only can a slow release be obtained by amino acid substitution but also a faster mode of action. Two rapid-acting insulin analogues, that is Humalog® (Eli Lilly) and NovoRapid® (Novo Nordisk), were also developed by using amino acid substitution to control the rapidly postprandial release of glucose. Humalog® is an insulin analogue with a lysine residue at B28 and proline residue at B29, while in NovoRapid® the proline residue at B28 is replaced with an aspartic acid [47, 48]. Both analogues have a more rapid onset of action when compared to human insulin.

Recombinant human interleukin-2 (IL-2) is another example in which the therapeutic effect is improved through molecular design. The commercial available non-glycosylated IL-2 analogue aldesleukin (Proleukin®, Chiron) is a des-alanyl-1 analogue of human IL-2 in which the cysteine residue in position 125 has been replaced by serine [43].

2.3.5 Protein Fusion

Fusion of proteins can be created through genetic engineering of a fusion gene. This typically involves removing the stop codon from a complementary DNA sequence coding for the first protein (e.g. a protein of therapeutic interest), then appending the complementary DNA sequence of the second protein (e.g. an endogenous protein with a long serum persistence) in frame by either ligation or overlap extension polymerase chain reaction (PCR). That DNA sequence will then be expressed by a cell as a single protein. Fusion proteins are designed to reduce clearance by increasing the size of the protein. Fc-domain of antibodies, human serum albumin (HSA) and transferrin are commonly used endogenous proteins for protein fusion. The prolonged half-lives can also be attributed to the long serum persistence properties of these endogenous proteins.

Etanercept, marketed under the trade name Enbrel®, is a fusion protein which links human soluble tumour necrosis factor (TNF) receptor to the Fc-domain of human immunoglobulin G1 (IgG1), and it acts as a TNF inhibitor. It has a greatly extended half-life *in vivo*, and a more profound and long-lasting biologic effect compared to a naturally occurring soluble TNF receptor [49]. Infliximab (Remicade®) is another fusion protein drug used to treat autoimmune disorders; it is administered by intravenous infusion, typically at 6–8 week intervals. One of advantages of the Fc-fusion protein is that the half-life can further be increased by engineering antibodies with increased affinity for the FcRn receptor [50]. Others have also attempted to fabricate single-chain Fc fusion protein to minimize size (compared to Fc fusion protein) as a mechanism for achieving better bioavailability and increased tumour penetration [51, 52].

Human serum albumin and transferrin are long-lived proteins in plasma. Proteins resulting from HSA and transferrin gene fusions have the long circulating life of albumin and transferrin while retaining their biological and therapeutic properties. Albuleukin (HSA-IL2) and Albutropin (HSA-hGH) are two HSA fusion proteins being clinically explored by Human Genome Sciences [53, 54]. Albulin is a long-acting insulin analogue obtained by direct gene fusion of a single-chain human insulin to HSA, which shows a relatively peak-less and 24 hour fashion of normalizing blood glucose level in diabetic mice [55].

BioRexis is developing transferrin fusion proteins to achieve improved pharmacokinetics, efficacy and tolerability. As one of its lead candidates, transferrin fused glucagon-like peptide 1 (GLP-1) displays increased serum half-life when compared to a GLP-1 peptide in an unfused state [56].

The systemic exposure of protein of therapeutic interest can be enhanced through protein fusion technology. However, it requires the protein to be amenable to modification at the N or C terminus. In addition, the absorption rate from the site of injection would be significantly decreased due to the enlarged size.

2.4 Depot Delivery Systems

Depot delivery systems are designed to extend or prolong the effect of protein and peptide drugs without modification of the biomacromolecules. It could provide continuous protein and peptide delivery after a single administration, thus reducing injection frequency, decreasing adverse side effects, saving cost and increasing patient compliance. Approaches for rate control and target site specific delivery by the parenteral route can be achieved by several different technologies similar to those used for low molecular weight drugs. Currently there are mainly three types of depot delivery systems providing sustained release of biopharmaceuticals *in vivo*, that is micro- and nanoparticulate systems, *in situ* depot-forming systems (gels) and implant systems. However, relatively few products based on more advanced formulation principles have reached the market [57]. For designing a rational depot delivery system of proteins and peptides it is important to understand both the properties of the drug molecule, such as its physicochemical properties, pharmacokinetics and pharmacodynamics, and also the characteristics of depot system, such as release mechanism and properties of carriers

2.4.1 Micro- and Nanoparticulate Systems

Micro- and nanoparticulate systems are the primary type of parenteral controlled-release protein and peptide products marketed in the past few years. They can be injected either subcutaneously or intramuscularly for systemic effect or injected into a specific body site for localized treatment. Further, because of their micron or sub-micron size range they can also be injected intravenously to achieve long circulation in blood. In terms of the materials used in these systems they can be divided in two groups, that is polymeric and lipid micro- and nanoparticulates.

2.4.1.1 Polymeric Micro- and Nanoparticulates

The polymers used in micro- and nanoparticulates systems should be biodegradable or biocompatible due to the non-retrievable nature of micro- and nanoparticulates after administration. A variety of synthetic and naturally occurring biodegradable polymers are available that can be used to fabricate polymeric micro- and nanoparticulates for protein delivery. Among them poly (lactic-*co*-glycolic acid) (PLGA) is the most commonly used biodegradable polymer due to its long safety history of application in medical devices. Many currently approved polymeric micro- and nanoparticulate systems for protein and peptide drugs are derived from PLGA.

Nutropin Depot® was the first approved protein (recombinant human growth hormone, rHGH) loaded PLGA microsphere product developed by Alkermes and Genetech in 1999 [58]. After a single injection it could provide sustained release of rHGH over a two or four week period. But in 2004 it was discontinued by the both companies due to the significant resources required for manufacturing and commercializing the product [59].

Compared to large protein molecules PLGA-based micro- and nanoparticulate systems are more commercially successful in delivering small peptide drugs, such as luteinizing hormone-releasing hormone (LHRH) agonists (Lupron Depot®, Trelstar™, Decapeptyl®) and octreotide (Suprecur®, Sandostatin LAR®, Somatuline® LA) [60]. The release of these peptides can be prolonged to 1–4 months [61]. The *in vitro* release study shows that the PLGA micro- and nanoparticulate possess the typical triphasic profile, that is initial burst release, lag phase and followed by zero order release. The release of the drug is mainly controlled by diffusion of drug from the polymer matrix and erosion of polymer.

Extensive works are still under progress to expand the application of PLGA-based micro- and nanoparticulates to the delivery of protein and peptide drugs [62–69]. To enhance desirable properties of PLGA for protein delivery many attempts have been made to modify the hydrophilicity of PLGA by synthesizing PLGA block copolymers with poly (ethylene oxide) [70] or poly (ethylene glycol) [71, 72]. Meanwhile, research in parenteral protein delivery by using micro- and nanoparticulates has been fuelled by the advent of novel carriers in recent years. Besides PLGA many other synthetic biodegradable polymers, such as polycaprolactone [73, 74], polyanhydride [75], polyorthoesters [76–78], polyphosphazenes [79] and so on, and other natural biodegradable polymers, like gelatine [80], albumin, dextran and so on, have been explored to develop micro- and nanoparticulate systems for parenteral delivery of protein and peptide drugs. But not so many of these micro- and nanoparticulate formulations have gained clinical or commercial success yet.

A number of methods can be used to fabricate polymeric micro- and nanoparticulates of proteins and peptides, for example spray drying [81], the double emulsion method [82–84], the phase separation technique [85], spray freeze-drying, ink jet technology, supercritical fluid technology, the emulsion solvent diffusion method, chemical and thermal cross-linking [86] and so on. For an ideal preparation technique it should meet the following requirements: high loading efficiency, high yield, optimal stability of biomacromolecules, stable batch reproducibility, low burst effect, adjustable release profiles and the ability to be scaled up. None of the aforementioned techniques can fulfil all these requirements. At present spray drying, the double emulsion method and phase separation technique are the most widely used techniques for polymeric micro- and nanoparticulates preparation. However, maintenance of protein and peptide stability is still one of the challenges during fabrication of micro- and nanoparticulates using these techniques. For example, in many cases organic solution has to be used to dissolve polymer, which may be detrimental to protein structure integrity. The selection of a relevant preparation method will be very much dependant on the physicochemical properties of both biomacromolecules and polymers, release mechanism of drug from carriers, stability of biomacromolecules and also factors related to application of micro- and nanoparticulates.

There are a number of challenges involved in developing polymeric micro- and nanoparticulates for protein and peptide drugs: (i) 'dose dumping' is always a potential risk when administering micro- and nanoparticulate systems due to the non-retrievable nature of this system as mentioned above; (ii) the attribute of 'burst effect' in micro- and nanoparticulates restricts the application of this system to proteins and peptides having a wide therapeutic index; (iii) the drug loading is generally low in micro- and nanoparticulates; (iv) normally large gauge needles have to be used to inject micro- and nanoparticulates; (v) the last but not least challenge is maintaining the integrity of protein and peptide in the formulation *in vivo*.

One of the principles of designing polymeric micro- and nanoparticulate systems for protein and peptide drugs is to protect these biomacromolecules from degradation *in vivo*. However, chemical degradation of the incorporated protein or peptide has been reported in these formulations. For example, degradation of lactides and glycolides in PLGA polymer will result in an acidic microenvironment that is detrimental to the stability of proteins [87]. The release of the aggregated or denatured protein many cause an unwanted immune response [88]. Therefore, much work still remains to meet the need for parenteral delivery of protein and peptide drugs by using polymeric micro- and nanoparticulates system though many advances have been made in recent years.

2.4.1.2 Lipid Micro- and Nanoparticulates

One of the few potential successful lipid-based sustained release formulations for peptides and proteins is the Depofoam™ concept by SkyePharma. Compared to conventional liposomes it has a larger available aqueous phase and has been shown to successfully encapsulate high amounts of various peptides and proteins, for example insulin [89, 90].

New lipid-based delivery systems may be developed based on the progress in nanosciences. It is well known that amphiphilic lipids such as phospholipids can self-assemble in water to form structures such as micelles, liposomes and tubes. By using carbon nanotubes as templates for polymerized lipid assemblies it may in future be possible to construct lipid-based drug delivery with unique geometries and release properties for drugs as for instances peptide and proteins [91].

2.4.2 *In Situ* Depot-Forming Systems

In situ depot-forming systems are usually viscous solutions or suspensions containing both biodegradable carrier and drug. The drug can be either dissolved or suspended in the system. When injected subcutaneously or intramuscularly the biodegradable carrier can form a depot of the drug by different mechanisms [92, 93]. The release and effect of the drugs can then be extended.

Eligard®, marketed by Sanofi-Aventis, is an injectable polymeric depot-forming system that delivers leuprolide acetate for palliative treatment of advanced prostate cancer. It was developed by QLT with the Atrigel® technology that can lead to a prolonged drug release profile of leuprolide *in vivo* with a duration of one or three months [94]. In the system, PLGA is used as a biodegradable carrier dissolved in N-methyl-2-pyrrolidone. Since N-methyl-2-pyrrolidone is miscible with water, it can, after injection, dissipate and water can diffuse into the system and result in precipitation of PLGA, thereby entrapping the drug.

SABER® is another similar system that has potential application for delivery of proteins and peptides [95]. It uses sucrose acetate isobutyrate as a carrier; this is dissolved in ethanol, benzyl alcohol or other water-miscible organic solvent. The selection of the organic solvent is rather crucial since it has impact on the initial drug burst and may raise toxicity.

ReGel®, an aqueous system, has also been developed and shown promising for protein and peptide delivery [96, 97]. It consists of a copolymer of hydrophobic PLGA blocks and hydrophilic PEG blocks. The copolymer is water soluble at ambient temperature, but it forms a gel at body temperature. Other biodegradable carriers, such as PEG-based copolymer with multiple thio [98], chitosan [99] and so on, were also attempted for delivering protein and peptide drugs by forming *in situ* depot system.

Compared to the micro- and nanoparticulate systems, the manufacturing process for *in situ* depot-forming systems is less complex; it mainly involves a simple dispersion or dissolution step. Furthermore, it may also allow for a higher drug loading and administration with a smaller gauge needle. However, an issue with this system is that the release profile of the drug may be influenced by the morphology of the depot forming on the site of injection [100].

2.4.3 Implant Systems

Implant systems require minor surgical procedures for administration. A second surgical procedure is needed to remove the implant if non-biodegradable carriers are used. The surgical nature of this system leads to less compatibility with patients compared to other depot delivery systems. But it is still valuable in enhancing systemic exposure of protein and peptide drugs by reducing the frequency of injection.

Zoladex®, marketed by AstraZeneca, is a biodegradable implant system containing goserelin acetate that can be administered subcutaneously monthly for treatment of prostate cancer [101]. Zoladex® 3.6 mg was first approved by the FDA in 1989; a 10.8 mg subcutaneous three-month biodegradable implant and other indications were subsequently approved. This implant is composed of PLGA as the biodegradable carrier.

Non-biodegradable carriers can also be used in implant systems to deliver peptide drugs. Viadur® was the first titanium osmotic implant using the Duros technology developed by Durect and Bayer [102]. It was used to deliver leuprolide acetate continuously for a full 12 months to manage the symptoms associated with advanced prostate cancer. However, in December 2007 Bayer discontinued marketing of Viadur® due to the limited long-term market viability of this product. Another example of non-biodegradable implant is Vantas®, marketed by Indevus Pharmaceuticals, that comprises cross-linked copolymer of hydroxypropyl methacrylate and 2-hydroxyl methacrylate. It is a 12-month subcutaneous implant used to deliver histrelin acetate, a nonapeptide analog of GnRH [103].

The retrievable nature of implant systems is an advantage over the other depot delivery systems.

2.5 Other Delivery Systems

2.5.1 Protein Crystallization and/or Precipitation

Besides chemical modification of proteins and depot delivery systems, protein crystallization or precipitation approaches have also been used to enhance systemic exposure of protein and peptide drugs [104, 105]. Insulin is the first crystalline protein to be approved for therapeutic use. In 1946, Nordisk (now Novo Nordisk A/S) developed isophane insulin (NPH), a suspension of crystalline zinc insulin combined with the positive charged protamine. When injected subcutaneously NPH has a prolonged action compared to regular insulin. In 1953, the company also launched Lente, a long-acting insulin zinc suspension. This is an approach that exploits an altered physical state of proteins and peptides, that is crystal or amorphous precipitate, to achieve improved stability and prolonged action [106, 107]. However, not all proteins can be readily designed into these forms. Insulin is still the only protein drug successful in this approach, though many other proteins and peptides have been attempted [108–110]. Issues such as the heterogeneity of crystal, protein agglomeration, lack of expertise in scaling-up crystallization and so on prevent wider application of this approach in delivery of proteins and peptides.

2.5.2 Injection Devices

Although non-invasive alternative routes are promising, all protein and peptide drugs are still administered though needles and syringes. The advances in injection device technology have provided patients with improved compliance and convenience. For example, the pen and auto-injector can help patients overcome the hesitation when self-injecting and make the needle-based device more 'user friendly' [111–113]. The first pens were introduced in 1984 for daily insulin injection. Auto-injectors are designed for use with fillable or prefilled syringes. During the last decade the use of prefilled syringe has exploded, especially for proteins, peptides and vaccines. The prefilled syringe is primarily used for liquid formulations. For lyophilized product a dual-chamber cartridge is also available.

The quality of today's needles can let patients feel less pain or even be painless. Now a needle-less injector is also available for parenteral insulin delivery, that is a gas jet auto-injector. It contains a cylinder of pressurized gas and propels a fine jet of liquid through the skin without the use of a needle [114]. One of the advantages of this device is obvious, it prevents needle-stick injury. Instead of delivering liquid Glide Pharma is developing a unique needle-free system (Glide SDI™) that delivers a solid dose through the skin [115]. It may render protein and peptide drugs with a sustained effect *in vivo* in the solid form. Device technologies may play a more and more important role in the delivery of protein and peptide drugs in the foreseeable future.

2.6 Conclusions

Protein and peptides drugs will become an even more important class of therapeutic agents than they are today. However, the full potential for the therapeutic use of these drugs is challenged by their intrinsic properties, such as the lack of physical and chemical stability, and the lack of optimal physicochemical properties for transport across various biomembranes. Thus, pharmaceutical scientists are faced with the challenge of formulating these drugs into safe, stable and therapeutic efficient formulations including the design of drug delivery systems. A success in this field to the benefit for the patients calls for a thoroughly interdisciplinary approach combining a number of disciplines from nano-/material science to clinical research.

References

1. Brange, J. (2000) Physical stability of proteins, in *Pharmaceutical Formulation Development of Peptides and Proteins* (eds S. Frokajer and L. Hovgaar), Taylor & Francis Limited, London, pp. 89–112.
2. Jorgensen, L., Kim, D.H., Vermehren, C. *et al.* (2004) Micropipette manipulation – a technique to evaluate the stability of water-in-oil emulsions containing proteins. *J. Pharm. Sci.*, **93**, 2994–3003.
3. Caliceti, P. and Veronese, F.M. (2003) Pharmacokinetic and biodistribution properties of poly(ethylene glycol)-protein conjugates. *Adv. Drug Del. Rev.*, **55**, 1261–1277.
4. Chakravarti, V.S., Borns, P., Lobell, J. and Douglas, S.D. (1991) Chondroosseous dysplasia in severe combined immunodeficiency due to adenosine deaminase deficiency (chondroosseous dysplasia in ADA deficiency SCID). *Pediatr. Radiol.*, **21**, 447–448.
5. Graham, M.L. (2003) Pegaspargase: a review of clinical studies. *Adv. Drug Deliv. Rev.*, **55**, 1293–1302.
6. Bukowski, R.M., Tendler, C., Cutler, D. *et al.* (2002) Treating cancer with PEG Intron: pharmacokinetic profile and dosing guidelines for an improved interferon-alpha-2b formulation. *Cancer*, **95**, 389–396.
7. Goffin V. and Touraine, P. (2004) Pegvisomant Pfizer/Sensus. *Curr. Opin. Invest. Drugs*, **5**, 463–468.
8. Reddy, K.R. (2004) Development and pharmacokinetics and pharmacodynamics of pegylated interferon alfa-2a (40kD). *Semin. Liver Dis.*, **24**, 33–38.
9. Molineux, G. (2004) The design and development of pegfilgrastim (PEG-rmetHuG-CSF, Neulasta). *Curr. Pharm. Des.*, **10**, 1235–1244.
10. Moshfegh , A.A. and Puliafito, C.A. (2005) Pegaptanib sodium for the treatment of neovascular age-related macular degeneration. *Expert Opin. Invest. Drugs*, **14**, 671–682.
11. Ryan, S.M., Mantovani, G., Wang, X. *et al.* (2008) Advances in PEGylation of important biotech molecules: delivery aspects. *Expert Opin. Drug Deliv.*, **5**, 371–383.
12. Pasut, G. and Veronese, F.M. (2007) Polymer drug conjugation, recent achievements and general strategies. *Porg. Olym. Sci.*, **32**, 933–961.
13. Roberts, M.J., Bentley, M.D. and Harris, J.M. (2002) Chemistry for peptides and protein PEGylation. *Adv. Drug Deliv. Rev.*, **54**, 459–476.

14. Herman, S., Hooftman, G. and Schacht, E. (1995) Poly(ethylene glycol) with reactive end groups: I modification of proteins. *J. Bioact. Biocomp. Polym.*, **10**, 145–187.
15. Veronese, F.M. (2001) Peptide and protein PEGylation: a review of problems and solutions. *Biomaterials*, **22**, 405–417.
16. Greenwald, R.B. (2001) PEGdrugs an overview. *J. Control. Release*, **74**, 159–171.
17. Coodson, R.J. and Katre, N.V. (1990) Site-directed PEGylation of recombinant interleukin-2 at its glycosylation site. *Biotechnology*, **8**, 343–346.
18. Zalipsky, S. and Menon-Rudolph, S. (1997) Hydrazide derivatives of poly (ethylene glycol) and their bioconjugates, in *Poly(ethylene glycol) Chemistry and Biological Applications*, ACS Symposium Series, vol. **680** (eds J.M. Harris and S. Zalipsky), ACS Books, Washington, DC, pp. 318–340.
19. Gaertner, H.F. and Offord, R.E. (1996) Site-specific attachement of functionalized poly(ethylene glycol) to the amino termimus of proteins. *Bioconjug. Chem.*, **7**, 38–44.
20. Greenwald, R.B., Pendri, A. and Bolikal, D. (1995) Highly soluble taxal derivatives: 7-polyethylene glycol carbomates and carbonates. *J. Org. Chem.*, **60**, 331–336.
21. Yankeelov, J.A. (1972) Modification of arginines by diketones. *Methods Enzymol.*, **25**, 566–579.
22. Akihito, S., Maeda, H., Kai, Y. and Ono, K. (1989) Polyethylene glycol derivatioves, modified peptides and production thereof. E.P. 0340741.
23. Garman, A.J. and Kalindjian, S.B. (1987) The preparation and properties of novel reversible polymer-protein conjugates. 2-omega-methoxypolyethylene (5000) glycoxymethylene-3-methylmaleyl conjugates of plasminogen activators. *FEBS Lett.*, **223**, 361–365.
24. Lee, S., Greenwald, R.B., McGuire, J. *et al.* (2001) Drug delivery systems employing 1,6-elimination: releasable poly(ethylene glycol) conjugates of proteins. *Bioconjug. Chem.*, **12**, 163–169.
25. Monkarsh, S.P., Spence, C., Porter, J.E. *et al.* (1997) Isolation of positional isomers of monopoly(ethylene glycol)ylated interferon-2α and the determination of their biochemical and biological characteristics, in *Poly(ethylene glycol) Chemistry and Biological Applications*, ACS Symposium Series, vol. **680** (eds J.M. Harris and S. Zalipsky), ACS Books, Washington, DC, pp. 207–216.
26. Na, D.H. and Lee, K.C. (2004) Capillary electrophoretic characterization of PEGylated human parathyroid hormone with matrix-assisted laser desorption/ionization time-of-flight mass spectrometry. *Anal. Biochem.*, **331**, 322–328.
27. Pesek, J.J., Matyska, M.T. and Krishnamoorthi, V. (2004) Separation of polyethylene glyco-modified proteins by open tubular capillary electrochromatography. *J. Chromatogr. A*, **1004**, 317–322.
28. Elliott, S., Lorenzini, T., Asher, S. *et al.* (2003) Enhancement of therapeutic protein *in vivo* activities through glycoengineering. *Nat. Biotechnol.*, **21**, 414–421.
29. Runkel, L., Meier, W., Pepinsky, R.B. *et al.* (1998) Structural and functional differences between glycosylated and non-glycosylated forms of human interferon-beta (IFN-beta). *Pharm. Res.*, **15**, 641–649.
30. Arakawa, T., Yphantis, D.A., Lary, J.W. *et al.* (1991) Glycosylated and unglycosylated recombinant-derived human stem cell factors are dimeric and have extensive regular secondary structure. *J. Biol. Chem.*, **266**, 18 942–18 948.
31. Langer, B.G., Weisel, J.W., Dinauer, P.A. *et al.* (1988) Deglycosylation of fibrinogen accelerates polymerization and increases lateral aggregation of fibrin fibers. *J. Biol. Chem.*, **263**, 15 056–15 063.
32. Sinclair, A.M. and Elliott, S. (2005) Glycoengineering: the effect of glycosylation on the properties of therapeutic proteins. *J. Pharm. Sci.*, **94**, 1626–1635.
33. Egrie, J.C., Dwyer, E., Browne, J.K. *et al.* (2003) Darbepoetin alfa has a longer circulating half-life and greater *in vivo* potency than recombinant human erythropoietin. *Exp. Hemato.*, **31**, 290–299.
34. Hartley, C., Elliott, S., Begley, C.G. *et al.* (2003) Kinetics of haematopoietic recovery after dose-intensive chemo/radiotherapy in mice: optimozed erythroid support with darbepoetin alpha. *Br. J. Haematol.*, **122**, 623–636.

35. Beals, J.M. and Shanafelt, A.B. (2006) Enhancing exposure of protein therapeutics. *Drug Discov. Today Technol.*, **3**, 87–94.
36. Parekh, R.B. (1994) Site-specific protein glycosylation. *Adv. Drug Del. Rev.*, **13**, 251–266.
37. Jenkins, N. and Curling, E.M.A. (1994) Glycosylation of recombinant proteins: problems and prospects. *Enzyme. Microb. Technol.*, **16**, 354–364.
38. Havelund, S., Plum, A., Ribel, U. *et al.* (2004) The mechanism of protraction of insulin detemir, a long-acting, acylated analog of human insulin. *Pharm. Res.*, **21**, 1498–1504.
39. Yuan, L., Wang, J. and Shen, W.C. (2005) Reversible lipidization prolongs the pharmacological effect, plasma duration, and liver retention of octreotide. *Pharm. Res.*, **22**, 220–227.
40. Knudsen, L.B., Nielsen, P.F., Huusfeldt, P.O. *et al.* (2000) Potent derivatives of glucagon-like peptide-1 with pharmacokinetic properties suitable for once daily administration. *J. Med. Chem.*, **43**, 1664–1669.
41. Foldvari, M., Attah-Poku, S., Hu, J. *et al.* (1998) Palmitoyl derivatives of interferon α: potential for cutaneous delivery. *J. Pharm. Sci.*, **87**, 1203–1208.
42. Wang, J., Shen, D. and Shen, W.C. (1999) Preparation, purification, and characterization of a reversible lipidized desmopressin with potential anti-diurectic activity. *Pharm. Res.*, **16**, 1674–1679.
43. Frokjaer, S. and Otzen, D.E. (2005) Protein drug stability: a formulation challenge. *Nat. Rev. Drug Discov.*, **4**, 298–306.
44. Gillies, P.S., Figgitt, D.P. and Lamb, H.M. (2000) Insulin glargine. *Drugs*, **59**, 253–260.
45. Luzio, S.D., Beck, P. and Owens, D.R. (2003) Comparison of the subcutaneous absorption of insulin glargine (Lantus) and NPH insulin in patients with Type 2 diabetes. *Horm. Metab. Res.*, **35**, 434–438.
46. Brown, L.R. (2005) Commercial challenges of protein drug delivery. *Expert Opin. Drug Deliv.*, **2**, 29–42.
47. DeFelippis, M.R., Bakaysa, D.L., Bell, M.A. *et al.* (1998) Preparation and characterization of cocrystalline suspension of [LysB28, ProB29]-human insulin analogue. *J. Pharm. Sci.*, **87**, 170–176.
48. Home, P.D., Barriocanal, L. and Lindholm, A. (1999) Comparative pharmacokinetics and pharmacodynamics of the novel rapid-acting insulin analogue, insulin aspart, in healthy volunteers. *Eur. J. Clin. Pharmacol.*, **55**, 199–203.
49. Madhusudan, S., Muthuramalingam, S.R., Braybrooke, J.P. *et al.* (2005) Study of etanercept, a tumor necrosis factor-alpha inhibitor, in recurrent ovarian cancer. *J. Clin. Oncol.*, **23**, 5950–5959.
50. Ghetie, V., Popov, S., Borvak, J. *et al.* (1997) Increasing the serum persistence of an IgG fragment by random mutagenesis. *Nat. Biotechnol.*, **15**, 637–640.
51. Reff, M.E. and Heard, C. (2001) A review of modifications to recombinant antibodies: attempt to increase efficacy in oncology applications. *Crit. Rev. Oncol. Hematol.*, **40**, 25–35.
52. Jazayeri, J.A. and Carroll, G.J. (2008) Fc-based cytokines: prospects for engineering superior therapeutics. *BioDrugs*, **22**, 11–26.
53. Osborn, B.L., Sekut, L., Corcoran, M. *et al.* (2002) Albutropin: a growth hormone-albumin fusion with improved pharmacokinetics and pharmacodynamics in rats and monkeys. *Eur. J. Pharmacol.*, **456**, 149–158.
54. Sung, C., Nardelli, B., LaFleur, D.W. *et al.* (2003) N-beta-albumin fusion protein that displays improved pharmacokinetic and pharmacodynamic properties in nonhuman primates. *J. Interferon Cytokine Res.*, **23**, 25–36.
55. Duttaroy, A., Kanakaraj, P., Osborn, B.L. *et al.* (2005) Development of a long-acting insulin analog using albumin fusion technology. *Diabetes*, **54**, 251–258.
56. Prior, C.P. (2007) Modified transferrin fusion proteins. United States Patent, 7176278.
57. Degim, I.T. and Celebi, N. (2007) Controlled delivery of peptides and proteins. *Curr. Pharm. Design*, **13**, 99–117.
58. Brown, L.R. (2005) Commercial challenges of protein drug delivery. *Expert Opin. Drug Deliv.*, **2**, 29–42.
59. Shi, Y. and Li, L.C. (2005) Current advances in sustained-release systems for parenteral drug delivery. *Expert Opin. Drug Deliv.*, **2**, 1039–1058.

60. Mundargi, R.C., Babu, V.R., Rangaswamy, V. *et al.* (2008) Nano/micro technologies for delivering macromolecular therapeutics using poly(D,L-lactide-co-glycolide) and its derivatives. *J. Control. Release*, **125**, 193–209.
61. Akers, M.J., Nail, S.L. and Saffell-Clemmer, W. (2007) Top ten hot topics in parenteral science and technology. *PDA J. Pharm. Sci. Technol.*, **61**, 337–361.
62. Zambaux, M.F., Bonneaux, F., Gref, R. *et al.* (1999) Preparation and characterization of protein C-loaded PLA nanoparticles. *J. Control. Release*, **60**, 179–188.
63. Blanco-Prieto, M.J., Besseghir, K., Zerbe, O. *et al.* (2000) *In vitro* and *in vivo* evaluation of a somatostatin analogue released from PLGA microspheres. *J. Control. Release*, **67**, 19–28.
64. Jackson, J., Liang, L., Hunter, W. *et al.* (2002) The encapsulation of ribozymes in biodegradable polymeric matrices. *Int. J. Pharm.*, **243**, 43–47.
65. Kostanski, J.W., Thanoo, B.C. and DeLuca, P.P. (2000) Preparation, characterization, and *in vitro* evaluation of 1- and 4-month controlled release orntide PLA and PLGA microspheres. *Pharm. Dev. Technol.*, **5**, 585–596.
66. Cho, S.W., Song, S.H. and Choi, Y.W. (2000) Effects of solvent selection and fabrication on the characteristics of biodegradable poly(lactide-co-glycolide) microspheres containing ovalbumin. *Arch. Pharm. Res.*, **23**, 385–390.
67. Zhu, K.J., Jiang, H.L., Du, X.Y. *et al.* (2001) Preparation and characterization of hCG loaded polylactide or poly(lactide-co-glycolide) microspheres using a modified water-in-oil-in water (w/o/w) emulsion solvent evaporation technique. *J. Microencapsul.*, **18**, 247–260.
68. Schrier, J.A. and DeLuca, P.P. (1999) Recombinant human bone morphogenetic protein-2 binding and incorporation in PLGA microsphere delivery systems. *Pharm. Dev. Technol.*, **4**, 611–621.
69. Prabhu, S., Sullivan, J.L. and Betageri, G.V. (2002) Comparative assessment of *in vitro* release kinetics of calcitonin polypeptide from biodegradable microspheres. *Drug Deliv.*, **9**, 195–198.
70. Kissel, T., Li, Y. and Unger, F. (2002) ABA-triblock copolymers from biodegradable polyester A-blocks and hydrophilic poly(ethylene oxide) B-blocks as a candidate for in situ forming hydrogel delivery systems for proteins. *Adv. Drug Deliv. Rev.*, **54**, 99–134.
71. Otsuka, H., Nagasaki, Y. and Kataoka, K. (2003) PEGylated nanoparticles for biological and pharmaceutical applications. *Adv. Drug Deliv. Rev.*, **55**, 403–419.
72. Choi, S., Baudys, M. and Kim, S.W. (2004) Control of blood glucose by novel GLP-1 delivery using biodegradable triblock copolymer of PLGA-PEG-PLGA in type 2 diabetic rats. *Pharm. Res.*, **21**, 827–831.
73. Sinha, V.R., Bansal, K., Kaushik, R. *et al.* (2004) Poly-epsilon-caprolactone microspheres and nanospheres: an overview. *Int. J. Pharm.*, **278**, 1–23.
74. Shenoy, D.B., D'Souza, R.J., Tiwari, S.B. and Udupa, N. (2003) Potential applications of polymeric microsphere suspension as subcutaneous depot for insulin. *Drug Dev. Ind. Pharm.*, **29**, 555–563.
75. Kumar, N., Langer, R.S. and Domb, A.J. (2002) Polyanhydrides: an overview. *Adv. Drug Deliv. Rev.*, **54**, 889–910.
76. Heller, J. and Barr, J. (2005) Biochronomer technology. *Expert Opin. Drug Deliv.*, **2**, 169–183.
77. Rothen-Weinhold, A., Schwach-Abdellaoui, K., Barr, J. *et al.* (2001) Release of BSA from poly (ortho ester) extruded thin strands. *J. Control. Release*, **71**, 31–37.
78. Bai, X.L., Yang, Y.Y., Chung, N.T.S. *et al.* (2001) The effect of polymer compositions on controlled release of protein from poly(ortho esters) microspheres. *J. Appl. Polym. Sci.*, **80**, 1630–1642.
79. Xu, X., Yu, H., Gao, S. *et al.* (2002) Polyphosphoester microspheres for sustained release of biologically active nerve growth factor. *Biomaterials*, **23**, 3765–3772.
80. Sinha, V.R. and Trehan, A. (2003) Biodegradable microspheres for protein delivery. *J Control. Release*, **90**, 261–280.
81. Bittner, B., Morlock, M., Koll, H. *et al.* (1998) Recombinant human erythropoietin (rHEPO) loaded poly(lactide-co-glycolide) microspheres: influence of the encapsulation technique and polymer purity on microspheres characteristics. *Eur. J. Pharm. Biopharm.*, **45**, 295–305.

82. Cui, F., Cun, D., Tao, A. *et al.* (2005) Preparation and characterization of melittin-loaded poly (DL-lactic acid) or poly (DL-lactic-co-glycolic acid) microspheres made by the double emulsion method. *J. Control. Release*, **107**, 310–319.
83. Nam, Y.S., Song, S.H., Choi, J.Y. and Park, T.G. (2000) Lysozyme microencapsulation within biodegradable PLGA microspheres: urea effect on protein release and stability. *Biotechnol. Bioeng.*, **70**, 270–277.
84. Yang, Y.Y., Chung, T.S. and Ng, N.P. (2001) Morphology, drug distribution and *in vitro* release profiles of biodegradable polymeric microspheres containing protein fabricated by double emulsion solvent extraction/evaporation method. *Biomaterials*, **22**, 231–241.
85. Redding, T.W., Schally, A.V., Tice, T.R. and Meyers, W.E. (1984) Long acting delivery systems for peptides: inhibition of rat prostate tumors by controlled release of DTrp6 leutinizing hormone releasing hormone injectable microcapsules. *Proc. Natl. Acad. Sci. USA.*, **81**, 5845–5848.
86. Stenekes, R.J., Franssen, O., van Bommel, E.M. *et al.* (1999) The use of aqueous PEG/dextran phase separation for the preparation of dextran microspheres. *Int. J. Pharm.*, **183**, 29–32.
87. Houchin, M.L. and Topp, E.M. (2008) Chemical degradation of peptides and proteins in PLGA: a review of reactions and mechanisms. *J. Pharm. Sci.*, **97**, 2395–2404.
88. Cleland, J.L., Powell, M.F. and Shire, S.J. (1993) Development of stable protein formulations: a close look at protein aggregation, deamidation, and oxidation. *Crit. Rev. Ther. Drug Carrier Syst.*, **10**, 307–377.
89. Ye, Q., Ashherman, J., Stevenson, M. *et al.* (2000) DepToam™ technology: a vehicle for controlled delivery of protein and peptide drugs. *J. Control. Release*, **64**, 155–166.
90. Howell, S.B. (2001) Clinical application of a novel sustained-release injectable drug delivery system: DepoFoam™ technology. *Cancer J.*, **7**, 219–227.
91. Thauvin, C., Rickling, S., Schultz, P. *et al.* (2008) Carbon nanotubes as templates for polymerized lipid assemblies. *Nat. Nanotechnol.*, **3**, 743–748.
92. Packhaeuser, C.B., Schnieders, J., Oster, C.G. and Kissel, T. (2004) *In situ* forming parenteral drug delivery systems: an overview. *Eur. J. Pharm. Biopharm.*, **58**, 445–455.
93. He, C., Kim, S.W. and Lee, D.S. (2008) *In situ* gelling stimuli-sensitive block copolymer hydrogels for drug delivery. *J. Control. Release*, **127**, 189–207.
94. Berges, R. and Bello, U. (2006) Effect of a new leuprorelin formulation on testosterone levels in patients with advanced prostate cancer. *Curr. Med. Res. Opin.*, **22**, 649–655.
95. Okumu, F.W., le Dao, N., Fielder, P.J. *et al.* (2002) Sustained delivery of human growth hormone from a novel gel system: SABER. *Biomaterials*, **23**, 4353–4358.
96. Choi, S. and Kim, S.W. (2003) Controlled release of insulin from injectable biodegradable triblock copolymer depot in ZDF rats. *Pharm. Res.*, **20**, 2008–2010.
97. Zentner, G.M., Rathi, R., Shih, C. *et al.* (2001) Biodegradable block copolymers for delivery of proteins and water-insoluble drugs. *J. Control. Release*, **72**, 203–215.
98. Qiu, B., Stefanos, S., Lalloo, A. *et al.* (2002) A hydrogel prepared by *in situ* cross-linking of a thiol-containing poly(ethylene glycol)-based copolymer: a new biomaterial for protein drug delivery. *Biomaterials*, **24**, 11–18.
99. Molinaro, G., Leroux, J.C., Damas, J. and Adam, A. (2002) Biocompatibility of thermosensitive chitosan-based hydrogels: an *in vivo* experimental approach to injectable biomaterials. *Biomaterials*, **23**, 2717–2722.
100. Cleland, J.L., Daugherty, A. and Mrsny, R. (2001) Emerging protein delivery methods. *Curr. Opin. Biotechnol.*, **12**, 212–219.
101. Perry, C.M. and Brogden, R.N. (1996) Goserelin: a review of its pharmacodynamic and pharmacokinetic properties, and therapeutic use in benign gynaecological disorders. *Drugs*, **51**, 319–346.
102. Moul, J.W. and Civitelli, K. (2001) Managing advanced prostate cancer with Viadur (leuprolide acetate implant). *Urol. Nurs.*, **21**, 385–388.
103. Chertin, B., Spitz, I.M., Lindenberg, T. *et al.* (2000) An implant releasing the gonadotropin hormone-releasing hormone agonist histrelin maintains medical castration for up to 30 months in metastatic prostate cancer. *J. Urol.*, **163**, 838–844.

104. Basu, S.K., Govardhan, C.P., Jung, C.W. and Margolin, A.L. (2004) Protein crystals for the delivery of biopharmaceuticals. *Expert Opin. Biol. Ther.*, **4**, 301–317.
105. Pechenov, S., Shenoy, B., Yang, M.X. *et al.* (2004) Injectable controlled release formulations incorporating protein crystals. *J. Control. Release*, **16**, 149–158.
106. Margolin, A.L. and Navia, M.A. (2001) Protein crystals as novel catalytic materials. *Angew. Chem. Int. Ed.*, **40**, 2204–2222.
107. Hallas-Møller, K., Petersen, K. and Schlichtkrull, J. (1952) Crystalline and amorphous insulin-zinc compounds with prolonged action. *Science*, **116**, 394–399.
108. Peterson, G.E. (2006) Intermediate and long-acting insulins: a review of NPH insulin, insulin glargine and insulin detemir. *Curr. Med. Res. Opin.*, **22**, 2613–2619.
109. Ayvazian, J.H. and Ayvazian, L.F. (1960) Effect of intravenous administration of crystalline pancreatic desoxyribonuclease in patients with gout. *N. Engl. J. Med.*, **263**, 999–1002.
110. Yang, M.X., Shenoy, B., Disttler, M. *et al.* (2003) Crystalline monoclonal antibodies for subcutaneous delivery. *Proc. Natl. Acad. Sci. USA.*, **100**, 6934–6939.
111. Brunton, S. (2008) Initiating insulin therapy in type 2 diabetes: benefits of insulin analogs and insulin pens. *Diabetes Technol. Ther.*, **10**, 247–256.
112. Haak, T., Edelman, S., Walter, C. *et al.* (2007) Comparison of usability and patient preference for the new disposable insulin device Solostar versus Flexpen, lilly disposable pen, and a prototype pen: an open-label study. *Clin. Ther.*, **29**, 650–660.
113. Pfützner, A., Asakura, T., Sommavilla, B. and Lee, W. (2008) Insulin delivery with FlexPen: dose accuracy, patient preference and adherence. *Expert Opin. Drug Deliv.*, **5**, 915–925.
114. Imoto, J. and Konishi, E. (2005) Needle-free jet injection of a mixture of Japanese encephalitis DNA and protein vaccines: a strategy to effectively enhance immunogenicity of the DNA vaccine in a murine model. *Viral Immunol.*, **18**, 205–212.
115. Bennett, S. and Potter, C. (2008) Solid dose injection of therapeutics and vaccines: effective, convenient and cost-effective alternative to needles. *OndrugDelivery*, **1**, 11–13.

3

Novel Non-Injectable Formulation Approaches of Peptides and Proteins

Shirui Mao,[1] *Dongmei Cun*[1,2] *and Yoshiaki Kawashima*[3]

[1] Shenyang Pharmaceutical University, School of Pharmacy, Shenyang, China

[2] University of Copenhagen, Department of Pharmaceutics and Analytical Chemistry, Faculty of Pharmaceutical Sciences, Copenhagen, Denmark

[3] Aichi Gakuin University, School of Pharmaceutical Science, Nagoya, Japan

3.1 Introduction

Rapid developments in the field of molecular biology and gene technology resulted in generation of many macromolecular drugs including peptides, proteins, polysaccharides and nucleic acids in great numbers. They are highly evolved molecules responsible for carrying out major biological processes and reactions. The future for proteins and peptides as therapeutic and prophylactic drugs seems very bright, considering the sequencing of the entire human genome and the huge strides being made in proteomics. However, when taken orally, peptides and/or proteins are degraded by the proteolytic enzymes in the gastrointestinal tract, and might be impermeable to the intestinal mucosa due to their hydrophilicity and large molecular size, leading to poor oral bioavailability [1]. As a result, systemic delivery of these macromolecular drugs has been limited to the parenteral route. Parenteral delivery of peptide/protein drugs has several drawbacks, including an invasive delivery method, a sterile dosage form, systemic side effects, inconvenient to patients and rapid clearance.

Over the last few decades, pharmaceutical scientists throughout the world have been trying to find alternative ways of administering peptides and proteins other than injection. So far, many efforts have been devoted to investigating the feasibility of nasal, pulmonary,

Delivery Technologies for Biopharmaceuticals: Peptides, Proteins, Nucleic Acids and Vaccines Edited by Lene Jorgensen and Hanne Mørck Nielsen

buccal, transdermal, ocular, rectal, sublingual, vaginal and controlled release oral drug delivery systems for the administration of proteins and peptides [1]. These alternative non-invasive routes potentially have lower peptide degrading enzyme activity. However, use of these alternative routes has had limited success in the absence of external stimuli to facilitate absorption. Therefore, various strategies have been implemented to promote the bioavailability of these drugs. The focus of this review is on the novel non-parenteral formulation approaches to improve the absorption of peptide and protein-based drugs, mainly including intranasal, pulmonal, peroral, transdermal and buccal routes. The parenteral formulation approaches are introduced elsewhere in this book.

3.2 Intranasal Delivery of Peptide/Protein Drugs

3.2.1 Properties of Intranasal Delivery

The intranasal route has been attracting a lot of interest as a non-invasive alternative to current parenteral injections for the systemic delivery of peptide and protein drugs [2]. This is due to its special organization of the nasal mucosae including a highly vascularized epithelium mucosae, a relatively large surface area (160–180 cm^2, epithelium covered with microvilli) and lower enzymatic activity than the gastrointestinal tract. Drugs absorbed from the nasal cavity pass directly into the systemic circulation, thereby avoiding first-pass liver metabolism. Thus, it has the potential to provide for a more rapid and complete drug absorption into systemic circulation and to allow for increased bioavailability. Examples of marketed intranasal administered peptide products include salmon calcitonin (e.g. Miacalcin®), desmopressin (e.g. DDAVP) and nafarelin (e.g. Synarel®).

Moreover, intranasal administration has a promising application in improving immunization. It is well recognized that the development of effective mucosally delivered vaccines is one strategy for potentially ameliorating some of the problems associated with the present vaccines [3, 4]. More importantly, nasal mucosa is rich in lymphoid tissue, named nasal-associated lymphoid tissue (NALT), which will ease the vaccination, and nasal administration allows the vaccine to be transported over a shorter distance before gaining access to the lymphoid tissue. Also, it has been shown that antibodies resulting from parenteral immunization do not necessarily reach mucosal surfaces from where most infectious agents enter the host [5]. One of the most effective methods to induce mucosal immunity in the upper respiratory tract is intranasal immunization [6].

The blood–brain barrier (BBB), segregating the brain interstitial fluid from the circulating blood, and the blood–cerebrospinal fluid barrier (BCB), separating the blood from the cerebrospinal fluid (CSF) that encircles the brain, provide efficient barriers to the diffusion of drugs from the blood stream into the central nervous system (CNS), especially of polar drugs such as peptides and proteins. Intranasal administration offers a non-invasive alternative route to the central nervous system for drug delivery, effectively bypassing the BBB [7]. Olfactory sensory neurons are the only first-order neurons whose cell bodies are located in a distal epithelium. Their dendritic processes are directly exposed to the external environment in the upper nasal passage, while their axons project through perforations in the cribriform plate of the ethmoid bone to synaptic glomeruli in the olfactory bulb. These unique anatomic and physiologic properties of the olfactory region provide both extracellular and intracellular pathways into the CNS that bypass the BBB [8].

The past few years have witnessed a sharp increase in the amount of research on the nasal pathway for CNS drug delivery [7]. Improved delivery to the brain via the intranasal route has been reported for some therapeutic peptides and proteins, such as insulin [9], vascular endothelial growth factor (VEGF) [10] and vasoactive intestinal peptide [11]. However, the total amount of drugs accessing the brain were reported to be low, and they were poorly absorbed and highly susceptible to the harmful environment of the nasal cavity [11–13]. Formulation approaches are used to solve this problem, as discussed in the following parts. On the other hand, it should be noted that there are also cases for which there was no evidence found for preferential delivery to the brain via intranasal dosing [14, 15]. Therefore, the potential for preferential brain delivery for intranasal dosing may be drug specific, or may depend on the study methods employed [12, 16].

3.2.2 Strategies to Enhance Intranasal Absorption

Despite the advantages of the intranasal route, the administration of peptides and proteins by this route is hampered by the chemical and physical instability of these molecules, the high metabolic activity, limited permeability of the mucosal barriers due to the high molecular weight and high hydrophilicity, and nasal mucociliary clearance system [17]. So far, many formulation approaches have been studied in order to improve the absorption of peptides and proteins through the nasal mucosa, such as using absorption enhancers, enzyme inhibitors, solutions of bioadhesive polymers or bioadhesive microspheres, nanoparticles and liposomes.

3.2.2.1 Absorption Enhancers

Adding absorption enhancers in the formulation is the most commonly used approach to enhance intranasal absorption. Commonly used intranasal absorption enhancers are listed in Table 3.1. They are capable of improving the transport/absorption of drugs with low bioavailability by different mechanisms [18]. They may help solubilize or stabilize the drug, they may alter properties of the mucus layer by opening tight junctions between the cells, or they may increase membrane fluidity. It is evident that the choice of an enhancer must be based on the following characteristics: effectiveness, safety, chemical inertness, lack of biological activity and rapid reversibility. The absorption enhancers, which increase the permeability of drugs through the epithelial membranes without causing any tissue damage, are especially useful for the delivery of peptide and protein drugs [18]. Although bile salts, fatty acids and surfactants exhibited permeation enhancing effects, they also produced membrane damage [38]. Therefore, development of new kinds of penetration enhancers without unacceptable side effects is an important approach for successful nasal delivery of peptide drugs.

Chitosan and its derivatives are among the most well known absorption enhancers with a good safety profile [18, 25, 39]. It has been shown that chitosan-based polymers are mucoadhesive and capable of opening the tight junctions between epithelial cells. So far, chitosan has been used extensively for the nasal delivery of peptides and proteins as well as, more recently, for the nasal delivery of vaccines at acidic pH [40]. *N*-Trimethyl chitosan chloride (TMC), a partially quaternized chitosan derivative, has excellent absorption enhancing effects even at neutral pH [39]. Among TMC with different degrees of quaternization (DQ), TMC with a DQ of 50–60% showed strong mucoadhesive and absorption enhancing properties [39]. Recently, a viable approach to enhance nasal drug delivery was exploited by using AT1002, a synthetic peptide, H-FCIGRL-OH, to regulate the intercellular tight junctions in order to enhance paracellular drug transport [41].

It was noticed that the absorption enhancing effect of the penetration enhancers is also influenced by polymer structure, molecular weight and charge density. The enhancing effects of cationized gelatins with different number of amino groups on the nasal absorption of insulin in rats depended on the amino group content [29]. Natsume *et al.* examined the effect of molecular weight and charge density on the enhancing effect of cationic polymers and found that the charge density was the most important parameter as far as the enhancement of the nasal absorption of FD4 was concerned [33, 42]. Spermine, a polyamine having four amino groups, could be useful to increase the charge density and, therefore, the nasal absorption of insulin in rats [31]. The combined use of N-acetyl-L-cysteine (NAC), which is a potent mucolytic agent, and a non-ionic surfactant could remarkably increase the nasal absorption of large molecular weight compounds [43].

3.2.2.2 Delivery Systems with Increased Retention Time

Mucoadhesive approach has been developed to improve the intranasal drug absorption because it could prolong the intimate contact time of the formulation on the nasal mucosa by adhering to the surface of the mucus layer. Different strategies can be used to achieve mucoadhesion.

Surface Modification. The bioadhesive delivery system can be obtained by means of surface modification of drug carriers with biological ligands, which recognize and adhere to specific chemical structures on the surface of cells, thus facilitating absorption of the delivery system. Good examples are lectins, proteins or glycoproteins of non-immunological origin, which specifically recognize sugar molecules and bind to the glycosylated membrane components. Lectin surface modification opened a novel pathway to improve the brain uptake of agents loaded by biodegradable poly(ethylene glycol)-poly(lactic acid) (PEG-PLA) nanoparticles after nasal administration. UEA I, specifically binding to l-fucose, which is largely located in the olfactory epithelium in the nasal cavity, might serve as a promising targeting ligand for the brain drug delivery system following intranasal administration [19]. A recent study suggested that wheat germ agglutinin (WGA) modification on the surface of PEG-PLA nanoparticles facilitated the uptake of an encapsulated fluorescence tracer in the CNS following intranasal administration [19].

Bioadhesive Microsphere Delivery System. To avoid rapid clearance due to ciliary beating and prolong the residence time in the nasal mucosa, different kinds of microspheres have been used in intranasal delivery of several peptide and protein drugs including insulin, such as the cross-linked dextran microspheres, starch microspheres and aminated gelatine microspheres [20–23]. Due to the bioadhesive nature of microspheres, the formulation is retained in the nasal cavity for an extended period. As reported recently, the antitoxin titres induced with tetanus toxoid dextran microsphere in rabbits by intranasal immunization was 175 times higher than the protective levels [24]. Several mechanisms have been reported for absorption enhancement effects of polymer-based microspheres, including [40]:

- that microspheres are deposited in the less or non-ciliated anterior part of the nasal cavity and slower nasal clearance;
- that the gelled system provides a local high drug concentration in close contact with the epithelial absorptive surface;
- that absorption of water by the microspheres from the mucus layer may induce reversible shrinking of the epithelial cells and widening of the tight junctions transiently and therefore the transport of hydrophilic compounds could be increased.

Bioadhesive Excipients. In addition to designing bioadhesive microspheres, mucoadhesion can simply be achieved by using bioadhesive excipients, such as gellan gum, pectin, chitosan, hydropropylmethylcellulose (HPMC) or methylcellulose (MC). B. Jansson *et al.* [26] have shown increased residence time and drug transfer across the nasal epithelium in rats using a gellan gum formulation. Low methylated pectins were shown to gel rapidly after administration into the nasal cavity of sheep and were retained for an extended period resulting in an increased half-time of clearance of the formulation [4, 27]. Chitosan has similarly been shown to have bioadhesive characteristics and to be retained in the nasal cavity for prolonged periods [12, 28]. As described in Section 3.2.2.1, chitosan has also been proven to be an efficient absorption enhancer that can transiently open tight junctions between epithelial cells and thereby significantly enhance the absorption of even large molecular weight polar drugs [27, 30, 32]. Particles consisting of peptides and proteins associated with polymeric chitosan and chitosan derivatives have been shown to enhance the absorption of these macromolecules across mucosal epithelia [34–37]. Recently, it was reported that sodium hyaluronate as a mucoadhesive component in nasal formulation enhances delivery of Fluorescein isothiocyanate-labelled 4 kDa dextran to brain tissue [8].

Insoluble Powder Formulations. Schipper *et al.* were probably the first to point out the difference in the efficacy of nasal absorption from powdery versus liquid formulations containing similar components [44]. They reported that the powder dosage form was far superior to the liquid formulation for insulin absorption. Similarly, it was reported that although the dry powder formulation using water-soluble compound as a drug carrier did not increase the absorption of polar molecules via the nasal route [45], a powder formulation using a water-insoluble compound was demonstrated be effective as a drug carrier to improve nasal bioavailability. A powder formulation using calcium carbonate ($CaCO_3$) improved the nasal bioavailability of elcatonin in the nasal cavity by increasing the residence time [46]. Moreover, a recent study demonstrated that, in dry powder formulations, the physicochemical and micromeritical characteristics of the powder material to be used as the filler can affect the nasal absorbability of peptide drugs [47]. Among the various water-insoluble pharmaceutical excipients, a less wettable or less water-absorbable powder such as ethylcellulose is the preferred candidate for nasal peptide delivery to assure improved bioavailability and low toxicity [47]. When a hydrophilic material is used as filler, it quickly absorbs the nasal fluid, hindering its use as a solvent for the drug. Even if the drug can still barely be dissolved in the nasal fluid, the filler traps the drug solution into its matrix, preventing the drug molecules from diffusing into

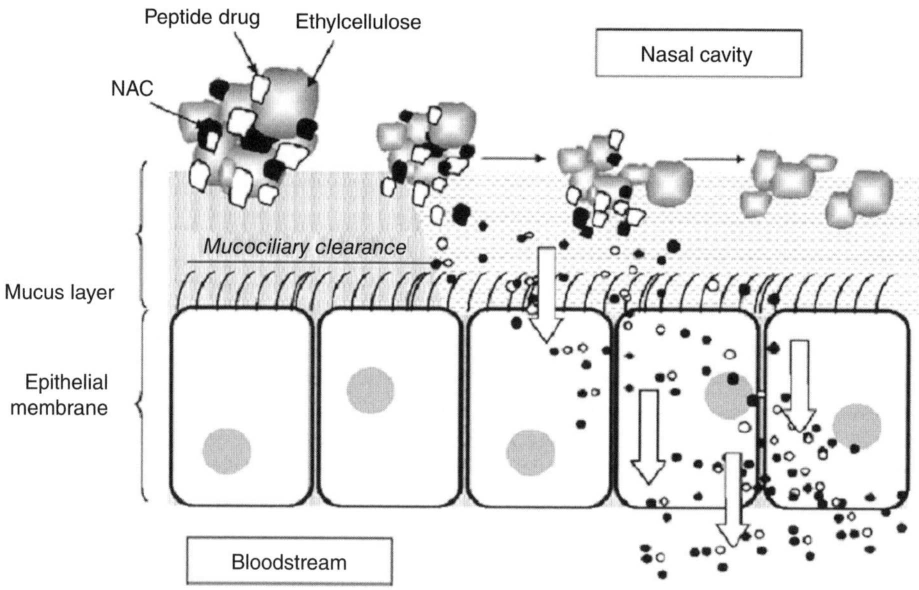

Figure 3.1 *Possible mechanism for the improved nasal bioavailability of a peptide drug delivered in a powder formulation with ethylcellulose and N-acetyl-l-cysteine [37] (Reprinted with permission from Elsevier)*

the epithelial membrane. In contrast, when the filler is not hydrophilic, the nasal fluid is exclusively used for dissolving the drug. Furthermore, when a water-soluble low molecular weight compound such as lactose is used as filler, the osmotic pressure generated may hinder drug permeation across the epithelial membrane. The osmotic pressure could induce the influx of fluid through the epithelial membrane, and this hydrodynamic flow would no doubt hinder the diffusion of the drug molecule into the absorption site. The possible mechanism for the improved nasal bioavailability of peptide drug delivered in a powder formulation with ethylcellulose and N-acetyl-L-cysteine is presented in Figure 3.1.

However, it is generally difficult to achieve a satisfactory effect for macromolecular drugs by applying a bioadhesive approach alone because it has to overcome the physical barrier of the epithelium for the drug to permeate into the circulation. The combined use of penetration enhancers and mucoadhesion should be an effective means to promote the nasal absorption of macromolecular drugs such as peptides and proteins.

3.2.2.3 Nanoparticle/Nanocomplex Drug Carriers

Nanoparticles. Nanoparticle drug carriers consist of solid biodegradable particles in sizes ranging from 10 to 1000 nm (generally, 50–300 nm). Since biological drugs such as peptides, proteins and DNA are poorly absorbed and highly susceptible to the harmful environment of the nasal cavity, the incorporation of these drugs into nanoparticles might be a promising approach to enhance their intranasal absorption [48]. PEG-g-chitosan nanoparticles noticeably improved the nasal absorption of insulin compared to an insulin-PEG-g-chitosan suspension and a control insulin solution [49]. Trimethyl chitosan nanoparticles were able to enhance the uptake of the loaded albumin in the rat nasal mucosa and induce strong systemic and local anti-haemagglutinin antibody responses in intranasally immunized mice with encapsulated influenza haemagglutinin [37, 50].

Nanocomplexes. It is well appreciated that the delivery of naked DNA to mucosal surfaces will normally give rise to little or no transfection of epithelial cells and poor immunological responses since naked DNA is a large, charged macromolecule that has little tendency to be absorbed across the mucosal surfaces [51]. A suitable delivery vehicle is required that will lead to an improved presentation of DNA to antigen presenting cells [52]. Using the traditional method to prepare nanoparticles is not appropriate due to the stability issue. Based on the negative charge of DNA, the addition of positively charged lipids and polymers leads to the compaction of the DNA to form nanoparticulate complexes, which has been shown to be advantageous for the efficiency of DNA vaccines [53]. Cationic polymeric materials such as polylysine, poly(ethylene imine) and chitosan have all been shown to be suitable for DNA intranasal delivery. As reported recently, chitosan/the plasmid pCR-X8-HBc-CETP (pCETP) nanoparticles could significantly attenuate the progression of atherosclerosis by intranasal immunization [54]. PEG-g-trimethyl chitosan insulin nanocomplexes noticeably improved the nasal absorption of insulin compared to a control insulin solution [55]. The nanocomplexes can be formulated to have different particle sizes and surface charge [56]. Moreover, in the nasal cavity, a positively charged nanoparticle is expected to demonstrate mucoadhesion and be cleared more slowly than a neutrally or negatively charged particle [57], leading to an improved opportunity for contact with lymphoid tissue such as the NALT [52]. While a variety of cationic polymers are available for the delivery of DNA to mucosal surfaces, it is important to choose a material that forms stable nanocomplexes and that is able to subsequently disassemble in the cell (endosomal release and escape) to allow the DNA to reach the nucleus. Besides, the resident time of nanoparticles in nasal cavity is limited because of mucociliary clearance, which is not available for the complete absorption of the formulation. This makes bioadhesive nanoparticle formulations with enhanced permeability a better alternative approach.

3.2.2.4 Liposomes

Development of effective delivery systems for the presentation of antigens to mucosal surfaces is critical for the success of these vaccines and liposomes have found extensive application in this area. In fact, liposomes have been identified as effective immunological adjuvants for the intranasal delivery of protein antigens for a long time [3]. The use of liposomes as a drug delivery system is an effective way of inducing the mucosal immune responses by increasing the systemic and mucosal immune responses against the nasally administered tetanus toxoid [58]. Intranasal administration of liposome-encapsulated haemagglutinin-encoding DNA was effective at eliciting a mucosal, cellular and humoral immune response [59]. This may be explained by the fact that liposomes can effectively entrap various drugs and biologicals, which are slowly released over an extended period *in vivo*, and are also efficiently taken up by the antigen presenting cells.

The physicochemical properties of liposomes can influence their utility as a delivery system and vaccine adjuvant. The composition of the lipids and their properties in liposome formulations are of special importance, such as the liquid–crystalline phase transition temperature (Tc) of the lipids, the charge of the lipids, inclusion of cholesterol and liposome size [3]. Enhanced vaccine efficacy was achieved using liposomes prepared from either whole *D. radiodurans* lipids or from a single isolated phosphoglycolipid previously identified as α-galactosylphosphatidylglyceroylalkylamine [60].

3.2.2.5 Microemulsions

Microemulsions, optically isotropic and thermodynamically stable systems of water, oil, surfactant and cosurfactant, have been studied as drug delivery systems on account of their solubilization capacity for poorly water-soluble drugs as well as their enhancement effect on topical and systemic drug availability. Microemulsions by virtue of their lipophilic nature and low globule size are widely explored as a delivery system to enhance uptake across the nasal mucosa [17] and for gene transfer [13]. Vyas *et al.* [61] have demonstrated increased brain uptake of clonazepam from a mucoadhesive microemulsion.

3.3 Pulmonary Delivery of Peptide/Protein Drugs

3.3.1 Properties of Pulmonary Delivery

Over the last few years, absorption of therapeutic macromolecules such as proteins and peptides administered by the pulmonary route has received great attention. Pulmonary delivery offers a number of advantages favourable for drug absorption. Unlike the nasal cavity (approximately 180 cm^2), an adult's lung offers a larger surface area for drug absorption (approximately 100 m^2) (Figure 3.2). In addition, good vascularization and the ultra-thinness of the alveolar epithelium (approximately 0.1–0.5 μm) can facilitate rapid drug absorption. Therefore, the lungs are far more permeable to macromolecules than any other portal of entry into the body [63]. Moreover, in contrast to oral delivery, where a drug can be heavily metabolized and altered by the enzymes of the gut and liver, the lungs have only a small fraction of the drug-metabolizing and efflux transporter activity [62], leading to a high bioavailability. Another benefit is the direct targeting of drugs to the lungs for treatment of pulmonary diseases, which usually requires smaller doses and reduces the potential for unwanted side effects. Importantly, pulmonarily administered vaccines can induce both systemic and local immune responses [64].

3.3.2 Strategies to Enhance Pulmonary Delivery

Despite the advantages of pulmonary delivery, pulmonary-delivered large macromolecular drugs may yield low bioavailability due to enzymatic degradation and/or the relative impermeability of the lungs to many of these drugs. The use of absorption enhancers and proteolytic enzyme inhibitors and suitably designed formulations are all useful approaches to increase the bioavailability of peptide and protein drugs, as described in the following.

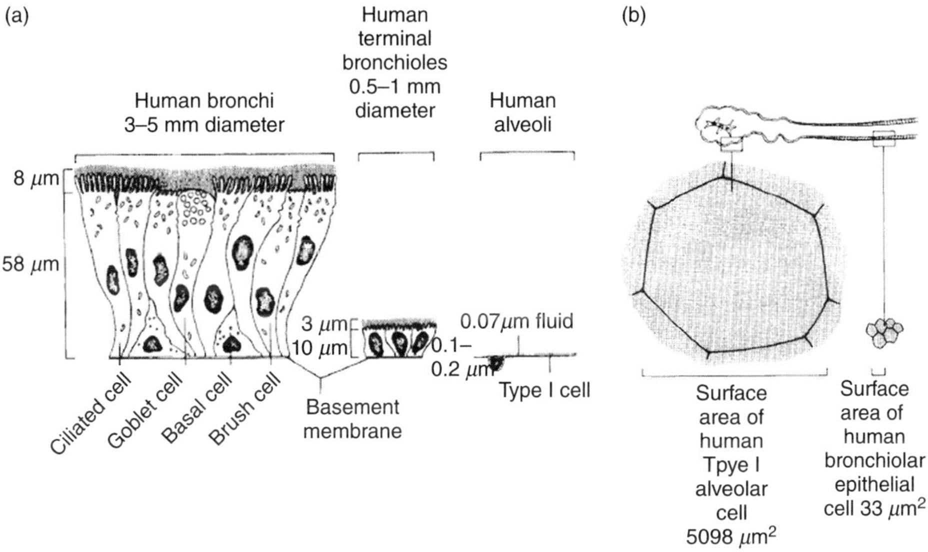

Figure 3.2 *Microstructure of pulmonary mucosae. (a) Lateral view of relative epithelial cell size and surface fluid thickness in different regions of the human lung; (b) Alveolar type I cells have very large surface area compared with airway cells [62] (Reprinted with permission from Elsevier)*

3.3.2.1 Protease Inhibitors and Permeation Enhancers

To improve the stability of proteins and peptides in the pulmonary microenvironment, the concept of including protease inhibitors in pulmonary drug formulations has been researched extensively [65]. Studies have shown that when nafamostat mesilate, which strongly inhibits a variety of proteases such as trypsin, plasmin and kallikaren, was co-administered with insulin, the relative bioavailability of insulin was approximately twice that obtained when the peptide was administered alone [66]. However, the bioavailability of some peptide and protein drugs from the pulmonary route is still poor even when co-administrated with protease inhibitors. Absorption enhancers are usually required to enhance the pulmonary absorption of such macromolecular drugs. Until now, various absorption enhancers have been used for improving the systemic availability of protein and peptide drugs after pulmonary administration alone or together with protease inhibitors [67, 68]. Most of the absorption enhancers used for intranasal delivery can also be used for the pulmonary route (Table 3.1). Recent advances made in the field of pulmonary protein delivery using absorption enhancers have been reviewed in detail by Hussain *et al.* [68]. The absorption enhancing effect is also influenced by the structure and charge density of the enhancers [69]. Similar to that of intranasal delivery, safe absorption enhancers are needed as suitable pulmonary delivery systems for peptide and protein drugs. However, the mechanism of action of most of the absorption enhancers could be due to an irreversible distortion of the alveolar epithelial cell layer, which could potentially make the lungs susceptible to the entry of exogenous allergens and dust particles

Table 3.1 Commonly used absorption enhancers for intranasal absorption of protein and peptide drugs

Type of absorption enhancer	Examples	Mechanism of function	Safety[a]	References
Bile salts and acids	Sodium glycocholate, sodium deoxycholate, sodium taurocholate, sodium glycodeoxycholate (GDC), taurocholic acid, alkyl glycosides	Disrupt membranes, open tight junctions, enzyme inhibition, mucolytic activity.	–	[18, 19]
Fatty acids	Sodium caprylate (C8), sodium caprate (C10) and sodium laurate (C12),	Increase fluidity of phospholipid domains.	–	[20]
	Oleic acid		+	[21]
Surfactants	Laureth-9, lysophosphatidylcholine (LPC), Span 85, sodium taurodihydrofusidate (STDHF)	Perturbation of intercellular lipids, protein domain integrity.	–	[22]
Acylcarnitines	Lauroylcarnitine chloride (LCC)	Member disruption.	–/+	[23]
	Palmitoylcarnitine chloride (PCC)		–	[23]
Phospholipids	Lysophospholipid, Didecanoylphosphatidyl-choline (DDPC)	Member disruption.	–	[24]
			–	
Cyclodextrins	Hydroxypropyl-beta-cyclodextrin	Disrupt membranes, open tight junctions.	+	[25, 26]
	Dimethyl-beta-cyclodextrin		+	[26]
	Methylated beta-cyclodextrins			[27]
	A-cyclodextrin (α-CD).		+	[24]

Cationic polymers	Chitosan and its derivatives	A combination of bioadhesion for preventing mucociliary clearance and a transient widening of the tight junctions in the nasal membrane.	++	[12, 25, 28]
	Gelatine		+	[29, 30]
	Spermine (SPM), spermidine (SPD) sperminated pullulans		+	[31, 32]
	Poly-L-arginines	Increase paracellular transport by inducing a transient and reversible widening of cell–cell junctions.	–/+	[33]
	Cetylpyridinium chloride (CPCL)	Disrupt membranes.	–	[33]
	Protamine			[34]
Anionic polymers	Sodium hyaluronate	Excellent mucoadhesive capacity, enhance the absorption of drugs and proteins via mucosal tissues.	–	[35]
	Alginate		–	[36]
	N-acetyl-L-cysteine (NAC)	Effectively reduce the viscosity of the mucous layer by breaking down the disulfide bonding of mucoproteins.	–	[37]
Synthetic peptide	ZOT-derived peptide AT-1002(H-FCIGRL-OH)	Enhance paracellular drug transport.	+	[31]
Chelators	EDTA	Interfere with Ca^{2+}.	+	[18]

[a] '–' unsafe; '+' safe.

inhaled during respiration. Thus, the use of absorption promoters in pulmonary drug delivery has generated safety concerns regarding possible long-term effects [68]. Therefore, there is an immediate need for discovering or developing novel excipients that could be considered safe and effective in chronic use. A recent study indicated that polyamines, especially spermine (SPM) and spermidine (SPD), can effectively improve the pulmonary absorption of insulin and other macromolecules without any membrane damage to the lung tissues [70].

3.3.2.2 Microparticle Delivery System

In addition to enzymatic degradation and low absorption from the respiratory epithelial barriers, the major factors limiting pulmonary absorption of the proteins are the poor deposition of the protein formulations at the alveolar region, and that the mucociliary escalator rapidly removes protein solution or particles from the central respiratory tract, and thereby prevents the access of the protein to the alveoli. Therefore, a suitable delivery system is required for efficient administration of protein deeply into the lungs. It is well recognized that lung deposition of inhaled aerosols is primarily controlled by their aerodynamic diameter. In humans, orally inhaled aerosols greater than 5–10 μm are mostly trapped by oropharyngeal deposition and are incapable of reaching the lung. In contrast, smaller aerosols $\leq$5 μm can penetrate into the lung, while the finest aerosols $\leq$1 μm are mostly exhaled without deposition. The fraction of particles with an aerodynamic diameter ranging from 1 to 5 μm should be as high as possible to guarantee a maximum deposition in the deep lung [71]. However, differences in aerosol characteristics, even within this 'respirable' size range of 1–5 μm, can still impact the lung-regional distribution. Improved bioavailability can be achieved by efficient targeting of the alveolar region of the lung with mono-dispersed particle formulations with aerodynamic diameters of 2–3 μm [72]. In this respect, microspheres have been proposed, since they can be tailored to exhibit appropriate aerodynamic properties [73] and possess a very narrow range of aerodynamic diameter [71, 74]. Furthermore, even within this 'respirable' size range of 1–5 μm, scintigraphy data showed that 30–50% of the inhaled dose impacts at airway bifurcations [75] and is subsequently excluded from systemic absorption by mucociliary clearance. Therefore, microspheres with mucoadhesive properties are preferred for the prolonged disposition. Also, it is suggested that microspheres having soft surfaces are easily accessible to alveolar macrophages, and effectively trapped by macrophages [76].

3.3.2.3 Nanoparticles

Nanoparticles have recently been proposed as valuable vehicles for efficient peptide and protein transport to the lung epithelium while avoiding unwanted mucociliary clearance and phagocytic mechanisms [77–79]. However, the utility of nanoparticles for pulmonary application is severely hindered because of their low inertia due to their excessively small dimensions and mass, which cause them to escape from lung deposition and be predominantly exhaled [80, 81]. Furthermore, their small size leads to particle–particle aggregation, making physical handling of nanoparticles difficult in liquid and dry powder forms. To solve these limitations, the production of spray-dried powders containing nanoparticles, which dissolve in the lungs into polymeric nanoparticles whose dimensions are sufficiently small to avoid mucociliary and phagocytic clearance until the particles have delivered their therapeutic payload, has recently been reported [82, 83]. It has been demonstrated that chitosan nanoparticle-loaded microspheres, obtained by spray-drying a suspension of nanoparticles in mannitol, exhibited adequate aerodynamic properties for lung delivery [79] and was demonstrated to be biocompatible with respiratory epithelial cell layers [84]. An important objective in the design of biodegradable polymeric materials is to avoid acute toxicity and accumulation of nanocarriers in the lung after administration [85]. SiRNA polyplexes formed with biodegradable poly(amino ester glycol urethane) were shown to display low cytotoxicity [86].

3.3.2.4 Bioadhesive System for Sustained Release

Although a number of high molecular weight drugs have been absorbed successfully through the lungs, such as insulin, heparin and granulocyte colony-stimulating factor (G-CSF) [87], most of the peptides have relatively short half-lives and the development of dosage forms with sustained pharmacological action is essential. However, the drug solution or microdrug particles ($<5\,\mu$m) are eliminated rapidly from the lung by ciliary movement and phagocytosis with macrophages [88]. Therefore, even if a sustained release dosage form is applied to pulmonary delivery, a prolonged pharmacological action cannot always be obtained as expected. To achieve sustained release, recently mucoadhesive PLGA nanospheres that were modified with chitosan (CS) were successfully developed to improve the pulmonary delivery of peptide by nebulization. The elimination rate of CS-modified nanospheres from the lungs was decreased significantly due to the mucoadhesive property after pulmonary administration compared to that of the unmodified nanospheres. The pharmacological action of CS-modified nanospheres loaded with elcatonin was prolonged significantly compared with that of the unmodified nanospheres. Furthermore, it was confirmed that CS on the surface of the nanospheres enhances the drug absorption by opening the intercellular tight junctions in the lung epithelium [89].

3.3.2.5 Liposomes

Liposomes have been investigated for the systemic absorption of different proteins after intratracheal delivery [90, 91]. The use of liposomes for pulmonary drug administration offers several advantages. They are versatile and tend to be relatively innocuous; they also provide protection to the encapsulated materials, minimize adverse side effects and prevent local irritations in the lung. Their organized structure permits the association of drugs to both the aqueous and lipid phase, and drug release can usually be controlled depending on the bilayer's number and composition [1, 92], or whether a drug-loaded vesicle was further incorporated inside a second vesicle, the encapsulation efficiency of particulate matter inside lipid vesicles, or the adsorption of lipid bilayers onto polyelectrolyte-coated capsules [93]. A recent study indicated that the liposomes have the potential to improve vasoactive intestinal peptide (VIP) inhalation therapy by providing a 'dispersible peptide depot' in the bronchi [94]. Animal studies showed that plasma glucose level was effectively reduced when liposomal insulin was delivered by inhalation, indicating that liposome-mediated pulmonary drug delivery promotes an increase in drug retention time in the lungs [95].

The mechanism of absorption enhancement by liposomes may be attributed to a possible interaction between the components of the liposomes and the pulmonary surfactants secreted at the surface of the alveolar membrane [96, 97]. To act as potential carrier for peptides to the lung, the delivery system has to meet certain criteria. Firstly, the liposomal formulation has to be stable and of optimal size to reach the lower bronchioles [98]. Secondly, the liposomes have to accommodate sufficient amounts of peptide and shield their payload from enzymatic degradation in the pulmonary microenvironment. Moreover, the suitability of liposomes for inhalation strongly depends on the drug to be delivered as well as on the lipid composition [99] or charge of the liposomal formulation and the method of application [100]. In general, drug retention is higher for smaller liposomes, which additionally are less rapidly opsonized than their larger counterparts [101]. Thus, liposomes below 200 nm in diameter are suggested to be optimal for pulmonary applications [98]. Similarly, Li and Mitra [102] demonstrated that the ability of liposomes to promote pulmonary protein absorption depends on the concentration, charge and length of the acyl chain of the phospholipids.

Moreover, the formulation of liposomal protein for pulmonary delivery as an aerosol must not cause adverse pulmonary reactions such as a cough or brochoconstriction. Toxicological studies indicated that charge inducing agents, for example stearylamine and dicetylphosphate, might cause an apparent disruption of pulmonary epithelial cells. The formulation must also stabilize the protein sufficiently to ensure that the protein survives the rigors of the aerosol generation process. The engineering of an effective liposomal drug formulation for inhalation therapy must take into consideration the leakage problem associated with the nebulization process, vesicle stability and release kinetics within the pulmonary milieu, and the altered pharmacokinetics of the encapsulated drugs [103]. The formulation components should not interfere with the generation of the aerosol. Moreover, successful drug delivery might be hampered by disintegration and instability of the liposomes caused by the high shear forces exerted by nebulizers leading to unwarranted drug release [104, 105]. The application of liposomes has also been suggested for sustained lung release of several drugs and their interaction with the endogenous phospholipids was proposed as a contribution to the prolonged retention of peptides within the lung [106]. However, liposomal formulations have been shown to be poor siRNA carriers for pulmonary gene delivery [107], possibly due to the surfactants inherent in the lung [108, 109].

Despite the many challenges faced by pulmonary drug delivery systems, several peptide and protein drugs are currently being investigated for potential systemic absorption through pulmonary system, including insulin, calcitonin, luteinizing-hormone-releasing hormone (LHRH) analogues, granulocyte colony-stimulating factor (G-CSF), and human growth hormone (hGH).

3.4 Buccal Administration of Peptide/Protein Drugs

3.4.1 Properties of Buccal Delivery

Buccal administration is a promising alternative to insufficient oral delivery and inconvenient parenteral delivery of hydrophilic macromolecular drugs, such as peptides and proteins. The buccal mucosa has an expanse of smooth muscle, which is relatively immobile, and it has abundant vascularization, thereby providing direct entry into the systemic circulation. The hepatic first-pass effect and degradation in the gastrointestinal tract is thereby avoided. Moreover, buccal delivery offers ease of administration and the possibility to terminate delivery easily if required [110].

The oral mucous membranes have a total surface area of 100 cm^2 (Figure 3.3). A division into different regions can be made: the floor of the mouth (sublingual), the buccal mucosa (cheeks), the gums (gingiva), the palatal mucosa and the lining of the lips. Buccal drug delivery implies the absorption of a drug from a suitable formulation through the mucosa of the cheeks, preferentially excluding other areas of the oral cavity. For buccal adhesive drug delivery system design, a surface area of 1–3 cm^2 and a daily dose of 25 mg or less are preferable. The maximal duration (i.e. contact time) of buccal delivery is approximately 4–6 h [110]. The delivery of peptide drugs across the buccal mucosa is a more convenient and safer approach than most other delivery methods.

To date, a wide variety of polypeptidic drugs have been evaluated for buccal absorption (reviewed by Junginger *et al.* and Sudhakar *et al.*) [110, 111]. It was shown that the buccal administration of proteins like insulin, interferons and interleukins has some advantages, and it reduces many related side effects. For example, buccal delivery provides constant and predictable levels of drug concentration in the blood. Another interesting advantage is its tolerance (in comparison with the nasal mucosa and skin) to delivery being dependent on the drug release from the vehicle as the rate limiting step rather than across the epithelium membrane.

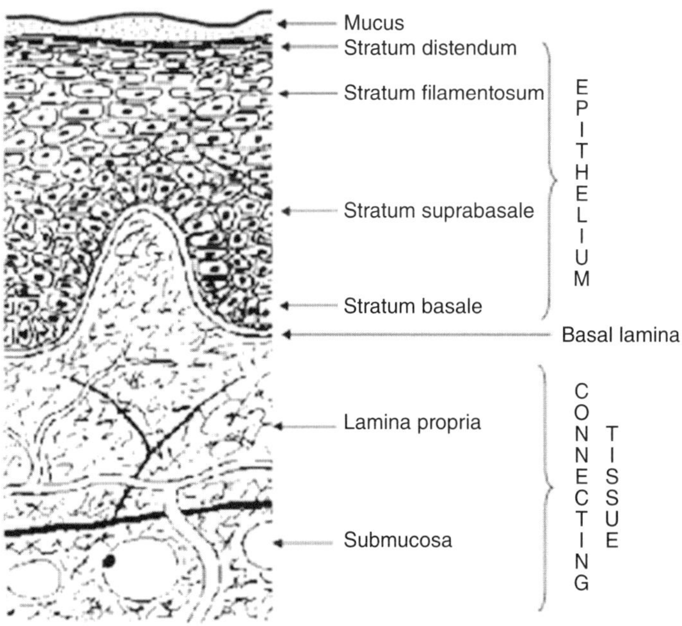

Figure 3.3 *Microstructure of buccal mucosae [111] (Reprinted with permission from Elsevier)*

3.4.2 Strategies to Improve Buccal Absorption

Despite the advantages of buccal delivery, peptides are generally not well absorbed through buccal mucosae because of their molecular size, hydrophilicity and the low permeability of the membrane. Peptide transport across buccal mucosa occurs via passive diffusion and is often accompanied by varying degrees of metabolism. Therefore, various approaches have been studied to improve the buccal absorption of peptides as described in the following.

3.4.2.1 Chemical Enhancers and Enzyme Inhibitors

Similar to that for intranasal and pulmonary drug delivery, it has been shown that buccal absorption can be improved by using various classes of penetration enhancers, such as bile salts, surfactants, fatty acids, chelators, cyclodextrins, chitosans and thiolated polymers [112, 113]. The degree of enhancement depended on a number of factors, including the characteristics of the permeant, the composition of the delivery vehicle and whether the tissue was pretreated with an enhancer. Even for the same kind of penetration enhancer, the difference in structure will influence its function. For example, trimethyl chitosan derived from the lower molecular weight chitosan and characterized by the highest degree of quaternization showed the best mucoadhesive and penetration enhancement properties for large peptides and proteins when administered via the buccal route [114]. Penetration enhancement of the buccal membrane is also drug specific [115]. It is fundamental for the penetration enhancer to be non-irritant, non-toxic and physiologically inactive. However, on account of the buccal mucosa's ability of rapid recovery, it seems to be less critical with the buccal epithelium than with other mucosal tissues. Still, the most efficient absorption promoters – bile acids salts – have a strong bitter taste, so regular use of compositions containing bile acids is hardly acceptable. Recently, lysalbinic acid was shown to increase significantly oral mucosa permeability for interferon and insulin and can be applied as an absorption enhancer for the buccal delivery of peptide drugs [116].

In addition, some enzyme inhibitors have been tested and shown to stabilize peptides against buccal mucosal enzymes [113]. Both enzyme inhibitors and permeation enhancers have been added to transbuccal tablet formulations of insulin and calcitonin to promote absorption [117].

3.4.2.2 Chemical Modification

Peptide absorption is mainly limited because of its instability, low diffusivity and rapid metabolism. Chemical modification can be essential in order to improve stability by minimizing enzymatic cleavage. Another important factor for the passage of peptide through biological membranes is the hydrophobicity. Generally lipophilic drugs absorb through the transcellular route, whereas hydrophilic drugs absorb through the paracellular route. Increasing the non-ionized fraction of ionizable drugs may increase drug penetration through the transcellular route. A potential useful approach is the chemical modification of peptides to produce prodrugs and analogues, which possess favourable physicochemical properties in comparison to the parent compound by rendering the peptide more lipophilic, thus facilitating absorption [118, 119]. This approach is more useful for small peptides with fewer than ten amino acid residues than for the larger peptides [120]. The idea behind such conversion is to protect the peptides against degradation by enzymes present at the mucosal barrier and to render them more lipophilic, thereby improving peptide transport [121]. Anyhow, the improvement in lipophilicity should be appropriate. Veillard *et al.* evaluated a lauryl-TRH tripeptide derivative in order to increase buccal absorption *in vitro* and *in vivo* [122]. In that study, it was shown that a lauryl derivative of a tripeptide (RP-56142) was absorbed from a buccal patch. In contrast, a myristoylated model dipeptide was found to be so lipophilic that it accumulated in the buccal epithelium and was unable to cross the entire mucosal barrier [123].

3.4.2.3 *Bioadhesive Delivery System*

The most successful approach for buccal mucosal delivery of peptides has been a bioadhesive formulation. Bioadhesive systems enhance buccal peptide absorption by providing an intimate contact between the drug dosage form and the buccal mucosa, maintaining a high drug concentration at the absorptive surface for an extended period. The mechanisms of bioadhesion and methods of evaluation have been reviewed recently [110]. Many mucoadhesive biocompatible polymers, including chitosan, hyaluronic acid, cellulose derivatives, poly(acrylates), gelatine and agarose, have been described for buccal drug delivery [124]. Potential bioadhesive polymers for buccal delivery of proteins and peptides are listed in Table 3.2. These systems exhibited increased peptide delivery when applied directly to a freshly excised rabbit mucosa. However, the duration of bioadhesion is largely limited due to the fast turnover rate of the mucus layer. Furthermore, the non-specificity of most bioadhesion processes can be strongly affected by various factors such as fluid pH, food intake and composition. Therefore a concomitant use of penetration enhancers were employed in some cases, such as in the case of buserelin and LHRH [125, 126]. Venugopalan *et al.* have shown that buccally administered insulin loading bioadhesive nanoparticles provided a significant hypoglycemic response without any detectable fluctuation in blood glucose profile and risk of hypoglycemia [127]. In addition to the possession of bioadhesive properties, some bioadhesive polymers have multiple functions, such as enzyme inhibition, permeation enhancing effect and so on. For example, bioadhesive polymer polycarbophil has also been shown to stabilize peptides by inhibiting proteolytic activity [128]. Carbopol 934P, 971P and 974P strongly inhibit proteolysis of insulin, calcitonin and insulin-like growth factor-I [129]. Chitosan has a strong permeation enhancing capacity [130].

3.4.2.4 *Liposomes*

Liposomal formulations with encapsulated drugs have been investigated for buccal administration [131]. However, it was reported that no significant difference in the blood glucose level profile was observed after buccal administration of liposome vesicles containing insulin (LEV-INS) compared with liposome vesicles alone in streptozocin-diabetic rats [132]. Petelin *et al.* showed that liposomes limit the transport of hydrophilic substances to the superficial layer of the epithelium [133]. Incorporation of protease inhibitors is one approach to improve the performance of less effective liposome peptide delivery systems [129].

Table 3.2 *Potential mucoadhesive biocompatible polymers for buccal absorption of protein and peptide drugs [111] (Reprinted with permission from Elsevier)*

Type of polymer	Examples	Characteristics
Chitosan and its derivatives	Chitosan, trimethyl chitosan, thiolated chitosan	Mucoadhesive agent due to either secondary chemical bonds such as hydrogen bonds or ionic interactions between the positively charged amino groups of chitosan and the negatively charged sialic acid residues of mucus glycoproteins or mucins.
Alginate and its derivatives	Sodium alginate	Excellent gel formation properties, Microstructure and viscosity are dependent on the chemical composition.
Cellulose derivatives	Hydroxypropyl cellulose	It is inert and shows no evidence of skin irritation or sensitization; compatible with most water-soluble gums and resins; synergistic with carboxymethyl cellulose (CMC) and sodium alginate.
	Hydroxypropylmethyl cellulose (HPMC)	Methocel E5, E15, E50, E4M, F50, F4M, K100, K4M, K15M, K100M, used for increasing viscosity and controlled release.
	Hydroxyethyl cellulose	Shows good viscosity stability over the range pH 2–12; synergistic with CMC and sodium alginate.
	Sodium carboxymethyl cellulose (CMC-Na)	Most CMC solutions are thixotropic; some are strictly pseudoplastic; with good bioadhesive strength.
Polycarbophil		Entangle the polymer with mucus on the surface of the tissue; hydrogen bonding between the non-ionized carboxylic acid and mucin.
Carbopol/carbomer	934 P, 940 P, 971 P and 974 P.	Common component in bioadhesive dosage forms, may mask the undesirable taste of the formulation.
Gums	Xanthan gum	It is more compatible with most non-ionic and anionic gums, featuring useful synergism with galactomannans.
	Guar gum	Forms viscous colloidal solution when hydrated in cold water. The optimum rate of hydration is between pH 7.5 and 9.0; FDA recognizes as safe.

Table 3.2 (continued)

Type of polymer	Examples	Characteristics
	Hydroxypropyl guar	Better compatibility with minerals than guar gum; good viscosity stability in the pH range 2–13.
	Carrageenan	All solutions are pseudoplastic with some degree of yield value; excellent thermoreversible properties.
Copolymers	Poly(hydroxy butyrate) Poly(e-caprolactone) Poly(ortho esters) Poly(cyano acrylates) Polyphosphazenes Poly(vinyl alcohol) Poly(ethylene oxide) Poly(hydroxytheyl methacrylate) Poly(ethylene oxide-b-propylene oxide)	Biocompatible and biodegradation ability is polymer structure dependent.

3.5 Oral Delivery of Peptide/Protein Drugs

3.5.1 Challenges of Oral Delivery

Among all the alternative routes of parental administration, oral delivery is undoubtedly the most acceptable route for delivering peptides and proteins, offering the advantages of convenience, self-administration without the pain and potential infections associated with injection and a high degree of patient compliance. However, successful oral delivery of peptides/proteins is strongly limited by their poor permeation capacity across the intestinal mucosa and enzymatically rapid degradation in the gastrointestinal (GI) tract [134].

The initial function of the GI tract is to digest and adsorb nutrients and water, but it also works as a fence to prevent the entry of pathogens, toxins and undigested macromolecules [135]. Peptides and proteins could be transported across the GI epithelium via either the transcellular or paracellular pathway. However, the hydrophilic characteristic and the large size of most therapeutic peptides and proteins prevent them from being transported by the transcellular route of absorption by means of passive diffusion [136]. Alternatively, the paracellular route is an aqueous extracellular route across the intercellular spaces between adjacent epithelial cells [134], and this pathway may be interesting for peptide and protein delivery due to the possible deficiency in proteolytic enzymes. But, unfortunately, it was demonstrated that molecules with a radius above 11–15 Å can hardly pass through the paracellular pathway due to the region's tight junction [136]. Apart from the physical barriers, the extremely low oral bioavailability is also attributed to the rapid presystemic enzymatic degradation caused by secreted proteases presented in gastric and intestinal fluids and also to membrane bound peptidases localized on the surface of various mucosal tissues [137].

3.5.2 Strategies to Enhance Oral Absorption

Despite these big challenges, there are some peptide-based drugs, such as desmopressin and cyclosporine A, that are available on the market as oral dosage forms, and some other oral administration products like GLP-1 formulated using Eligen® technology is being evaluated in phase I clinic trials (Emisphere, www.emisphere.com). These facts indicated that it is not absolutely impossible to delivery peptides and proteins orally. To date, huge efforts have been dedicated to the development of oral formulations and improvement of their absorption across the intestinal mucosa.

3.5.2.1 Absorption Enhancer and Enzyme Inhibitor

It has been demonstrated that co-administration with enzyme inhibitors can significantly improve the oral bioavailability of peptides and proteins [138]. However, the safety issue of the co-administration with enzyme inhibitors, especially in long-term therapy, is of concern because they can bring out intestinal mucosal damage, disturbance of the digestion of nutritive protein and stimulate protease secretion due to feedback regulation. It has been proposed that these problems can be solved by using novel drug delivery systems to realize simultaneous release of the inhibitory agent and the therapeutic peptides, thereby restricting the enzyme inhibitors in the small area where they are needed [139]. In addition, a conjugation of enzyme inhibitor and mucoadhensive polymer also has been developed, which not only confers inhibitory properties to polymer but also decreases the side effects by immobilization of the enzyme inhibitor in the delivery systems [140].

Formulating peptides or proteins with membrane permeation enhancing excipients is another effective way to improve their oral bioavailability, since the critical issue of poor permeation across the intestinal epithelial membrane can be potentially addressed by the use of some absorption enhancers [141]. To date, certain surfactants, bile salts, fatty acids and sodium salts of fatty acid, nitrogen-containing rings, salicylates, chelating agents, and toxins and venom extracts like melittin have shown permeation enhancement effects *in vitro* or *in vivo* [134]. Critical premises for successfully adapting this approach are: (i) reproducible intestinal permeation, (ii) low toxicity, and (iii) understanding the mechanisms of permeation enhancement [142]. Generally, permeation enhancers facilitate drug absorption via two different mechanisms: (i) the transcellular route – perturbation of the cell membrane integrity – and (ii) the paracellular route – opening the tight junctions between epithelial cells. Structure–function relationships for permeation enhancers have been established by understanding the mechanisms and it was shown that fatty esters exhibited the most paracellular behaviour while cationic and zwitterionic surfactants displayed the most transcellular behaviour [143]. Apart from those low molecular weight permeation enhancers mentioned above, some polymers, such as chitosan and poly(acrylic acid) derivatives, also show great permeation enhancing effect, which can be further enhanced by thiolation and combination with reduced glutathione (GSH) [144]. Thiol groups of the polymer can inhibit protein tyrosine phosphatase, a key enzyme, which is involved in the closing process of tight junctions, through a GSH-mediated mechanism. Just like the use of enzyme inhibitors, toxicity is also a big concern associated with permeation enhancers. It is worthy to mentioning that the results of some studies indicated the opening of the cellular barrier by chitosan is transient [145]. This special manner makes a great difference in toxicity between chitosan and the classic penetration enhancers, which are known to lead to irreversible epithelial damage.

3.5.2.2 *Particulate Delivery Systems*

Encapsulation of peptides and proteins in particulate delivery systems, such as polymeric nanoparticles (10–1000 nm in size) and microspheres (1–1000 μm), liposomes, emulsions and solid lipid nanoparticles (SLNs), has been developed extensively over the past few years. Particulate delivery systems represent an exciting approach to protect the fragile peptides and proteins against enzymatic degradation in the GI tract, to control the rate or site of entrapped macromolecules release and to enhance the uptake and transport of orally administered macromolecules.

However, the absorption mechanism of particles through the gastrointestinal epithelial cells has not been fully understood yet. Theoretically, particles can be transported across the intestinal epithelium by the transcellular or paracellular pathway [146]. Under physiological conditions, the absorption of particles through the paracellular route is very limited due to the narrow intercellular space and the tight junction between the epithelial cells. Some data from SEM and TEM observation confirmed that nanocapsules smaller than 150 nm can be absorbed via the paracellular pathway through the intestinal mucosa [147]. But, to some extent, paracellular permeation can be enhanced by surfactants and some polymers, such as chitosan and thiolated polymers, transiently opening the tight junctions. Alternatively, uptake of particulate can mainly occur via transcytosis by a general absorptive cell, that is, enterocytote or microfold cell (M cell). M cells, the highly specialized epithelial cells, are located in the human follicle-associated epithelium (FAE) which overlays the intestinal lymphatic tissues, that is the Peyer's patch [148] (Figure 3.4). M-cells transport particles from the gut by endocytosis at the apical part of the cell to the basolateral hollow cavity, where they are released by exocytosis. Although M-cells only represent a very low percentage of the intestinal epithelium (5% of the FAE) [149], it is generally believed that FAE and M-cells are the privileged places for particles absorption.

It is generally agreed that the physicochemical properties of particles, including particle size, surface hydrophobicity, surface charge, shape, elasticity and stability of particles in the GI tract, play a critical role in the extent and mechanism of particle uptake at mucosal surfaces, but there are also other factors that can control the final absorption rate. It has been known for a long time that particles with a size up to 15 μm can be taken up by M-cells, and that small particles were taken up to a greater extent than large particles [150]. Obviously, it is reasonable to enhance the absorption efficiency of peptide or protein loaded particles by optimizing the formulations in terms of size and surface hydrophobicity [146].

Polymeric Particles. Over the past decades, there has been a considerable growth in the development of polymeric particulate delivery systems for oral delivery of peptides and proteins. Biodegradable lactide-based polymers PLA and PLGA, pH sensitive complexation hydrogel poly(methacrylic acid-grafted-ethylene glycol)) (P(MAA-g-EG)), natural polymers such as chitosan, pectins, alginate and cyclodextrins, and some other polymers like poly(alkylcyanoacrylate) and poly(ε-caprolactone) have been extensively

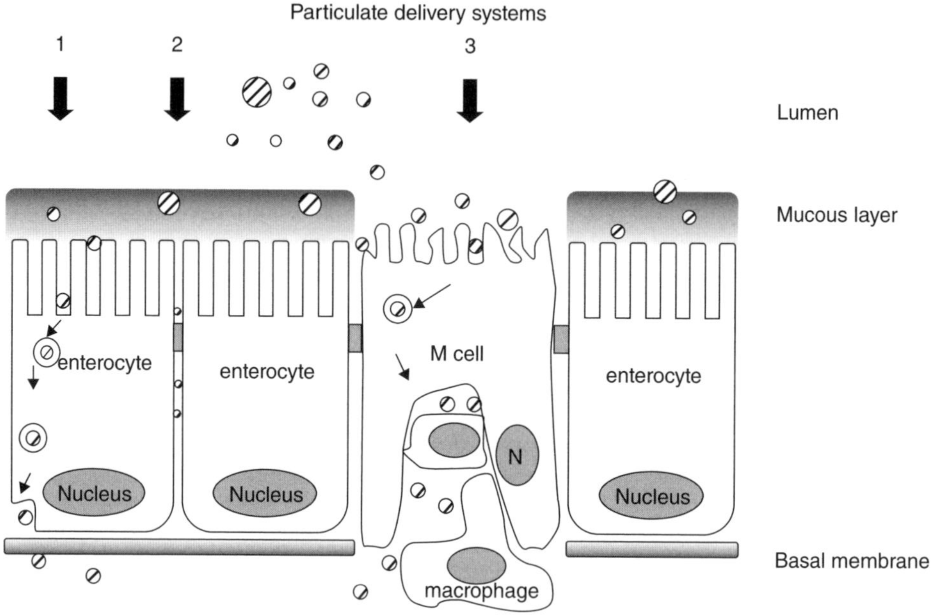

Figure 3.4 *Schematic representation of the three possible mechanisms for the intestinal absorption of particles containing peptides or proteins: 1 transcytosis by enterocyte; 2 paracellular transport; 3 transcellular transport by M cells*

explored as oral delivery carriers of peptides and proteins. Although PLA and PLGA have the advantages of excellent biodegradability and biocompatibility, there are some intrinsic shortcomings, such as the undesired initial burst release and the denaturation of protein and peptide during preparation, storage and release [151].

Chitosan, a natural biodegradable polymer, has attracted increasing attention. Chitosan-based microspheres and nanoparticles have been shown to be promising as an oral delivery carrier due to their penetration enhancement ability and mucoadhensive properties [145]. In addition, there is plenty of possibility to modify chitosan to improve its features or add additional functionality into it. The synthesis and properties of some representative modified chitosans, such as trimethylated chitosan, chitosan-inhibitor conjugates, thiolated polymer, hydrophobically modified chitosan and PEGylated chitosan, have been reviewed by Martin *et al.* [152].

In recent years, considerable efforts has been devoted to the development of pH sensitive hydrogel, which can protect entrapped peptides from the harsh acid environment of the stomach and release them in the intestine [153]. The representative pH-sensitive graft copolymer is synthesized by UV-initiated free radical polymerization of methacrylic acid (MAA) and methoxy-terminated poly(ethylene glycol) monomethacrylate (PEGMA) and consists of cross-linked PMAA grafted with PEG chains, henceforth designated as P(MAA-g-EG) [154]. P(MAA-g-EG)-based particles were considered as the most promising delivery device for oral insulin delivery because of the pH-sensitive swelling property, the proteolytic enzyme inhibition and mucoadhensive properties [155].

Lipid Particles. Liposomes are the most common form of lipid particles. Liposomes have good permeation properties, since their bilayer structures are similar to a cell membrane, but they are easily destroyed by bile salts present in the small intestine, so oral administration of liposomes is limited to very stable formulations that can resist the low pH of the stomach, the bile salts and hydrolytic enzymes of the intestine [156]. Approaches such as double liposome [157] and PEG-coated liposome [158] have been developed to improve the stability of liposomes. Furthermore, oral administration of peptide incorporated into solid lipid nanoparticles [159] or microemulsion [160] also showed some promising results.

3.5.2.3 *Mucoadhesive Oral Drug Delivery Systems*

The concept of mucoadhesion is used to describe the adhesion between an adhesive polymeric device and the gastrointestinal mucosa. Mucoadhesive oral drug delivery systems have attracted increasing attention over the last two decades. It was assumed that this delivery pathway could circumvent the problems associated with oral delivery devices, including the physiological turnover at the intestinal wall and the rapid transit of the devices in intestinal lumen. The mechanism is to localize and prolong the drug at a specific target site, and maintain direct contact with the intestinal cells, thereby increasing the drug concentration gradient.

Enhanced mucoadhesive property is commonly achieved by using some mucoadhesive polymers, such as polyacrylate, cellulose derivatives, chitosan derivatives and alginate, as the component of particles or using them as coating materials to modify the surface of particles [161–163]. Recently, self-assembling pectin-liposome nanocomplexes (PLNs) have been evaluated *in vitro* and *in vivo*. It was shown that the multiple-unit mucoadhesive dosage forms are able to adhere and penetrate through the mucus layer of the GI tissues [164]. Furthermore, coating with poly(vinyl alcohol) (PVA) or Vitamin E succinylated PEG 1000 also confers adhesion capability to the particles [165]. The binding is mainly based on non-specific interactions (hydrogen bond and ionic interaction) between the particles and the intestinal mucosa. It is worth pointing out that the period of these non-specific mucoadhesive systems remaining at the site of attachment is limited partly due to the high turnover and sloughing rate of the mucus and the possible displacement of the delivery systems by soluble mucus in the luminal content [166]. Furthermore, interaction with food and other intestinal contents may also significantly decrease the efficiency of the mucoadhesion.

Apart from this, modifying the surface of the particles by coupling various cytoadhesive ligands is a more specific and effective approach to gain mucoadhesive properties, which was recognized as the second new generation of mucoadhesion-specific site-directed bioadhesive. The decorated particles can be recognized by epithelial cells or M-cells and consequently attached to a specific site at the mucosal surface [146]. Some important cytoadhesive ligands that have been extensively explored included lectins, bacterial adhesins, amino acid sequences and monoclonal antibody [167]. Lectin is a carbohydrate-binding protein of non-immune origin from either plant or bacteria that agglutinates cells or precipitates polysaccharides or glycoconjugates. The specificity, multivalent nature and their non-immunogen properties make lectins attractive candidates for the development of specific mucoadhesive oral peptide and protein delivery system [166].

3.5.2.4 Colon Specific Drug Delivery

The colon has attracted much attention as a site for delivery of peptides and proteins because of the low activity of proteolytic enzyme there. In addition, other advantages, including greater responsiveness to absorption enhancers and protease inhibitors, near neutral pH and longer residence time, also facilitate the adsorption of those macromolecules [168]. Colon specific drug delivery systems were designed to be able to remain intact in the stomach and the small intestine, and release the peptides or proteins only after entering the large intestine in response to the time that the dosage form takes to transit to colon, the special microflora distribution and the distinct enzymes or the pH change along the GI tract. Correspondingly, those systems were called (i) time-controlled systems [169], (ii) enzymatically-controlled systems [170] and (iii) pH sensitive systems, respectively. However, the success with those systems is limited so far mainly due to the high variability between individuals. Combining various mechanisms could be a rational resolution to this issue [171].

3.5.2.5 Chemical Modification

Lipidization. Chemical modification via lipidization, aimed at improving lipophilicity and enzymatic stability of peptides and proteins, appears to be a reasonable approach to improve the oral bioavailability of peptides and proteins, since the high hydrophilic feature and enzymatic degradation are not favourable for the absorption of peptides and proteins through intestinal epithelium cells [136].

The lipidization can usually be done by forming an amine bond between a carboxyl group of a lipid molecular and an amino group of a peptide. Various fatty acids with different chain lengths, like butyric acid, palmitic acid, lauric acid and caprylic acid, have previously been used as lipid moieties to modify peptides such as insulin [172], desmopressin [173] and leucine enkephalin analogue DADLE [174]. It was shown that these lipophilic derivatives of peptides possessed enhanced permeability across intestinal membrane, higher stability against the enzymatic degradation and longer plasma half-life. However, the problems associated with the synthesis of acyl derivatives, such as low solubility and reduced biological activity of the lipidized product, have limited its extensive application [136]. In order to avoid these problems, a method termed reversible aqueous lipidization (REAL) has been developed. Several therapeutic peptides, such as desmopressin [175], salmon calcitonin [176] and Leu-enkephalin [177], have been lipidized by this technique, and those REAL peptides exhibit not only improved pharmacokinetic and pharmacodynamic properties but also an increase in epithelial absorption.

Furthermore, peptides have been chemically modified by conjugating with lipoamino acids (LAAs) or their oligomers called lipopeptides, which represent a class of compounds combining the structural features of lipids and amino acids [178]. The conjugate of LAAs and different peptides, such as LHRH, thyrotropin releasing hormone (TRH) [178], α-conotoxin MII [179] and a dipeptide L-Glu-L-Trp-OH [180], showed enhanced metabolic stability and also a higher rate of oral uptake.

Coupling with Targeting Ligands. Modification of proteins by coupling them with targeting endogenous ligands (transport-carrier molecules) is another promising strategy to enhance intestinal absorption of peptides and proteins [181]. The conjugated molecule can be specifically recognized by cellular receptors in the GI tract and successfully transported from the apical to the basolateral membrane of epithelium cells via a receptor-mediated endocytotic mechanism. Receptor recognizable ligands, including lectins, toxins, transferrin (Tf), cobalamin (vitamin B12), folate and viral haemagglutinins, have been explored to form a conjugate with peptide to improve the intestinal absorption [182]. As one of the examples, insulin was covalently linked to iron-loaded human Tf via a disulfide linkage and it was demonstrated that conjugation to Tf can markedly improve the hypoglycemic effect of insulin in streptozotocin induced diabetic rats [183].

The key point to successful use of this strategy is the need to ensure that the coupling does not block the active site both on the ligand and the therapeutic peptide. The ligand must still be recognized by the receptor and the peptide/protein must preserve its original biological activity to induce the desired effect. The disadvantage of this delivery system is the limited numbers of intrinsic-factor receptors on the intestinal epithelium membrane. Consequently, there is an upper limit on the amount of efficacious peptide that can be delivered to the circulation via the pathway, and the factional absorption will decrease with the increase in the dose. However, this could be counteracted by means of multiple dosing or by using ligand-tagged nanoparticles, which could significantly increase the loading of peptide per ligand [184].

3.6 Transdermal Delivery of Peptide/Protein Drugs

3.6.1 Properties of Transdermal Delivery

Transdermal delivery of peptides is an attractive route, since this route has many advantages, such as being painless, avoiding the hepatic first-pass effect, the skin has a lower proteolytic activity than the gastrointestinal tract, having the possibility of continuously delivering drugs with short half-lives, thus prolonging their effect [185], and encouraging patient compliance, since a transdermal delivery system would be easy to apply and remove. However, previous studies suggested that peptide drugs might undergo extensive enzymatic degradation in the viable skin [186]. In addition, the basic prerequisite of transdermal drug delivery is the ability of the drug to penetrate the outermost layer of the skin, that is the stratum corneum (Figure 3.5). This is comprised of keratin-rich cells embedded in multiple lipid bilayers, and it is an excellent barrier to the transport of large molecules and charged compounds. Therefore, it is desirable to increase the transport of peptides across the lipoidal barrier membranes by overcoming the problems of their permeability and stability.

3.6.2 Strategies to Improve Transdermal Absorption of Proteins/Peptides

Since the amount of macromolecules that can be administered transdermally is quite low, it is desirable to facilitate the transdermal transport using different strategies, usually chemical or physical approaches.

3.6.2.1 Chemical Enhancers

One approach to modify the barrier properties of the stratum corneum is to use certain adjuvants, including absorption enhancers and/or metabolic inhibitors [188–190]. It is

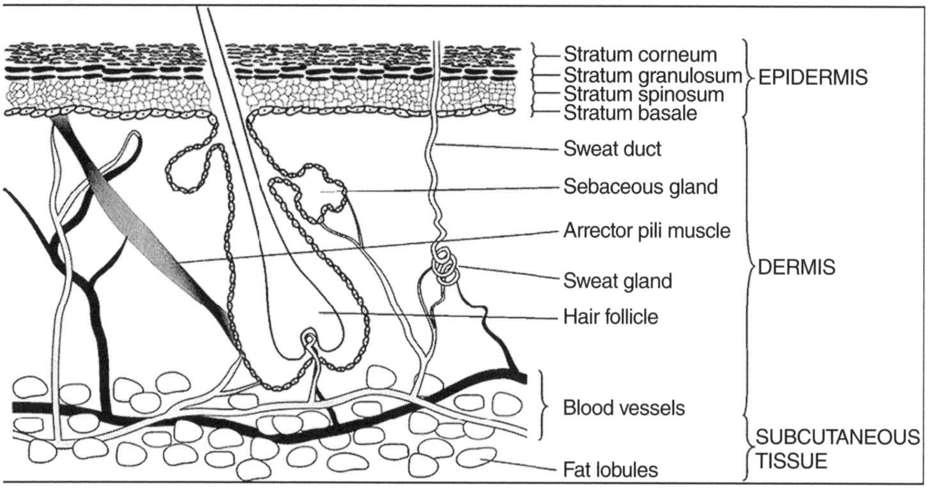

Figure 3.5 *Diagrammatic representation of a cross-section through human skin showing the different cell layers and appendages [187] (Reprinted with permission from Elsevier)*

noticed that the absorption enhancers commonly used in other mucosa, such as intranasal or pulmonary, cannot be applied to the transdermal route directly due to the difference in physiological structure of the application site. Some commonly used chemical enhancers for the transdermal absorption of proteins/peptides and their mechanism of action are summarized in Table 3.3 [191–196]. It was reported that in the presence of penetration enhancers such as terpenes and ethanol, the transdermal penetration of TRH and M-TRH through human epidermal membranes can be delivered in amounts sufficient to elicit biologic effects [187, 197]. A frequently reported effect is that penetration enhancers act by partitioning into and interacting with the outermost layer of the skin to ideally induce a reversible increase in skin permeability. These invasive approaches might prove to be useful, but there are still some problems in their safety and skin irritation.

3.6.2.2 Structure Modification

Another promising approach is to use prodrugs and analogues to improve the transdermal absorption of peptide and protein drugs. It was indicated that the transdermal delivery of tetragastrin was improved by lipophilic modification with fatty acids [198]. Similarly, the stability and permeability of phenylalanyl-glycine across the skin was improved by chemical modification with various fatty acids [199].

3.6.2.3 Liposomes

Although the majority of reports dealing with standard liposomes concentrate on improved drug deposition into skin and its appendages, some studies showed improved transdermal delivery from these nano-aggregates. Fresta and Puglisi [200] found that vesicles of unsaturated phospholipid (fluid liposomes), produced high percutaneous absorption and tissue distribution rather than skin accumulation. Employing a human skin graft model,

Table 3.3 *Representative absorption enhancers for transdermal absorption of protein and peptide drugs*

Type of transdermal absorption enhancer	Examples	Mechanism of action	References
Solvents	Ethanol	Cause dehydration of skin, lipid extraction and changes in skin proteins.	[187, 197]
	Dimethyl acetamide	Cause lipid extraction and changes in skin proteins.	[193]
	Ethyl acetate	Cause lipid extraction.	[193]
	Isopropyl myristate (IPM), isopropyl palmitate	Cause increase in lipid fluidity.	[193, 194]
	Propylene glycol	Cause dehydration of skin.	[193]
	Transcutol®	Act by improving drug solubility within the membrane rather than by increasing drug diffusivity across it.	[198]
	Sulfoxides	Induce significant structural perturbations.	[199]
Fatty acids and esters	Oleic acid	Be able to induce lipid fluidization as well as phase separation within the membrane.	[200]
	Dimethylaminohexanoic acid dodecyl ester	Directly decreased the skin barrier resistance.	[201]
Hydrocarbons (alkanes and alkenes)	Alkyldisiloxanes, decanol	A reversible decrease of the barrier resistance.	[202, 203]
Terpenes	Menthol, cineole	Able to affect the intercellular packing of the stratum corneum lipids but the disruption is entirely reversible.	[192, 197, 204]
	Thiomenthol derivatives		[205]
Surfactants	Sodium lauryl sulfate, Span	Disruption of the densely packed lipids that fill the extra cellular spaces of the stratum corneum.	[206]
	Sugar esters		[207]
Azone (1-dodecylaza-cycloheptan-2-one)	Laurocapram	Act by lipid fluidization as well as ion pairing.	[199]
Urea		A combination of increasing stratum corneum water content and through the keratolytic activity.	[190]
Cyclodextrin	β-cyclodextrin (β-CD), randomly methylated β-cyclodextrin (RAMEB) and 2-hydroxypropyl β-cyclodextrin (HPβ-CD)	Extract significant amounts of cholesterol from isolated stratum corneum.	[208]

a liposomal formulation augmented the uptake of γ-interferon into the epidermis of viable human skin compared with aqueous solution [201]. Liposome-encapsulated antibodies were distributed rapidly into the deep cutaneous regions of piglet skin with a clearly raised percutaneous absorption compared to aqueous solutions [202]. To overcome the stratum corneum barrier of the skin without carrying out any skin penetration, cationic liposomes, DC-Chol/DOPE and DOTAP, were employed as vehicles for the transdermal antigen DNA delivery. The developed lipoplex-patch DNA vaccines have proven to be simple and non-invasive, by which the antibodies incurred provide marked therapeutic effects in test animals [203]. It was suggested that flexible liposomes are more efficient in transdermal drug delivery. The presence of a surfactant increases the elasticity of the lipid bilayers [202].

3.6.2.4 Micro/Nano Particles

Some other dosage forms, such as nanospheres [204], can also be used to enhance transdermal absorption of protein or peptide drugs. It has been demonstrated that particles with a diameter up to 10 mm are able to penetrate into the annexes of the skin, that is sweat and sebaceous glands and hair follicles [205]. A recent study by Tahara *et al.* indicated that solid-in-oil nanodispersion shows the possibility of transdermal protein delivery [206].

3.6.2.5 Physical Approaches

Peptides typically contain acidic and basic functional groups that are usually charged at physiological conditions. Therefore, passive transdermal diffusion of these large molecular weight ions across the lipophilic outermost layers of the skin is extremely difficult. Based on their properties, some physical approaches are reported to be able to enhance the transdermal absorption. A number of studies have demonstrated that electroporation-mediated transdermal delivery of peptides, polysaccharides, oligonucleotides and genes may be possible at clinically relevant rates [207]. Iontophoresis or sonophoresis techniques are extensively studied [208–210]. *In vitro* and animal *in vivo* iontophoretic delivery of several small peptides, in the molecular weight range 1000–5000 Da, have been reported widely and outlined in the literature [211]. Transdermal absorption of peptide drugs, insulin and calcitonin under iontophoresis using pulsed current was enhanced by the switching technique [212]. The use of pulsed direct current profile resulted in an enhanced transport of LHRH and Nafarelin across human epidermis and with a slightly better stability of LHRH than during the iontophoresis [213]. The pulse depolarization iontophoresis (PDPIP) system is a unique system which provides a method of depolarizing the skin surface after application of each pulse and could enhance transdermal absorption of peptide drugs such as salmon calcitonin (sCT) [214]. A recently study indicated that the erbium:yttrium-aluminium-garnet (Er:YAG) laser can be effective for transdermal delivery of peptides and protein-based drugs [215].

3.7 Conclusions

Nasal, pulmonary, buccal, oral and transdermal mucous membranes have all been evaluated as potential alternative routes for protein or peptide absorption. The advantages and

Table 3.4 *Advantages and disadvantages of different administration routes [216] (Reprinted with permission from Elsevier)*

Route	Approximate surface area	Physiological properties	Enzymatic barrier properties	Unenhanced bioavailability (%)	Technological challenges
Nasal	150 cm^2	Mucus, ciliated, columnar pseudostratified epithelium (10 μm)	Medium	2–20	Small surface area; permeation
Pulmonary	80–140 m^2	Bronchi and bronchioles: mucus, ciliated columnar pseudostratified epithelium (10–60 μm); alveoli: squamous epithelial monolayer (<1 μm)	Low	20–80	Precision of delivery
Buccal	100 cm^2 (non-keratinized)	Mucus, stratified, partly keratinized epithelium (500–600 μm), hydrated	Medium	0–5	Permeation
Oral	>200 m^2	Thick mucus layer, columnar epithelial monolayer (10 μm), pH variations	High	0–1	Enzymatic stability; permeation; efflux
Trans-dermal	1.8 m^2	Keratinized stratified epithelium (100–200 μm)	Very low	0–1	Permeation

disadvantages of different administration routes are summarized in a recent review [216], as shown in Table 3.4. It is clear that these alternative approaches have major limitations of variable bioavailability and the safety concerns of the enhancers used in the formulations. Among them, protein and peptide drugs stand to benefit from nasal delivery (with bioavailability of 2–20%) and pulmonary delivery (20–80%) the most, especially for drugs that require a quick onset of action. However, compared to the pulmonary route, the intranasal route is less demanding in terms of formulation technology, such as particle size requirements and complexity of delivery device, and has a very promising future for the delivery of peptides and proteins with appropriate formulation design.

References

1. Cryan, S.A. (2005) Carrier-based strategies for targeting protein and peptide drugs to the lungs. *AAPS J.*, **7**, 20–41.
2. Pontiroli, A.E. (1998) Peptide hormones: review of current and emerging uses by nasal delivery. *Adv. Drug Deliv. Rev.*, **29**, 81–87.

3. Alpar, H.O., Bowen, J.C. and Brown, M.R.W. (1992) Effectiveness of liposomes as adjuvants of orally and nasally administered tetanus toxoid. *Int. J. Pharm.*, **88**, 335–344.
4. Illum, L. (2003) Nasal drug delivery-possibilities, problems and solutions. *J. Control. Release*, **87**, 187–198.
5. Bowersock, T.L., HogenEsch, H., Suckow, M. *et al.* (1996) Oral vaccination with alginate microsphere systems. *J. Control. Release*, **39**, 209–220.
6. Rebelatto, M.C., Guimond, P., Bowersock, T.L. and HogenEsch, H. (2001) Induction of systemic and mucosal immune response in cattle by intranasal administration of pig serum albumin in alginate microparticles. *Vet. Immunol. Immunopathol.*, **83**, 93–105.
7. Graff, C.L. and Pollack, G.M. (2005) Nasal drug administration: potential for targeted central nervous system delivery. *J. Pharm. Sci.*, **94**, 1187–1195.
8. Horvát, S., Fehér, A., Wolburg, H. *et al.* (2009) Sodium hyaluronate as a mucoadhesive component in nasal formulation enhances delivery of molecules to brain tissue. *Eur. J. Pharm. Biopharm.*, **72**, 252–259.
9. Sigurdsson, P., Thorvaldsson, T. and Gizuararson, S. (1997) Olfactory absorption of insulin to the brain. *Drug Deliv.*, **4**, 195–200.
10. Yang, J.P., Liu, H.J., Cheng, S.M. *et al.* (2009) Direct transport of VEGF from the nasal cavity to brain. *Neurosci. Lett.*, **449**, 108–111.
11. Dufes, C., Olivier, J.C., Gaillard, F. *et al.* (2003) Brain delivery of vasoactive intestinal peptide (VIP) following nasal administration in rats. *Int. J. Pharm.*, **255**, 87–97.
12. Illum, L. (2004) Is nose-to-brain transport of drugs in man a reality? *J. Pharm. Pharmacol.*, **56**, 3–17.
13. Kim, T.M., Chung, H., Kwon, I.C. *et al.* (2000) *In vivo* gene transfer to the mouse nasal cavity mucosa using a stable cationic lipid emulsion. *Mol. Cells*, **10**, 142–147.
14. van den Berg, M.P., Verhoef, J.C., Romeijn, S.G. and Merkus, M.W. (2004) Uptake of estradiol or progesterone into the CSF following intranasal and intravenous delivery in rats. *Eur. J. Pharm. Biopharm.*, **58**, 131–135.
15. Yang, Z., Huang, Y., Gan, G. and Sawchuk, R.J. (2005) Microdialysis evaluation of the brain distribution of stavudine following intranasal and intravenous administration to rats. *J. Pharm. Sci.*, **94**, 1577–1588.
16. Charlton, S.T., Davis, S.S. and Illum, L. (2007) Nasal administration of an angiotensin antagonist in the rat model: Effect of bioadhesive formulations on the distribution of drugs to the systemic and central nervous systems. *Int. J. Pharm.*, **338**, 94–103.
17. Ugwoke, M.I., Verbeke, N. and Kinget, R. (2001) The biopharmaceutical aspects of nasal mucoadhesive drug delivery. *J. Pharm. Pharmacol.*, **53**, 3–21.
18. Davis, S.S. and Illum, L. (2003) Absorption enhancers for nasal drug delivery. *Clin. Pharmacokinet.*, **42**, 1107–1128.
19. Gao, X., Tao, W., Lu, W. *et al.* (2006) Lectinconjugated PEG-PLA nanoparticles: preparation and brain delivery after intranasal administration. *Biomaterials*, **27**, 3482–3490.
20. Pereswetoff-Morath, L. and Edman, P. (1995) Dextran microspheres as a potential nasal drug delivery system for insulin - *in vitro* and *in vivo* properties. *Int. J. Pharm.*, **124**, 37–44.
21. Illum, L., Farraj, N.F., Davis, S.S. *et al.* (1990) Investigation of the nasal absorption of biosynthetic human growth hormone in sheep – use of a bioadhesive microsphere delivery system. *Int. J. Pharm.*, **63**, 207–211.
22. Oechslein, C.R., Fricker, G. and Kissel, T. (1996) Nasal delivery of octreotide: absorption enhancement by particulate carrier systems. *Int. J. Pharm.*, **139**, 25–32.
23. Wang, J., Tabata, Y. and Morimoto, K. (2006) Aminated gelatin microspheres as a nasal delivery system for peptide drugs: Evaluation of *in vitro* release and *in vivo* insulin absorption in rats. *J. Control. Release*, **113**, 31–37.
24. Sajadi Tabassi, S.A., Tafaghodi, M. and Jaafari, M.R. (2008) Induction of high antitoxin titers against tetanus toxoid in rabbits by intranasal immunization with dextran microspheres. *Int. J. Pharm.*, **360**, 12–17.
25. Sinswat, P. and Tengamnuay, P. (2003) Enhancing effect of chitosan on nasal absorption of salmon calcitonin in rats: comparison with hydroxypropyl- and dimethyl-beta-cyclodextrins. *Int. J. Pharm.*, **257**, 15–22.

26. Jansson, B., Hagerstrom, H., Fransen, N. *et al.* (2005) The influence of gellan gum on the transfer of fluorescein dextran across rat nasal epithelium *in vivo*. *Eur. J. Pharm. Biopharm.*, **59**, 557–564.
27. Illum, L. (2000) Transport of drugs from the nasal cavity to the central nervous system. *Eur. J. Pharm. Sci.*, **11**, 1–18.
28. Soane, R.J., Frier, M., Perkins, A.C. *et al.* (1999) Evaluation of clearance characteristics of bioadhesive systems in humans. *Int. J. Pharm.*, **78**, 55–65.
29. Seki, T., Kanbayashi, H., Nagao, T. *et al.* (2005) Effect of aminated gelatin on the nasal absorption of insulin in rats. *Biol. Pharm. Bull.*, **28**, 510–514.
30. Hinchcliffe, M. and Illum, L. (1999) Intranasal insulin delivery and therapy. *Adv. Drug Deliv. Rev.*, **35**, 199–234.
31. Seki, T., Kanbayashi, H., Chono, S. *et al.* (2007) Effects of a sperminated gelatin on the nasal absorption of insulin. *Int. J. Pharm.*, **338**, 213–218.
32. Smith, J.M., Dornish, M. and Wood, E.J. (2004) Involvement of protein kinase C in chitosan glutamate-mediated tight junction disruption. *Biomaterials*, **26**, 3269–3276.
33. Natsume, H., Iwata, S., Ohtake, K. *et al.* (1999) Screening of cationic compounds as an absorption enhancer for nasal drug delivery. *Int. J. Pharm.*, **185**, 1–12.
34. Fernandez-Urrusuno, R., Calvo, P., Remunan-Lopez, C. *et al.* (1999) Enhancement of nasal absorption of insulin using chitosan nanoparticles. *Pharm. Res.*, **16**, 1576–1581.
35. van der Lubben, I.M., Kersten,G., Fretz, M.M. *et al.* (2003) Chitosan microparticles for mucosal vaccination against diphtheria: oral and nasal efficacy studies in mice. *Vaccine*, **28**, 1400–1408.
36. Vila, A., Sanchez, A., Janes, K. *et al.* (2004) Low molecular weight chitosan nanoparticles as new carriers for nasal vaccine delivery in mice. *Eur. J. Pharm. Biopharm.*, **57**, 123–131.
37. Amidi, M., Romeijn, S.G., Borchard, G. *et al.* (2006) Preparation and characterization of proteinloaded N-trimethyl chitosan nanoparticles as nasal delivery system. *J. Control. Release*, **111**, 107–116.
38. Schiepper, F.W.H.M., Merkus, N.G.M., Hermens, W.A.J.J. *et al.* (1993) Absorption enhancers in nasal drug delivery: efficacy and safety. *J. Control. Release*, **24**, 201–208.
39. Hamman, J.H., Stander, M. and Kotze, A.F. (2002) Effect of the degree of quaternization of N-trimethyl chitosan chloride on absorption enhancement: *in vivo* evaluation in rat nasal epithelia. *Int. J. Pharm.*, **232**, 235–242.
40. Illum, L., Jabbal-Gill, I., Hinchcliffe, M. *et al.* (2001) Chitosan as a novel nasal delivery system for vaccines. *Adv. Drug Carrier. Rev.*, **51**, 81–96.
41. Song, K., Fasano, A. and Eddington, N.D. (2008) Enhanced nasal absorption of hydrophilic markers after dosing with AT1002, a tight junction modulator. *Eur. J. Pharm. Biopharm.*, **69**, 231–237.
42. Miyamoto, M., Natsume, H., Iwata, S. *et al.* (2001) Improved nasal absorption of drugs using poly-l-arginine: effects of concentration and molecular weight of poly-l-arginine on the nasal absorption of fuorescein isothiocyanate-dextran in rats. *Eur. J. Pharm. Biopharm.*, **52**, 21–30.
43. Matsuyama, T., Morita, T., Horikiri, Y. *et al.* (2006) Enhancement of nasal absorption of large molecular weight compounds by combination of mucolytic agent and nonionic surfactant. *J. Control. Release*, **110**, 347–352.
44. Schipper, N.G.M., Romeijn, S.G., Verhoef, J.C. and Merkus, F.W.H.M. (1993) Nasal insulin delivery with dimethyl-beta-cyclodextrin as an absorption enhancer in rabbits – powder more effective than liquid formulations. *Pharm. Res.*, **10**, 682–686.
45. Ugwoke, M.I., Kaufmann, G., Verbeke, N. and Kinget, R. (2000) Intranasal bioavailability of apomorphine from carboxymethylcellulose based drug delivery systems. *Int. J. Pharm.*, **202**, 125–131.
46. Ishikawa, F., Katsura, M., Tamai, I. and Tsuji, A. (2001) Improved nasal bioavailability of elcatonin by insoluble powder formulation. *Int. J. Pharm.*, **224**, 105–114.
47. Matsuyama, T., Morita, T., Horikiri, Y. *et al.* (2007) Influence of fillers in powder formulations containing N-acetyl-L-cysteine on nasal peptide absorption. *J. Control. Release*, **120**, 88–94.

93. McPhail, D., Tetley, L., Dufes, C. and Uchegbu, I.F. (2000) Liposomes encapsulating polymeric chitosan based vesicles – a vesicle in vesicle system for drug delivery. *Int. J. Pharm.*, **200**, 73–86.
94. Hajos, F., Stark, B., Hensler, S. *et al.* (2008) Inhalable liposomal formulation for vasoactive intestinal peptide. *Int. J. Pharm.*, **357**, 286–294.
95. Huang, Y. and Wang, C. (2006) Pulmonary delivery of insulin by liposomal carriers. *J. Control. Release*, **113**, 9–14.
96. Liu, F.Y., Shao, Z., Kildsig, D.O. and Mitra, A.K. (1993) Pulmonary delivery of free and liposomal insulin. *Pharm. Res.*, **10**, 228–232.
97. Meyboom, A., Maretzki, D., Stevens, P.A. and Hofmann, K.P. (1999) Interaction of pulmonary surfactant protein A with phospholipid liposomes: a kinetic study on head group and fatty acid specificity. *BBA Mol. Cell Biol. Lipids*, **1441**, 23–35.
98. Labiris, N.R. and Dolovich, M.B. (2003) Pulmonary drug delivery. Part II: the role of inhalant delivery devices and drug formulations in therapeutic effectiveness of aerosolized medications. *Br. J. Clin. Pharmacol.*, **56**, 600–612.
99. Stark, B., Debbage, P., Andreae, F. *et al.* (2007) Association of vasoactive intestinal peptide with polymer-grafted liposomes: Structural aspects for pulmonary delivery. *Biochimica. et Biophysica. Acta*, **1768**, 705–714.
100. Waldrep, J.C., Keyhani, K., Black, M. and Knight, V. (1994) Operating characteristics of 18 different continuous-flow jet nebulizers with beclomethasone dipropionate liposome aerosol. *Chest*, **105**, 106–110.
101. Niven, R.W., Speer, M. and Schreier, H. (1991) Nebulization of liposomes. II. The effects of size and modeling of solute release profiles. *Pharm. Res.*, **8**, 217–221.
102. Li, Y. and Mitra, A.K. (1996) Effect of phospholipids chain length, concentration, charge, and vesicle size on pulmonary insulin absorption. *Pharm. Res.*, **13**, 76–79.
103. Shek, P.N., Suntres, Z.E. and Brooks, J.I. (1994) Liposomes in pulmonary applications: physicochemical considerations, pulmonary distribution and antioxidant delivery. *J. Drug Target*, **2**, 431–442.
104. Niven, R.W. and Schreier, H. (1990) Nebulization of liposomes. I. Effects of lipid composition. *Pharm. Res.*, **7**, 1127–1133.
105. Niven, R.W., Carvajal, T.M. and Schreier, H. (1992) Nebulization of liposomes. III. The effects of operating conditions and local environment. *Pharm. Res.*, **9**, 515–520.
106. Sharma, A. and Sharma, U.S. (1997) Liposomes in drug delivery: progress and limitations. *Int. J. Pharm.*, **154**, 123–140.
107. Griesenbach, U., Kitson, C., Escudero, S. *et al.* (2006) Inefficient cationic lipid-mediated siRNA and antisense oligonucleotide transfer to airway epithelial cells *in vivo*. *Respir. Res.*, **7**, 26.
108. Ernst, N., Ulrichskotter, S., Schmalix, W.A. *et al.* (1999) Interaction of liposomal and polycationic transfection complexes with pulmonary surfactant. *J. Gene Med.*, **1**, 331–340.
109. Rudolph, C., Lausier, J., Naundorf, S. *et al.* (2000) *In vivo* gene delivery to the lung using polyethylenimine and fractured polyamidoamine dendrimers. *J. Gene Med.*, **2**, 269–278.
110. Sudhakar, Y., Kuotsu, K. and Bandyopadhyay, A.K. (2006) Buccal bioadhesive drug delivery – a promising option for orally less efficient drugs. *J. Control. Release*, **114**, 15–40.
111. Junginger, H.E., Hoogstraate, J.A. and Verhoef, J.C. (1999) Recent advances in buccal drug delivery and absorption – *in vitro* and *in vivo* studies. *J. Control. Release*, **62**, 149–159.
112. Senel, S. and Hýncal, A.A. (2001) Drug permeation enhancement via buccal route: possibilities and limitations. *J. Control. Release*, **72**, 133–144.
113. Veuillez, F., Kalia, Y.N., Jacques, Y. *et al.* (2001) Factors and strategies for improving buccal absorption of peptides. *Eur. J. Pharm. Biopharm.*, **51**, 93–109.
114. Sandri, G., Rossi, S., Bonferoni, M.C. *et al.* (2005) Buccal penetration enhancement properties of *N*-trimethyl chitosan: influence of quaternization degree on absorption of a high molecular weight molecule. *Int. J. Pharm.*, **297**, 146–155.
115. Shojaei, A.H. (1998) Buccal mucosa as a route for systemic drug delivery: a review. *J. Pharm. Pharmaceut. Sci.*, **1**, 15–30.

116. Starokadomskyy, P.L. and Dubey, I.Ya. (2006) New absorption promoter for the buccal delivery: preparation and characterization of lysalbinic acid. *Int. J. Pharm.*, **308**, 149–154.
117. Heiber, S.J., Ebert, C.D., Dave, S.C. *et al.* (1994) *In vivo* buccal delivery of calcitonin. *J. Control. Release*, **28**, 269–271.
118. Kahns, A.H., Buur, A. and Bundgaard, H. (1993) Prodrugs of peptides. 18. Synthesis and evaluation of various esters of desmopressin (dDAVP). *Pharm. Res.*, **10**, 68–74.
119. Bundgaard, H. and Moss, J. (1990) Prodrugs of peptides. 6. Bioreversible derivatives of Thyrotropin-Releasing Hormone (TRH) with increased lipophilicity and resistance to cleavage by the TRH-specific serum enzyme. *Pharm. Res.*, **7**, 885–892.
120. Zhou, X.H. (1994) Overcoming enzymatic and absorption barriers to nonparenterally administered protein and peptide drugs. *J. Control. Release*, **29**, 239–252.
121. Hughes, R.A., Toth, I., Ward, P. *et al.* (1991) Lipidic peptides. III: lipidic amino acid and oligomer conjugates of morphine. *J. Pharm. Sci.*, **80**, 1103–1105.
122. Veillard, M.M., Longer, M.A., Martens, T.W. and Robinson, J.R. (1987) Preliminary studies of oral mucosal delivery of peptide drugs. *J. Control. Release*, **6**, 123–131.
123. Veuillez, F., Ganem-Quintanar, A., Deshusses, J., Falson-Rieg, F. and Buri, P. (1998) Comparison of the *ex vivo* oral mucosal permeation of tryptophan-leucine (Trp-Leu) and its myristoyl derivative. *Int. J. Pharm.*, **170**, 85–91.
124. Shojaei, A.H. and Li, X. (1997) Mechanisms of buccal mucoadhesion of novel copolymers of acrylic acid and polyethylene glycol monomethylether monomethacrylate. *J. Control. Release*, **47**, 151–161.
125. Hoogstraate, J.A., Verhoef, J.C., Pijpers, A. *et al.* (1996) *In vivo* buccal delivery of the peptide drug buserelin with glycodeoxycholate as an absorption enhancer in pigs. *Pharm. Res.*, **13**, 1233–1237.
126. Nakane, S., Kakumoto, M., Yukimatsu, K. and Chien, Y.W. (1996) Oramucosal delivery of LHRH: pharmacokinetic studies of controlled and enhanced transmucosal permeation. *Pharm. Dev. Technol.*, **1**, 251–259.
127. Venugopalan, P., Sapre, A., Venkatesan, N. and Vyas, S.P. (2001) Pelleted bioadhesive polymeric nanoparticles for buccal delivery of insulin: preparation and characterization. *Pharmazie*, **56**, 217–219.
128. Li, C., Bhatt, P.P. and Johnston, T.P. (1997) Transmucosal delivery of oxytocin to rabbits using a mucoadhesive buccal patch. *Pharm. Dev. Technol.*, **2**, 265–274.
129. Wang, W. (1996) Oral protein drug delivery. *J. Drug Target*, **4**, 195–232.
130. Senel, S., Kremer, M.J., Kas, S. *et al.* (2000) Enhancing effect of chitosan on peptide drug delivery across buccal mucosa. *Biomaterials*, **21**, 2067–2071.
131. Sveinsson, S.J. and Holbrook, W.P. (1993) Oral mucosal adhesive ointment containing liposomal corticosteroid. *Int. J. Pharm.*, **95**, 105–109.
132. al-Achi, A. and Greenwood, R. (1993) Buccal administration of human insulin in streptozocin-diabetic rats. *Res. Commun. Chem. Pathol. Pharmacol.*, **82**, 287–306.
133. Petelin, M., Sentjurc, M., Stolic, Z. and Skaleric, U. (1998) EPR study of mucoadhesive ointments for delivery of liposomes into the oral mucosa. *Int. J. Pharm.*, **173**, 193–202.
134. Hamman, J.H., Enslin, G.M. and Kotze, A.F. (2005) Oral delivery of peptide drugs – barriers and developments. *Biodrugs*, **19**, 165–177.
135. Christiane, D., Reis, C.P. and Maincent, P. (2008) Nanoparticle strategies for the oral delivery of insulin. *Expert. Opin. Drug Deliv.*, **5**, 45–68.
136. Pauletti, G.M., Gangwar, S., Knipp, G.T. *et al.* (1996) Structural requirements for intestinal absorption of peptide drugs. *J. Control. Release*, **41**, 3–17.
137. Woodley, J.F. (1994) Enzymatic barriers for gi peptide and protein delivery. *Crit. Rev. Ther. Drug Carrier Syst.*, **11**, 61–95.
138. Yamamoto, A., Taniguchi, T., Rikyuu, K. *et al.* (1994) Effects of various protease inhibitors on the intestinal absorption and degradation of insulin in rats. *Pharm. Res.*, **11**, 1496–1500.
139. Bernkop-Schnnrch, A. (1998) The use of inhibitory agents to overcome the enzymatic barrier to perorally administered therapeutic peptides and proteins. *J. Control. Release*, **52**, 1–16.
140. Bernkop-Schnnrch, A. and Kast, C.E. (2001) Chemically modified chitosans as enzyme inhibitors. *Adv. Drug Deliv. Rev.*, **52**, 127–137.

141. Lee, V.H.L., Yamamoto, A. and Kompella, U.B. (1991) Mucosal penetration enhancers for facilitation of peptide and protein drug absorption. *Crit. Rev. Ther. Drug Carrier Syst.*, **8**, 91–192.
142. Aungst, B.J., Saitoh, H., Burcham, D.L. *et al.* (1996) Enhancement of the intestinal absorption of peptides and nonpeptides. *J. Control. Release*, **41**, 19–31.
143. Whitehead, K. and Mitragotri, S. (2008) Mechanistic analysis of chemical permeation enhancers for oral drug delivery. *Pharm. Res.*, **25**, 1412–1419.
144. Bernkop-Schnnrch, A., Kast, C.E. and Guggi, D. (2003) Permeation enhancing polymers in oral delivery of hydrophilic macromolecules: thiomer/GSH systems. *J. Control. Release*, **93**, 95–103.
145. Prego, C., Torres, D. and Alonso, M.J. (2005) The potential of chitosan for the oral administration of peptides. *Expert. Opin. Drug Deliv*, **2**, 843–854.
146. des Rieux, A., Fievez, V., Garinot, M. *et al.* (2006) Nanoparticles as potential oral delivery systems of proteins and vaccines: a mechanistic approach. *J. Control. Release*, **116**, 1–27.
147. Damgé, C., Michel, C., Aprahamian, M. *et al.* (1990) Nanocapsules as carriers for oral peptide delivery. *J. Control. Release*, **13**, 233–239.
148. Florence, A.T. (1997) The oral absorption of micro- and nanoparticulates: neither exceptional nor unusual. *Pharm. Res.*, **14**, 259–266.
149. Giannasca, P.J., Giannasca, K.T., Leichtner, A.M. and Neutra, M.R. (1999) Human intestinal M cells display the sialyl Lewis A antigen. *Infect. Immun.*, **67**, 946–953.
150. Christiane, D., Reis, C.P. and Maincent, P. (2008) Nanoparticle strategies for the oral delivery of insulin. *Expert. Opin. Drug Deliv.*, **5**, 45–68.
151. Singh, R., Singh, S. and Lillard, J.W. (2008) Past, present, and future technologies for oral delivery of therapeutic proteins. *J. Pharm. Sci.*, **97**, 2497–2523.
152. Martin, W., Hirofumi, T. and Andreas, B.-S. (2009) Modified chitosans for oral drug delivery. *J. Pharm. Sci.*, **98**, 1648–1659.
153. Kim, B. and Peppas, N.A. (2002) Synthesis and characterization of pH-sensitive glycopolymers for oral drug delivery systems. *J. Biomater. Sci. Polym. Ed.*, **13**, 1271–1281.
154. Kavimandan, N.J., Losi, E. and Peppas, N.A. (2006) Novel delivery system based on complexation hydrogels as delivery vehicles for insulin-transferrin conjugates. *Biomaterials*, **27**, 3846–3854.
155. Babu, V.R., Patel, P., Mundargi, R.C. *et al.* (2008) Developments in polymeric devices for oral insulin delivery. *Expert. Opin. Drug Deliv.*, **5**, 403–415.
156. Zhou, F. and Neutra, M.R. (2002) Antigen delivery to mucosa-associated lymphoid tissues using liposomes as a carrier. *Biosci. Rep.*, **22**, 355–369.
157. Yamabe, K., Kato, Y., Onishi, H. and Machida, Y. (2003) Potentiality of double liposomes containing salmon calcitonin as an oral dosage form. *J. Control. Release*, **89**, 429–436.
158. Li, H., Song, J.H., Park, J.S. and Han, K. (2003) Polyethylene glycol-coated liposomes for oral delivery of recombinant human epidermal growth factor. *Int. J. Pharm.*, **258**, 11–19.
159. AnJ, A. and Souto, E. (2007) Solid lipid nanoparticles as a drug delivery system for peptides and proteins. *Adv. Drug Deliv. Rev.*, **59**, 478–490.
160. Zheng, J.Y. and Fulu, M.Y. (2006) Decrease of genital organ weights and plasma testosterone levels in rats following oral administration of leuprolide microemulsion. *Int. J. Pharm.*, **307**, 209–215.
161. Thongborisute, J., Takeuchi, H., Yamamoto, H. and Kawashima, Y. (2006) Visualization of the penetrative and mucoadhesive properties of chitosan and chitosan-coated liposomes through the rat intestine. *J. Liposome Res.*, **16**, 127–141.
162. Cui, F., Qian, F. and Yin, C. (2006) Preparation and characterization of mucoadhesive polymer-coated nanoparticles. *Int. J. Pharm.*, **316**, 154–161.
163. Takeuchi, H., Matsui, Y., Yamamoto, H. and Kawashima, Y. (2003) Mucoadhesive properties of carbopol or chitosan-coated liposomes and their effectiveness in the oral administration of calcitonin to rats. *J. Control. Release*, **86**, 235–242.
164. Thirawong, N., Thongborisute, J., Takeuchi, H. and Sriamornsak, P. (2008) Improved intestinal absorption of calcitonin by mucoadhesive delivery of novel pectin-liposome nanocomplexes. *J. Control. Release*, **125**, 236–245.

165. Win, K.Y. and Feng, S.S. (2005) Effects of particle size and surface coating on cellular uptake of polymeric nanoparticles for oral delivery of anticancer drugs. *Biomaterials*, **26**, 2713–2722.
166. Kompella, U.B. and Lee, V.H.L. (2001) Delivery systems for penetration enhancement of peptide and protein drugs: design considerations. *Adv. Drug Deliv. Rev.*, **46**, 211–245.
167. Vasir, J.K., Tambwekar, K. and Garg, S. (2003) Bioadhesive microspheres as a controlled drug delivery system. *Int. J. Pharm.*, **255**, 13–32.
168. Sinha, V.R., Singh, A., Kumar, R.V. *et al.* (2007) Oral colon specific drug delivery of protein and peptide drugs. *Crit. Rev. Ther. Drug Carrier Syst.*, **24**, 63–92.
169. Gazzaniga, A., Maroni, A., Sangalli, M.E. and Zema, L. (2006) Time-controlled oral delivery systems for colon targeting. *Expert. Opin. Drug Deliv.*, **3**, 583–597.
170. Sinha, V.R. and Kumria, R. (2003) Microbially triggered drug delivery to the colon. *Eur. J. Pharm. Sci.*, **18**, 3–18.
171. Yang, L., Chu, J.S. and Fix, J.A. (2002) Colon-specific drug delivery: new approaches and *in vitro/in vivo* evaluation. *Int. J. Pharm.*, **235**, 1–15.
172. Asada, H., Douen, T., Mizokoshi, Y. *et al.* (1994) Stability of acyl derivatives of insulin in the small intestine: relative importance of insulin association characteristics in aqueous solution. *Pharm. Res.*, **11**, 1115–1120.
173. Kahns, A.H., Buur, A. and Bundgaard, H. (1993) Prodrugs of peptides. 18. Synthesis and evaluation of various esters of desmopressin (dDAVP). *Pharm. Res.*, **10**, 68–74.
174. Uchiyama, T., Kotani, A., Tatsumi, H. *et al.* (2000) . Development of novel lipophilic derivatives of DADLE (leucine enkephalin analogue): intestinal permeability characteristics of DADLE derivatives in Rats. *Pharm. Res.*, **17**, 1461–1467.
175. Wang, J., Shen, D. and Shen, W.C. (1999) Preparation, purification, and characterization of a reversibly lipidized desmopressin with potentiated anti-diuretic activity. *Pharm. Res.*, **16**, 1674–1679.
176. Wang, J., Chow, D., Heiati, H. and Shen, W.C. (2003) Reversible lipidization for the oral delivery of salmon calcitonin. *J. Control. Release*, **88**, 369–380.
177. Wang, J., Hogenkamp, D.J., Tran, M. *et al.* (2006) Reversible lipidization for the oral delivery of leu-enkephalin. *J. Drug Target.*, **14**, 127–136.
178. Toth, I., Flinn, N., Hillery, A. *et al.* (1994) Lipidic conjugates of Luteinizing-Hormone-Releasing Hormone (LHRH)+ and Thyrotropin-Releasing-Hormone (TRH)+ that release and protect the native hormones in homogenates of human intestinal epithelial (Caco 2) cells. *Int. J. Pharm.*, **105**, 241–247.
179. Blanchfield, J.T., Gallagher, O.P., Cros, C. *et al.* (2007) Oral absorption and *in vivo* biodistribution of alpha-conotoxin MII and a lipidic analogue. *Biochem. Biophys. Res. Commun.*, **361**, 97–102.
180. Bergeon, J.A., Chan, Y.N., Charles, B.G. and Toth, I. (2008) Oral absorption enhancement of dipeptide L-Glu-L-Trp-OH by lipid and glycosyl conjugation. *Biopolymers*, **90**, 633–643.
181. Morishita, M. and Peppas, N.A. (2006) Is the oral route possible for peptide and protein drug delivery? *Drug Discov. Today*, **11**, 905–910.
182. Russell-Jones, G.J. (2004) Use of targeting agents to increase uptake and localization of drugs to the intestinal epithelium. *J. Drug Target.*, **12**, 113–123.
183. Xia, C.Q., Wang, J. and Shen, W.C. (2000) Hypoglycemic effect of insulin-transferrin conjugate in streptozotocin-induced diabetic rats. *J. Pharmacol. Exp. Ther.*, **295**, 594–600.
184. Chalasani, K.B., Russell-Jones, G.J., Jain, A.K. *et al.* (2007) Effective oral delivery of insulin in animal models using vitamin B12-coated dextran nanoparticles. *J. Control. Release*, **122**, 141–150.
185. Ogiso, T., Iwaki, M., Tanino, T. *et al.* (2000) *In vitro* skin penetration and degradation of peptides and their analysis using a kinetic model. *Biol. Pharm. Bull.*, **23**, 1346–1351.
186. Steinstrtisser, I. and Merkle, H.P. (1995) Dermal metabolism of topically applied drugs: pathways and models reconsidered. *Pharm. Acta Helv.*, **70**, 3–24.

187. Magnusson, B.M., Runn, P., Karlsson, K. and Koskinen, L.-O.D. (1997) Terpenes and ethanol enhance the transdermal permeation of the tripeptide thyrotropin releasing hormone in human epidermis. *Int. J. Pharm.*, **157**, 113–121.
188. Ruland, A., Kreuter, J. and Rytting, H.J. (1994) Transdermal delivery of the tetrapeptide hisetal (melanotropin(6–9)). I. Effect of various penetration enhancers: *in vitro* study across hairless mouse skin. *Int. J. Pharm.*, **101**, 57–61.
189. Choi, H., Flynn, G.L. and Amidon, G.L. (1990) Transdermal delivery of bioactive peptides: the effect of n-decylmethylsulfoxide, pH and inhibitors on enkephalin metabolism and transport. *Pharm. Res.*, **7**, 1099–1105.
190. Williams, A.C. and Barry, B.W. (2004) Penetration enhancers. *Adv. Drug Deliv. Rev.*, **56**, 603–618.
191. Williams, A.C. and Barry, B.W. (1992) Skin absorption enhancers. *Crit. Rev. Ther. Drug Carrier Syst.*, **9**, 305–353.
192. Asbill, C.S. and Michniak, B.B. (2000) Percutaneous penetration enhancers: Local versus transdermal activity. *Pharm. Sci. Technol. Today (PSTT)*, **3**, 36–41.
193. Pillai, O., Nair, V. and Panchagnula, R. (2004) Transdermal iontophoresis of insulin: IV. Influence of chemical enhancers. *Int. J. Pharm.*, **269**, 109–120.
194. Kogan, A. and Garti, N. (2006) Microemulsions as transdermal drug delivery vehicles. *Adv. Colloid Interface Sci.*, **123–126**, 369–385.
195. He, W., Guo, X. and Zhang, M. (2008) Transdermal permeation enhancement of *N*-trimethyl chitosan for testosterone. *Int. J. Pharm.*, **356**, 82–87.
196. Novotný, M., Hrabálek, A., Janůšová, B. *et al.* (2009) Dicarboxylic acid esters as transdermal permeation enhancers: Effects of chain number and geometric isomers. *Bioorg. Med. Chem. Lett.*, **19**, 344–347.
197. Magnusson, B.M. and Runn, P. (1999) Effect of penetration enhancers on the permeation of the thyrotropin releasing hormone analogue pGlu-3-methyl-His-Pro amide through human epidermis. *Int. J. Pharm.*, **178**, 149–159.
198. Setoh, K., Murakami, M., Araki, N. *et al.* (1995) Improvement of transdermal delivery of tetragastrin by lipophilic modification with fatty acids. *J. Pharm. Pharmacol.*, **47**, 808–811.
199. Yamamoto, A., Setoh, K., Murakami, M. *et al.* (2003) Enhanced transdermal delivery of phenylalanyl-glycine by chemical modification with various fatty acids. *Int. J. Pharm.*, **250**, 119–128.
200. Fresta, M. and Puglisi, G. (1996) Application of liposomes as potential cutaneous drug delivery systems. *In vitro* and *in vivo* investigation with radioactively labelled vesicles. *J. Drug Target.*, **4**, 95–101.
201. Short, S.M., Paasch, B.D., Turner, J.H. *et al.* (1996) Percutaneous absorption of biologically-active interferon-gamma in a human skin graft-nude mouse model. *Pharm. Res.*, **13**, 1020–1027.
202. El Maghraby, G.M., Barry, B.W. and Williams, A.C. (2008) Liposomes and skin: From drug delivery to model membranes. *Eur. J. Pharm. Sci.*, **34**, 203–222.
203. Cheng, J., Huang, H., Tseng, W. *et al.* (2009) Transcutaneous immunization by lipoplex-patch based DNA vaccines is effective vaccination against Japanese encephalitis virus infection. *J. Control. Release*, **135**, 242–249.
204. Nicoli, S., Santi, P., Couvreur, P. *et al.* (2001) Design of triptorelin loaded nanospheres for transdermal iontophoretic administration. *Int. J. Pharm.*, **214**, 31–35.
205. Rolland, A., Wagner, N., Chatelus, A. *et al.* (1993) Site-specific drug delivery to pilosebaceous structures using polymeric microspheres. *Pharm. Res.*, **10**, 1738–1744.
206. Tahara, Y., Honda, S., Kamiya, N. *et al.* (2008) A solid-in-oil nanodispersion for transcutaneous protein delivery. *J. Control. Release*, **131**, 14–18.
207. Banga, A.K. and Prausnitz, M.R. (1998) Assessing the potential of skin electroporation for the delivery of protein- and gene-based drugs. *Trends Biotechnol.*, **16**, 408–412.
208. Tachibana, K. (1992) Transdermal delivery of insulin to alloxan/diabetic rabbits by ultrasound exposure. *Pharm. Res.*, **9**, 952–954.

209. Langkjaer, L., Brange, J., Grodsky, G.M. and Guy, R.H. (1998) Iontophoresis of monomeric insulin analogues *in vitro*: effects of insulin charge and skin pretreatment. *J. Control. Release*, **51**, 47–56.
210. Dujardin, N., Van Deà Smissen, P. and Préat, V. (2001) Topical gene transfer into rat skin using electroporation. *Pharm. Res.*, **18**, 61–66.
211. Green, P.G. (1996) Iontophoretic delivery of peptide drugs. *J. Control. Release*, **41**, 33–48.
212. Tomohira, Y., Machida, Y., Onishi, H. and Nagai, T. (1997) Iontophoretic transdermal absorption of insulin and calcitonin in rats with newly-devised switching technique and addition of urea. *Int. J. Pharm.*, **155**, 231–239.
213. Raiman, J., Koljonen, M., Huikko, K. *et al.* (2004) Delivery and stability of LHRH and Nafarelin in human skin: the effect of constant/pulsed iontophoresis. *Eur. J. Pharm. Sci.*, **21**, 371–377.
214. Nakamura, K., Katagai, K., Mori, K. *et al.* (2001) Transdermal administration of salmon calcitonin by pulse depolarization-iontophoresis in rats. *Int. J. Pharm.*, **218**, 93–102.
215. Fang, J., Lee, W., Shen, S. *et al.* (2005) Transdermal delivery of macromolecules by erbium:YAG laser. *J. Control. Release*, **100**, 75–85.
216. Moeller, E.H. and Jorgensen, L. (2008) Alternative routes of administration for systemic delivery of protein pharmaceuticals. *Drug Discov. Today: Technol.*, in press.

4.2 Barriers to Delivery of Nucleic Acids

4.2.1 Anatomical

In systemic circulation, the first barrier encountered is the blood stream. The blood contains serum proteins, nucleases and macrophages that can aggregate, degrade or phagocytose the delivery vector, respectively, even before it reaches the tissue of interest. Naked plasmids have a significantly reduced half-life as compared to when complexed with cationic liposomes [2].

Clearance by the reticuloendothelial system (RES) is of great concern, and particle size often determines by what organ (liver, kidney or spleen) and how quickly the particle will be cleared from the blood. Extended plasma circulation is desired to allow a majority of the injected dose time to reach the target tissue. Poly(ethylene glycol) (PEG), a hydrophilic inert polymer, has been conjugated to many chemical vectors to extend circulation time, such as liposomes, patented by Lasic's group, later termed Stealth® liposomes [3–6]. Once in the vasculature of the target tissue, the delivery vector must leave the blood stream and transverse into the interstitial space of the tissue.

Tumour vasculature is inherently disorganized, immature and tortuous due to the presence of discontinuous endothelium and an incomplete basement membrane. The blood vessels in the tumour attempt to supply the rapidly growing tumour cells with nutrients and cannot grow fast enough to fully develop. The Enhanced Permeability and Retention (EPR) effect is often used in cancer therapy to deliver drugs to the interstitum of tumours, skipping the usual barrier of transversing the endothelium [7–10]. However, this permeability is not universal in tumour vasculature. Heterogeneous permeability of tumour endothelium and interstitum was shown in window chamber experiments using by rhodamine-labelled stealth liposomes [11]. Doxil® (a commercial formulation of doxorubicin encapsulated in stealth liposomes), showed increased accumulation in tumour tissue compared to normal tissue in humans [12].

While the EPR effect is useful in many tumour microenvironments, other methods have been developed to further enhance the delivery. For efficient delivery, the vector may be targeted to the specific tissue of interest by ligand attachment on the vector, by targeting a receptor on the cells in the tissue of interest or by triggering release at the desired site. These two topics are discussed in depth later. Being able to discriminate between target and non-target cells is crucial to ensure specific expression of the nucleic acids delivered.

4.2.2 Cellular

Once at the plasma membrane, the chemical vector must be endocytosed to enter the cell. Endocytosis can occur by many routes including macropinocytosis, clathrin-mediated and non-clathrin-mediated endocytosis and entry via caveolae [13]. Polyplexes of poly(ethylene imine) (PEI) complexed with pDNA were shown to attach to cell surfaces, migrating into clumps, and then endocytosed [14]. It has been postulated that cationic lipid carriers can enter cells through fusion with the cell membrane; however, much more evidence exists for endocytotic pathways [15]. However, Douglas *et al.*, have recently published results arguing that endocytosis is cell-line dependent. The authors report that after clathrin-mediated endocytosis, complexes follow the endo-lysosomal pathway, which initiates the proton-sponge effect for release (discussed under heading 4.5.2). Further, the authors report that caveolin-mediated endocytosis leaves the complexes trapped in vesicles that become 'transfection-incompetent' [16]. For successful gene delivery to occur, the nanoparticle or naked plasmid must escape the endosome before being degraded upon fusion of a lysosome with the current endosome [17].

Once the particle or naked plasmid is in the cytoplasm, it faces additional intracellular barriers. Diffusional mobility of plasmid DNA is limited due to the cytoskeleton, particularly actin [18] and molecular crowding in the cytoplasm. Intracellular mobility of DNA was shown to be size dependent. Double stranded DNA >250 bp had significantly decreased diffusion compared to DNA of 100 bp or less [19]. Active cytosolic nucleases can degrade the plasmid. However, if the nucleic acids being delivered function via an RNAi mechanism, they need travel no further than the cytoplasm where crucial enzymes such as RISC and Dicer reside. Entry into the nucleus by the plasmid is rate limiting and can only be achieved when the cell divides or through nuclear membrane pores, which have size limitations on what can enter or exit. Nucleus localization sequences (NLSs) can be conjugated to pDNA to facilitate entry through the nuclear pores, [20] which can increase in size from 9 to 39 nm from docking of appropriate NLSs [21]. However, most delivery systems rely on non-specific entry to the nucleus [19].

4.3 Major Classes of Delivery Vectors

4.3.1 Introduction

Cationic moieties are able to condense DNA, RNAs and other oligonucleotides through the ionic interaction of the nucleic acid's negatively charged phosphate backbone with cationic charges, which can come from lipids, polymers or peptides. These different classes of delivery vectors have been developed with individual limitations on size, charge ratio, nucleic acid capacity and toxicity.

4.3.2 Cationic Lipids

Cationic liposomes were first used for DNA transfection by Felgner *et al.* in 1987 with the introduction of the cationic lipid, N-[1-(2,3-dioleyloxy)propyl]-N,N,N-trimethylammonium chloride (DOTMA) [22]. Soon after, DOTMA's ester analogue DOTAP (1,2-dioleoyl-3-trimethylammonium-propane) also saw popularity as a cationic lipid used in drug delivery [23, 24].

Lipoplexes are a popular chemical vector delivery technique, due to the relatively low immune response accompanying their delivery. These vectors combine cationic lipids with DNA to produce a complex, a product of the interaction of the negatively charged phosphate backbone and the positively charged lipid head groups. Transfection has been shown to be a function of charge ratio (lipid:DNA or (+ : −)). Using a 4 : 1 charge ratio, the *in vivo* transfection of IL-12 plasmids was improved by conjugating albumin to the exterior of the lipoplex in combination with intratumoral (i.t.) delivery [25]. The lipids used were 1-palmitoyl-2-oleoyl-*sn*-glycero-3-ethylphosphocholine (EPOPC) and cholesterol (chol) in a 1 : 1 ratio. Albumin was added to the outside of the membrane to enhance the escape of DNA from the endocytotic pathway, since it has been described to undergo a low-pH induced conformational change, acquiring fusogenic properties [26].

Similar to IL-18, IL-12 enhances the proliferation and cytotoxic activity of both T and NK cells and induces the production of IFN-γ. Also delivered (in a separate experiment) by this albumin conjugated lipoplex was the 'suicide' gene, HSV-tk, which when delivered with the prodrug ganciclovir (GCV), converts into a toxic active metabolite that causes cell death. The attachment of albumin greatly increased the transfection efficiency of the lipoplex at all charge ratios (4 : 1, 2 : 1 and 1 : 1). This study showed that four i.t. injections of these plasmids over several days at 40 μg each (with HSV-tk accompanied by intraperitoneal (i.p.) injections of GCV) showed significant tumour growth delay in a subcutaneous TSA model (a mouse mammary adenocarcinoma, which is highly aggressive, immunogenic and has a high proliferation capacity) compared to the control injection of buffer. Complete tumour regression was seen in one member of the immunogene therapy group (one of six). Even after four i.t. injections of the control plasmid, some growth delay was seen, the authors attributing this to cytosine-phosphate-guanine (CpG) motifs in the plasmid that elicited a slight immune response. Higher T-lymphocyte infiltration was seen from control plasmids verses untreated mice, showing the immune response from even an empty plasmid. However, albumin coated lipoplexes containing the IL-12 plasmid (40 μg) showed significant tumour regression, high T-lymphocyte infiltration and extensive necrotic and hemorrhagic areas. Also, with both formulations (with and without albumin), this study demonstrates a frequent problem of high transfection in the lung, rather than the target tissue.

LPD, which stands for 'liposome-polycation-DNA', was developed by Huang's group and uses cationic liposomes differently than other groups. While most cationic liposomes condense nucleic acids in layers, with the cationic bilayers associating with negatively charged nucleic acids, making a multilamellar liposome, LPD is different. LPD first condenses nucleic acids with a polycation such as polylysine or protamine, making a solid particle, and then mixes that solution with charged liposomes, which effectively coat the surface of the solid DNA-polycation core. LPDII complexed DNA and polylysine at a ratio of 1 : 0.75, then trapped it in folate-targeted anionic pH-sensitive liposomes. LPDII is coated with PEG conjugated to folate, to help the particle evade RES detection and target folate over-expressing cells such as KB cells [27]. The particle's charge depends on the lipid/DNA ratio, and at high lipid/DNA ratios the anionic particles are formed, while at low lipid/DNA ratios cationic particles are formed. The cationic particles were found to have high transfection but to not be tissue specific, whereas the anionic particles showed a receptor dependent transfection activity.

In another, more recent formulation of LPD, three different siRNA sequences were co-formulated in each LPD to cause simultaneous silencing of three important proteins in cancer survival: MDM2, c-myc and vascular endothelial growth factor (VEGF) [28]. The LPD were made with DOTAP as the cationic lipid, protamine as the polycation, protected against RES uptake by a PEG coating and targeted using anisamide to the sigma receptor on murine lung metastasis. A 70–80% reduction in cancerous nodules in the lung was seen in the targeted treatment group compared to control.

4.3.3 Cationic Polymers

Poly(ethylene imine) is a cationic polymer commonly used in gene delivery to create 'polyplexes', indicating the complexation of DNA and a polymer. Linkers made of PEG, an inert hydrophilic polymer, can also be added to prevent uptake by the RES [29]. Because of the enhanced release from the endosome due to the proton sponge effect, cationic polymers have been popular in gene delivery. However, the toxicity of the vector is often an issue at high molecular weights. Polyplexes have also been conjugated to targeting ligands, either directly to the PEI or to a PEG linker on the PEI such as peptides with the RGD (Arg-Gly-Asp) motif [29].

Focal adhesion kinase (FAK) is a protein kinase in cell cycle progression and cell migration. Modified PEI (M-PEI) was used to transfect tumour cells and tissues with a FAK siRNA plasmid (psiFAK) to downregulate the tendency of metastases forming in the mouse model from left hind footpad model using B16F10 (mouse melanoma). After i.t. delivery of psiFAK, the tumour showed no metastases in the lungs and lymph nodes in any mouse ($n = 6$), compared to phosphate buffered saline (PBS) or either control plasmid (pLuc or pKH3). Primary tumours also experienced a statistically significant growth delay and survival time when treated with psiFAK as compared to the controls [30].

In addition to modified PEI, linear and branched PEI have also been used in intratumoral (i.t.) gene delivery. Coll *et al.* [31] compared three routes of delivery (i.v., intracystic and i.t.) to human non-small cell lung carcinoma (NSCLC) using a Luciferase plasmid to discover the best method of polyplex administration with linear PEI. A charge ratio of 10 polymer:DNA provided the best *in vivo* transfection, causing increasing expression from 24 to 96 hours. When delivering intravenously to solid tumours, minimal expression was seen in the highly vascularized tumours, independent of size. However, a nearly threefold increase in expression was seen in the normal lung tissue, making these nanoparticles useless for solid tumour delivery if injected i.v., not even benefiting noticeably from the EPR effect. A common lung metastasis model was used here that consists of i.v. injection of tumour cells, which are caught in the lung and create many nodules therein. After an i.v. injection of these PEI particles, only tumour-free mice expressed Luciferase and no expression was seen in tumour loaded lungs. However, using other delivery routes, these particles were effective. In several cancers, including some ovarian and gliomas, the tumour wall expands to create a liquid filled cavity. In a cyst-forming tumour mouse model H358NL, after draining the internal liquid (up to 5 ml), a small volume of polyplexes were delivered into the cyst cavity [31]. This method of delivery showed high transfection and long lasting expression and could also be used post-surgery, after the bulk of the tumour had been removed and some residual tumour cells remained. The authors found that the volume injected i.t. with a normal syringe travelled to the periphery of the tumour. A micropump technique was developed for i.t. delivery at a rate of 20 μl/min. This method showed high and long lasting expression of the transgene, where normal syringe injection was not effective. However, free plasmid injection was inefficient with the micropump and more effective with the normal syringe injection, indicating that the hydrostatic pressure induced from normal i.t. injection possibly aided in delivery.

Rozema *et al.* introduced a multifuntional polymer for delivery of siRNA to hepatocytes [32]. Named 'Dynamic PolyConjugates', these particles use an endosomolytic polymer, an amphipathic poly(vinyl ether) termed PBAVE, composed of butyl and amino vinyl ethers. The polymer can target cells of interest, evade the RES and deliver a therapeutic payload by binding to targeting ligands, PEG and siRNA. Once inside the endosome and the pH drops, a pH-labile bond connecting the targeting ligand and PEG is cleaved, exposing many positive charges and inducing the proton sponge effect, drawing water into the endosome until it bursts, allowing the siRNA to escape into the cytoplasm. This vector has a mechanism for many of the challenges of delivery and lives up to its dynamic name, able to change as each event presents itself.

The Davis group introduced a cationic cyclodextran-containing polymer for condensing pDNA in 1999 showing comparable transfection to PEI and Lipofectamine formulations *in vitro* [33]. The aims was to decrease toxicity during transfection, and the cyclodextran formulation showed no signs of toxicity at a 70:1 (+/−) charge ratio in the presence of 10% serum. Later development of this system included conjugation of an adamantane-terminated modifier to PEG to stabilize the particle and maximize plasma circulation time, as well as conjugating the modifier to targeting ligands such as transferrin [34–36]. Characterization of the particle's formulation showed that size can be 'dialed in' between 60 and 150 nm, similarly with a zeta potential ranging between 10 and 30 mV depending on how the formulation is made and modified [36].

Water-soluble lipopolymers (WSLPs) are another class of polymers used for nucleic acid delivery [37]. Branched PEI was conjugated to a hydrophobic lipid anchor. Single or repeated i.t. injections of WSLP containing IL-12 plasmid were compared to see the tolerance of the method and the efficacy on tumour regression and survival in a CT26 colorectal model. The studies also showed that the complexes could be retained in the tumour for up to 24 hours. The WSLP/pIL-12 plasmid complexes performed better against the controls of no treatment, free plasmid and 25K PEI/pIL-12 in both four and eight day repeat i.t. injections. After 45 days, 60% of the treatment group that received WSLP/pIL-12 every four days were alive, compared to 40% of the group that was treated with WSLP/pIL-12 every eight days, while all members of the single injection and control groups had reached study endpoints by day 33 and 26, respectively. However, the repeat injection method did cause some side effects. After six weeks of injections (either every four or eight days), the majority of mice in these groups could no longer receive injections because of increased tumour necrosis.

4.3.4 Cationic Dendrimers

Cationic dendrimers are built of positively charged molecules, usually amino groups, branching out from a central core. Each increased addition of branches is termed a new 'generation'. Several consistencies are seen across dendrimer transfection research. Increased generation yields increased transfection, but also increased toxicity. Even dendrimers delivered alone (not complexed with DNA) can induce host cell DNA damage [38]. Also, because cationic dendrimers can contain many positive charges, as they are formed from cationic monomers polymerized outwards from a central core, the proton sponge effect can be observed with dendrimers as well [39]. Endocytosis of dendrimers complexed with nucleic acids (dendriplexes) varies across cell lines [40, 41]. Dendriplexes have been shown to internalize through a cholesterol-dependent mechanism [41], but in other cells, cholesterol depletion does not seem to have an effect in cells which express few caveolae (HeLa and HepG2) [40]. These cells, as well as endothelial cells, endocytosed dendrimers complexed with DNA (dendriplexes) even with inhibition of clathrin-independent, phagocytic and macropinocytic pathways. It was discovered by the same group (Manunta *et al.*) that an increased rate of uptake occurred with over-expression of caveolin 1 [39].

Issues with cationic dendrimers such as transfection efficiency and cytotoxicity increasing in tandem, have been solved somewhat by Zhang *et al.* The research group conjugated poly(amido amine) (PAMAM) dendrimers to the surface of poly(D,L-lactide-*co*-glycolide) (PLGA) microspheres. Plasmid DNA was trapped by these particles by incubating PAMAM-coated PLGA microspheres (positively charged) with negatively charged pDNA. Unmodified PLGA microspheres were made by double emulsion and solvent evaporation of a mixture of PLGA to pDNA [42].

The core of the dendrimer can be modified, giving a different branching pattern and thus amino group density, dictated by the number of reaction centres available on the core [43]. The core that builds a less dense dendrimer, has been shown to require a higher generation to effectively condense DNA. A trimesyl core gives three places for dendrimer building to begin, while pentaerythritol gives four and inositol gives six. Trimesyl core dendrimers required generation six dendrimers to effectively condense DNA, while pentaerythritol and inositol only required generation five dendrimers. However, interestingly, the trimesyl core generation six dendrimers were smaller (100–300 nm) than the pentaerythritol or inositol generation five dendrimers (>600 nm), as well as showing greater transfection. In trimesyl core dendrimers, as commonly reported with all dendrimers, cytotoxicity did increase with generation; however, the toxicity was much less than that of PEI [43].

4.3.5 Cell Penetrating Peptides (CPP) and DNA Mimics

Cell penetrating peptide-based drug delivery systems have been developed to administer genes to tumours. Morris *et al.* have developed an amphipathic peptide, Pep-3, suitable for either i.t. or i.v. delivery, with slight modifications to overcome systemic barriers when delivered intravenously [44]. Plasmids that target antisense cyclin B1 (HypNA-pPNA) were delivered to a xenograft animal tumour model (PC3, human prostate carcinoma) as a DNA mimic.

The studies also showed that peptide binding of DNA mimics and uptake of the complex into cells are dependent on the hydrophobicity and charge of the peptide, respectively, with a minimum of four cationic charges required within the hydrophobic domain. The greatest growth delay was seen with i.t. delivery of 5 μg of the Pep-3 DNA mimic delivered. To evade the immune system in systemic circulation, Pep-3 was coated with poly(ethylene glycol). The newly coated Pep-3/DNA mimic/PEG nanoparticle was delivered intravenously at 10 μg and saw similar tumour growth delay as 5 μg Pep-3/DNA mimic delivered i.t. As not exemplified in previous studies, this chemical vector had similar growth delay effects from i.t. and i.v. delivery after some minor vector modifications.

The rabies virus glycoprotein (RVG) has been conjugated to siRNA to transverse the blood–brain barrier [45]. RVG targets the nicotinic acetylcholine receptor (AchR) on neuronal cells, enabling the entry of virus particles. The virus protein (RVG) was conjugated to a chain of nine D-arginine residues as a spacer, then to siRNA, which allowed transfection specifically in neuronal cells (Neuro-2a) *in vitro*. *In vivo*, after i.v. delivery, particles were found in the brain but not the liver or spleen and showed a high level of fluorescein isothiocyanate (FITC) fluorescence. Finally, after infecting non-obese diabetic severe combined immunodeficiency (NOD/SCID) mice with flaviviral encephalitis and injecting an antiviral siRNA sequence conjugated to RVG, survival was significantly extended.

4.3.6 Naked DNA

Degradation of free plasmids by rat plasma has been shown to have a half-life of 1.2 minutes for supercoiled pDNA, 21 minutes for open circular pDNA and 11 minutes for linear pDNA [2]. However, in the same study, complexing pDNA with cationic liposomes (1,2-dioleoyl-3-trimethylammonium-propane : L-alpha-dioleoyl-phosphatidylethanolamine (DOTAP : DOPE, 1 : 1 w/w)), allowed the supercoiled plasmid to be detected after 5.5 hours.

Other than hydrodynamic injection, (a large volume quickly injected, efficient in delivery to the murine liver [46]), naked DNA plasmid delivery has seen little success in i.v. delivery. Eliminating many barriers encountered in systemic circulation, it makes sense to assume that i.t. injection would show greater transfection efficiency and anti-tumour effects for naked plasmids. Intratumoral delivery of a naked interleukin (IL-18) plasmid was delivered to CT26 tumours established in the liver [47]. CT26 is an undifferentiated murine adenocarcinoma that models colorectal cancer, the most common site of metastasis being the liver. Tumours were established by injection of CT26 into the lower left lobe of the liver. IL-18 protein was found in all subjects injected with either the control vector or IL-18 plasmid, but significantly higher levels in the 50 μg plasmid-injected group. Mice treated with the control vector did not generate a CT26-specific T cell response, while those treated with the IL-18 producing plasmid contained a significant number of IFN-γ (interferon gamma) producing $CD4^+$ T cells and had significantly lower tumour growth. While most studies use solid tumour models in the hind flank of mice, this study shows the versatility of i.t. injection as it can also be administered to a solid tumour within a major organ within the body cavity.

4.4 Targeted Delivery

4.4.1 Introduction

Attaching a ligand specific to the tissue of interest increases localization of the nanoparticle to that tissue. Many different receptors are over-expressed in tumour tissues making the corresponding ligands desired as targeting agents. Also, hepatocytes express the asialoglycoprotein receptor in high amounts, making galactose and asialofetuin desired ligands for targeting in that application.

4.4.1.1 Galactose

Galactose can be used as a targeting agent to the asialoglycoprotein receptor (ASGP-R) on mammalian hepatocytes [48]. Asialofetuin, the receptor's natural ligand can also be used. N-acetylgalactose (NAG) has been conjugated to the end of endosomolytic polymers to target the vehicle to hepatocytes for pDNA delivery [32]. Galactose has also been conjugated to liposomes for delivery of oligodeoxynucleotides (ODNs) [49].

4.4.2 Targeting Solid Tumours and/or Metastasis

4.4.2.1 Transferrin

Transferrin is an iron-binding glycoprotein often used as a targeting ligand. Expression of transferrin receptors on the cell surface is upregulated in quickly dividing cells due to their need for iron. Transferrin receptor density can be increased by treating cells with the cell-permeable iron chelator desferrioxamine [50]. Transferrin targeting was used to deliver a reporter plasmid using polylysine or protamine to human leukaemic cells, K-562, termed 'transferrinfection' [50]. Other systems targeting immune cells have used transferrin as a ligand, such as liposome targeting to rabbit bone marrow precursor cells [51]. In addition to K-562 cells, transferrin targeting has been efficient with neuroblastoma Neuro-2A cells, melanoma H225 and 16F10 cells, with increased expression in all cell lines, some as high as 1000–10000-fold [1]. Joshee *et al.* showed that adding pDNA to a mixture of transferrin and DOTAP/DOPE (1 : 1) cationic liposomes yielded significantly higher transfection in Panc-1 cells than the formulation where pDNA and cationic liposomes were mixed first, then mixed with transferrin, although zeta potentials of the formulations were similar (around −33 mV). When conjugating gold nanoparticles to transferrin and adding pDNA to the gold-transferrin and cationic liposome mixture, both the labelled transferrin and pDNA colocalized at the perinuclear space and in the nucleus, suggesting cotransport intracelluarly and into the nucleus [52]. Transferrin has been conjugated to many different chemical vectors including: cationic cyclodextran polymers [34, 35], polylysine and protamine [50], PEI [1] and liposomes [51, 53]. Anti-transferrin receptor single chain antibody fragments have similarly been used to target the upregulation of these receptors [47]. The cationic liposome decorated with these anti-transferrin receptor single chain antibody fragments was able to target solid tumours and metastasis.

4.4.2.2 Folate

Folate receptors are over-expressed in many tumours including ovary, kidney, uterus, testis, brain, colon, lung, myelocytic blood cells, as well as KB cells, a nasopharyngeal epidermal carcinoma cell line [27, 54–61]. Published by Lee and Huang, LPDII used polylysine to complex DNA, then was coated with a pH sensitive anionic lipid, as well as the folate ligand as a targeting mechanism to KB cells [27]. It was shown that transfection was controlled by charge for the cationic particles (low lipid/DNA ratio) and by the folate ligand for the anionic particles (high lipid/DNA ratio). KB cells also showed high uptake of neutral liposomes encapsulating doxorubicin targeted using folate conjugated to a PEG linker, secured to the lipid bilayer with a cholesterol anchor [62]. Wang and Low indicated that folate should be conjugated to the carrier at the γ-carboxyl group terminus, so as not to interfere with the receptor-ligand binding pocket [63]. Folate has also been conjugated to PEI [64] and polylysine [65] for gene delivery.

4.4.2.3 Sigma Receptor

Sigma receptors are expressed on many normal tissues within the body, including the heart, liver, endocrine glands, kidneys, lungs, gonads, central nervous system and ovaries [66–68]. Novakova *et al.* showed that in a rat heart, gene expression of sigma receptors can be upregulated by intense stress stimuli, with hypoxia and immobilization showing the greatest upregulation, and exposure to cold showing no effect on gene expression at all [67]. Hypoxia is an environment found in many tumours, thus the upregulation of sigma receptors under these conditions correlate with an upregulation of sigma receptors found in several cancers. Across both human and rodent cancer lines, tumours show an upregulation of the sigma receptors, making substrates for the receptor efficient targeting moieties. Vilner *et al.* tested 13 different rodent and human cell lines, and found that all tumours except one (human MCF-7 breast adenocarcinoma) expressed some level of sigma receptor 1 and 2. MCF-7 did not express the sigma receptor 1 at all. Other cell lines tested include: T47D breast ductal carcinoma, NCI-H727 lung carcinoid, A375 melanoma (amelanotic), ThP-1 leukaemia, U-138MG glioblastoma, SK-N-SH neuroblastoma, LNCaP.FGC prostate, C6 glioma (rat), NB41A3 neuroblastoma (mouse), N1E-115 neuroblastoma (mouse), NG108–15 mouse neuroblastoma X rat glioma (hybrid) and S-20Y neuroblastoma (mouse) [69]. Other studies have discovered other cancer cell lines lacking the sigma receptors, including HeLa, KB, HepG2 and Chinese hamster ovary cells [17].

Ligands for the sigma receptor are diverse and include haloperidol, reduced haloperidol, fluphenazine, perphenazine, trifluoperazine, BD737, LR172, BD1008, SH344, trifluperidol, thioridazine and (–)-butaclamol [70]. Often these ligands can be conjugated to delivery vectors to target cells with over-expressing sigma receptors. Haloperidol has been conjugated to the distal end of PEG linked to a phospholipid and used to deliver a reporter gene to MCF-7 cells, showing a 10-fold increase in expression, compared to the untargeted lipoplex. Administration of spironolacetone (a sigma receptor downregulator) with the lipoplex showed a 10-fold lower gene expression than those cells not treated with spironolacetone [17]. Benzamides have been shown to be sigma receptor ligands [71]. Huang's group first used anisamide, a benzamide ligand that has shown great affinity for sigma receptors, in targeting stealth liposomes loaded with doxorubicin to DU-145, a human prostate adenocarcinoma cell line, by conjugating anisamide to the distal end of PEG [72]. The group has also used anisamide to target LPD to the human NCI-H460 lung cancer line [73, 74] and has shown targeting effects in B16F10 (murine melanoma cells) as a metastatic cancer model [75].

4.4.2.4 RGD

The sequence of RGD (arginine-glycine-aspartic acid) attaches to many proteins in the extracellular matrix and on the cell surface, and specifically is a ligand for integrin receptor $\alpha_v\beta_3$ [76]. The RGD motif plays a role in many processes, including cell growth, migration and differentiation [77]. When targeted with RGD-PEG-PEI or RGD-PEI complexes, saturable binding was shown in Mewo human melanoma cells, and low binding was shown in A549 human lung carcinoma cells [76]. When the PEG spacer was not used, transfection of the Mewo cells was increased 1–2-fold; however, attachment of

RGD to polyplexes using a PEG spacer delivering pDNA against vascular endothelial growth factor showed increased delivery to tumours, growth delay and increased survival rate [29].

4.4.2.5 Intratumoral Delivery

Local injection is a promising treatment option especially in the case of solid tumours and has been taken advantage of for gene delivery with chemical vectors, owing to the inherent lower transfection ability seen with chemical vectors compared to viruses. Intratumoral delivery evades many of the barriers that i.v. delivery faces, as the delivery vector starts in the tumour interstitum. However, in i.t. delivery there is rapid plasmid DNA clearance from the tumour interstitium that can be attributed to the leaky vasculature, and local drug diffusion may cause non-specific toxicity [37]. Also, i.t. delivery is only practical where direct injection is possible, such as solid tumours and to an excised tumour cavity or, in the case of cystic fluid filled tumours that have been drained, to treat the residual tumour cells as suggested by Coll *et al.* [31]. Many of the genes delivered by i.t. delivery centre around immunogene therapy, producing large amounts of cytokines to recruit T and NK (natural killer) cells to the tumour (and only the tumour, hence the direct injection method), to produce a localized immune response.

Intratumoral injection may not be the most realistic treatment for all tumour types, but for solid tumours it is able to deliver nearly the entire volume of injection to the tumour interstitum without depending on the EPR effect. Although i.t. gene therapy is not a viable method for treating metastasis, unless the metastasis can be localized, as in the liver [47]. Most i.t. delivered drugs have no cancer cell targeting mechanism or mechanism to evade RES detection, thus are quickly cleared from the bloodstream. As shown in previous sections, many of the same vectors employed in i.t. delivery of nucleic acid-based drugs have been used in i.v. delivery, with slight modifications, increasing the targeting properties and extending systemic circulation time. Intratumorally administered gene therapy can become a combined therapy when incorporated with a mechanical delivery enhancement method, such as electroporation, ultrasound or use of a micropump injector, showing superior transfection of tumour cells *in vivo*.

4.5 Triggered Release

4.5.1 Introduction

Triggered release is an important aspect of gene delivery. Triggered release can include anything from facilitated escape from endosomes, temperature or pH sensitive carriers or a mechanical method such as electroporation or ultrasound. Mechanical methods can trigger release by permeating cell membranes or breaking open delivery vectors and are often used in concert with intratumoral delivery due to the local application of the mechanical method [78–83]. Using mechanical methods to facilitate gene delivery is a broad subject and outside the scope of this chapter; however, many quality reviews have been written on the subject [84–86].

4.5.2 Proton Sponge Effect

Boussif *et al.* first coined the term for the 'proton sponge effect' when describing PEI in endosomes, but it has been shown to be true for many different chemical delivery vectors [87]. This phenomenon occurs when cationic delivery vectors (reported for polymers, lipids and dendrimers) enter an endosome. The large buffering capacity of non-protonated amino groups or other proton acceptors causes a large influx of hydrogen ions. As protons enter the endosome, so do chloride anions to maintain the electro-neutrality. Water also diffuses in via osmosis to keep the osmotic pressure the same inside and outside the endosome. This influx of water causes the endosome to burst, releasing the delivery vector into the cytosolic space. This mechanism has been verified in cultured cells [88].

Cationic lipids can have interactions with negatively charged endosomal lipids (such as phosphatidylserine) and create non-lamellar structures, possibly disrupting endosomes and allowing for escape into the cytoplasm [89]. Also, greater generation dendrimers conjugated to PLGA particles showed greater buffering capacity, which the authors pointed towards increased proton sponge effect, or triggered release from endolysosomal compartments by lysis of endolysosomes as water diffuses through the membrane to compensate for the large number of positive charges and the lower pH [42].

Sakae *et al.* used PEI/pDNA complexes coated with PEG-Suc (PEG-Succinic acid), using negatively charged carboxylate groups as proton acceptors, and showed an increase in the buffering capacity compared to polyplexes coated with another anionic modified PEG, called PEG-C. While PEG-C accepted 41 μmol of protons between pH 7.4 and 5.5, PEG-Suc accepted 66 μmol. The pK_{a1} and pK_{a2} values of PEG-Suc were 4.00 and 5.24, respectively. The addition of PEG-Suc reduced the zeta potential from +37.8 mV for the PEI/pDNA complexes to −24 to −34 mV. A greater amount of PEG-C was needed to achieve the same negative zeta potential as with PEG-Suc [90]. While much focus has been given to cationic chemical vectors for gene delivery, this paper shows, in a new light, the potential of negatively charged chemical vectors to deliver *in vivo*, based on the proton sponge effect.

4.5.3 Magnetofection™ (Magnetic Delivery)

Buerli *et al.* presented a protocol for magnetofection reporting procedures used in the laboratories of Fritschy (University of Zürich, Switzerland), Medina (INMED/INSERM, Marseille, France) and Fuhrer (Univeristy of Zürich, Switzerland and Mount Sinai Medical Centre, New York, USA). The magnetofection reagent used by all three groups is CombiMag (OZ Biosciences) with a magnetic plate for delivery of cDNA and shRNA into rat hippocampal neurons (embryonic day 18/19). The protocol shows effective double-magnetofection of pyramidal and GABAergic interneurons, with expression of two genes in one neuron [91].

Scherer *et al.* showed that using TransMAG^{PEI} associated with chemical vectors (as well as viral vectors and naked plasmids), increased expression of the luciferase reporter gene delivered. TransMAG^{PEI} are superparamagnetic iron oxide nanoparticles coated with PEI, obtained from Chemicell in Berlin, Germany. Particle sizes ranged from 400–1000 nm. All the chemical vectors used (PEI, AVET-PEI, GenePorter and Lipofectamine) showed increased luciferase activity after 10 minutes of magnetofection with a magnetic field compared to standard transfection (4 hours incubation) in both NIH3T3 and CHO-K1 cell lines. Magnetofection (use of transMAG^{PEI} particles associated with the carrier) without a magnetic field never produced significantly greater gene expression over the standard transfection. Of the chemical vectors used, Lipofectamine™ showed the greatest enhancement in expression, with 971-fold greater expression with magnetofection and applied magnetic field compared to standard transfection [92].

Xenariou *et al.* used TransMAG^{PEI} to deliver Lipofectamine™ 2000 or a cationic lipid (GL67) complexed with pDNA *in vitro*, showing a 300-fold and 30-fold increases in Luciferase gene expression, respectively, compared to the vectors/pDNA without TransMAG^{PEI} after exposure to a magnetic field. However, delivering to nasal epithelium of mice *in vivo*, the control particles, GL67/pDNA, had significantly higher luciferase gene expression (50-fold) than TransMAG^{PEI}-GL67/pDNA complexes; also, the formulation precipitated *in vivo*. When the precipitate was removed, and treatment with the supernatant was also completed, no significant difference existed between the groups of treatment with the supernatant and the precipitates, or between use of the magnet and without. *In vivo* magnetofection was also low in TransMAG^{PEI}-pDNA complexes with and without the magnet compared to naked pDNA [93].

Lee *et al.* used PolyMAG, a superparamagnetic iron oxide nanoparticle, for transfecting mouse embryonic stem cells. PolyMAG consists of superparamagnetic nanoparticles (tsMAG-PEI) coated with PEI, complexed to pDNA with free PEI. Significant increases in transfection efficiency of green fluorescent protein (GFP) were seen with magnetofection (PolyMAG) compared to the control vector, FuGENE6. A magnetic field was applied by placing the plates on the MagnetoFactor plate device for 15 minutes at 37 °C, then returned to incubate at 37 °C. After 50 subpassages in culture, the mouse embryonic stem cells that underwent magnetofection still expressed GFP [94].

4.5.4 Hyperthermic Delivery

Hyperthermia is often used in drug delivery to increase vasodilation in the tumour, increasing the EPR effect. Clinically, temperatures in the range of 40–44 °C are used for local, regional, part body or full body hyperthermia [95]. Chemical vectors have been made with thermosensitive materials such as lipids and polymers to facilitate release of nucleic acids (and other drugs) before, during or after heat is applied.

Chen *et al.* showed a reduced transition temperature of K_8-ELP(1–60) block copolymer particles after complexing with pDNA from 71.5 to 44.9 °C, bringing the particle into a clinically relevant hyperthermic temperature range. ELP is an elastin-like polypeptide with 60 repetitive pentapeptide units, which aggregates above its transition temperature facilitating release, while K_8 signifies a cationic oligolysine used to complex and condense the pDNA. However, while the development of a temperature sensitive block copolymer gene delivery vehicle was produced, and showed delivery of GFP plasmids in MCF cells *in vitro*, hyperthermic triggered release was not established. Release profiles were determined by concentrations of heparin sodium, an anionic protein found in the body. Still, the future of this block copolymer is bright, having a clinically relevant transition temperature and functional activity [96].

Zintchenko *et al.* produced a block copolymer with PEI and poly(N-isopropylacrylamide) (PNIPAM), which when complexing pDNA made uncharged particles of about

150–200 nm that aggregated to sizes of a few microns after 15 minutes of incubation, releasing pDNA at temperatures between 37 and 42 °C. Neuro2A cells were transfected and showed almost two orders of magnitude higher gene expression after several cycles through the polymer's transition temperature. The studies showed that hyperthermia was able to facilitate the rate at which the particles aggregated, but not the rate of uptake into cells. The enhanced gene expression was attributed to the larger size of the aggregated thermosensitive particles in comparison to the control PEI-PEG copolymers, which leads to a greater osmotic pressure inside the endosomes, a greater amount of PEI to facilitate the protein sponge effect and, thus, more DNA released into the cytoplasm [97].

4.5.5 Reductive Agents

Sulfhydryl cross-linked chemical gene delivery systems have been developed to facilitate protection of DNA and triggered release. By creating peptides with one to four cystine residues, McKenzie *et al.* reported a greater ability to create sulfhydryl bonds, and by including many lysine residues the peptides could achieve an appropriate cationic charger to complex with DNA [98]. Five peptides presented in the paper (such as Cys-Trp-Lys_{18}) were able to spontaneously oxidize after binding to pDNA, causing interpeptide disulfide bonds, decreasing the particle size and preventing dissociation of DNA in the presence of concentrated sodium chloride. These five peptides fully condensed DNA at a peptide to DNA ratio of 0.3–0.4 nmol of peptide/μg of DNA, with a charge ratio of 2 : 1 (+/−). Release of DNA was triggered by the reducing environment of the endosome and cytosol, reducing the sulfide bond, showing a 5–60-fold increase in DNA expression compared to uncross-linked peptides. Kwok *et al.* showed *in vivo* delivery of sulfhydryl linked PEG-peptide/glycopeptide DNA nanoparticles with large biodistribution to the liver. In this study, peptides with the most positive zeta potentials (18.1–21.5 mV) showed greatest distribution to the hepatocytes, and those with the most negative zeta potential (−16.1 to −13.0 mV) had the greatest accumulation in Kupffer cells (macrophages of the liver) [99].

Lee and colleagues used degradable nanogels made of hyaluronic acid (HA) to deliver siRNA, triggering release as a function of glutathione (GSH) concentrations [100]. Glutathione is a major reducing agent found in the cytoplasm and triggered the release of siRNA from the hyaluronic acid nanogels that were cross-linked with disulfide linkages. When 10 mM of GSH was used, 100% of the siRNA was released in 60 minutes, compared to only 5.8% siRNA released when 0 mM GSH was used. The HA/siRNA nanogels were also shown to be non-toxic, have GFP expression in HCT-116 silenced comparable to the silencing of PEI/siRNA complexes *in vitro* and, finally, significantly better silencing of GFP as compared to PEI/siRNA complexes in the presence of serum *in vitro*. The hyaluronic acid nanogels also had specific endocytosis mediated by the CD44 receptor [100].

4.5.6 Biotin–Avidin

The strong affinity of biotin for avidin has been well established. Xiong and colleagues developed a nanoparticle that takes advantage of this to trigger release of DNA. The carrier consisted of avidin conjugated to PEG, and biotin conjugated to PEI. PEI complexed the plasmid DNA, the biotin bound the avidin and the PEG was included to protect the carrier from clearance by the RES. The group found that conjugating PEI to biotin decreased the binding affinity to avidin, making it possible for dissociation of biotin from avidin *in vivo* in a biotin-rich environment. Localization *in vivo* to tumour interstitium was attributed to the enhanced permeability and retention effect. After polyplexes had accumulated in tumour interstitium, additional biotin could be delivered either orally or intravenously to bind with the avidin-PEG molecules, triggering release of the PEG shell from the PEI-pDNA complex. This also allows the polyplex to associate with the cell membrane, encouraging endocytosis [101].

4.5.7 Electrochemical

DNA conjugated to gold nanoparticles has elicited interest in electrochemically triggered release. Wang *et al.* in a proof-of-concept experiment showed that it was possible to immobilize non-thiolated layers of dsDNA or ssDNA on gold nanoparticles, and release the DNA by controlling the electric potential [102].

4.6 Recent Clinical Trials Using Naked and Chemically Complexed Nucleic Acids

4.6.1 Introduction

When searching for 'gene therapy' clinical trials on www.clinicaltrials.gov, a website maintained by the National Institutes of Health (NIH) as a 'registry of federally and privately supported clinical trials conducted in the United States and around the world', nearly 1500 trials were returned. However, the majority of these trials had been flagged because genotyping, gene expression analysis or gene expression profiling was included in the studies' aims. Other trials included delivery vehicles such as adenoviruses, adeno-associated viruses, lentiviruses, retroviruses or cell-based therapy, where immune cells were transformed with a virus to express a certain protein and re-injected into the body. Although these fields are older and more developed than those of non-viral, chemical vectors, some chemical vectors appeared on this list in less than two percent of the clinical trials, ranging from phase 1 to phase 4. In the following section, clinical trials have an associated number (i.e. [NCT00000000]). These clinical trials can be found online by going to http://www.clinicaltrials.gov/show/NCT00000000, where NCT00000000 is the NCT identification number associated with the trial.

4.6.2 Naked Nucleic Acids in Clinical Trials

4.6.2.1 Naked DNA

In phase 1 studies at the Seoul National University Hospital, Korea, IL-2 plasmids are being used as an immune adjuvant to vaccinate patients with HIV antigens. A human mutant plasmid called GX-12 contains both human IL-2 and HIV-1 antigen genes. An intramuscular injection of GX-12 is delivered in combination with highly active antiretroviral therapy (HAART) [NCT00517569].

A plasmid encoding vascular endothelial growth factor (VEGF) has been delivered to the heart to promote angiogenesis and vasculogenesis to treat myocardial ischaemia and angina pectoris. In phase 1 and 2 trials, at the Cardiac Catheterization Laboratory, The Heart Centre, University Hospital, Rigshospitalet, Copenhagen, Denmark, VEGF-A165 plasmid was used to treat chronic myocardial ischaemia. This treatment was also combined with granulocyte colony-stimulating factor (G-CSF) therapy, recruiting bone marrow derived stem cells to regenerate the cardiac tissue at the sight of damage, promoting angiogenesis and vasculogenesis [NCT00135850]. Similarly, at Northwestern University, Evanston, IL, USA, in phase 1 trials, phVEGF165 is being delivered by intramyocardial injection into the heart, and G-CSF is also administered treating patients with ischaemic heart failure. [NCT00279539]. In an effort to enhance vasculogenesis, VEGF has been delivered with another growth factor, basic fibroblast growth factor (bFGF). In phase 2 trials at the Institute of Cardiology, Warsaw, Poland, plasmid coding human VEGF-A165/bFGF was injected into ischaemic myocardium of refractory coronary artery disease patients. These two important angiogenic growth factors, bFGF and VEGF-A were combined into one plasmid (bicistronic). This therapy aims at stimulating neoangiogenesis to increase myocardial perfusion [NCT00620217].

In phase 2 trials at 35 locations across the United States, recombinant plasmid DNA coding for VEGF2 (pVGI.1) is administered using an experimental cardiac direct injection catheter (Stiletto™) system for patients with angina pectoris [NCT00090714].

DNA encoding VEGF has also been used to treat critical limb ischaemia, leg ulcers, peripheral artery disease and diabetic neuropathy. At over 30 study locations, in a phase 3 trial, plasmid XRP0038/NV1FGF (delivering phVEGF165) is being used for critical limb ischaemia (CLI) patients with skin lesions [NCT00566657]. At three locations across the United States, in a phase 1 study, a VEGF plasmid (pVGI.1) is injected intramuscularly into leg muscles to treat moderate to high risk critical limb ischaemia and leg ulcers. VEGF treatment is being tested to see if it improves rest, pain and/or heals ulcers in the legs of patients with peripheral artery disease [NCT00304837].

At centres in Boston and New York, USA, VEGF DNA is being delivered in the form of pVGI.1 to express VEGF2 to treat diabetic neuropathy in the legs. The study is in phase 1 and 2 trials [NCT00056290]. In phase 1 trials by Diabetes and Glandular Disease Research Associates, San Antonio, USA, plasmid hVEGF-A is used to treat diabetic neuropathy. The plasmid, named SB-509, by Sangamo BioSciences, is administered as an intramuscular injection [NCT00110500].

A phase 1 and 2 trial at Memorial Sloan-Kettering Cancer Centre, New York, USA, is using sargramostim, also known as granulocyte-macrophage colony-stimulating factor (GM-CSF). DNA is used as an adjuvant for a multi-epitope vaccine to treat melanoma. DNA is injected subcutaneously on day 0 and the vaccine delivered on day 5, also subcutaneously. This study is treating stage IIB, stage IIC, stage III or stage IV melanoma [NCT00085137].

In Connecticut, USA, a phase 1 trial is underway to treat leaky bladder problems. The plasmid, hMaxi-K will be given as a single administration into the bladder through a catheter. Changes in incontinence episodes, maximum bladder capacity, urgency episodes and participant quality of life will be analysed, among other variables [NCT00495053].

Turner syndrome and SHOX (short stature homeobox) deficiency is being treated at a phase 3 trial conducted at Eli Lilly, Philadelphia, PA, USA. A recombinant DNA plasmid for somatropin, also known as HGH, human growth hormone, is being administered and 'height velocity' of those on this treatment and those on placebo will be measured. In this study, treatment will be continued until patients reach normal adult height [NCT00190658].

At two locations, the Dubowitz Neuromuscular Centre, Hammersmith Hospital and the Clinical Trails Unit, St Mary's Hospital, both in London, United Kingdom, Duchenne Muscular Dystrophy (DMD) is being treated by intramuscular administration of antisense therapy with the use of antisense oligonucleotides (AONs), which has the potential to restore the production of dystrophin, the defective protein, in >70% of DMD. The intramuscular administered morpholino oligomer is directed against exon 51 (AVI-4658 PMO) [NCT00159250].

At locations in six European countries, phase 2 trials are underway to treat patients with severe peripheral arterial occlusive disease, Fontaine's stage IV. The pCOR plasmid, named XRP0038 (NV1FGF), was constructed by inserting the gene coding for the fibroblast growth factor. FGF plays a role in angiogenesis [NCT00368797].

4.6.2.2 Naked DNA with Electroporation

In the first clinical trial to combine electroporation with free plasmid delivery, free IL-12 plasmid is injected into melanoma tumours near the surface of the skin then electroporation is applied. This phase 1 study is being carried out at the H. Lee Moffitt Cancer Centre & Research Institute, in Tampa, USA [NCT00323206].

4.6.2.3 Hybrid Oligonucleotides

In phase 1 trial at Albert Einstein Comprehensive Cancer Centre, New York, USA, an 18-mer hybrid oligonucleotide is being used to study the effectiveness of combining docetaxel and GEM 231 in treating patients who have recurrent or refractory solid tumours. GEM-231 (a mixed backbone oligonucleotide) is a strand of synthetic DNA which has been modified with 2′-O-methyl ribose at both ends in order to resemble RNA [103] targeted against the RIalpha subunit of protein kinase A (PKA) [104].

4.6.3 Lipid-Based Nucleic Acid Carriers in Clinical Trials

Cationic liposomes composed of DMRIE/DOPE, have been used to delivering pGT-1 (cystic fibrosis gene) to cells lining the nose of cystic fibrosis patients. DMIRE is a cationic lipid, an abbreviation for 1,2-dimyristyloxypropyl-3-dimethyl-hydroxy ethyl ammonium bromide, while DOPE is a neutral lipid, an abbreviation for that stands for L-alpha-dioleoyl phosphatidylethanolamine. Delivery was facilitated by syringe instillation over thirty minutes to the right inferior nasal turbinate. Studies are in phase 1 trials at the University of Alabama, Birmingham, USA [NCT00004471].

At the M.D. Anderson Cancer Centre, Houston, USA, DOTAP:chol liposomes were complexed with the fus1 gene to be delivered intravenously in the treatment of lung cancer [NCT00059605].

To treat patients with recurrent or refractory stage III or stage IV head and neck cancer, cationic liposomes (DOTMA:chol) were used to deliver the interleukin-2 gene. The

17. Mukherjee, A., Prasad, T.K., Rao, N.M. and Banerjee, R. (2005) Haloperidol-associated stealth liposomes: a potent carrier for delivering genes to human breast cancer cells. *J. Biol. Chem.*, **280**, 15619–15627.
18. Dauty, E. and Verkman, A.S. (2005) Actin cytoskeleton as the principal determinant of size-dependent DNA mobility in cytoplasm: a new barrier for non-viral gene delivery. *J. Biol. Chem.*, **280**, 7823–7828.
19. Lukacs, G.L., Haggie, P., Seksek, O. *et al.* (2000) Size-dependent DNA mobility in cytoplasm and nucleus. *J. Biol. Chem.*, **275**, 1625–1629.
20. Hebert, E. (2003) Improvement of exogenous DNA nuclear importation by nuclear localization signal-bearing vectors: a promising way for non-viral gene therapy? *Biol. Cell*, **95**, 59–68.
21. Pante, N. and Kann, M. (2002) Nuclear pore complex is able to transport macromolecules with diameters of about 39 nm. *Mol. Biol. Cell*, **13**, 425–434.
22. Felgner, P.L., Gadek, T.R., Holm, M. *et al.* (1987) Lipofection: a highly efficient, lipid-mediated DNA-transfection procedure. *Proc. Natl. Acad. Sci. USA*, **84**, 7413–7417.
23. Walker, C., Selby, M., Erickson, A. *et al.* (1992) Cationic lipids direct a viral glycoprotein into the class I major histocompatibility complex antigen-presentation pathway. *Proc. Natl. Acad. Sci. USA*, **89**, 7915–7918.
24. Quattrone, A., Papucci, L., Morganti, M. *et al.* (1994) Inhibition of MDR1 gene expression by antimessenger oligonucleotides lowers multiple drug resistance. *Oncol. Res.*, **6**, 311–320.
25. Faneca, H., Cabrita, A.S., Simoes, S. and Pedroso de Lima, M.C. (2007) Evaluation of the antitumoral effect mediated by IL-12 and HSV-tk genes when delivered by a novel lipid-based system. *Biochim. Biophys. Acta*, **1768**, 1093–1102.
26. Schenkman, S., Araujo, P.S., Dijkman, R. *et al.* (1981) Effects of temperature and lipid composition on the serum albumin-induced aggregation and fusion of small unilamellar vesicles. *Biochim. Biophys. Acta*, **649**, 633–647.
27. Lee, R.J. and Huang, L. (1996) Folate-targeted, anionic liposome-entrapped polylysine-condensed DNA for tumor cell-specific gene transfer. *J. Biol. Chem.*, **271**, 8481–8487.
28. Li, S.D., Chono, S. and Huang, L. (2008) Efficient oncogene silencing and metastasis inhibition via systemic delivery of siRNA. *Mol. Ther.*, **16**, 942–946.
29. Kunath, K., Merdan, T., Hegener, O. *et al.* (2003) Integrin targeting using RGD-PEI conjugates for *in vitro* gene transfer. *J. Gene Med.*, **5**, 588–599.
30. Li, S., Dong, W., Zong, Y. *et al.* (2007) Polyethylenimine-complexed plasmid particles targeting focal adhesion kinase function as melanoma tumor therapeutics. *Mol. Ther.*, **15**, 515–523.
31. Coll, J.L., Chollet, P., Brambilla, E. *et al.* (1999) *In vivo* delivery to tumors of DNA complexed with linear polyethylenimine. *Hum. Gene Ther.*, **10**, 1659–1666.
32. Rozema, D.B., Lewis, D.L., Wakefield, D.H. *et al.* (2007) Conjugates for targeted *in vivo* delivery of siRNA to hepatocytes. *Proc. Natl. Acad. Sci. USA*, **104**, 12982–12987.
33. Gonzalez, H., Hwang, S.J. and Davis, M.E. (1999) New class of polymers for the delivery of macromolecular therapeutics. *Bioconjug. Chem.*, **10**, 1068–1074.
34. Bellocq, N.C., Pun, S.H., Jensen, G.S. and Davis, M.E. (2003) Transferrin-containing, cyclodextrin polymer-based particles for tumor-targeted gene delivery. *Bioconjug. Chem.*, **14**, 1122–1132.
35. Hu-Lieskovan, S., Heidel, J.D., Bartlett, D.W. *et al.* (2005) Sequence-specific knockdown of EWS-FLI1 by targeted, nonviral delivery of small interfering RNA inhibits tumor growth in a murine model of metastatic Ewing's sarcoma. *Cancer Res.*, **65**, 8984–8992.
36. Bartlett, D.W. and Davis, M.E. (2007) Physicochemical and biological characterization of targeted, nucleic acid-containing nanoparticles. *Bioconjug. Chem.*, **18**, 456–468.
37. Yockman, J.W., Maheshwari, A., Han, S.O. and Kim, S.W. (2003) Tumor regression by repeated intratumoral delivery of water soluble lipopolymers/p2CMVmIL-12 complexes. *J. Control. Release*, **87**, 177–186.

38. Omidi, Y., Hollins, A.J., Drayton, R.M. and Akhtar, S. (2005) Polypropylenimine dendrimer-induced gene expression changes: the effect of complexation with DNA, dendrimer generation and cell type. *J. Drug Target*, **13**, 431–443.
39. Behr, J.P. (1997) The proton sponge: a trick to enter cells the viruses did not exploit. *Chimia*, **51**, 34–36.
40. Manunta, M., Nichols, B.J., Tan, P.H. *et al.* (2006) Gene delivery by dendrimers operates via different pathways in different cells, but is enhanced by the presence of caveolin. *J. Immunol. Methods*, **314**, 134–146.
41. Manunta, M., Tan, P.H., Sagoo, P. *et al.* (2004) Gene delivery by dendrimers operates via a cholesterol dependent pathway. *Nucleic Acids Res.*, **32**, 2730–2739.
42. Zhang, X.Q., Intra, J. and Salem, A.K. (2007) Conjugation of polyamidoamine dendrimers on biodegradable microparticles for nonviral gene delivery. *Bioconjug. Chem.*, **18**, 2068–2076.
43. Zhang, X.Q., Wang, X.L., Huang, S.W. *et al.* (2005) *In vitro* gene delivery using polyamidoamine dendrimers with a trimesyl core. *Biomacromolecules*, **6**, 341–350.
44. Morris, M.C., Gros, E., Aldrian-Herrada, G. *et al.* (2007) A non-covalent peptide-based carrier for *in vivo* delivery of DNA mimics. *Nucleic Acids Res.*, **35**, e49.
45. Kumar, P., Wu, H., McBride, J.L. *et al.* (2007) Transvascular delivery of small interfering RNA to the central nervous system. *Nature*, **448**, 39–43.
46. Liu, F., Song, Y.K. and Liu, D. (1999) Hydrodynamics-based transfection in animals by systemic administration of plasmid DNA. *Gene Ther.*, **6**, 1258–1266.
47. Pirollo, K.F., Zon, G., Rait, A. *et al.* (2006) Tumor-targeting nanoimmunoliposome complex for short interfering RNA delivery. *Hum. Gene Ther.*, **17**, 117–124.
48. Wu, J., Nantz, M.H. and Zern, M.A. (2002) Targeting hepatocytes for drug and gene delivery: emerging novel approaches and applications. *Front Bio. Sci.*, **7**, d717–725.
49. Yu, F., Jiang, T., Zhang, J. *et al.* (2007) Galactosylated liposomes as oligodeoxynucleotides carrier for hepatocyte-selective targeting. *Pharmazie*, **62**, 528–533.
50. Cotten, M., Langle-Rouault, F., Kirlappos, H. *et al.* (1990) Transferrin-polycation-mediated introduction of DNA into human leukemic cells: stimulation by agents that affect the survival of transfected DNA or modulate transferrin receptor levels. *Proc. Natl. Acad. Sci. USA*, **87**, 4033–4037.
51. Stavridis, J.C., Deliconstantinos, G., Psallidopoulos, M.C. *et al.* (1986) Construction of transferrin-coated liposomes for *in vivo* transport of exogenous DNA to bone marrow erythroblasts in rabbits. *Exp. Cell Res.*, **164**, 568–572.
52. Joshee, N., Bastola, D.R. and Cheng, P.W. (2002) Transferrin-facilitated lipofection gene delivery strategy: characterization of the transfection complexes and intracellular trafficking. *Hum. Gene Ther.*, **13**, 1991–2004.
53. Cheng, P.W. (1996) Receptor ligand-facilitated gene transfer: enhancement of liposome-mediated gene transfer and expression by transferrin. *Hum. Gene Ther.*, **7**, 275–282.
54. Campbell, I.G., Jones, T.A., Foulkes, W.D. and Trowsdale, J. (1991) Folate-binding protein is a marker for ovarian cancer. *Cancer Res.*, **51**, 5329–5338.
55. Coney, L.R., Tomassetti, A., Carayannopoulos, L. *et al.* (1991) Cloning of a tumor-associated antigen: MOv18 and MOv19 antibodies recognize a folate-binding protein. *Cancer Res.*, **51**, 6125–6132.
56. Franklin, W.A., Waintrub, M., Edwards, D. *et al.* (1994) New anti-lung-cancer antibody cluster 12 reacts with human folate receptors present on adenocarcinoma. *Int. J. Cancer Suppl.*, **8**, 89–95.
57. Garin-Chesa, P., Campbell, I., Saigo, P.E. *et al.* (1993) Trophoblast and ovarian cancer antigen LK26. Sensitivity and specificity in immunopathology and molecular identification as a folate-binding protein. *Am. J. Pathol.*, **142**, 557–567.
58. Holm, J., Hansen, S.I., Hoier-Madsen, M. *et al.* (1994) Folate receptor of human mammary adenocarcinoma. *Apmis*, **102**, 413–419.
59. Rettig, W.J., Garin-Chesa, P., Beresford, H.R. *et al.* (1988) Cell-surface glycoproteins of human sarcomas: differential expression in normal and malignant tissues and cultured cells. *Proc. Natl. Acad. Sci. USA*, **85**, 3110–3114.

60. Ross, J.F., Chaudhuri, P.K. and Ratnam, M. (1994) Differential regulation of folate receptor isoforms in normal and malignant tissues *in vivo* and in established cell lines. Physiologic and clinical implications. *Cancer*, **73**, 2432–2443.
61. Weitman, S.D., Lark, R.H., Coney, L.R. *et al.* (1992) Distribution of the folate receptor GP38 in normal and malignant cell lines and tissues. *Cancer Res.*, **52**, 3396–3401.
62. Zhao, X.B., Muthusamy, N., Byrd, J.C. and Lee, R.J. (2007) Cholesterol as a bilayer anchor for PEGylation and targeting ligand in folate-receptor-targeted liposomes. *J. Pharm. Sci.*, **96**, 2424–2435.
63. Wang, S. and Low, P.S. (1998) Folate-mediated targeting of antineoplastic drugs, imaging agents, and nucleic acids to cancer cells. *J. Control. Release*, **53**, 39–48.
64. Guo, W. and Lee, R.L. (1999) Receptor-targeted gene delivery via folate-conjugated polyethylenimine. *AAPS Pharm. Sci.*, **1**, E19.
65. Mislick, K.A., Baldeschwieler, J.D., Kayyem, J.F. and Meade, T.J. (1995) Transfection of folate-polylysine DNA complexes: evidence for lysosomal delivery. *Bioconjug. Chem.*, **6**, 512–515.
66. Wolfe, S.A. Jr, Culp, S.G. and De Souza, E.B. (1989) Sigma-receptors in endocrine organs: identification, characterization, and autoradiographic localization in rat pituitary, adrenal, testis, and ovary. *Endocrinology*, **124**, 1160–1172.
67. Novakova, M., Bruderova, V., Sulova, Z. *et al.* (2007) Modulation of expression of the sigma receptors in the heart of rat and mouse in normal and pathological conditions. *Gen. Physiol. Biophys.*, **26**, 110–117.
68. Hellewell, S.B., Bruce, A., Feinstein, G. *et al.* (1994) Rat liver and kidney contain high densities of sigma 1 and sigma 2 receptors: characterization by ligand binding and photoaffinity labeling. *Eur. J. Pharmacol.*, **268**, 9–18.
69. Vilner, B.J., John, C.S. and Bowen, W.D. (1995) Sigma-1 and sigma-2 receptors are expressed in a wide variety of human and rodent tumor cell lines. *Cancer Res.*, **55**, 408–413.
70. Vilner, B.J., de Costa, B.R. and Bowen, W.D. (1995) Cytotoxic effects of sigma ligands: sigma receptor-mediated alterations in cellular morphology and viability. *J. Neuro. Sci.*, **15**, 117–134.
71. Mach, R.H., Huang, Y., Freeman, R.A. *et al.* (2004) Conformationally-flexible benzamide analogues as dopamine D3 and sigma 2 receptor ligands. *Bioorg. Med. Chem. Lett.*, **14**, 195–202.
72. Banerjee, R., Tyagi, P., Li, S. and Huang, L. (2004) Anisamide-targeted stealth liposomes: a potent carrier for targeting doxorubicin to human prostate cancer cells. *Int. J. Cancer*, **112**, 693–700.
73. Li, S.D. and Huang, L. (2006) Surface-modified LPD nanoparticles for tumor targeting. *Ann. NY Acad. Sci.*, **1082**, 1–8.
74. Li, S.D., Chen, Y.C., Hackett, M.J. and Huang, L. (2008) Tumor-targeted delivery of siRNA by self-assembled nanoparticles. *Mol. Ther.*, **16**, 163–169.
75. Li, S.D., Chono, S. and Huang, L. (2008) Efficient gene silencing in metastatic tumor by siRNA formulated in surface-modified nanoparticles. *J. Control. Release*, **126**, 77–84.
76. Kunath, K., Merdan, T., Hegener, O. *et al.* (2003) Integrin targeting using RGD-PEI conjugates for *in vitro* gene transfer. *J. Gene Med.*, **5**, 588–599.
77. Ruoslahti, E. (1996) RGD and other recognition sequences for integrins. *Ann. Rev. Cell Dev. Biol.*, **12**, 697–715.
78. Henshaw, J.W., Zaharoff, D.A., Mossop, B.J. and Yuan, F. (2007) Electric field-mediated transport of plasmid DNA in tumor interstitium *in vivo*. *Bioelectrochemistry*, **71**, 233–242.
79. Zaharoff, D.A., Barr, R.C., Li, C.Y. and Yuan, F. (2002) Electromobility of plasmid DNA in tumor tissues during electric field-mediated gene delivery. *Gene Ther.*, **9**, 1286–1290.
80. Tamura, T., Nishi, T., Goto, T. *et al.* (2001) Intratumoral delivery of interleukin 12 expression plasmids with *in vivo* electroporation is effective for colon and renal cancer. *Hum. Gene Ther.*, **12**, 1265–1276.
81. Takahashi, Y., Nishikawa, M. and Takakura, Y. (2006) Suppression of tumor growth by intratumoral injection of short hairpin RNA-expressing plasmid DNA targeting beta-catenin or hypoxia-inducible factor 1alpha. *J. Control. Release*, **116**, 90–95.

82. McCray, A.N., Ugen, K.E., Muthumani, K. *et al.* (2006) Complete regression of established subcutaneous B16 murine melanoma tumors after delivery of an HIV-1 Vpr-expressing plasmid by *in vivo* electroporation. *Mol. Ther.*, **14**, 647–655.
83. Duvshani-Eshet, M., Benny, O., Morgenstern, A. and Machluf, M. (2007) Therapeutic ultrasound facilitates antiangiogenic gene delivery and inhibits prostate tumor growth. *Mol. Cancer Ther.*, **6**, 2371–2382.
84. Niidome, T. and Huang, L. (2002) Gene therapy progress and prospects: nonviral vectors. *Gene Ther.*, **9**, 1647–1652.
85. Mir, L.M., Moller, P.H., Andre, F. and Gehl, J. (2005) Electric pulse-mediated gene delivery to various animal tissues. *Adv. Genet.*, **54**, 83–114.
86. Favard, C., Dean, D.S. and Rols, M.P. (2007) Electrotransfer as a non viral method of gene delivery. *Curr. Gene Ther.*, **7**, 67–77.
87. Boussif, O., Lezoualc'h, F., Zanta, M.A. *et al.* (1995) A versatile vector for gene and oligonucleotide transfer into cells in culture and *in vivo*: polyethylenimine. *Proc. Natl. Acad. Sci. USA*, **92**, 7297–7301.
88. Sonawane, N.D., Thiagarajah, J.R. and Verkman, A.S. (2002) Chloride concentration in endosomes measured using a ratioable fluorescent Cl- indicator: evidence for chloride accumulation during acidification. *J. Biol. Chem.*, **277**, 5506–5513.
89. Lewis, R.N. and McElhaney, R.N. (2000) Surface charge markedly attenuates the nonlamellar phase-forming propensities of lipid bilayer membranes: calorimetric and (31)P-nuclear magnetic resonance studies of mixtures of cationic, anionic, and zwitterionic lipids. *Biophys. J.*, **79**, 1455–1464.
90. Sakae, M., Ito, T., Yoshihara, C. *et al.* (2008) Highly efficient *in vivo* gene transfection by plasmid/PEI complexes coated by anionic PEG derivatives bearing carboxyl groups and RGD peptide. *Biomed. Pharmacother.*, **62**, 448–453.
91. Buerli, T., Pellegrino, C., Baer, K. *et al.* (2007) Efficient transfection of DNA or shRNA vectors into neurons using magnetofection. *Nat. Protoc.*, **2**, 3090–3101.
92. Scherer, F., Anton, M., Schillinger, U. *et al.* (2002) Magnetofection: enhancing and targeting gene delivery by magnetic force *in vitro* and *in vivo*. *Gene Ther.*, **9**, 102–109.
93. Xenariou, S., Griesenbach, U., Ferrari, S. *et al.* (2006) Using magnetic forces to enhance non-viral gene transfer to airway epithelium *in vivo*. *Gene Ther.*, **13**, 1545–1552.
94. Lee, C.H., Kim, E.Y., Jeon, K. *et al.* (2008) Simple, efficient, and reproducible gene transfection of mouse embryonic stem cells by magnetofection. *Stem Cells Dev.*, **17**, 133–141.
95. Schlemmer, M., Lindner, L.H., Abdel-Rahman, S. and Issels, R.D. (2004) [Principles, technology and indication of hyperthermia and part body hyperthermia]. *Radiologe*, **44**, 301–309.
96. Chen, T.H., Bae, Y., Furgeson, D.Y. and Biosynthetic, I. (2008) Nanobiomaterials (IBNs) for hyperthermic gene delivery. *Pharm. Res.*, **25**, 683–691.
97. Zintchenko, A., Ogris, M. and Wagner, E. (2006) Temperature dependent gene expression induced by PNIPAM-based copolymers: potential of hyperthermia in gene transfer. *Bioconjug. Chem.*, **17**, 766–772.
98. McKenzie, D.L., Kwok, K.Y. and Rice, K.G. (2000) A potent new class of reductively activated peptide gene delivery agents. *J. Biol. Chem.*, **275**, 9970–9977.
99. Kwok, K.Y., Park, Y., Yang, Y. *et al.* (2003) *In vivo* gene transfer using sulfhydryl cross-linked PEG-peptide/glycopeptide DNA co-condensates. *J. Pharm. Sci.*, **92**, 1174–1185.
100. Lee, H., Mok, H., Lee, S. *et al.* (2007) Target-specific intracellular delivery of siRNA using degradable hyaluronic acid nanogels. *J. Control. Release*, **119**, 245–252.
101. Xiong, M.P., Forrest, M.L., Karls, A.L. and Kwon, G.S. (2007) Biotin-triggered release of poly(ethylene glycol)-avidin from biotinylated polyethylenimine enhances *in vitro* gene expression. *Bioconjug. Chem.*, **18**, 746–753.
102. Wang, J., Jiang, M. and Mukherjee, B. (2000) On-demand electrochemical release of DNA from gold surfaces. *Bioelectrochemistry*, **52**, 111–114.
103. Blay, J. (2000) Technology evaluation: GEM-231. Hybridon Inc. *Curr. Opin. Mol. Ther.*, **2**, 468–472.
104. Agrawal, S., Kandimalla, E.R., Yu, D. *et al.* (2002) GEM 231, a second-generation antisense agent complementary to protein kinase A RIalpha subunit, potentiates antitumor activity of

irinotecan in human colon, pancreas, prostate and lung cancer xenografts. *Int. J. Oncol.*, **21**, 65–72.
105. Kaushik, A. (2001) Leuvectin Vical Inc. *Curr. Opin. Investig. Drugs*, **2**, 976–981.
106. Gleich, L.L., Gluckman, J.L., Armstrong, S. *et al.* (1998) Alloantigen gene therapy for squamous cell carcinoma of the head and neck: results of a phase-1 trial. *Arch. Otolaryngol. Head Neck Surg.*, **124**, 1097–1104.
107. Bailey, C., Negus, R., Morris, A. *et al.* (2007) Chemokine expression is associated with the accumulation of tumour associated macrophages (TAMs) and progression in human colorectal cancer. *Clin. Exp. Metastasis*, **24**, 121–130.

5

Viral Technology for Delivery of Nucleic Acids

Shervin Bahrami and Finn Skou Pedersen
Aarhus University, Department of Molecular Biology, Aarhus, Denmark

5.1 Introduction

The therapeutic potential for medicine based on DNA or RNA is enormous, ranging from the correction of mutated genes for treating inherited diseases to the silencing of the expression of harmful genes by RNA interference. The fields of gene and nucleic acid therapy have been explored for years with variable success.

One, or perhaps the biggest obstacle, for the use of nucleic acid-based therapy as an effective treatment is the efficient and directed delivery to the cells of interest. Among several methods used for nucleotide delivery, virally-based systems hold the most promising perspectives in terms of efficiency. Viruses are naturally evolved nucleotide delivery packages. Although there are significant differences between virus families, they have two characteristics in common. They all contain a genetic blueprint in the form of either DNA or RNA (and that in either single or double-stranded forms) that can use the cellular machinery for gene expression and a protein machinery for the efficient and *active* delivery of these into their host cells. The key phrase here is of course 'active'. Since cells are surrounded by membranes that are impermeable to most molecules, among them polynucleotides, an active mechanism for penetrating this barrier has evolved in every known virus.

Viruses make perfect gene delivery vehicles for biomedical purposes if: (i) their cargo can be manipulated to include the heterologous nucleic acids of interest; (ii) their delivery machinery can be exploited to deliver their cargo to predetermined cells of interest; and (iii) their harmful side effects can be controlled.

Delivery Technologies for Biopharmaceuticals: Peptides, Proteins, Nucleic Acids and Vaccines Edited by Lene Jorgensen and Hanne Mørck Nielsen

All viruses contain polynucleotides and proteins. Some, such as retroviruses and influenza viruses, are enveloped in a membrane composed of a lipid bilayer while others, such as adenoviruses or the parvovirus adeno-associated virus (AAV), are not. The mechanisms of viral entry can be divided into two main groups depending on the presence or absence of a lipid envelope. Topologically, the core of enveloped viruses belongs to the same space as the cytosol of target cells. Therefore, fusion of the viral and cellular membranes is sufficient for the entry of the viral core into the cytosol. On the contrary, non-enveloped viral cores are in a topological sense part of the extracellular space. Entry of such viruses therefore depends on traversing the plasma membrane of host cells.

The basis for viral technology for delivery of polynucleotides is the fact that virus genomes contain two main categories of information, *cis*-acting and *trans*-acting elements. *Cis*-acting sequences exert their function at the nucleotide level. They are involved in the regulation of transcription, translation and replication (depending on the viral life cycle) as well as in packaging of the viral genome into new particles. The *trans*-acting elements, on the other hand, consist of the protein-coding genes that direct the production of the structural proteins and enzymes that make up a virion. Hence, it is possible to construct artificial viral genomes, commonly referred to as viral *vectors*, in which part or all of the protein-coding genes are replaced by heterologous sequences, that is the polynucleotide cargo. When the viral structural proteins and enzymes are supplied *in trans*, such a vector will be packaged in a virion just like the normal viral genome, ready to be introduced into target cells (Figure 5.1).

Once the viral carrier has entered the target cell, the nucleic acid cargo comes into action. Depending on what the cargo is meant to do, different viral families can be used. Retroviruses, including lentiviruses, are endowed with an enzymatic mechanism for integrating their genomes into the host cells' DNA. Through genetic engineering, heterologous genes (also called transgenes) can be inserted into the retroviral genomes, making them ideal vectors for the stable and permanent expression of genes. Other viruses, such as adenoviruses, express their genes from genomes that are not integrated into cellular DNA and, therefore, if not replicated are sequestered and lost through cell division. Thus, manipulated adenoviral genomes offer means for transient expression of heterologous genes. However, in some non-dividing cells *in vivo*, adenovirus vectors may be maintained for months to years. Even viruses of bacteria or insects that do not themselves replicate in vertebrate hosts can be modified to deliver genetic cargo into mammalian cells. In the following sections, these different aspects of viral entry and gene expression are explained and the perspectives for their use as delivery vectors discussed. As an alternative to strategies based upon the encapsidation of genetically engineered viral genomes in cellular production systems, virus particles may in special cases be loaded with chemically or enzymatically synthesized nucleic acid. Finally, components of viruses may be used in conjugation with synthetic delivery vehicles, one example being the incorporation of a viral envelope protein capable of membrane fusion into a chemically defined liposome.

Attenuated viruses have a long history in medicine, mainly for vaccination purposes but also in tumour therapy. While this development initially used natural isolates, modern technology allows the specific design of modified viruses for, for example, vaccination, immunotherapy and tumour therapy by oncolytic viruses. There is, therefore, no longer a clear border between virus therapy/vaccination and the use of viral vectors for gene

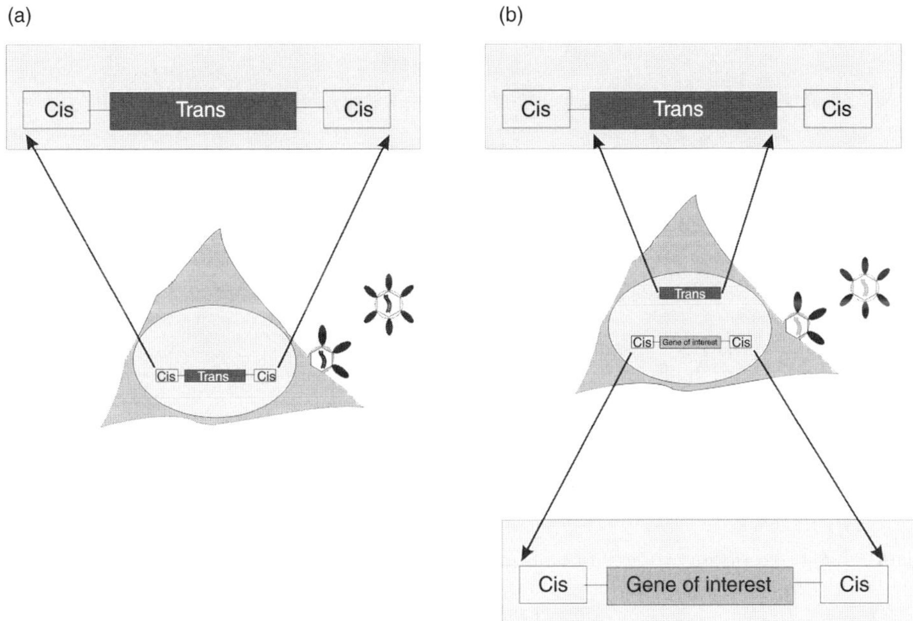

Figure 5.1 *Packaging of viral vectors: The coupling of a gene of interest to viral* cis-acting *sequences promotes its incorporation into a viral particle when* trans-acting *factors are provided by the cell. (a) A cell producing viruses. (b) A cell producing viral particles carrying a vector*

therapy/vaccination. In this chapter, the focus is on simple vehicles for single-cycle delivery of mainly heterologous nucleic acid cargo. For information on modified viruses that undergo multiple rounds of replication in target tissues to destroy tumour cells, or viruses that stimulate an immune response through the expression of a number of viral genes in addition to a transgene, recent reviews can be referred to [1–4].

5.2 Barriers and Topological Requirements to Cellular Entry

As mentioned above, when it comes to entry into host cells, viruses fall into two different categories depending on whether or not they contain a lipid envelope. The mechanisms for entry of these two groups are radically different: Enveloped viruses enter cells through fusion of their own membrane with that of the host cell, whereas non-enveloped viruses must penetrate the membrane of the target cells (Figure 5.2).

5.2.1 Enveloped Viruses and Membrane Fusion

Many important biological activities in eukaryotic cells involve the fusion of membranes. Some of these involve fusion of vesicles inside a cell, such as endo- and exocytosis or intra-Golgi transport, others involve intercellular membranes, such as the fusion of sperm

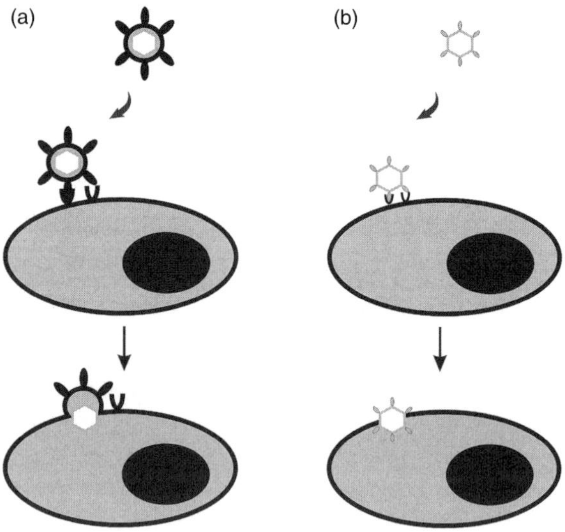

__Figure 5.2__ Two distinct modes of viral entry. (a) The nucleoprotein core of an enveloped virus enters by membrane fusion. (b) A non-enveloped virus enters by membrane penetration. The figure depicts entry at the plasma membrane. However, both modes of viral entry may alternatively take place through an intracellular (e.g. endosomal) membrane after endocytosis

and egg cells and viral-cell fusion [5]. The systems that mediate these functions, although different, share some mechanistic characteristics. This is not surprising since the fusion mechanism is dictated by the chemistry of lipids, the main component of membranes. Lipid molecules are amphipathic. They contain a hydrophilic, and sometimes charged end, and a hydrophobic tail of hydrocarbons. When mixed with water, the hydrophobic tails cluster together spontaneously, leaving the hydrophilic ends exposed to the aqueous environment. Biological membranes are basically bilayers of lipid molecules with entangled hydrophobic tails. This arrangement constitutes the thermodynamically most stable conformation of amphipathic molecules by exposing the hydrophilic ends of the molecules to the aqueous environment while masking the hydrophobic parts from it. For two membranes to fuse, they must first come into close proximity before merging into one. Thus, membrane fusion requires the expelling of water from a hydrophilic surface and bringing two equally charged surfaces together, both of which are energetically unfavourable processes. Because of this high kinetic barrier, membrane fusion does not occur spontaneously and requires an active mechanism [6, 7].

Much of the current knowledge and thinking about the mechanism of fusion of viral and cellular membranes is based upon the function of influenza haemagglutinin (HA) [8]. This is the protein device that mediates the fusion of influenza virus with its host cells. Haemagglutinin consists of two subunits: HA1 and HA2 (Figure 5.3a). The smaller subunit HA1 is anchored within the viral membrane and folded into an energetically unstable conformation, which is in turn locked in place by interaction with the larger HA2

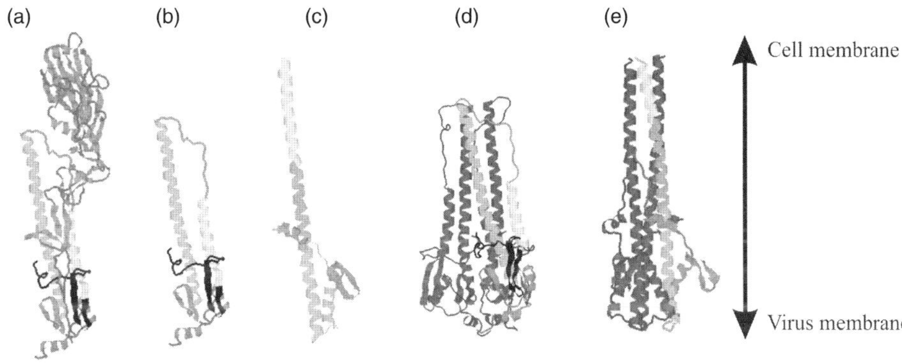

Figure 5.3 *Activation of the membrane-fusion machinery of an enveloped virus, the haemagglutinin (HA) protein of influenza A. (a) HA1 and HA2 in complex before fusion (non-acidic pH). (b) and (d) The HA1 part of (a), in monomeric (b) and trimeric (d) forms. (c) and (e) HA1 in the activated, extended helix form at acidic pH in monomeric (c) and trimeric (e) forms. Note that the black residues of (b) and (d) are not part of the structures shown in (c) and (e)*

subunit (Figure 5.3a). At the extracellular pH the carboxy terminal of HA1 forms a long α-helix that is connected through a loop to an anti-parallel short N-terminal α-helix. This is in turn connected to a hydrophobic 'fusion peptide', which is buried in the trimeric structure (parts of the black) (Figure 5.3b,d). The HA2 subunit has the ability to bind to the cellular receptor, in this case sialic acid. When this happens, the virus is internalized into the endosomal compartment encountering acidic pH. Upon pH change, the connective loop folds into an α-helix extending the triple coiled coil structure. This results in insertion of the formerly buried fusion peptide (a stretch of hydrophobic amino acids) into the target membrane. Subsequently, a hairpin structure forms near the viral membrane at the C-terminal of the protein, resulting in tilting of the HA protein in the plane of the membranes, pulling them closer together (Figure 5.3c,e). What happens next is not completely understood, but it is thought that with ends of the HA protein in opposite membranes, and their close proximity, lipid mixing and fusion can occur.

5.2.2 Non-Enveloped Viruses

Non-enveloped viruses enter cells by binding to surface receptors and subsequently penetrating the cellular membranes either at the surface or in the endosome. Subgroup C adenoviruses, for example, bind to the Coxsackie Adenovirus Receptor (CAR) and are endocytosed, after which it is thought that their interaction with $\alpha_V\beta_5$ integrins at the low pH of the endosomes results in permeabilization of the membrane, followed by endosomal escape into the cytosol. Although the exact mechanism is not known, it has been shown that the adenovirus capsid has the ability to lyse lipid bilayer membranes at a pH optimum corresponding to the endosomal pH [9].

5.3 Routes of Administration

In most cases, the effector molecules delivered by a viral vector are needed at a certain site in an organism (and usually in a certain type of cell). Therefore, it is necessary for a delivery vector to be efficiently targeted to its destination. For certain types of cells, such as cells of the blood or skin, targeting issues can to some extent be addressed through *ex vivo* delivery in which the target cells are removed, genetically modified, and then reintroduced into the organism. However, the ideal usage of viral vectors is by delivery of cargo *in vivo*. To this end, the natural variation of the tropism of viruses adds to the toolbox. Viruses differ in tropism, that is the ability to infect different kinds of cells. Not every virus can infect every cell.

During natural infections, viruses use a variety of strategies to be transmitted from host to host and to overcome host barriers to entry. The strategy used by a given virus is reflected by the structure of the viruses that spread, and how the virus is shed from its previous host. Some viruses are stable, some very unstable. Animal viruses have six primary points of entry into a host organism. These include the epithelial surfaces of skin, respiratory tract, gastrointestinal tract, genitourinary tract, conjunctiva and, finally, the interface between mother and foetus/embryo [10]. These entry portals of specific viruses are potentially valuable as points of delivery of viral vectors.

However, at the current level of technology this natural entry repertoire of viral vehicles has been exploited only to a limited extent in clinical trials. Delivery has mostly been by injection into the blood stream for systemic delivery or vascular delivery to organs such as liver or heart, or by local injection into tumour tissue, muscle or brain, or by intradermal injection. For example vectors derived from adenoviruses and adeno-associated virus 2 have been used for delivery in the upper respiratory tract in clinical trials [11].

5.4 Delivery Vector Requirements

5.4.1 Targeting

There are different determinants for the variety of tropisms that are displayed by wild-type viruses, but the single most important one is the ability to use specific molecules on the surface of a target cells as binding and entry receptors. The choice of receptor is dependent on the affinity of the viral surface proteins (HA2 in influenza, SU in retroviruses, etc.) for these receptor molecules present on the host cell. Therefore, it is theoretically possible to change the tropism of viruses by manipulating their surface proteins. It seems like a simple idea to introduce binding domains for alternative receptors into the viral surface proteins in order to change their tropism, and indeed this has been tried many times with varying success [12, 13].

For example the insertion of stromal derived factor-1α (SDF-1α) into the surface protein of a retrovirus (murine leukaemia virus) resulted in significant transduction of CXCR-4 expressing cells, albeit at an efficiency that was around three orders of magnitude lower than that of the wild-type virus on its natural mouse cell hosts. Interestingly, the same study showed that the insertion of epidermal growth factor (EGF) into the surface protein of the same virus did not alter its receptor usage [14]. Other peptide insertions have resulted in similar results [15–19]; for example, chimaeric surface proteins containing apelin, the ligand for the G-protein coupled receptor APJ, show a small but significant ability to transform APJ expressing cells [16], while a peptid ligand for $\alpha_v\beta_3$ integrin inserted into the same type of virus resulted in virions that could bind to the integrin molecules but were unable to infect cells expressing these [17].

Targeting of adenovirus vectors has been attempted using fibre protein mutants impaired in binding to the broadly expressed CAR receptor [20]. Insertion of endothelial cell-binding peptides, isolated through the phage-display technique, into such adenoviral vectors have similarly resulted in a significant but low increase in gene transfer to endothelial cells (Figure 5.4b) [21].

A similar idea is to use a second molecule as a bridge between the viral vector and the cell receptor. For example, the coat protein of AAV has been biotinylated so that it can bind to an adaptor protein made by fusion of streptavidin to different protein ligands. Incubation of the biotinylated virus and the adaptor molecule resulted in increase of infection efficiency of cells with appropriate receptors by several orders of magnitude (Figure 5.4c) [22].

It is also possible to change the cell specificity of a virus of interest by incorporating the surface protein of another virus into it. This procedure is restricted in the case of non-enveloped viruses since the protein structure of the two viruses involved must be compatible. On the other hand, incorporation of a different surface protein into most enveloped viruses is possible even between widely different viral families. This is because there is little interaction between the viral core and the surface protein, which is embedded into the lipid envelope of the virus. Incorporation of a heterologous surface protein in enveloped viruses is referred to as pseudotyping (Figure 5.4d).

As seen from the mentioned examples, redirecting a virus' tropism may not be a trivial matter. In almost all cases, the engineered viral proteins bind perfectly to the intended receptor, but viruses either do not enter the cells or do so with a much reduced efficiency. In the case of enveloped viruses, this is because the membrane fusion process is under meticulous control and it seems as if the slightest change to the structure of viral surface proteins disables it. Efficient use of viruses for gene delivery requires better design of

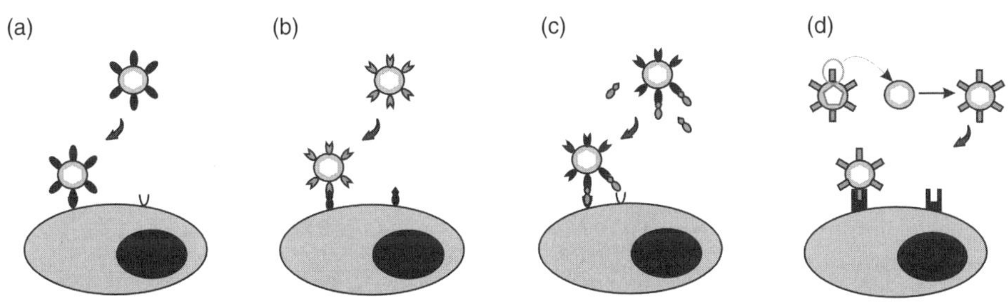

Figure 5.4 Targeting strategies by surface modification of viral vector particles. (a) Wild-type surface protein. (b) Genetically engineered surface protein exposing a peptide ligand. (c) Surface protein coated in vitro *with a chemically synthesized ligand. (d) Coating by a surface protein of another virus in the cellular production system*

retargeted surface proteins. Targeting attempts with non-enveloped viruses have generally been more successful, probably because no complicated process like fusion activation must occur prior to infection. This, in turn, has the disadvantage of making non-enveloped viral targeting less specific. Residual binding to non-targeted cells can cause off-target infections. One important block to efficiently targeted infection seems to be degradation resulting from retarded endosomal escape caused by the introduced targeting modifications [23, 24].

5.4.2 Expression

After entry of a viral vector to the designated target cells, the nucleic acid cargo is inside the cytoplasm. Depending on the nature of the cargo, it might be necessary for the DNA or RNA cargo to be translocated into the nucleus of the cell. In many settings, the cargo is a coding gene that needs to be transcribed. If the viral vector lacks its own RNA-polymerase to make mRNAs, as is the case with, for example, retroviruses and adenoviruses, it is vital for the cargo to get access to the cellular transcription machinery in the nucleus in order to achieve transgene expression. Correct expression of the transgene in such cases is dependent on the presence of appropriate *cis*-elements in the polynucleotide construct such as promoter, enhancer and so on. In the case of retroviruses, stable integration of the vector DNA into the host genome may result in sustained expression of the transgene [25]. However, in some cases integrated retroviral vectors are subject to epigenetic silencing *in vivo* [26]. Silencing is influenced by vector design as well as target cell and chromosomal integration site. Retroviral vectors that are less sensitive to silencing have been developed, and in the case of lentiviral vectors the silencing of integrated proviruses is not an important issue [26].

Design of the vector can benefit from regulatory elements naturally found in the target cell, which adds another dimension to the control over gene expression from the vector. Transgene cassettes may be constructed to direct the expression of mRNAs for protein production or non-coding RNAs with regulatory effects. Examples of non-coding RNAs are ribozymes or short hairpin RNAs (shRNAs) designed for the knockdown of specific messenger RNAs (mRNAs). Robust and efficient expression cassettes for such small RNAs have been designed and are suited for use in a variety of viral vectors [27]. Further developments aim at the generation of expression cassettes that can be regulated by small ligands such as tetracyclines [28].

Transgene expression might depend on factors other than vector design and *cis*-elements. For example, some retroviruses, such as murine leukaemia viruses (MLVs), are only capable of infecting dividing cells. This is because these viruses do not have any active means of entering the nucleus and must wait for the nuclear membrane to be degraded during cell division. Thus MLV-based vectors can be used for the expression of transgenes in tumours of tissues that do not normally undergo cell division, such as brain [29]. Other more genetically complex members of the retrovirus family, such as lentiviruses, do not suffer from this limitation and hence make good delivery vectors to, for example, neurons [30].

In some cases, such as the delivery of small interfering RNA (siRNA), the polynucleotide cargo does its job directly and does not need to be transcribed. In cases like this, no *cis*-elements or elaborate vector designs are needed. It suffices to package the siRNA into a delivery vehicle that takes it to the cytoplasm.

5.4.3 Safety

For all their beneficial characteristics, viruses are natural parasites evolved to spread in the host organism, often becoming pathogens. This raises important issues of safety in using viruses as delivery vehicles with risks yet to be fully defined. Moreover, the current generations of genuine animal viral vectors cannot be assembled from purified components in a test tube. Instead, they must be produced in the complex biological systems of cultured animal cells, which places extra emphasis on issues such as quality control and standardized scale-up of production for clinical use.

Viruses exploit the gene-expression machinery of the host cells for their own propagation. To achieve this, many viruses encode proteins that control basic regulatory processes, such as DNA replication or transcription of the host cell in addition to viral replication. While *trans*-acting proteins may be absolutely required in the cells used to produce viral vector particles, their expression in the target cell may be undesired to avoid possible adverse effects on the cell and to limit further spread of the vector. Hence, a major task in the development of viral vector technology is the separation of genes encoding viral *trans*-acting factors from the viral *cis*-elements needed for gene transfer (Figure 5.5).

Using naturally occurring infectious agents such as viruses as delivery vehicles draws attention to possible interactions between virus and vector. Indirect effects on the vector may come from an immune reaction against the virus and direct effects on the vector may involve rescue by or recombination with a naturally infecting virus. In some case, such as the lentivirus HIV-1, vectors may be modified to lack *cis*-elements needed for rescue or recombination in a target cell infected by a virus [31].

Insertional mutagenesis is another critical safety issue associated with vectors such as retroviral vectors that are inserted into the genomic DNA of the target cell. When a retrovirus genome becomes integrated into the host cell genome, the viral promoter/enhancer elements that are meant to drive the viral gene expression can act on nearby host genes and disturb the regulation of their transcription. From studies in mice and chicken, it is well established that multiple rounds of insertional mutagenesis by retroviruses can cause the development of cancer [32]. Recently, insertional mutagenesis by the gene transfer vector was found to be associated with the development of T-cell leukaemia in a clinical trial, as discussed below.

In vivo delivery raises its own concerns regarding an innate or adaptive immune response to vector particles as well as other systemic effects. The intravenous administration of adenovirus vectors results in delivery mainly to the liver. High doses of first and second generation adenovirus vectors have been observed to cause an acute inflammatory response and liver toxicity in mice and larger animals such as baboons [33]. In a clinical trial for ornithine transcarbamylase deficiency, high doses of a second generation vector caused a fatal acute inflammatory response in one patient [34]. The latest generation of fully deleted adenovirus vectors that lack coding sequences for all viral proteins causes less acute toxicity following the administration of large doses in animals [33].

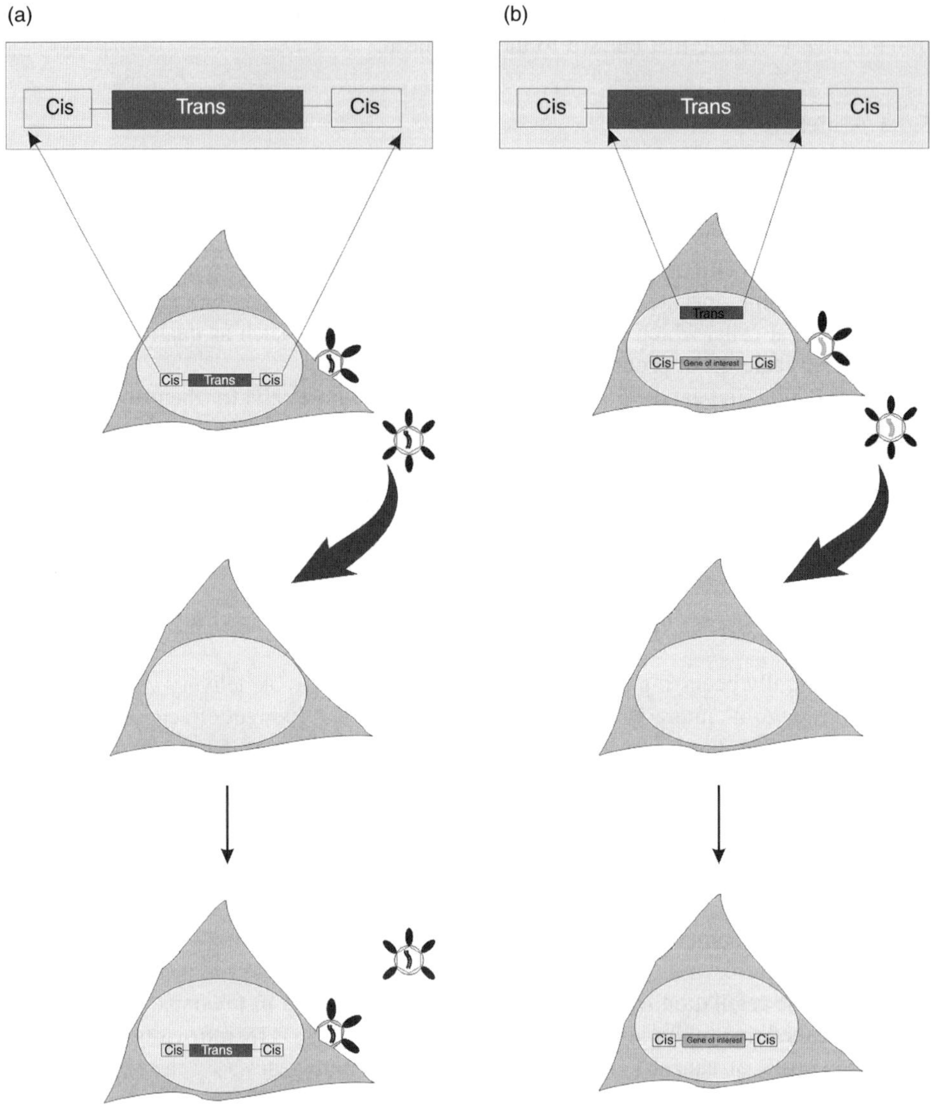

Figure 5.5 *Gene transfer by viral vector particle. (a) The virus itself harbours all* cis-*elements and encodes all viral proteins; it is therefore able to replicate in multiple rounds. (b) The viral vector particle produced from a specialized packaging cell line transfers only the viral* cis*-elements and the gene of interest; it lacks the genes for viral proteins and undergoes only one round of replication*

5.5 Examples of Viral Delivery Technology

Below the focus is firstly on four types of viral vectors based upon retroviruses, lentivurses, adenoviruses and adeno-associated viruses, respectively. These systems are most highly developed in terms of transfer vectors devoid of viral coding sequences, and have all been taken into clinical trials (Table 5.1). Important developments not detailed here are also ongoing towards the making of vectors from other viruses, such as large DNA viruses of the herpesviruses and poxviruses families. Finally, at the end of the chapter attention is drawn to vectors exploiting viruses of non-vertebrate origin as well as possibilities for *in vitro* loading of delivery vehicles.

5.5.1 Retroviral and Lentiviral Vectors

Two types of therapeutic vectors are derived from the viral family *Retroviridae*. The term retroviral vectors covers vectors derived from murine leukaemia virus, which is a gammaretrovirus, whereas the term lentiviral vectors refers to vectors derived from human immunodeficiency virus type 1 (HIV-1), which is a lentivirus. These vectors have an insert capacity of 5–6 kilobases (kb) of DNA, and both retroviral and lentiviral vectors sustain stable insert of their DNA into chromosomal DNA of the host cell [25].

Some groups of enveloped viruses are unstable and difficult to purify and concentrate. Retroviral vector half-lives are in the range of a few hours at 37 °C and one to a few months at −80 °C, depending on formulation [35, 36]. Part of this instability is associated with the loss of a functional envelope protein required for fusion. One approach to improve stability involves the formulation of the vector core by another envelope protein expressed in the producer cell. Much improvement has been gained by such substitution of the envelope protein from the rhabdovirus vesicular stomatitis virus (VSV) for a retroviral or lentiviral envelope protein. This pseudotyping significantly improves vector stability and allows for the concentration and purification by ultracentrifugation [37]. However, one further bottleneck for the efficient use of VSV envelope pseudotyped vectors is their rapid inactivation by serum complement proteins. To overcome this hurdle, monomethoxypoly(ethylene glycol) has been conjugated to vector particles, which had no effect on the transduction efficiency. This treatment reduced the rate of vector inactivation by non-heat-inactivated serum by a factor of 1000 and prolonged the circulation half-life of active vectors, after a single intravenous dose, by a factor of five [38].

Table 5.1 *Frequently used viral vectors*

Virus	Enveloped	Stable integration	Insert capacity (kb[a])	Genome	Particle diameter (nm)
Retrovirus	Yes	Yes	5–6	RNA	100
Lentivirus	Yes	Yes	5–6	RNA	100
Adenovirus	No	No	36	DNA	90
Adeno-associated virus	No	No	5–6	DNA	23

[a] kb is 1000 base pairs.

The choice of producer cell line also has an influence on the composition of vector particles. One well studied example is the difference among mammalian species of (α1-3) galactosyltransferase, an enzyme that adds sugar residues on the envelope glycoproteins. This enzyme is not encoded by the genomes of humans and other primates, but is expressed in common vector producer cell lines from other species such as mice. Humans have high levels of antibodies to Gal(α)1,3Gal (Gal) epitopes, and viruses harbouring this on their envelope glycoproteins are rapidly inactivated in human serum [39].

By another strategy, directed evolution of a retrovirus in multiple steps led to the identification of stability-enhancing mutations of the viral protease. These mutations enhanced infectivity half-lives in viral supernatant at 37 °C from 5–8 h up to around 40 h. During the life cycle of the virus, the protease is responsible for the proteolytic maturation of released viral particles towards infectivity [40].

In principle, a vector that inserts its DNA covalently into the chromosomal DNA of its host cell causes a mutation. Depending upon the site of insertion and the nucleotide sequence of the vector, such a mutation may be innocent or harmful to the cell. Retroviruses including lentiviruses insert their genome into target DNA without any preference at the sequence level and are able to target all regions of the genome. It is well known that replication-competent retroviruses can cause cancer in animals by multiple rounds of insertional mutagenesis through mostly gain-of-function type of mutations, such as the dysregulation of oncogene expression [32]. In contrast, there is no evidence that such gain-of-function mutations are a common result of insertional mutagenesis by lentiviruses. During the early years of gene therapy by retroviral vectors, theoretical considerations had led to the conclusion that the risk of oncogenic transformation from a single dose of replication-defective retroviral vectors *ex vivo* was negligible, since combinations of rare hits at multiple genome targets would have to occur in the same cell [41]. The primary safety concern in conjunction with retroviral *ex vivo* gene therapy has, therefore, been to avoid the presence of replication-competent viruses in the vector stocks that would sustain multiple rounds of replication and cause additional insertional mutations [42].

A gene therapy protocol recently led to the restoration of a functional immune system in X-SCID patients who are born without a common γ receptor (γ-c) needed for a functional immune system [43]. A functional copy of the γ-c gene was transferred to haematopoietic stem cells *ex vivo* using a Moloney murine leukaemia virus-derived vector. Some of the treated infants subsequently developed clonal T-cell leukemias in which a copy of the gene therapy vector was detected at positions near LMO-2, a gene previously implicated as a translocation partner in T-cell leukaemias [44]. Together with further analyses this provided strong evidence that vector insertion at LMO-2 led to the dysregulated expression of this gene and had a causative role in the development of malignant disease. This first case of an adverse effect caused by insertional mutagenesis in a human gene therapy trial raised the question of whether there was anything particular to this protocol that made this effect so frequent. Subsequent investigations point to an oncogenic effect of the vector-expressed γ-c gene in itself, the role of the X-SCID background and the high doses of vector used as well as the type of vector [45]. The particular vector harboured the complete region of the transcriptional promoter-enhancer region of Moloney murine leukaemia virus known to be a potent insertional mutagen that causes rapid T-cell lymphomas in mice. Recent studies emphasize that the risk of insertional mutagenesis can be significantly reduced through improved vector design [46]. Self-inactivating (SIN) vectors lack the viral transcriptional control elements and are less potent gene activators. Another possibility to improve the protocol is to use lentiviral vectors as delivery vehicles. A final strategy that is being explored is to endow the vector with an integration machinery that targets it preferentially to non-coding regions of the genome or to specific locations. Work towards this difficult task takes advantage of heterologous DNA recombinases and tethering proteins with specific DNA-recognition [47, 48].

The most developed lentiviral vectors are derived from HIV-1, a virus, which has been extensively characterized, as it is the pathogenic agent responsible for AIDS. The development of lentiviral vectors followed the basic principles established by the construction of retroviral vector from murine leukaemia viruses. However, the elimination of accessory genes of HIV-1 without compromising gene transfer and transgene efficiencies posed additional challenges [37]. Lentiviral vectors based upon HIV-1 have recently been introduced in clinical trials, in all cases based upon *ex vivo* gene transfer, and targeting HIV-infection, cancer or inherited diseases [11].

Lentiviral vectors are commonly used in a formulation where they are pseudotyped with a heterologous envelope protein such as the envelope protein of VSV [25]. Good results were recently obtained by pseudotyping HIV-1 derived vectors by the envelope glycoproteins of the measles virus [49]. In contrast to VSV, which enters by membrane fusion at a low pH after endocytosis, the measles virus enters by fusion directly at the plasma membrane in a pH-independent manner. Interestingly, chimaeric measles envelope proteins engineered to harbour specific receptor ligands, such as epidermal growth factor or a single chain antibody, to CD20 were found to sustain efficient entry of lentiviral vectors into cells expressing the cognate receptors [49].

Compared to the standard retroviral vectors, the nucleophilic preintegration complex of lentiviral vectors allows them to transduce non-dividing as well as dividing cells, which holds promise for the stable *in vivo* delivery to many tissues. In contrast to murine leukaemia virus vectors, lentiviral vectors are less prone to epigenetic silencing and, as mentioned above, have not been reported to induce cancer by insertional mutagenesis. To fully avoid insertional mutagenesis, non-integrating lentiviral vectors have been developed and found to sustain long-term expression in non-dividing cells [50].

5.5.2 Adenovirus Vectors

Adenoviruses are non-enveloped, medium sized DNA viruses. Adenovirus vectors have been used in clinical trials for both *ex vivo* and *in vivo* delivery [11]. Moreover, oncolytic adenoviruses that replicate under restricted conditions have been used in clinical trials for cancer treatment [1, 2]. Several features of adenoviruses make them attractive vehicles for *in vivo* delivery. Their molecular biology is well-understood, they are stable, can be produced at high titres, are able to infect many cell types, have a very large DNA-insert capacity and are easy to manipulate (Table 5.1) Adenovirus infections are common in humans, but they are not life-threatening in immunocompetent individuals. Most vectors have been based on serotype 5 adenovirus (Ad5). However, adenoviruses come in several serotypes that are being exploited for differences in receptor usage and cell-type preferences and to circumvent restrictions to delivery caused by prior immunity. For example, Ad5 uses CAR as its main receptor, but CAR is absent from some candidate target cells. Other serotypes, such as Ad35 and Ad11, may infect cells in the absence of CAR [51].

For delivery purposes, expression of viral protein or multiple rounds of replication may cause an acquired immune response and result in the elimination of vector-bearing cells. While this may suffice and actually be preferred for vaccination strategies [3], other applications require sustained vector expression for longer periods. The first generation of adenovirus vectors was deleted of the regulatory genes E1 and E3, the former being supplied *in trans* for vector production. The second generation vectors have additional defects in regulatory genes, but as the first generation vectors they have retained most viral genes. In contrast, the latest generation of fully deleted vectors are devoid of all viral protein-coding genes and carry only *cis*-acting sequences needed for gene transfer [52]. All viral proteins needed to produce fully deleted vectors are provided *in trans* in packaging cells. First and second generations of adenoviral vectors were found to sustain vector expression in baboons for 1–2 weeks while vector expression from fully deleted vectors without any viral protein-coding sequences lasted for nearly two years [53].

The chemical modification of adenovirus vectors with synthetic polymers such as poly(ethylene glycol) (PEG) or poly-N-(2-hydroxypropyl) methacrylamide (poly-HPMA) holds promise for several critical aspects of *in vivo* delivery [54, 55]. Polymers can be attached to adenoviral vector particles without destroying capsid integrity and gene transfer activity. Heavy polymer modification using HPMA or PEG can ablate the normal tropism of the virus and serve as a platform for retargeting strategies [56–58]. Several critical limitations of *in vivo* delivery by adenovirus vectors may be overcome by polymer coating. For example PEG coating caused a strong reduction in the innate immune response and liver toxicity following the intravenous administration of vector particles to mice. Moreover, also in mouse studies, PEGylation led to a significant reduction in the induction of an adaptive humoral and cellular immune response to the initial vector dose and aided the vector to escape neutralization from prior immunity [55, 59].

5.5.3 Adeno-Associated Virus Vectors

Parvoviruses have a single-stranded DNA genome and are among the smallest mammalian DNA viruses. Adeno-associated viruses (AAVs) are non-pathogenic parvoviruses of man and other species that have been developed into a promising gene therapy vector technology [60]. Natural isolates of AAV need regulatory proteins from other viruses such as adenoviruses to replicate. The cellular production systems for AAV vector particles therefore include *trans*-acting factors from adenovirus as well as AAV. The vector particles of DNA and protein can be produced to high titres, withstand purification and are stable under various conditions [60]. The AAV genome is around 4.7 kb, of which the necessary *cis*-elements to be retained in vector DNA include only the two 145 nucleotide long inverted terminal repeats. The insert capacity has been reported to be somewhat above genome size, up to 6 kb, in standard vectors derived from the human AAV serotype 2 (AAV2) [60, 61] (Table 5.1).

AAV2 has several favourable features for *in vivo* delivery. Its common presence in the human population without known pathogenic effects adds to the safety profile. Although AAV vector DNA appears to reside in an episomal, non-integrated state, persistent transgene expression for up to two years has been demonstrated in non-dividing cells of a variety of tissues in small and large animals [61]. Hence, the risk of insertional mutagenesis by AAV vectors is considered low. Conversion of the single-stranded genome to its double-stranded form is believed to be a major bottleneck for adequate vector gene expression after *in vivo* delivery. Modified self-complementary vectors with partially double-stranded genomes have been found to exhibit enhanced transduction efficiencies in several tissues [62].

The AAV capsid protein of the delivery vehicle may induce a humoral immune response, whereas little cytotoxic T-cell responses have been observed in animals models [63]. Moreover, a large part of the human population has antibodies against AAV2. As described above for adenovirus vectors, polymer coating has also been successfully applied to AAV to ablate virus tropism, enable retargeting and protect against neutralizing antibodies [64]. AAVs have broad cellular tropism. The natural variation among over 100 AAV isolates from different species offers a rich source for the generation of mosaic or chimaeric vectors of improved properties [65]. Currently, vectors developed from more than ten other AAV subtypes present a favourable repertoire of features for gene transfer to diverse target tissues *in vivo* [65]. The molecular basis of the target specificity of AAV serotypes is poorly understood. Not only receptor and co-receptor interactions, but also post-entry steps appear to play a role [65]. Subtype differences may potentially be used to overcome prior immunity, or induced immunity in case of a need for repeated administration. AAV vectors, mostly AAV2, but also AAV6, are involved in many clinical trials for inherited or acquired diseases [11, 60]. Finally, shRNA expression cassettes are commonly used in AAV vectors with good results [66].

5.5.4 Mixed Delivery Systems

Prokaryotic viruses, such as filamentous single-stranded DNA viruses of *E. coli*, have had wide biotechnological impact. Among other things, their ease of genetic manipulation and production at high titres offer affinity based selection of individual display mutants from complex peptide display libraries. Selection from such libraries permits the identification of viruses adapted to bind to mammalian receptors. Hence, bacterial viruses offer the prospect of specific cell targeting, but are not in themselves suited for the delivery of a gene of interest into mammalian cells [67]. To promote transgene delivery, bacterial viruses such as the single-stranded DNA virus M13 can be equipped with features that improve its ability to deliver genetic cargo into mammalian cells. A promising case is provided by a recent integrin-directed hybrid vector containing elements of two single-stranded DNA viruses:, the filamentous bacterophage M13 and the mammalian parvovirus AAV. When tested in a mouse model, this hybrid vector had retained the ligand-directed targeting properties of M13 to a mammalian receptor as well as its transgene delivery and expression features as an AAV-vector [68, 69].

Baculoviruses are a group of insect viruses widely used as vectors for heterologous protein production in insect cells. While baculovirus vectors are unable to replicate in mammalian cells, they are capable of delivering functional transgenes into cultured mammalian cells, in some cases with high efficiencies [70]. The highly developed technology of producing high titres of baculovirus vectors in insect cells makes it attractive to explore these vehicles for further applications in cultured mammalian cells or *in vivo*. In a recent report the ability of baculovirus vectors to knock-down target genes by RNA interference was tested and vectors equipped with transgene cassettes directing the expression of targeted shRNAs were found to mediate efficient knock-down of lamin A/C in cultured human cells [71].

5.5.5 *In Vitro* Assembled Delivery Vehicles

The ability to efficiently load cargo into virus particles *in vitro* would add advantages in terms of versatility as well as product control. An efficient *in vitro* process of packaging DNA into pseudovirions has been developed for polyomaviruses, a family of small, non-enveloped mammalian DNA viruses. While the pseudovirions deliver their DNA cargo efficiently into the cytosol of mammalian cells, transgene expression is restricted by limited nuclear import. A recent study based on the polyomavirus simian virus 40 showed that pseudovirions could also package small interfering RNAs (siRNAs) and deliver these into their site of action in the cytosol where they acted efficiently as mediators of gene knock-down by RNA interference [72]. This technology has been brought forward into a clinical cancer trial [11].

The term virosome has been used for decades to denote a liposome that displays a viral envelope protein on its surface. As delivery vehicles these semisynthetic particles may combine the versatility of cargo and ease of loading into liposomes with the efficient receptor-dependent membrane fusion triggered by viral envelopes. Virosomes are generated by adding extracts of viral particles during liposome production or by fusing liposomes and viruses. Virosomes have exploited fusogenic envelope proteins from various virus families, including paramyxoviruses such as Sendai virus, influenza viruses and rhabdoviruses such as vesicular stomatitis virus [73]. siRNAs are promising cargoes for virosome delivery because they do not need to migrate beyond the cytoplasm once they have been delivered through the lipid bilayer membrane of the cell by the viral envelope-mediated fusion. Complexed to cationic lipid, siRNA can be efficiently encapsidated into virosomes [74]. Recent work has used virosomes with the influenza haemagglutinin protein as an efficient delivery vehicle for siRNAs *in vitro* and *in vivo* [75]. Such virosomes may be particularly suited for respiratory tract delivery.

5.6 Conclusion

Delivery vehicles have been made from diverse animal viruses that allow the controlled single-cycle transfer of genetic material. These are, in all cases, biosynthetic products made by specialized cells. The field faces challenges of targeting and increasing efficacy of delivery, in the ability to control harmful side effects as well as innate and adaptive immune responses. Furthermore, the potential interaction and recombination with replication-competent viruses in the environment remains an inherent safety issue. Hence, no viral vectors are yet part of common medical practice. Gene transfer can take place to isolated cells outside the body (*ex vivo*) or by viral vectors in the body (*in vivo*). For *in vivo* delivery the selective targeting of predetermined cells remains a difficult issue that is the focus of much investigation. Tools to modify the virus surface to direct it to cells harbouring a given receptor or to avoid an immune reaction include genetic engineering and chemical modification, as well as phenotypic mixing with heterologous virus proteins.

The stable chromosomal maintenance of *ex vivo* or *in vivo* transferred genetic material creates a risk of insertional mutagenesis, as seen in a recent clinical trial using *ex vivo* gene transfer by a retroviral vector. However, latest developments in non-integrating vectors (adenovirus, adeno-associated virus and non-integrating lentivirus) demonstrate long-term persistence of trans-gene expression of months to years in non-dividing cells *in vivo*. New developments that may bypass some of the safety issues raised by using modified vertebrate viruses in man include delivery vehicles based upon viruses of distant species as well as procedures for *in vitro* loading of virus particles or virosomes with, for example, small interfering RNAs.

References

1. Vaha-Koskela, M.J., Heikkila, J.E. and Hinkkanen, A.E. (2007) Oncolytic viruses in cancer therapy. *Cancer Lett.*, **254**, 178–216.
2. Russell, S.J. and Peng, K.W. (2007) Viruses as anticancer drugs. *Trends Pharmacol. Sci.*, **28**, 326–333.
3. Tatsis, N. and Ertl, H.C. (2004) Adenoviruses as vaccine vectors. *Mol. Ther.*, **10**, 616–629.
4. Liniger, M., Zuniga, A. and Naim, H.Y. (2007) Use of viral vectors for the development of vaccines. *Expert Rev. Vaccines*, **6**, 255–266.
5. Sapir, A., Avinoam, O., Podbilewicz, B. and Chernomordik, L.V. (2008) Viral and developmental cell fusion mechanisms: conservation and divergence. *Dev. Cell*, **14**, 11–21.
6. Chernomordik, L.V., Zimmerberg, J. and Kozlov, M.M. (2006) Membranes of the world unite! *J. Cell Biol.*, **175**, 201–207.
7. Chernomordik, L.V. and Kozlov, M.M. (2003) Protein-lipid interplay in fusion and fission of biological membranes. *Ann. Rev. Biochem.*, **72**, 175–207.
8. Bentz, J. and Mittal, A. (2003) Architecture of the influenza hemagglutinin membrane fusion site. *Biochim. Biophys. Acta*, **1614**, 24–35.
9. Leopold, P.L. and Crystal, R.G. (2007) Intracellular trafficking of adenovirus: many means to many ends. *Adv. Drug Deliv. Rev.*, **59**, 810–821.
10. Virgin, S. (2007) Pathogenesis of viral infection, in *Fields' Virology*, 5th edn (eds D.M. Knipe and P.M. Howley), Lippincott Williams and Wilkins, Philadelphia, USA, pp. 327–388.
11. Gene Therapy Clinical Trials Worldwide (2008) Provided by the *J. Gene Med.* Update September 2008. http://www.wiley.co.uk/genmed/clinical/ (accessed 1 November 2008).
12. Bartosch, B. and Cosset, F.L. (2004) Strategies for retargeted gene delivery using vectors derived from lentiviruses. *Curr. Gene Ther.*, **4**, 427–443.
13. Silman, N.J. and Fooks, A.R. (2000) Biophysical targeting of adenovirus vectors for gene therapy. *Curr. Opin. Mol. Ther.*, **2**, 524–531.
14. Katane, M., Takao, E., Kubo, Y. *et al.* (2002) Factors affecting the direct targeting of murine leukemia virus vectors containing peptide ligands in the envelope protein. *EMBO Rep.*, **3**, 899–904.
15. Qi, P., Han, J.X., Lu, Y.Q. and Wang, C.X. (2007) Redirecting retroviral tropism by insertion of hepatitis B virus PreS1 core peptide into the envelope. *Arch. Virol.*, **152**, 1721–1730.
16. Bahrami, S., Duch, M. and Pedersen, F.S. (2007) Ligand presentation on a synthetic flexible hinge in Moloney murine leukemia virus SU supports entry via a heterologous receptor. *Virology*, **363**, 303–309.
17. Wu, B.W., Lu, J., Gallaher, T.K. *et al.* (2000) Identification of regions in the Moloney murine leukemia virus SU protein that tolerate the insertion of an integrin-binding peptide. *Virology*, **269**, 7–17.
18. Gollan, T.J. and Green, M.R. (2002) Selective targeting and inducible destruction of human cancer cells by retroviruses with envelope proteins bearing short peptide ligands. *J. Virol.*, **76**, 3564–3569.

19. Gollan, T.J. and Green, M.R. (2002) Redirecting retroviral tropism by insertion of short, non-disruptive peptide ligands into envelope. *J. Virol.*, **76**, 3558–3563.
20. Kirby, I., Davison, E., Beavil, A.J. *et al.* (1999) Mutations in the DG loop of adenovirus type 5 fiber knob protein abolish high-affinity binding to its cellular receptor CAR. *J. Virol.*, **73**, 9508–9514.
21. Nicklin, S.A., Von Seggern, D.J., Work, L.M. *et al.* (2001) Ablating adenovirus type 5 fiber-CAR binding and HI loop insertion of the SIGYPLP peptide generate an endothelial cell-selective adenovirus. *Mol. Ther.*, **4**, 534–542.
22. Ponnazhagan, S., Mahendra, G., Kumar, S. *et al.* (2002) Conjugate-based targeting of recombinant adeno-associated virus type 2 vectors by using avidin-linked ligands. *J. Virol.*, **76**, 12900–12907.
23. Meier, O. and Greber, U.F. (2004) Adenovirus endocytosis. *J. Gene Med.*, **6** (Suppl. 1), S152–S163.
24. Everts, M. and Curiel, D.T. (2004) Transductional targeting of adenoviral cancer gene therapy. *Curr. Gene Ther.*, **4**, 337–346.
25. Pedersen, F.S. and Duch, M.R. (2005) Retroviruses in human gene therapy, in *Encyclopedia of Life Sciences*, John Wiley & Sons, Ltd. www.els.net.
26. Ellis, J. (2005) Silencing and variegation of gammaretrovirus and lentivirus vectors. *Hum. Gene Ther.*, **16**, 1241–1246.
27. Aagaard, L. and Rossi, J.J. (2007) RNAi therapeutics: principles, prospects and challenges. *Adv. Drug Deliv. Rev.*, **59**, 75–86.
28. Aagaard, L., Amarzguioui, M., Sun, G. *et al.* (2007) A facile lentiviral vector system for expression of doxycycline-inducible shRNAs: knockdown of the pre-miRNA processing enzyme Drosha. *Mol. Ther.*, **15**, 938–945.
29. Tai, C.K. and Kasahara, N. (2008) Replication-competent retrovirus vectors for cancer gene therapy. *Front. Biosci.*, **13**, 3083–3095.
30. Wong, L.F., Goodhead, L., Prat, C. *et al.* (2006) Lentivirus-mediated gene transfer to the central nervous system: therapeutic and research applications. *Hum. Gene Ther.*, **17**, 1–9.
31. Grunwald, T., Pedersen, F.S., Wagner, R. and Uberla, K. (2004) Reducing mobilization of simian immunodeficiency virus based vectors by primer complementation. *J. Gene Med.*, **6**, 147–154.
32. Pedersen, F.S. and Sørensen, A.B., Pathogenesis of oncoviral infections, in *Retroviruses: Molecular Microbiology and Genomics* (ed. R. Kurth), Horizon Press, in press.
33. Muruve, D.A. (2004) The innate immune response to adenovirus vectors. *Hum. Gene Ther.*, **15**, 1157–1166.
34. Raper, S.E., Chirmule, N., Lee, F.S. *et al.* (2003) Fatal systemic inflammatory response syndrome in a ornithine transcarbamylase deficient patient following adenoviral gene transfer. *Mol. Genet. Metab.*, **80**, 148–158.
35. Cruz, P.E., Silva, A.C., Roldao, A. *et al.* (2006) Screening of novel excipients for improving the stability of retroviral and adenoviral vectors. *Biotechnol. Prog.*, **22**, 568–576.
36. Wikstrom, K., Blomberg, P. and Islam, K.B. (2004) Clinical grade vector production: analysis of yield, stability, and storage of gmp-produced retroviral vectors for gene therapy. *Biotechnol. Prog.*, **20**, 1198–1203.
37. Cockrell, A.S. and Kafri, T. (2007) Gene delivery by lentivirus vectors. *Mol. Biotechnol.*, **36**, 184–204.
38. Croyle, M.A., Callahan, S.M., Auricchio, A. *et al.* (2004) PEGylation of a vesicular stomatitis virus G pseudotyped lentivirus vector prevents inactivation in serum. *J. Virol.*, **78**, 912–921.
39. Takeuchi, Y., Porter, C.D., Strahan, K.M. *et al.* (1996) Sensitization of cells and retroviruses to human serum by (alpha 1-3) galactosyltransferase. *Nature*, **379**, 85–88.
40. Vu, H.N., Ramsey, J.D. and Pack, D.W. (2008) Engineering of a stable retroviral gene delivery vector by directed evolution. *Mol. Ther.*, **16**, 308–314.
41. Moolten, F.L. and Cupples, L.A. (1992) A model for predicting the risk of cancer consequent to retroviral gene therapy. *Hum. Gene Ther.*, **3**, 479–486.

42. Vanin, E.F., Kaloss, M., Broscius, C. and Nienhuis, A.W. (1994) Characterization of replication-competent retroviruses from nonhuman primates with virus-induced T-cell lymphomas and observations regarding the mechanism of oncogenesis. *J. Virol.*, **68**, 4241–4250.
43. Cavazzana-Calvo, M., Hacein-Bey, S., de Saint Basile, G. *et al.* (2000) Gene therapy of human severe combined immunodeficiency (SCID)-X1 disease. *Science*, **288**, 669–672.
44. Hacein-Bey-Abina, S., Von Kalle, C., Schmidt, M. *et al.* (2003) LMO2-associated clonal T cell proliferation in two patients after gene therapy for SCID-X1. *Science*, **302**, 415–419.
45. Shou, Y., Ma, Z., Lu, T. and Sorrentino, B.P. (2006) Unique risk factors for insertional mutagenesis in a mouse model of XSCID gene therapy. *Proc. Natl. Acad. Sci. USA*, **103**, 11730–11735.
46. Ryu, B.Y., Evans-Galea, M.V., Gray, J.T. *et al.* (2008) An experimental system for the evaluation of retroviral vector design to diminish the risk for proto-oncogene activation. *Blood*, **111**, 1866–1875.
47. Lombardo, A., Genovese, P., Beausejour, C.M. *et al.* (2007) Gene editing in human stem cells using zinc finger nucleases and integrase-defective lentiviral vector delivery. *Nat. Biotechnol.*, **25**, 1298–1306.
48. Tan, W., Dong, Z., Wilkinson, T.A. *et al.* (2006) Human immunodeficiency virus type 1 incorporated with fusion proteins consisting of integrase and the designed polydactyl zinc finger protein E2C can bias integration of viral DNA into a predetermined chromosomal region in human cells. *J. Virol.*, **80**, 1939–1948.
49. Funke, S., Maisner, A., Muhlebach, M.D. *et al.* (2008) Targeted cell entry of lentiviral vectors. *Mol. Ther.*, **16**, 1427–1436.
50. Apolonia, L., Waddington, S.N., Fernandes, C. *et al.* (2007) Stable gene transfer to muscle using non-integrating lentiviral vectors. *Mol. Ther.*, **15**, 1947–1954.
51. Stone, D., Liu, Y., Li, Z.Y. *et al.* (2007) Comparison of adenoviruses from species B, C, E, and F after intravenous delivery. *Mol. Ther.*, **15**, 2146–2153.
52. Volpers, C. and Kochanek, S. (2004) Adenoviral vectors for gene transfer and therapy. *J. Gene Med.*, **6** (Suppl. 1), S164–S171.
53. Morral, N., O'Neal, W., Rice, K. *et al.* (1999) Administration of helper-dependent adenoviral vectors and sequential delivery of different vector serotype for long-term liver-directed gene transfer in baboons. *Proc. Natl. Acad. Sci. USA*, **96**, 12816–12821.
54. O'Riordan, C.R., Lachapelle, A., Delgado, C. *et al.* (1999) PEGylation of adenovirus with retention of infectivity and protection from neutralizing antibody *in vitro* and *in vivo*. *Hum. Gene Ther.*, **10**, 1349–1358.
55. Kreppel, F. and Kochanek, S. (2008) Modification of adenovirus gene transfer vectors with synthetic polymers: a scientific review and technical guide. *Mol. Ther.*, **16**, 16–29.
56. Fisher, K.D., Stallwood, Y., Green, N.K. *et al.* (2001) Polymer-coated adenovirus permits efficient retargeting and evades neutralising antibodies. *Gene Ther.*, **8**, 341–348.
57. Stevenson, M., Hale, A.B., Hale, S.J. *et al.* (2007) Incorporation of a laminin-derived peptide (SIKVAV) on polymer-modified adenovirus permits tumor-specific targeting via alpha6-integrins. *Cancer Gene Ther.*, **14**, 335–345.
58. Wortmann, A., Vohringer, S., Engler, T. *et al.* (2008) Fully detargeted polyethylene glycol-coated adenovirus vectors are potent genetic vaccines and escape from pre-existing anti-adenovirus antibodies. *Mol. Ther.*, **16**, 154–162.
59. Croyle, M.A., Le, H.T., Linse, K.D. *et al.* (2005) PEGylated helper-dependent adenoviral vectors: highly efficient vectors with an enhanced safety profile. *Gene Ther.*, **12**, 579–587.
60. Coura Rdos, S. and Nardi, N.B. (2007) The state of the art of adeno-associated virus-based vectors in gene therapy. *Virol. J.*, **4**, 99.
61. Zhong, L. and Srivastava, A. (2007) Adeno-associated viral vectors in gene therapy, in *Encyclopedia of Life Sciences*, John Wiley & Sons, Ltd. www.els.net.
62. Wu, J., Zhao, W., Zhong, L. *et al.* (2007) Self-complementary recombinant adeno-associated viral vectors: packaging capacity and the role of rep proteins in vector purity. *Hum. Gene Ther.*, **18**, 171–182.
63. Zaiss, A.K. and Muruve, D.A. (2005) Immune responses to adeno-associated virus vectors. *Curr. Gene Ther.*, **5**, 323–331.

64. Carlisle, R.C., Benjamin, R., Briggs, S.S. *et al.* (2008) Coating of adeno-associated virus with reactive polymers can ablate virus tropism, enable retargeting and provide resistance to neutralising antisera. *J. Gene Med.*, **10**(4), 400–411.
65. Wu, Z., Asokan, A. and Samulski, R.J. (2006) Adeno-associated virus serotypes: vector toolkit for human gene therapy. *Mol. Ther.*, **14**, 316–327.
66. Grimm, D., Pandey, K. and Kay, M.A. (2005) Adeno-associated virus vectors for short hairpin RNA expression. *Methods Enzymol.*, **392**, 381–405.
67. Clark, J.R. and March, J.B. (2006) Bacteriophages and biotechnology: vaccines, gene therapy and antibacterials. *Trends Biotechnol.*, **24**, 212–218.
68. Hajitou, A., Rangel, R., Trepel, M. *et al.* (2007) Design and construction of targeted AAVP vectors for mammalian cell transduction. *Nat. Protoc.*, **2**, 523–531.
69. Hajitou, A., Trepel, M., Lilley, C.E. *et al.* (2006) A hybrid vector for ligand-directed tumor targeting and molecular imaging. *Cell*, **125**, 385–398.
70. Kost, T.A., Condreay, J.P. and Jarvis, D.L. (2005) Baculovirus as versatile vectors for protein expression in insect and mammalian cells. *Nat. Biotechnol.*, **23**, 567–575.
71. Nicholson, L.J., Philippe, M., Paine, A.J. *et al.* (2005) RNA interference mediated in human primary cells via recombinant baculoviral vectors. *Mol. Ther.*, **11**, 638–644.
72. Kimchi-Sarfaty, C., Brittain, S., Garfield, S. *et al.* (2005) Efficient delivery of RNA interference effectors via *in vitro*-packaged SV40 pseudovirions. *Hum. Gene Ther.*, **16**, 1110–1115.
73. Felnerova, D., Viret, J.F., Gluck, R. and Moser, C. (2004) Liposomes and virosomes as delivery systems for antigens, nucleic acids and drugs. *Curr. Opin. Biotechnol.*, **15**, 518–529.
74. de Jonge, J., Holtrop, M., Wilschut, J. and Huckriede, A. (2006) Reconstituted influenza virus envelopes as an efficient carrier system for cellular delivery of small-interfering RNAs. *Gene Ther.*, **13**, 400–411.
75. Huckriede, A., De Jonge, J., Holtrop, M. and Wilschut, J. (2007) Cellular delivery of siRNA mediated by fusion-active virosomes. *J. Liposome Res.*, **17**, 39–47.

6

The Innate Immune Responses, Adjuvants and Delivery Systems

S. Moein Moghimi
University of Copenhagen, Department of Pharmaceutics and Analytical Chemistry, Faculty of Pharmaceutical Sciences, Copenhagen, Denmark

6.1 Introduction

Vaccination strategies are no longer restricted to the prevention of infectious diseases; they now include therapeutic intervention of chronic viral infections, cancer, autoimmunity and allergy [1–3]. While neutralizing antibody responses are apparently important to combat most extracellular pathogens, strong cell-mediated responses are critical for protection against malaria, tuberculosis, herpes simplex virus, human immunodeficiency virus (HIV), hepatitis C virus as well as cancer. Often, a combination of both humoral and cell-mediated responses is necessary for elimination of some pathogens and cancers. There is also a need for vaccine formulations capable of inducing tolerance or modulating T helper (T_H) cell responses (as in allergy and autoimmunity) [2, 3].

Immune responses against pathogens and antigens are initiated in secondary lymphoid organs (lymph nodes, mucosal-associated lymphoid tissues and spleen). For elucidation of immune responses, antigens entering at peripheral sites such as skin, gastrointestinal tract and blood must reach secondary lymphoid organs where antigen-specific naïve B and T cells circulate. Antigens may reach these sites either in native form (soluble or cell-associated), and often transported with bound antibodies and/or complement, or processed by antigen-presenting cells, primarily macrophages and dendritic cells (DCs) [2, 3]. However, not every antigen can elicit an immune response; immunogenicity will

Delivery Technologies for Biopharmaceuticals: Peptides, Proteins, Nucleic Acids and Vaccines Edited by Lene Jorgensen and Hanne Mørck Nielsen

vary with antigen type and tertiary structure, amount, extent of resistance to lysosomal degradation and the context in which they are detected by the innate immune system (e.g. propensity to activate complement system). It has now become apparent that the stimulation of the innate immune system (defined as a rapid burst of inflammatory cytokines and activation of macrophages and DCs) can have an important role in the evolution of the adaptive immune responses [4]. Thus, inclusion of immune potentiators (or adjuvants, Table 6.1), which trigger early innate immune responses that subsequently assist the generation of potent and persistent adaptive immune responses, is crucial to vaccine effectiveness [5, 6]. Vaccines based on attenuated live organisms seem to fulfill these criteria; their invasiveness provides efficient delivery to antigen-presenting cells and some components of the pathogens (such as pathogen-associated molecular patterns) have the inherent property of stimulating the innate immune system.

As a result of these observations, a plethora of organic and inorganic materials have been used to improve the immunogenicity of inactivated and subunit vaccines (Table 6.2) [6–11]. Among them, particulate adjuvants such as liposomes and polymeric microspheres can not only protect antigens against degradation and provide a depot effect, but

Table 6.1 *Adjuvants as immunomodulators and antigen delivery systems*

Mode of action through:

- Antigen protection from degradation and elimination
- Controlling the physical appearance of antigen in vaccine formulation and its presentation to antigen-presenting cells (DCs and macrophages)
- Providing similar signals associated with infections thus inducing inflammatory reactions
- Site-specific targeting of DC endocytic and pattern-recognition receptors
- Controlling intracellular distribution of antigens

Optimal requirements:

- Should be chemically defined for consistent manufacturing
- Should elicit protective immune responses with weak antigens at lower antigen doses and reduce the frequency of booster immunization
- Be effective in young children, elderly and immunocompromised individuals
- Should be biodegradable, exhibit high stability and be non-immunogenic itself

Adjuvant types (may also be classified in accordance to source, mode of action and physicochemical properties); see also Table 6.2:

- Mineral salts (crystalline aluminium oxyhydroxide, amorphous aluminium hydroxyphosphate, hydrated calcium phosphate)
- Oil-in-water emulsions (e.g. MF59, AS04) and micellar/surfactant-based formulations (e.g. purified saponins, Polysorbate 80)
- Particulate (e.g. ISCOMS, liposomes, virosomes, polymeric nano/microspheres, gold, iron oxide, carbon nanotube)
- Microbial derivatives (see TLR and NOD ligands in Table 6.3)
- Endogenous human immunomodulators (cytokines that can be administered either as protein or plasmid coded, such as hGM-CSF and IL-12, Immudaptin or C3d tandem array)

Table 6.2 *Selected examples of adjuvant complexes: formulation aspects and associated responses*

Adjumer™ or polyidi(carboxylatophenoxy)lphosphazene

In soluble form serves as an adjuvant for parenteral administration but can be cross-linked to form microsphere hydrogel for mucosal vaccination; Adjumer has been evaluated in Phase I and II clinical trials (well tolerated with no adverse reactions); Adjumer sustained antibody response in mice after a single parenteral immunization as well as IgG and IgA responses after mucosal immunization

Adju-Phos (aluminium phosphate gel)

For human (diphtheria, tetanus and pertussis) and veterinary vaccine applications; may cause mild erythema and/or mild transient swellings at the site of injection; stimulate IL-4 and T_H-2 responses in mice with enhanced IgG1 and IgE production

Alhydrogel (aluminium hydroxide gel; Alum)

Same applications and adjuvanticity as Adju-Phos

Antigen formulation (SPT)

A microfluidized vaccine adjuvant vehicle composed of Squalene, poloxyethylene sorbitan monooleate (Polysorbate 80), polyoxyethene-polyoxypropylene block copolymers such as Pluronic® L121, phosphate-buffered saline and the designated antigen; induces humoral and cellular immune response; its safety and potency has been demonstrated in Phase I/II trials; Pluronic L121 may up-regulate expression of MHC class II molecules in antigen-presenting cells

Avridine® or *N,N*-dioctadecyl-*N′,N′*-bis(2-hydroxyethyl)propanediamine

For incorporation into liposomal formulations or in saline suspension with alum-precipitated antigen; induces humoral and cellular immunity (proliferation of B and T lymphocytes, macrophage activation, induction of interferon, enhancement of mucosal immunity when administered orally with antigen, induction of IgG2a and IgG2b isotypes); intranasal administration to humans induces interferon in nasal secretions and protection against rhinovirus challenge

Calcium phosphate gel

Has been used as adjuvant in vaccine formulations against diphtheria, tetanus, pertussis and poliomyelitis; adsorbs soluble antigens and presents them in a particulate form to DCs; well tolerated

Dimethyl dioctadecylammonium bromide (DDA)

Formulated as in oil-based emulsions in association with liposomes (functions as antigen carrier by direct binding or modification at oil/water interface); DDA with trehalose 6,6′-dibehenate (a synthetic analogue of trehalose 6,6′-dimycolate, which is an immunostimulatory component of the mycobacterial cell wall) forms vesicular structures and the adjuvant monophosphoryl lipid A may also be added; stimulate both humoral and cell-mediated immune responses against a wide range of antigens in animals; delayed-type hypersensitivity reactions are augmented after subcutaneous or intradermal routes of injection

Deoxycholic acid sodium salt (DOC/alum complex)

Enhances immune response to membrane proteins

Gamma inulin

Included in adjuvant formulations as a primary adjuvant and also as the immune stimulant when combined as composite particles with alum in the adjuvant Algammulin (a gamma inulin/alum composite adjuvant); a potent activator of the alternative pathway of complement; enhances both humoral and cell-mediated immunity; non-pyrogenic and non-antigenic (low toxicity in experimental animals)

(continued overleaf)

Table 6.2 (continued)

ImmTher® (N-acetylglucosamine-N-acetylmuramyl-L-ala-D-isoGlu-L-Ala-glycerol dipalmitate)

Formulated as adjuvant in liposomes consisting of 1-palmitoyl-2-oleoyl phosphatidylcholine and 1,2-dioleoyl phosphatidylglycerol; induces high levels of TNF-α, IL-1 and IL-6; rather safe in humans with no observed haematological hepatic or neural toxicity

Immune stimulating complexes (ISCOMes)

A cage-like structure complex composed of Quillaja saponins, cholesterol, phospholipids and antigen in phosphate-buffered saline; ISCOM based vaccines promote both antibody and cell-mediated responses following subcutaneous or intramuscular routes of injection; non-parenteral administration (intranasal and oral routes) has also proven to be effective

(Homo- and copolymers) of lactic and glycolic acid (PLGA, PGA, PLA)

Antigens may be adsorbed or entrapped by microspheres formed from these materials (a depot system) and exhibit enhanced and prolonged antibody responses compared to equivalent doses of free antigen; such systems may eliminate or reduce the need for booster immunizations; antigen-loaded PLGA particles are generally formulated by the emulsion solvent evaporation method but this may result in low encapsulation efficiency of highly water-soluble proteins; instability of protein antigens, however, may occur during polymer degradation as this induces accumulation of acidic monomers and consequently generates a low pH inside the particles

MF59

An intramuscular adjuvant composed of squalene, poloxyethylene sorbitan monooleate (Polysorbate 80) and sorbitan trioleate (Span 85); intramuscular injection in combination with a variety of subunit antigens results in elevated humoral response and cytotoxic lymphocytes

Virosomes

Virosomes are unilamellar vesicles incorporating influenza haemagglutinin; a commercial example is Inflexal V®, a 150 nm unilamellar vesicle prepared by detergent removal of influenza surface glycoproteins and a mixture of natural and synthetic phospholipids (70% egg phosphatidylcholine/20% phosphatidylethanolamine/10% envelop phospholipids of influenza virus); phosphatidylethanolamine may directly stimulate B-cells; formulation contains haemagglutinin (HA)-1 (its globular head contains a sialic acid site that binds to antigen-presenting cell surface receptors and initiate internalization) and HA-2, which mediates fusion of virosome and endosomal membrane (due to conformational change of HA-2 at low pH) thus leading to antigen release into cytosol and processing via proteasome complex

Walter Reed liposomes

Vesicles composed of dimyristoyl phosphatidylcholine and dimyristoyl phosphatidylglycerol, cholesterol, Lipid A from *Salmonella Minnesota* R595; formed vesicles are then adsorbed to aluminium hydroxide gel; have undergone four Phase I and Phase I/IIa clinical trials (three containing recombinant antigens derived from *Plasmodium falciparum* sporozite and the other with gp120 of HIV envelope); no adverse reactions was reported in humans may also deliver antigens to professional antigen-presenting cells through endocytic/phagocytic mechanisms [6]. Although these empirical approaches have clearly demonstrated the adjuvant activity of many substances, several products were also shown to elicit significant adverse effects, both locally and systemically, that precluded their use in human vaccine development. As a result, currently licensed adjuvants for human use are a few and include aluminium phosphate or hydroxyl gel (alum, licensed in the 1920s), the squalene-based oil-in-water emulsions MF59 (first licensed in Europe for the flu vaccine FLUAD) and AS03 (approved for the pandemic flu vaccine Prepandrix), and AS04 or monophosphoryl Lipid A formulated in alum (first approved for an HBV vaccine, Fendrix); these are delivery/depot-type adjuvants, which predominantly induce antibody responses.

There is a clear need for improved understanding of adjuvant mechanisms of action in different species, thus providing a basis for the rational selection of adjuvants for use with new human vaccines, such as highly purified subunit proteins or synthetic peptides, as well as veterinary vaccines. In this respect and owing to their ability to orchestrate a repertoire of innate and adaptive immune responses, DCs are key targets for developing new therapies that bring about resistance (as in infection and cancer) or tolerance (silencing in autoimmunity, allergy and transplantation scenarios). In this chapter innate features of DCs are briefly examined in relation to adjuvant design and antigen delivery systems.

6.2 Dendritic Cell Immunobiology

Circulating DCs, arising from progenitors in the bone marrow, home to tissues as immature cells with high capacity for phagocytosis, endocytosis and macropinocytosis [3, 12]. After antigen capture, immature DCs migrate from peripheral tissues via the afferent lymphatics into the T cell area of the draining lymph node, where they mature to antigen-presenting (in the context of major histocompatability complex, MHC, displaying molecules and increased expression of co-stimulatory molecules) T-cell priming cells to bring about clonal selection (Figure 6.1) [12, 13]. Recent evidence also indicates that steady-state migration of DCs into the lymph node occurs in healthy organisms; this may continuously impart tolerance against self-antigens to T-cells [14, 15]. Thus adoptive responses are either immunogenic or tolerogenic, depending on the specific DC linage, maturational stage and a spectrum of microenvironmental activation signals. Additionally, DCs may further educate effector cells to home in to sites of tissue injury. Following task completion, mature DCs eventually die by apoptosis. Understanding of peripheral DC immunobiology has provided numerous macromolecular and particulate-based vaccination strategies that simultaneously exploit DCs' distinct endocytic receptors for efficient antigen capture as well as non-phagocytic pattern recognition receptors to stimulate DC maturation, thus leading to strong T_H-1 responses [3, 5, 6]. This is in contrast to T_H-2 responses achieved by aluminium salts (alum), which are believed to induce non-specific inflammation and enhance antigen uptake by peripheral DCs through a depot effect (but see also Section 6.4 for recent proposed mechanisms of alum adjuvanticity).

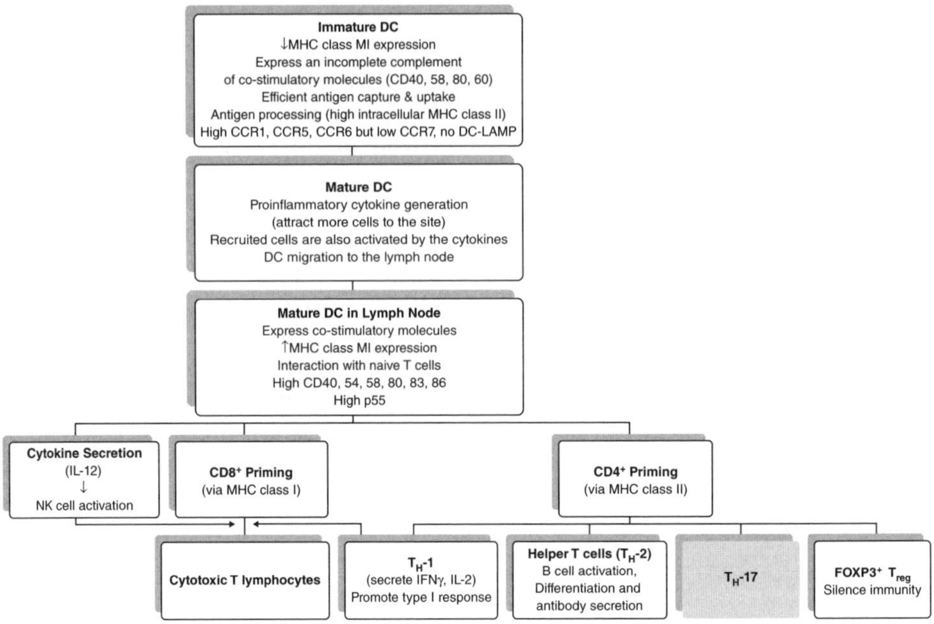

Figure 6.1 *The role of DC in immunity. CCR refers to chemokine (C-C motif) receptor. CCR6 is expressed only by Langerhans cells. CCR7 is up-regulated in mature DCs; this promotes responsiveness to macrophage inflammatory protein-3β and secondary lymphoid-tissue chemokine, the latter being a potent chemokine for both mature DCs and naïve T cells. LAMP refers to lysosome-associated membrane protein. Uncommitted CD4$^+$ cells can be induced to differentiate towards T_H-1 (accompanied by secretion of interleukin-2 and interferon-γ for cell-mediated responses and activation of CD8$^+$ cytotoxic T cells), T_H-2 (characterized by secretion of interleukins-4, 5 and 6 to help B-cells and induction of IgE and IgG4 in man but IgG1, IgE and IgA in mice), T_H-17 (interleukin-17 expressing cells that are linked to early inflammation and autoimmunity) and T_{reg} (cells for immune silencing) phenotypes according to local cytokine profile, however, there is no direct evidence for the existence of discrete T_H-17 cells in humans*

A second, but a less well defined, pathway of antigen delivery also exists, which functions independently of cellular trafficking along the lymphatics [16, 17]. Phenotypically and functionally resident DCs in lymph nodes are immature but constantly sample self-antigens and present them to T-cells for maintaining tolerance; however, following simultaneous exposure to an antigen and a maturation stimulus, such resident DCs activate T cells. These DCs can take up soluble antigens (smaller than 70–80 kDa) that enter the afferent lymphatics through a conduit system, which drains into the lumen of high endothelial venules [17]. The conduits are formed from collagen fibres and ensheathed by tightly interconnected non-haematopoietic fibroblastic reticular cells with occasionally embedded resident DCs, resulting in a compartment that is separate from T and B cell microenvironments [17]. A similar conduit system also exists in the spleen, where it delivers low molecular weight materials from the blood to the white pulp region [18].

Since immature DCs present in high numbers in lymph nodes when compared with peripheral tissues, they may serve as a valid target for vaccine delivery. However, this is a challenging idea since efficient co-delivery of antigen and adjuvant is remarkably difficult, but, if achievable, it could minimize the risk of premature antigen presentation that often leads to silencing.

A recent study [19] has remarkably claimed that interstitially injected polymeric nanoparticles of 20 nm size, which drain rapidly from the injection site due to their small size and hydrophilicity [20, 21], can physically target resident DCs in the subcapsular sinuses of regional draining lymph nodes, but not in T cell areas, and elicit their maturation (presumably through the generation of complement activation products, which may serve as 'danger signals'). These suggestions remain tantalizingly vague since DCs are known to be localized in both T cell and conduit areas and not much is known as how lymph nodes are populated by DCs in the steady state. Perinodal adipose tissue (PAT), however, is exquisitely equipped for local interactions with lymphoid tissues and, therefore, may provide a possible explanation for the above-mentioned observations [22]. Indeed, DCs are also found in abundance in PAT and cytokines, such as TNF-α and interleukin-6 (IL-6), to both of which PAT is highly responsive [23], are key modulators of immunity. These cytokines are known to promote DC migration into lymph nodes. In addition, PAT has a specialized fatty acid content and site-specific changes of fatty acids within PAT further exerts local immune modulatory effects to include DC migration and maturation [24, 25]. Therefore, the above observations may represent particle capture from subcapsular sinuses by migratory DCs arriving from PAT. Afferent lymphatic vessels also enter PAT and branch into numerous fine vessels before entering through the lymph node capsule into subcapsular sinus [22]. Nanoparticle extravasation from afferent lymphatic branches into PAT is a further possibility, although not investigated, that could account for DC uptake and migration.

6.3 Antigen Capture

Immature DCs express numerous uptake receptors and different receptors on the same DC subset may direct antigen to separate processing pathways (exogenous MHC class I pathway and MHC class II-rich compartments) [3, 12, 26, 27]. The bulk of these receptors are lectins with carbohydrate recognition capacity. For example, DEC-205 [28] and mannose receptor MRC1/CD206 [26, 29, 30] are type I transmembrane proteins with multiple contiguous lectin domains. The cytoplasmic tail of DEC-205 is highly efficient in antigen targeting to late endosomes, and the generated polypeptides are transported into the MHC class II-rich compartment for their loading onto the nascent MHC class II molecules while DCs mature. The mannose receptor CD206 binds a variety of mannosylated antigens and pathogens to include HIV, cytomegalovirus, mycobacteria and Candida; however, it can prevent DC maturation and contribute to immune evasion. Therefore, in the absence of additional stimuli, mannose receptor ligation might not lead to DC activation and enhanced immunogenicity. DC-SIGN (CD209) is another DC-restricted C-type lectin receptor (a type II transmembrane protein) that has been implicated in antigen internalization and presentation to T cells [31]. Other type II endocytic DC receptors include LOX-1 [32], langerin (CD207) [33], DCIR (CLEC4A) [34] and

DECTIN-2 [35]. These C-type lectins are differently expressed by distinct subsets of DCs and their expression can vary with the state of DC maturation. For example, myeloid DCs ($CD11c^+$) in skin epidermis (also known as Langerhans cells) express langerin and DEC-205 and are capable of inducing strong killer T cell responses, whereas DCs in the dermis express DC-SIGN and mannose receptor and can activate antibody-forming B cells [31, 36]. Also, distinct skin DC populations, such as dermal DCs, following migration colonize lymph node areas distinct from Langerhans cells [37]; the outcome of skin vaccination therefore depends on lectin and a DC subset that interacts with the antigen.

Other antigen internalizing receptors that further regulate DC responses include DORA [38], a member of the immunoglobulin superfamily, which is downregulated following CD40L engagement, immunoglobulin-like transcript-3 [39], which may potentially negatively regulate DC activation and maturation, and Fcγ receptor type I [40], which is responsible for antigen–antibody complexes internalization and capable of mediating both activating and inhibitory signals in DCs; for instance, immunity to antibody-coated tumour cells may be induced through selective engagement of activating Fc receptors on DCs.

DCs also express complement receptors (CRs) [41, 42]. Notably, stimulation of CR3 on DC populations induces immunosuppressive effects [42]. Thus, C3-conjugated products could be targeted to specific DC populations to induce both antigen uptake and tolerization. CR2, which interacts with C3d, expression is largely restricted to follicular DCs and B lymphocytes [43]. Conjugation of C3d to antigen dramatically lowers the threshold for cognate B lymphocytes activation through interaction with CR2 in conjugation with the B cell antigen receptor complex and CD19 [44]. Immobilized C1q has also been shown to induce monocyte-derived dendritic cell maturation and activation to include IL-12 and TNF-α secretion as well elevation of their T cell stimulating capacity [45]. C1q interaction was mediated through two receptors, cC1qR and gC1qR, recognizing the globular and collagenous regions of the molecule, respectively. These observations may indicate that C1q-opsonized antigens may play a role in induction and regulation of immune response.

6.4 Particulate Antigen Delivery Systems

A wide range of particulate systems (e.g. polymeric nanoparticles, polymerosomes, carbon nanotubes, liposomes, etc.) can aid antigen delivery to DCs [6, 9, 10, 19, 46–51]. The extent of particle/complex uptake and subsequent antigen processing and DC response strongly depends on physicochemical characteristics (such as particle size, shape and surface properties) and chemical composition of the carrier, as well as the mode of antigen association (surface adsorbed or grafted, entrapped). For instance, particle uptake and subsequent antigen processing may be enhanced following surface decoration with DC endocytic receptor ligands to include monoclonal antibodies, aptamers and carbohydrates. Surface functionalization of particulate carriers with targeting ligands is also beneficial if they are to be used for oral antigen delivery. The absorption of particles in the intestine following oral administration takes place at the Peyer's patches [52]. The epithelial cell layer overlaying Payer's patches contains specialized M cells, where their apical membranes bind and take up pathogens and particles from the lumen and transport them to underlying macrophages and DCs [52] for either T_H-2 or T_H-3 T_{reg} biased response. However, the efficiency of sampling orally delivered particles by M cells is very low as these cells are relatively rare in the epithelial lining. The use of M cell-specific probes, such as secretory IgAs and reovirus $\sigma 1$ protein, in conjugation with DC endocytic receptor ligands, may enhance the efficiency of particulate targeting and subsequent presentation to regional DCs [52, 53].

The efficiency of cross-presentation of exogenous antigens in DCs also appears to be particle size dependent. For example, antigens bound to 50 nm polystyrene nanoparticles are shuttled into late acidic endosomes and lysosomal compartments (pH 4.5–5.0), leading to rapid and unregulated antigen degradation and inefficient cross-presentation. On the other hand, DCs shuttle antigens bound to 500 nm or 3 μm polystyrene particles to early endosomal compartments with a more neutral environment (pH 6–7); this preserves the majority of antigens for subsequent generation of peptides to be loaded to the MHC class I pathway [54]. Alternatively, antigen delivery to cytosol, and hence channelling into the MHC class I route, can be dramatically enhanced by triggered release in late endosomes; this is achievable with pH sensitive and fusogenic carriers [55]. A pH sensitive carrier maintains a stable structure at neutral pH but destabilizes and subsequently disrupts endosomal integrity causing cargo release into cytoplasm. Apart from pH sensitive liposomes [55], other examples include polyketal nanoparticles [56] and block copolymer micelles of poly(methacrylic acid)-*b*-poly(ethylene oxide)-poly-L-lysine [57]. Alternatively, by mimicking the characteristics of certain toxins or viruses, cytoplasmic antigen delivery may also be enhanced. This requires antigen conjugation to toxins such as inactive derivatives of diphtheria toxin, anthrax toxin and B-subunit of *Escherichia coli* heat-labile enterotoxin, or toxin co-encapsulation (e.g. listeriolysin O, HA2 fusion peptide of influenza virus) with antigens in particulate vehicles [55, 58, 59]. Moreover, such approaches are highly desirable for cytoplasmic delivery of nucleic acid-based vaccines as well as plasmids encoding immunosuppressive cytokines; the latter strategy is to allow DCs to subsequently downregulate T cell responses (e.g. expansion of $FOXP3^+$ expressing T_{reg} cells) and has been demonstrated with poly(lactic-*co*-glycolic acid) nanoparticles modified with a histidine rich peptide and carrying IL-10 encoding plasmid *in vitro* [50].

The adjuvanticity of particulate antigen carriers may in part be related to their complement activating nature, as in carbon nanotubes [60, 61] and certain liposomes [62]. Similarly, some of the stabilizing components of an antigen carrier (as in non-ionic surfactants) may further generate complement split products by activating complement through all the three known pathways; such agents may also be used as adjuvants on their own right, and some are even capable in upregulating the surface expression of MHC class II antigens in antigen-presenting cells [63–66].

6.5 Signalling Receptors and the Role of Adjuvants

Different microbial products affect DC maturation in distinct ways, thereby launching alternative versions of host immunity. The mechanism responsible for this potentiation of immune responses is based on the action of microbial ligands particularly on pattern recognition receptors (PRRs) comprising Toll-like receptors (TLRs), cytoplasmic nucleotide-binding oligomerization domain proteins (NOD) and NALP3 (cytoplasmic NOD-like receptor protein, cryopyrin), Table 6.3 [5, 6, 67–69]. TLRs 1, 2, 4, 5, 6 and 11 are surface exposed, whereas TLRs 3, 7, 8 and 9 are endosomal. TLRs are also differently expressed by distinct subsets of DCs. For example, plasmacytoid DCs, isolated from human blood, only express TLR7 and 9; only myeloid DCs express TLR3 but TLR4 expression is observed in monocytes and sometimes in myeloid DCs, thus indicating different function of DC subsets [70–73]. Langerhans cells express a selective array of TLRs, notably TLR1, 2, 3, 5, 6 and 10 [74]. Following ligation, PRRs initiate signalling cascades, through the adaptor protein MyD88, which is used by all TLRs except TLR3, leading to activation of transcriptional factors such as NF-κB and MAP kinases that ultimately results in expression of pro-inflammatory cytokines (e.g. IL-6 and IL-12) [75, 76]. TLR3 (and also TLR4) signals through the adaptor protein TRIF [75, 77]. This activates interferon regulatory factor (IRF) 3, 7 and NF-κB, leading to production of antiviral type I interferon (IFN)-α and -β. IFN-α can also be induced through TLR7 and 9 signalling through activation of IRF-7. NOD (NOD1 and 2) ligation with their ligands activates NF-κB through the receptor-interacting serine/threonine kinase with the caspase recruitment domain, resulting in expression of numerous inflammatory cytokines and chemokines [78, 79]. PRRs can also function synergistically; synergy has been observed between NOD1/2 and sub-active doses of TLR4 agonists and also between NOD2 and TLR9 [80–82].

Thus, appreciation of PRR biology is beginning to offer rational approaches to discovery and development of vaccine adjuvants such as synthetic TLR agonists. A handful of first generation agents, such imidazoquinolines and unmethylated CpG oligonucleotides (e.g. as in combination with cationic liposomes), are now known to induce protective

Table 6.3 *TLR and NOD receptors expressed by DCs*

Pattern recognition receptor	Ligands
TLR1	Triacyl lipoproteins, Pam3Cys
TLR2 (also expressed on B cells in nasal associated lymphoid tissue)	Gram positive peptidoglycan; lipoprotein; lipoarabionmannan; zymosan yeast particles
TLR3 (endosomal)	dsRNA (viral); poly I:C
TLR4	Monophosphoryl lipid A; paclitaxel
TLR5	Bacterial flagellins
TLR6	Lipotechoic acid; lipopetides from mycoplasma
TLR7/8 (endosomal)	ssRNA; imidazoquinolines such as imiquimod and resiquimod
TLR9 (endosomal)	Unmethylated CpG oligonucleotides
TLR10	Unknown
TLR11	*Toxoplasma* profiling
NOD1/2 (cytosolic)	*meso*-diaminopimelic acid and muramyl dipeptide, respectively
NALP3 (cytosolic)	Uric acid crystals, bacterial RNA

T_H-1-like responses after sensing by TLR 3, 7, 8 and 9 and have shown some efficacy in animals and human trials [6]. Similarly, TLR 2 engagement by Pam3Cys leads to a preferential production of T_H-2 cytokines [6]. Recent studies have also shown that *in vitro* alum can induce direct activation of NALP3 inflammasome complex (apoptosis-associated speck-like protein and caspase-1 protease) in macrophages and synergize with lipopolysaccharide stimulation of TLR4 for the secretion of several proinflammatory cytokines including IL-1β, IL-18 and IL-33 [83]. Presumably, part of *in vivo* mechanisms of alum adjuvanticity may arise through induction of necrosis in unidentified cells, resulting in production of uric acid (a NALP3 substrate). Other potential adjuvants, such as chitosan, silica and Quil A (a component of ISCOMS), also seem to activate NALP3 inflammasome [84, 85]. The interaction between TLR4 and C1q binding structures on DCs has also been investigated. For instance Baruah et al. [86] have shown that in the presence of TLR4 ligands C1q increases the phagocytosis of apoptotic cells by immature DCs and the release of IL-12, whereas other studies concluded that the selective inhibition of TLR4-induced IL-12 production could be achieved after cross-linking of gC1qR on the surface of macrophages and DCs [87].

PRRs therefore offer a wide range of potential targets for vaccine adjuvants and could rationalize adjuvant discovery, particularly through high-throughput screening, combinatorial peptide synthesis and immunogenomics. Ultimately, such approaches may enhance adjuvant activity and safety by combining delivery vehicles and immunostimulants.

6.6 Conclusions

DCs are unique antigen-presenting cells; they induce, sustain and modulate immune responses. Maturation stage, route of antigen entry (systemic, mucosal, epidermal and intradermal routes), antigen quantity and its presentation time (prior to or after DC activation), and cytokine microenvironment all influence DC effector functions. Further understanding of DC subset pathophysiology and function at different stages of disease will be essential for developing effective vaccination strategies against pathogens, cancer and autoimmune diseases. For instance, in HIV there is a progressive loss and dysfunction of blood myeloid and plasmacytoid DCs; monocyte-derived DCs from HIV infected patients are apparently impaired in their ability to stimulate T cells and produce cytokines. Therefore, only through advances of immunological knowledge may efficient carriers and adjuvants be designed to target either all DCs or a specific subtype and deliver antigens together with DC activators or inhibitors. Such approaches may also be applied for rational *ex-vivo* modulation of DC activities for mechanistic research as well as for immunotherapy protocols involving DC injection [88].

Acknowledgements

Financial support from the Danish Agency for Science, Technology and Innovation (Det Strategiske Forskningsråd), reference 2106-08-0081, is gratefully acknowledged.

References

1. Holmgren, J. and Czerkinsky, C. (2005) Mucosal immunity and vaccines. *Nat. Med.*, **11**, S45–S53.
2. Ulmer, J.B., Valley, U. and Rappuoli, R. (2006) Vaccine manufacturing: challenges and solutions. *Nat. Biotechnol.*, **24**, 1377–1383.
3. Steinman, R.M. and Banchereau, J. (2007) Taking dendritic cells into medicine. *Nature*, **449**, 419–426.
4. Hoebe, K., Janssen, E. and Beutler, B. (2004) The interface between innate and adaptive immunity. *Nat. Immunol.*, **5**, 971–974.
5. Pashine, A., Valiante, N.M. and Ulmer, J.B. (2005) Targeting the innate immune response with improved vaccine adjuvants, *Nat. Med.*, **11**, S63–S68.
6. Guy, B. (2007) The perfect mix: recent progress in adjuvant research. *Nat. Rev. Microbiol.*, **5**, 505–517.
7. Linblad, E.B. (2004) Aluminium adjuvants – in retrospect and prospect. *Vaccine*, **22**, 3658–3668.
8. Linblad, E.B. (2004) Aluminium compounds for use in vaccines. *Immunol. Cell Biol.*, **82**, 497–505.
9. Combadiere, B. and Mahe, B. (2008) Particle-based vaccines for transcutaneous vaccination. *Comp. Immunol. Microbiol. Infect. Dis.*, **31**, 293–315.
10. Gupta, R.K. and Siber, G.R. (1995) Adjuvants for human vaccines – current status, problems and future prospects. *Vaccine*, **13**, 1263–1276.
11. Morein, B., Sundquist, B., Hoglund, S. *et al.* (1984) Iscom, a novel structure for antigenic presentation of membrane proteins from enveloped viruses. *Nature*, **308**, 457–460.
12. Banchereau, J., Briere, F., Caux, C. *et al.* (2000) Immunobiology of dendritic cells. *Ann. Rev. Immunol.*, **18**, 767–811.
13. Randolph, G.J., Angeli, V. and Swartz, M.A. (2005) Dendritic-cell trafficking to lymph nodes through lymphatic vessels. *Nat. Rev. Immunol.*, **5**, 617–628.
14. Steinman, R.M., Hawiger, D. and Nussenzweig, M.C. (2003) Tolerogenic dendritic cells. *Ann. Rev. Immunol.*, **21**, 685–711.
15. Lutz, M.B. and Schuler, G. (2002) Immature, semi-mature and fully mature dendritic cells: which signals induce tolerance or immunity? *Trend Immunol.*, **23**, 445–449.
16. Ingulli, E., Ulman, D.R., Lucido, M.M. and Jenkins, M.K. (2002) *In situ* analysis reveals physical interactions between $CD11b^+$ dendritic cells and antigen-specific CD4 T cells after subcutaneous injection of antigen. *J. Immunol.*, **169**, 2247–2252.
17. Sixt, M., Kanazawa, N., Selg, M. *et al.* (2005) The conduit system transports soluble antigens from the afferent lymph to resident dendritic cells in T cell area of the lymph node. *Immunity*, **22**, 19–29.
18. Nolte, M.A., Belien, J.A., Schadee-Eestermans, I. *et al.* (2003) A conduit system distributes chemokines and small blood-borne molecules through splenic white pulp. *J. Exp. Med.*, **198**, 505–512.
19. Reddy, S.T., van der Vlies, A.J., Simenoi, E. *et al.* (2007) Exploiting lymphatic transport and complement activation in nanoparticle vaccines. *Nat. Biotechnol.*, **25**, 1159–1164.
20. Moghimi, S.M., Hawley, A.E., Christy, N.M. *et al.* (1994) Surface engineered nanospheres with enhanced drainage into lymphatics and uptake by macrophages of the regional lymph nodes. *FEBS Lett.*, **344**, 25–30.
21. Moghimi, S.M. (2003) Modulation of lymphatic distribution of subcutaneously injected poloxamer 407-coated nanospheres: the effect of the ethylene oxide chain configuration. *FEBS Lett.*, **540**, 241–244.
22. Knight, S.C. (2008) Specialized perinodal fat fuels and fashions immunity. *Immunity*, **28**, 135–138.
23. Tilg, H. and Moschen, A.R. (2006) Adipocytokines: mediators linking adipose tissue, inflammation and immunity. *Nat. Rev. Immunol.*, **6**, 772–783.
24. Thurnher, M. (2007) Lipids in dendritic cell biology: messengers, effectors, and antigens. *J. Leukoc. Biol.*, **81**, 154–160.

25. Pond, C.M. (2003) Paracrine relationships between adipose and lymphoid tissues: implications for the mechanism of HIV-associated adipose redistribution syndrome. *Trend Immunol.*, **24**, 13–18.
26. Sallusto, F., Cella, M., Danieli, C. and Lanzavecchia, A. (1995) Dendritic cells use macropinocytosis and the mannose receptor to concentrate macromolecules in the major histocompatibility complex class II compartment: downregulation by cytokines and bacterial products. *J. Exp. Med.*, **182**, 389–400.
27. Reis e Sousa, C., Stahl, P.D. and Austyn, J.M. (1993) Phagocytosis of antigens by Langerhans cells *in vitro*. *J. Exp. Med.*, **178**, 509–519.
28. Jing, W., Swiggard, W.J., Heufler, C. *et al.* (1995) The receptor DEC-205 expressed by dendritic cells and thymic epithelial cells is involved in antigen processing. *Nature*, **375**, 151–155.
29. Engering, A.J., Cella, M., Fluitsma, D. *et al.* (1997) The mannose receptor functions as a high capacity and broad specificity antigen receptor in human dendritic cells. *Eur. J. Immunol.*, **27**, 2417–2425.
30. Tan, M.C., Mommaas, A.M., Drijfhout, J.W. *et al.* (1997) Mannose receptor-mediated uptake of antigens strongly enhances HLC class II-restricted antigen presentation by cultured dendritic cells. *Eur. J. Immunol.*, **27**, 2426–2435.
31. Geijtenbeek, T.B., Torensma, R., van Vliet, S.J. *et al.* (2000) Identification of DC-SIGN, a novel dendritic cell-specific ICAM-3 receptor that supports primary immune response. *Cell*, **100**, 575–585.
32. Delneste, Y., Magistrelli, G., Gauchat, J.F. *et al.* (2002) Involvement of LOX-1 in dendritic cell-mediated antigen cross-presentation. *Immunity*, **17**, 353–362.
33. Valladeau, J., Duvert-Frances, V., Pin, J.J. *et al.* (1999) The monoclonal antibody DCGM4 recognizes langerin, a protein specific of Langerhans cells, and is rapidly internalized from the cell surface. *Eur. J. Immunol.*, **29**, 2695–2704.
34. Dudziak, D., Kamphorst, A.O., Heidkamp, G.F. *et al.* (2007) Differential antigen processing by dendritic cell subsets *in vivo*. *Science*, **315**, 107–111.
35. Ariizumi, K., Shen, G-L., Shikano, S. *et al.* (2000) Cloning of a second dendritic cell-associated C-type lectin (Dectin-2) and its alternatively spliced isoforms. *J. Biol. Chem.*, **275**, 11957–11963.
36. Mommaas, A.M., Mulder, A.A., Jordens, R. *et al.* (1999) Human epidermal Langerhans cells lack functional mannose receptors and a fully developed endosomal/lysosomal compartment for loading of HLA class II molecules. *Eur. J. Immunol.*, **29**, 571–580.
37. Kissenpfennig, A., Henri, S., Dubois, B. *et al.* (2005) Dynamics and function of Langerhans cells *in vivo*. *Immunity*, **22**, 643–654.
38. Bates, E.E., Dieu, M-C., Ravel, O. *et al.* (1998) CD40L activation of dendritic cells downregulate DORA, a novel member of immunoglobulin superfamily. *Mol. Immunol.*, **35**, 513–524.
39. Cella, M., Dohring, C., Samaridis, J. *et al.* (1997) A novel inhibitory receptor (ILT3) expressed on monocytes, macrophages, and dendritic cells involved in antigen processing. *J. Exp. Med.*, **185**, 1743–1751.
40. Fanger, N.A., Wardwell, K., Shen, L. *et al.* (1996) Type I (CD64) and type II (CD32) Fc γ receptor-mediated phagocytosis by human blood dendritic cells. *J. Immunol.*, **157**, 541–548.
41. Rozis, G., de Silva, S., Benlahrech, A. *et al.* (2005) Langerhans cells are more efficiently transduced than dermal dendritic cells by adenovirus vectors expressing either group C or group B fibre protein: implications for mucosal vaccines. *Eur. J. Immunol.*, **35**, 2617–2626.
42. Behrens, E.M., Sriram, U., Shivers, D.K. *et al.* (2007) Complement receptor e ligation of dendritic cells suppresses their stimulatory capacity. *J. Immunol.*, **178**, 6268–6279.
43. Zabel, M.D. and Weis, J.H. (2001) Cell-specific regulation of the CD21 gene. *Int. Immunopharmacol.*, **1**, 483–493.
44. Dempsey, P.W., Allison, M.E., Akkaraju, S. *et al.* (1996) C3d of complement as a molecular adjuvant: bridging innate and acquired immunity. *Science*, **271**, 348–350.
45. Csomor, E., Bajtaya, Z., Sandora, N. *et al.* (2007) Complement protein C1q induces maturation of human dendritic cells. *Mol. Immunol.*, **44**, 3389–3397.

46. O'Hagan, D.T., Singh, M. and Ulmer, J.B. (2004) Microparticles for the delivery of DNA vaccines. *Immunol. Rev.*, **199**, 191–200.
47. O'Hagan, D.T. and Rappuoli, R. (2004) Novel approaches to vaccine delivery. *Pharm. Res.*, **21**, 1519–1530.
48. Pantarotto, D., Partidos, C.D., Hoebeke, J. *et al.* (2003) Immunization with peptide-functionalized carbon nanotubes enhances virus-specific neutralizing antibody responses. *Chem. Biol.*, **10**, 991–996.
49. Akagi, T., Wang, X., Uto, T. *et al.* (2007) Protein direct delivery to dendritic cells using nanoparticles based on amphiphilic poly(amino acid) derivatives. *Biomaterials*, **28**, 3427–3436.
50. Jia, L., Kovacs, J.R., Zheng, Y. *et al.* (2008) Expansion of Foxp3-expressing regulatory T cells *in vitro* by dendritic cells modified with polymeric particles carrying a plasmid encoding interleukin-10. *Biomaterials*, **29**, 1250–1261.
51. Perrie, Y., Mohammed, A.R., Kiby, D.J. *et al.* (2008) Vaccine adjuvant systems: enhancing the efficacy of sub-unit protein antigens. *Int. J. Pharmaceut.*, **364**, 272–280.
52. Simecka, J.W. (1998) Mucosal immunity of gastrointestinal tract and oral tolerance. *Adv. Drug Deliv. Rev.*, **34**, 235–259.
53. Lavelle, E.C. (2001) Targeted delivery of drugs to the gastrointestinal tract. *Crit. Rev. Ther. Drug Carrier Syst.*, **18**, 341–386.
54. Tran, K.K. and Shon, H. (2009) The role of phagosomal pH on the size-dependent efficiency of cross-presentation by dendritic cells. *Biomaterials*. **30**, 1356–1362.
55. Drummond, D.C., Zignani, M. and Leroux, J.C. (2000) Current status of pH-sensitive liposomes in drug delivery. *Prog. Lipid. Res.*, **39**, 409–460.
56. Heffernan, M.J., Kasturi, S.P., Yang, S.C. *et al.* (2009) The stimulation of CD8+ T cells by dendritic cells pulsed with polyketal microparticles containing ion-paired protein antigen and poly(inosinic acid)-poly(cytidylic acid). *Biomaterials*, **30**, 910–918.
57. Boudier, A., Aubert-Pouessel, A., Louis-Plence, P. *et al.* (2009) The control of dendritic cell maturation by pH-sensitive polyion complex micelles. *Biomaterials*, **30**, 233–241.
58. Beauregard, K.E., Lee, K.D., Collier, R.J. and Swanson, J.A. (1997) pH-dependent perforation of macrophage phagosomes by Listeriolysin O from *Listeria monocytogenes*. *J. Exp. Med.*, **186**, 1159–1163.
59. Lee, K.D., Oh, Y.K., Portnoy, D.A. and Swanson, J.A. (1996) Delivery of macromolecules into cytosol using liposome containing hemolysin from Listeria monocytogenes. *J. Biol. Chem.*, **271**, 7249–7252.
60. Salvador-Morales, C., Flahaut, E., Sim, E. *et al.* (2006) Complement activation and protein adsorption by carbon nanotubes. *Mol. Immunol.*, **43**, 193–201.
61. Hamad, I., Hunter, A.C., Rutt, K.J. *et al.* (2008) Complement activation by PEGylated single-walled carbon nanotubes is independent of C1q and alternative pathway turnover. *Mol. Immunol.*, **45**, 3797–3803.
62. Moghimi, S.M. and Hamad, I. (2008) Liposome-mediated triggering of complement cascade. *J. Liposome Res.*, **18**, 195–209.
63. Hunter, R.L., McNicholl, J. and Lal, A.A. (1994) Mechanisms of action of nonionic block copolymer adjuvants. *AIDS Res. Hum. Retroviruses*, **10**, S95–S98.
64. Moghimi, S.M., Hunter, A.C., Dadswell, C.M. *et al.* (2004) Causative factors behind poloxamer 188 (Pluronic F68, Flocor™)-induced complement activation in human sera. A protective role against poloxamer-mediated complement activation by elevated serum lipoprotein levels. *Biochim. Biophys. Acta*, **1689**, 103–113.
65. Moghimi, S.M. and Szebeni, J. (2003) Stealth liposomes and long circulating nanoparticles: critical issues in pharmacokinetics, opsonization and protein-binding properties. *Prog. Lipid Res.*, **42**, 463–478.
66. Gaymalov, Z.Z., Yang, Z., Pisarev, V.M. *et al.* (2009) The effect of nonionic block copolymer pluronic P85 on gene expression in mouse and antigen-presenting cells. *Biomaterials*. **30**, 1232–1245.
67. Malmgaard, L. (2005) Dendritic cells, toll-like receptors, and T-cell responses: lessons from viral infections *in vivo*. *Viral Immunol.*, **18**, 584–594.

68. Barton, G.M. and Medzhitov, R. (2002) Toll-like receptors and their ligands. *Curr. Topic Microbiol. Immunol.*, **270**, 81–92.
69. Gordon, S. (2002) Pattern recognition receptors: doubling up for the innate immune response. *Cell*, **111**, 927–930.
70. Ito, T., Amakawa, R., Kaisho, T. *et al.* (2002) Interferon-alpha and interleukin-12 are induced differentially by Toll-like receptor 7 ligands in human blood dendritic cell subsets. *J. Exp. Med.*, **195**, 1507–1512.
71. Jarrossay, D., Napolitani, G., Colonna, M. *et al.* (2001) Specialization and complementarity in microbial molecule recognition by human myeloid and plasmacytoid dendritic cells. *Eur. J. Immunol.*, **31**, 3388–3393.
72. Uronen-Hansson, H., Allen, J., Osman, M. *et al.* (2004) Toll-like receptor 2 (TLR2) and TLR4 are present inside human dendritic cells, associated with microtubules and glogi apparatus but are not detectable on the cell surface: integrity of microtubules is required for interleukin-12 production in response to internalized bacteria. *Immunology*, **111**, 173–178.
73. Kadowaki, N., Ho, S., Antonenko, S. *et al.* (2001) Subsets of human dendritic cell precursors express different Toll-like receptors and respond to different microbial antigens. *J. Exp. Med.*, **194**, 863–869.
74. Flacher, V., Bouschbacher, M., Verronese, E. *et al.* (2006) Langerhans cells express a specific TLR profile and differently respond to viruses and gram-positive bacteria. *J. Immunol.*, **177**, 7959–7967.
75. Akira, S. and Takeda, K. (2004) Toll-like receptor signaling. *Nat. Rev. Immunol.*, **4**, 499–511.
76. Akira, S., Yamamoto, M. and Takeda, K. (2003) Role of adapters in Toll-like receptor signaling. *Biochem. Soc. Trans.*, **31**, 637–642.
77. Honda, K., Yanai, H., Negishi, H. *et al.* (2005) IRF-7 is the master regulator of type-I interferon-dependent immune responses. *Nature*, **434**, 772–777.
78. Chamaillard, M., Hashimoto, M., Horie, Y. *et al.* (2003) An essential role for NOD1 in host recognition of bacterial peptidoglycan containing diaminopimelic acid. *Nat. Immunol.*, **4**, 702–707.
79. Girardin, S.E., Boneca, I.G., Viala, J. *et al.* (2003) Nod2 is a general sensor of peptidoglycan through muramyl dipeptide (MDP) detection. *J. Biol. Chem.*, **278**, 8869–8872.
80. Napolitani, G., Rinaldi, A., Bertoni, F. *et al.* (2005) Toll-like receptor agonist combinations synergistically trigger a T helper type 1-polarizing program in dendritic cells. *Nat. Immunol.*, **6**, 769 776.
81. Fritz, J.H., Girardin, S.E., Fitting, C. *et al.* (2005) Synergistic stimulation of human monocytes and dendritic cells by Toll-like receptor 4 and NOD1- and NOD2-activating agonists. *Eur. J. Immunol.*, **35**, 2459–2470.
82. van Heel, D.A., Ghosh, S., Hunt, K.A. *et al.* (2005) Synergy between TLR9 and NOD2 innate immune responses is lost in genetic Crohn's disease. *Gut*, **54**, 1553–1557.
83. Franchi, L. and Nunez, G. (2008) The Nlrp3 inflammasome is critical for aluminium hydroxide-mediated IL-1β secretion but dispensable for adjuvant activity. *Eur. J. Immunol.*, **38**, 2085–2089.
84. Li, H., Willingham, S.B., Ting, J.P. and Re, F. (2008) Cutting edge: inflammasome activation by alum and alum's adjuvant effects are mediated by NLRP3. *J. Immunol.*, **181**, 17–21.
85. Dostert, C., Petrilli, V., Van Bruggen, R. *et al.* (2008) Innate immune activation through Nalp3 inflammasome sensing of asbestos and silica. *Science*, **320**, 674–677.
86. Baruah, P., Dumitriu, I.E., Peri, G. *et al.* (2006) The tissue pentraxin PTX3 limitsC1q-mediated complement activation and phagocytosis of apoptotic cells by dendritic cells. *J. Leukoc. Biol.*, **80**, 87–95.
87. Waggoner, S.N., Cruise, M.W., Kassel, R. and Hahn, Y.S. (2005) gC1q receptor ligation selectively down-regulatehuman IL-12 production through activation of the phosphoinositide 3-kinase pathway. *J. Immunol.*, **175**, 4706–4714.
88. Figdor, C.G., de Vries, I.J.M., Lesterhuis, W.J. and Melief, C.J.M. (2004) Dendritic cell immunotherapy: mapping the way. *Nat. Med.*, **10**, 475–480.

Delivery Technologies for Biopharmaceuticals

7

Lipid Nanoparticle-Based Systems for Delivery of Biomacromolecule Therapeutics

Susana Martins,[1] *Domingos C. Ferreira*[1] *and Eliana B. Souto*[2,3]

[1] University of Porto, Department of Pharmaceutical Technology, Faculty of Pharmacy, Porto, Portugal

[2] Fernando Pessoa University, Department of Pharmaceutical Technology, Faculty of Health Sciences, Porto, Portugal

[3] Institute of Biotechnology and Bioengineering, Centre of Genetics and Biotechnology, University of Trás-os-Montes and Alto Douro, Vila Real, Portugal

7.1 Introduction

The development of new drug entities (NDEs), namely peptides, proteins and nucleic acids, for the treatment of several life-threatening diseases is a multidisciplinary area that requires a continuous demonstration of the safety and efficiency of such NDEs, which should be as good as or even better than the classical ones.

The majority of the drugs developed nowadays belong to class III of the Biopharmaceutical Classification System (BCS), that is showing high aqueous solubility and low permeability due to their high molecular weight and low partition coefficient values [1]. Thus, the main prominent cause of failure in reaching the pharmaceutical market is associated with poor biopharmaceutical properties, that is drug absorption and distribution. Furthermore, the lack of significant efficacy, high toxicity and adverse side effects are additional reasons to drop the molecules in the early stages of the industrial development.

Delivery Technologies for Biopharmaceuticals: Peptides, Proteins, Nucleic Acids and Vaccines Edited by Lene Jorgensen and Hanne Mørck Nielsen

Examples of these NDEs include peptides, proteins and other molecules having molecular weights above 500, which suffer gastrointestinal (GI) degradation and/or poor intestinal permeability. Approaches to overcome these limitations, in order to maximize drug absorption and thus to improve oral bioavailability, may include developing drug formulations using chemical penetration enhancers and/or colloidal carriers. The former play an important role in changing transcellular absorption pathways, whereas the latter may specifically change the tight junctions between cells to facilitate paracellular uptake [2].

With regard to systemic administration of lipophilic drugs, co-solvents are often used as solubilizing enhancers, which can elicit pain at the site of injection, allergic reactions and drug precipitation. Many of the large molecular weight drugs (e.g. peptides, proteins, DNA, oligonucleotides, siRNA) are sensitive to enzyme-mediated degradation in the blood, turning drug delivery and active targeting into a huge challenge. A suitable alternative to achieve both *in vitro* and *in vivo* drug stability would be their formulation in colloidal carriers. In fact, liposomes, polymeric and lipid nanoparticles have been studied intensively so as to increase physicochemical stability of many drugs and to achieve a particular targeting site.

A major advantage of drug delivery and targeting to a specific site of action by means of colloidal carriers is the reduction of large drug dose required to achieve a therapeutic benefit. Furthermore, when drugs are administered systemically they suffer blood dilution and are taken up by various tissues and organs, which for several is the contributing factor to adverse side effects. Thus, favourable pharmacokinetic profiles from systemically administered colloidal carriers can lead to improvements in the therapeutic outcomes. It is also important to point out that coating the colloidal carriers with hydrophilic polymers, such as poly(ethylene glycol) (PEG), can change the size and the electrical charge of the carriers, thereby reducing their immunogenicity and improving the pharmacokinetic properties of the drug [3]. Moreover, the carrier uptake by the organs of the reticuloendothelial system (RES), that is liver and spleen, will be minimized or totally avoided, thus prolonging the circulation time of the incorporated drug. Opsonization is one of the main biological barriers that colloidal carriers need to overcome to achieve drug delivery and targeting. If these carriers are very hydrophobic, they will quickly bind to opsonin proteins present in the blood serum and through the complement-mediated pathway be recognized and removed by the macrophages of the mononuclear phagocytic system (MPS).

Nanotechnology is a novel subject addressed to the development of drug delivery systems within the nanometre size range (typically <500 nm), specifically designed to provide medical and pharmaceutical benefits to treat several life-threatening diseases, such as cancer and infections. Nanotechnology can achieve desirable pharmacokinetic and biopharmaceutical properties, and overcome some major toxicological concerns, allowing thereafter the development of novel formulations for several NDEs.

When selecting a carrier system for a specific drug, it is necessary to take into account several properties of the molecule being incorporated, for example lipophilicity, solubility and stability in the biological fluids, degree of ionization, permeability and gastrointestinal metabolism, as well as systemic pharmacokinetics and pharmacodynamics, and protein binding properties.

Nanoparticulate carriers composed of lipid materials have the advantages of being composed of biodegradable and biocompatible materials resembling human tissues. Such carriers are the solid lipid nanoparticles (SLNs) and the lipid-drug conjugates (LDCs).

These carriers have been exploited for the delivery and targeting of peptides and proteins [2], vaccines [4], antisense oligonucleotides [5], plasmid DNA [6–10] and short interfering RNA (siRNA) [11, 12].

7.2 Definitions and Properties of Solid Lipid Nanoparticles (SLNs)

Solid lipid nanoparticles (SLNs) are a relatively new class of drug carrier that has gained considerable interest over the last decade. SLNs have been successfully employed for efficient delivery of numerous NDEs by various delivery routes. During the last 15 years several reviews on the subject have been published [13–18].

SLNs combine advantages of traditional colloidal drug carrier systems like liposomes, polymeric nanoparticles and emulsions (e.g. physical stability, protection of labile drugs from degradation, controlled release and easy preparation [19]). Nevertheless, at the same time SLNs may avoid or minimize the drawbacks associated with them, for example, toxicity of materials upon biodegradation and difficulties in scaling-up [20, 21]. SLNs do not carry the relatively high costs usually associated with large scale production, and they have less storage and drug leakage problems compared to systems such as liposomes [16, 22]. The significant toxicity and acidity associated with a number of biodegradable polymeric materials are also not observed with SLNs [23, 24]. Furthermore, the main advantage offered by lipid carriers in drug delivery is the use of physiological lipids or lipid molecules with a history of safe use in human medicine, decreasing the risk of acute and chronic toxicity [16, 25].

SLNs are defined in the scientific literature as submicron sized carriers (50–1000 nm) composed of physiological, biodegradable and biocompatible lipids that remain in a solid state at room and body temperatures [26]. SLNs can suitably be prepared using a wide variety of lipids, including fatty acids (e.g. stearic acid, palmitic acid), mono-, di- or triglycerides (e.g. tristearin, tripalmitin, trilaurin), glyceride mixtures (e.g. glyceryl behenate) or waxes (e.g. cetyl palmitate) [13, 16], and stabilized by biocompatible surfactants of choice (non-ionic or ionic). SLNs are compatible with many surfactants (e.g. poloxamer 188, polysorbate 80, lecithin, sodium glycocholate) that are approved by drug regulatory agencies. Generally, lipids that can be employed as a matrix for SLNs are highly purified triglycerides, complex glyceride mixtures or even waxes. However, recently, in SLNs based on mixtures of solid and liquid lipids, the so-called nanostructured lipid carriers (NLCs), high amounts of lecithins, amphiphilic cyclodextrins and para-acyl-calix(4)-arenes have been investigated [27–29]. NLCs overcome limitations of conventional SLNs, such as unexpected dynamics of polymorphic transitions and inherently low drug loading, due to the crystalline structure of the solid lipid [17, 30]. NLCs can be prepared by mixing spatially different lipids or liquid lipids (oils) with solid lipids. If the proper amount of oil is mixed with the solid lipid, a phase separation and the formation of oily nanocompartments within the solid lipid matrix can occur. This leads to the possibility of incorporating a high amount of drug inside these systems, because liquid lipids solubilize drugs to a much higher extent than solid lipids. Moreover, control of drug release can still be observed because solid lipids are present in the NLC matrix.

SLNs are suitable for the incorporation of both hydrophobic and hydrophilic drugs in considerable amounts within the lipid matrix. In addition to the natural ability of SLNs to efficiently incorporate hydrophobic compounds, the more recently developed variations of SLNs (e.g. polymer-lipid hybrid nanoparticles, lipid-drug conjugate nanoparticles) further expand the role of SLNs for the encapsulation of hydrophilic and ionic compounds [13, 16, 17].

There have been an increasing number of methods described in the literature for SLN preparation, including hot and cold high pressure homogenization [13, 16], via microemulsion [31] or phase inversion-based [32] techniques, by means of solvent emulsification-based approaches [33] and by membrane extruder techniques [34, 35].

From the methods listed above, those based on solvent emulsification are usually the most applied for producing SLNs containing biopharmaceuticals, for example, peptides, proteins, nucleic acids and other genetic materials, because they do not require the melting of the solid bulk lipid to produce a hot nanoemulsion. Instead, lipid materials are dissolved in a suitable organic solvent as described below.

The emulsification/evaporation method makes the encapsulation of biomacromolecules feasible, and it is based on the preparation of nanoparticle dispersions from o/w emulsions. The lipophilic material is initially dissolved in a water-immiscible organic solvent (e.g. chloroform, dichloromethane, cyclohexane [36, 37]) that is emulsified in an aqueous phase. Depending on the composition and the concentration of the lipid in the organic phase (i.e. a certain amount of polar organic solvents can be added to this phase [12]), very small particle sizes can be obtained (30–100 nm). A clear disadvantage of this method is the use of organic solvents; as the residual solvent in the particles may give a toxicity that cannot be ignored.

A solvent displacement method also based on the production of an o/w emulsion but using polar organic solvents, such as acetone, methanol or ethanol [28, 38, 39], can also be employed. In this case, the organic phase is injected into an aqueous surfactant solution under magnetic stirring. A violent spreading occurs because of the miscibility of both phases. Droplets of solvent of nanometer size are torn from the o/w interface and are rapidly stabilized by the surfactant molecules present in the aqueous phase until diffusion of the solvent is complete and lipid precipitation has occurred. Removal of solvent can be performed by distillation. Lipid nanoparticles are formed after total evaporation of the water-miscible organic solvent.

The emulsification/diffusion technique was developed using non-toxic and physiologically compatible solvents and monoglycerides or waxes as components of the inner phase of o/w emulsions [40]. According to the moderate water solubility of the solvents employed (e.g. ethyl acetate, benzyl alcohol), the dilution of the emulsions determines the diffusion of the organic solvent from the droplets to the continuous phase with the consequent instant solidification of lipophilic material.

A highly attractive alternative to produce solvent-free SLNs would be the supercritical fluid technology (SCF) [41]. This process is based on the use of supercritical carbon dioxide (CO_2) showing the advantages of not being inflammable, it is of low toxicity and cost, and sterilization can be employed. Atomizing supercritical CO_2 at low critical temperature and pressure can produce colloidal carriers as dried powders in small and large scale [42]. Within the SCF there are several methods, namely raid expansion of supercritical solution (RESS), supercritical anti-solvent (SAS), particles from gas saturated solutions (PGSS), gas/supercritical anti-solvent (GAS/SAS), the aerosol solvent extraction system (ASES), solution enhanced dispersion by supercritical fluids (SEDS) and the particles from compressed anti-solvent (PCA), which are applied according to the drug solubility in the SCF. Several proteins have been processed by SCF techniques, mainly the SAS and PGSS methods [42]. Using a modified PGSS method to prepare insulin-loaded SLNs, the peptide was first dissolved in dimethylsulfoxide and then added to melted mixtures of tristearin, phosphatidylcholine and dioctyl sulfosuccinate [3]. The lipid mass was mixed with compressed CO_2, followed by atomization. SLNs of a mean size below 500 nm were showing sustained release properties and preserving insulin biological activity. Coating protein crystals of a given particle size with solid lipids dissolved in supercritical CO_2 is another alternative to avoid the use of organic solvents. The process is carried out under mild conditions, thereby preserving the protein integrity. Bovine serum albumin (BSA) crystals have been coated with tripalmitin using this process [43, 44].

The incorporation of drugs in SLNs can be described by three models, namely the homogenous matrix model (in which drug is either molecularly dispersed or present as amorphous clusters in the lipid matrix), the drug-enriched shell model (an outer lipid shell containing drug with a lipid core) and the drug-enriched core model (a drug core surrounded by a lipid layer or a reservoir type system) [13, 16, 17]. The type of system obtained will obviously be dependent on the properties of the drug and on the matrix composition (in terms of lipids and surfactants), as well as on the method selected for their production. In many cases, a mixture of the three types of system is obtained that depends on the coefficient partition of the drug between the lipid and the aqueous phases; the drug molecules can also be at the surface of the system and, therefore, not physically protected by the matrix.

7.3 Definitions and Properties of Lipid-Drug Conjugates (LDCs)

Lipid-drug conjugates (LDCs) have emerged as an approach to improve delivery of hydrophilic drugs. LDCs can be described as a special form of nanoparticle consisting 100% of conjugates between the lipid and the drug, or a mixture of LDCs with suitable lipids. LDCs are obtained by converting a hydrophilic drug to a lipophilic one, or a prodrug, mainly via esterification or amidation [15]. Usually, LDCs are synthesized to have a melting point in the range of 50 to 100 °C, so that they can be processed by high pressure homogenization (HPH) to achieve sizes in the nanometre range (<1 μm). The production of nanoparticles by means of HPH is easy and fast. Firstly, the lipid phase is melted at approximately 5–10 °C above its melting point, then stirred in a hot surfactant solution to disperse this lipid phase into oil droplets stabilized by a surfactant film (i.e. pre-emulsion). Secondly, the obtained pre-emulsion is homogenized by applying a pressure between 200 and 500 bar. Thirdly, after homogenization a hot nanoemulsion is obtained and cooled, which leads to recrystallization of the lipid and formation of lipid nanoparticles. The HPH technique is also highly suitable for large scale production and is of relatively low cost [14].

LDCs have been developed to improve the delivery and targeting of hydrophilic NDEs. LDC technology has been applied to improve delivery of the hydrophilic anti-trypanosomial drug diminazene diaceturate (DZA) for the treatment of the chronic phase of human African trypanosomiasis, where the targeting of the drug to brain is essential. Olbrich *et al.* reported that DZA-loaded LDCs could successfully be targeted to the brain due to selective adsorption of apolipoproteins (ApoE, ApoA – I and ApoA – IV) onto the surface of these lipid carriers [45, 46]. These mediating agents are the key factors for the delivery and direct targeting of drugs to the brain. Furthermore, *in vivo* studies showed localization of LDCs in the endothelial cells of the blood vessels in the brain, as observed by confocal laser scanning microscopy [45]. The authors also observed that the *in vivo* performance and cytotoxicity potential of LDCs was strongly governed by the type of lipid used for the conjugation and the drug to lipid ratio. Much exploration is required in the field of LDCs to prove its success in parasitic diseases. Nevertheless, the fact that LDCs can themselves be treated as a NDE might be a limiting factor in their rapid commercialization. LDCs as delivery systems have proven to be highly efficient in therapy.

7.4 Administration, Delivery and Targeting

The scientific literature reports studies on lipid nanoparticles exploited for several applications, such as oral, nasal, pulmonary, ocular, rectal and parenteral routes.

With regard to oral delivery, to increase the bioavailability of drugs these should be attached to suitable colloidal carriers protecting the drug molecules against degradation. Furthermore, if the carrier is composed of lipids, an enhanced permeability of the mucosa will be observed. The GI uptake and mucosal translocation of drug-loaded lipid nanoparticles to the systemic circulation might occur via the Peyer's patches [47], which is the principal mechanism of lipid particulate material translocation across the intestine. Bargoni *et al.* showed that SLNs based on stearic acid and phosphatidylcholine were evidenced in lymph and blood after duodenal administration to rats [48]. The small diameters of SLNs may possibly facilitate their uptake by the lymphatic system. Other transport mechanisms might also occur, for example across intact epithelial membranes, possibly by endocytosis (transcellular route), or through tight junctions (paracellular route); however, they depend on the carrier properties, such as particle size, electrical charge, hydrophobicity, shape, elasticity and affinity to lectins, as well as their physicochemical stability *in situ* [49–51].

Lipid nanoparticles might also reach the blood stream through intranasal administration. It has been reported that lipid particles (>1 μm) containing protein antigens were able to reach the lungs and be taken up by lung macrophages, which contributed to an increased specific mucosal immune response [52]. The nasal-associated lymphoid tissue is capable of taking up and transporting particulate and soluble antigens.

A larger mucosal absorption area is obtained via administration through the pulmonary route, being easily reached by means of aerosolization. Radiolabelled SLNs administered by aerosolization or by endotracheal delivery were monitored by gamma-scintigraphy and Videira *et al.* found that pulmonary delivery involves phagocytosis by alveolar macrophages [53, 54]. Inhaled SLNs were translocated to regional lymph nodes, being therefore an effective way to achieve the lymphatics, in particular for vaccine delivery.

Other routes involving mucosal uptake include ocular and rectal administrations. Nevertheless, these have solely been tested for the delivery of tobramycin [55] and diazepam [56], respectively.

Colloidal carriers offer great promise to improve the therapeutic effectiveness and safety profile of cancer chemotherapy. Since the 1990s, a larger number of SLNs for the delivery of cytotoxic drugs have successfully been developed. Different sub-classes of cytotoxic agents and their derivatives have been encapsulated [57]. Furthermore, a number of compounds (verapamil [58] and cyclosporine A [59–63]), which possess activities to overcome the multidrug resistance phenotype in cells resistant to cytotoxic drugs, have been loaded into these systems. Cytotoxic anticancer drugs are a class of compound with highly diverse molecular structures and physicochemical properties [64]. Consequently, carrier materials, such as polymers, that can bind one cytotoxic drug may not be compatible with another drug. The versatility of SLNs is thus extremely important for the incorporation and delivery of these drugs [65].

The popularity and industrial acceptability of any novel drug delivery system is mainly governed by its versatility in terms of modifications (either surface or matrix) so as to widen its application [66]. Modifications may also be required to minimize or overcome demerits associated with the parent delivery system. Similar to other colloidal carriers, lipid nanoparticles are rapidly cleared by the monocyte-phagocyte system (MPS), leading to passive targeting to the liver and spleen. This would be highly useful for the effective treatment of leishmaniasis. On the contrary, modification in SLN structures might be required to treat trypanosomiasis, malaria or other infectious diseases. Additionally, modifications would also be required to address issues like poor encapsulation efficiency of amphiphilic peptides or proteins (e.g. glycoproteins) and also in the case of highly hydrophilic ones (e.g. insulin and calcitonin).

There are a considerable number of examples in the literature on applications of SLNs to modify the drug release profiles according to the required purposes and the therapeutic needs. The so-called stealth or long circulating SLNs have successfully been applied to various anticancer drugs, such as doxorubicin and paclitaxel, to improve their therapeutic efficacy and tolerability [67, 68]. SLNs based on amphiphilic cyclodextrins and para-acyl-calix-(4)-arene, which are macrocyclic host compounds based on methylene-bridged phenolic units with lipophilic substitution, have been produced to enhance the loading capacity of these carriers [28, 29]. Schubert *et al.* have produced SLNs based on mixtures of solid lipids and considerable amounts of phospholipids, which are highly useful for thc incorporation of amphiphilic molcculcs, such as amphotcricin B or mclarsoprol [27]. Moreover, such SLNs would exhibit certain characteristics of liposomes, with the further advantage of improved biological and physicochemical stability. SLNs composed of pH sensitive lipids, such as phosphotidyl ethanolamine, might also be promising for similar purposes. Cationic SLNs based on solid lipids and charged inducers, such as steryl amine, and the surfactants cetyl trimethylammonium bromide (CTAB) and dioleoyl trimethylammonium propane (DOTAP), have also been described in the literature, and they were shown to be tolerated by CPH54A cell lines [9]. Interestingly, cationic SLNs were found to exhibit similar transfection potential to those of commercialized transfection kits based on liposomes [69]. This observation led to successful development of TransoPlex, a transfection kit based on cationic SLNs [70].

Cationic colloidal carriers have been found to be very effective in eradicating leishmanial and trypanosomial infections [66]. Moreover, they were also effective in delivering antisense oligonucleotides for the eradication of leishmanial infections. Cationic SLNs might therefore have great potential for the treatment of parasitic diseases, but their applications in this field are still not explored.

The ability to target desired sites of action either by itself or by suitable modification is one of the desired characteristics of the colloidal carriers. As compared to others, there are, relatively, fewer reports on the targeting ability of lipid nanoparticles. Nevertheless, SLNs have shown potential in improving liver targeting [71] and brain delivery of various molecules, such as idarubacin and camptothecin [72, 73]. The brain targeting potential of SLNs can be very useful for the treatment of central nervous system infections, such as trypanosomiasis and even cerebral malaria. To achieve brain targeting and delivery no particular modifications in the lipid nanoparticle structure is required. On the contrary, this mechanism is likely to be mediated by the phenomenon of differential protein adsorption, as observed with LDCs. Müller *et al.* have shown that brain targeting of SLNs can be achieved when polysorbate 80 is used as a stabilizer [74]. As expected, the targeting is mediated by adsorption of Apo E. The authors have also shown that the protein adsorption on the SLNs is not transient and the adsorption pattern is retained for a considerably long time, which could be sufficient to achieve successful transport of the therapeutic moiety at the desired target [75].

7.5 Toxicity and Safety

The hemo- and cytocompatibility of colloidal carriers has raised a novel issue – so-called 'nanotoxicology' – which needs to be taken into account when developing nanoparticles for drug delivery. From the past studies, polymeric nanoparticles have exhibited cytotoxic effects in peritoneal resident and thioglycolate-elicited macrophages. Thus, a possible upregulation of proinflammatory cytokines must also be addressed in the development of lipid nanoparticles. In the case of SLNs, the cytotoxicity can be expected to be minimal or completely absent, mostly due to the better physiological acceptability when compared to polymeric nanoparticles. Nevertheless, not all SLNs can be considered innocuous. SLNs based on stearic acid and dimethyl-dioctadecylammonium bromide exhibited marked cytotoxic effect on the murine peritoneal macrophages, though they were found to be absolutely safe when tested *in vivo* in acute toxicity studies [45]. On the other hand, SLNs based on triglyceride, cetyl palmitate and paraffin did not induce significant macrophage toxicity or cytokine production [76–78].

The literature reveals that cytocompatibility of SLNs is dependent of the lipid matrix and surfactant used in their production [76, 77]. Koziara *et al.* showed that conventional and stealth SLNs based on cetyl alcohol, polysorbate 80 and emulsifying wax are totally safe to various blood components at clinically relevant concentrations [79]. Olbrich *et al.* showed that SLNs consisting of glycerol behenate as lipid, and polysorbate 80 and poloxamer 188 as surfactants, showed no binding to erythrocytes (SLN binding 10.0%) [80, 81]. In contrast, when sorbitan trioleate was used, blood cell affinity of labelled SLN was increased, leading to significant aggregation of red blood cells (75.3%). Furthermore, from the cytotoxicity point of view, it has been shown that trimyristin SLNs, when produced with pharmaceutically accepted surfactants for intravenous injection, such as poloxamer 188, soy lecithin, sodium cholate and polysorbate 80, could be well tolerated by RAW 264.7 macrophages [82]. Besides, other studies on SLNs have shown that they are not likely to induce any allergic reactions, hypersensitivity or cytokine production [81].

Liu *et al.* studied the cytotoxic effects on the A549 cells of insulin loaded and non-loaded SLNs composed of stearic acid, palmitic acid and soybean phosphatidylcholine as lipid matrix, stabilized by a mixture of sodium cholate and poloxamer 188 as surfactants

[83]. These A549 cells have widely been applied in various studies to mimic the structural and biochemical characteristics of human alveolar type II cells. The results obtained showed no cytotoxic effects of SLNs. Nevertheless, when using poloxamer 188 as surfactant, a very low cytotoxicity that was dependent on the applied concentration could be observed [83].

7.6 Applications for Biopharmaceuticals

7.6.1 Peptides and Proteins

Being composed of a lipid matrix, it is expected that most of the peptides and proteins (which are hydrophilic) will show poor association or incorporation efficiencies, as well as loading capacities. During the production and/or shelf life of the carriers, peptides and proteins will partition to the water phase of the dispersions, unless they are chemically linked to the lipid matrix (e.g. LDCs). Nevertheless, several peptides and proteins have successfully been incorporated within SLNs, for example, insulin [2, 36, 50, 51, 83–85], calcitonin [2, 49, 86, 87] and cyclosporine A [59–63].

It has recently been shown that salmon calcitonin was successfully incorporated (approximately 86%) within trimyristin SLNs (stabilized by poloxamer 407) for oral administration [87]. The pharmacological activity of the protein was evaluated in rats revealing that SLNs lowered the basal blood calcium levels by up to 20% with 500 IU/kg dose and sustaining hypocalcaemia over eight hours.

Prolonged *in vitro* release and sustained *in vivo* effects have been observed for various pharmaceutical proteins formulated in SLNs [49, 60, 62]. The mechanism depends on the physicochemical characteristics of the protein, particle size, lipid matrix composition, surfactants used, drug distribution throughout the matrix, method of preparation and production parameters [2]. Burst release might also be observed, in particular for hydrophilic peptides and proteins that might be localized in the o/w interface [51]. This can be useful to achieve a suitable blood concentration at the beginning of the treatment, which will be prolonged by means of a sustained release. An incomplete release has been pointed out as a result of adsorption of the release peptide or protein onto the particle surface [88].

The adsorption properties shown by peptides and proteins when in contact with solid–liquid interfaces can be exploited to load colloidal carriers with these drugs. Adsorption onto preformed lipid particles can be used to measure their interaction with blood components and to assess the *in vitro* macrophage uptake, as well as to improve the half-life of the drug in the blood [68, 75, 89]. Furthermore, this property can also be exploited for vaccine administration onto mucosal surfaces.

7.6.2 Nucleic Acids

Unlike other classes of therapeutic agent, genetic material (i.e. DNA, RNA and antisense oligonucleotides) can be used for either 'gain of function' actions, that is gene therapy, or 'loss of function' activity, much like conventional antagonist drugs [6, 90, 91]. Thus, gene therapy has emerged as a promising way to treat genetic diseases, virus infections and cancer, as well as for vaccination [92].

Gene therapy depends on large nucleic acids encoding a needed protein, while gene-specific inhibition can be achieved with short nucleic acid oligonucleotides, requiring a match only to a short, unique sequence in the gene, such as antisense. The introduction of siRNA into cells in culture has proven to be highly selective and potent, being considered as a novel biological research tool replacing the antisense and ribozyme methods in recent latest years [90, 93].

Nowadays, there are mainly two groups of gene transfer system, namely viral and non-viral vectors. Both are currently employed to deliver and target therapeutic genes into somatic cells. Viral vectors are biological systems derived from naturally evolved viruses capable of transferring genetic material into the host cells, being therefore more effective than the non-viral ones. Nevertheless, their application is limited due to their potential oncogenicity, pathogenicity (viral recombination and the risk of an immune or inflammatory response) and also because they are solely suitable to carry very small nucleic acids. On the other hand, non-viral vectors are safer, of low cost, more reproducible and can carry higher molecular weight nucleic acids, protecting them from components of the biological medium and, fundamentally, from nuclease digestion [10]. The main problem of non-viral systems is their low transfection efficiency. The literature reports that non-viral vectors are usually composed of cationic peptides [94], polymers [95, 96] or lipids [9, 97–99], and also combinations of these [100–103]. Being positively charged, they can interact with the negatively charged nucleic acid through electrostatic interactions. Cationic liposomes have been among the most efficient synthetic gene delivery reagents [97].

Olbrich *et al.* introduced SLNs as a non-viral transfection system [69]. From the application point of view, SLNs have good stability [104], which facilitates the industrial production and the manipulation for different processes such as lyophilisation.

To produce non-viral lipid vectors (so-called 'lipoplexes'), positively charged lipids are required to bind the negatively charged DNA molecules. DNA molecules solely bind lipids in a condensed form, thus condensation increases as the charge ratio (+/–) increases, which obviously increases the DNA/cationic lipid ratio [10, 105]. Condensation is required to facilitate the mobility of DNA molecules, which is limited by their large size. Furthermore, it has been reported that formulations are able to bind and condense DNA, providing positively charged lipoplexes protecting DNA from external agents. However, if condensation is excessive, the DNA release from the complexes is more difficult and transfection could be limited. In fact, the release of DNA from the complexes may be one of the most crucial steps determining the optimal ratio for lipoplexes-mediated transfection [106]. Since it influences the superficial charge of the lipoplexes, DNA condensation is a crucial factor determining the transfection ability of SLNs, that is DNA cell entry, its delivery from SLNs, its protection from DNAses and its topology (i.e. supercoiled, open circular or linear [107]). For gene therapy the positive charge of lipoplexes is advantageous because it facilitates the interaction with the negative charged cell surface and the cell entry. Elouahabi and Ruysschaert have postulated that this positive charge could also improve the entry in the cell, facilitating the invagination of the cell plasma membrane and inducing the early steps of the endocytosis process [108]. Clathrin-mediated endocytosis is the main entry process for the cationic lipid formulations [109]; nevertheless, fusion is also possible where part of the DNA can be released to the medium and not to the cytoplasm [110].

Since the initial studies, hundreds of cationic lipids have been synthesized as candidates for non-viral gene delivery [111] and a few were used for clinical trials [112]. Cationic lipids are needed for the preparation of SLNs used in gene therapy for an additional reason, which is their surfactant activity. The surfactant activity is required to produce an initial emulsion, and the positive charge to provide the superficial charge to the SLNs. The positive superficial charge of the SLNs is needed for their further interaction with negatively charged DNA. However, cationic lipids can be toxic when used repeatedly, and can induce inflammatory reactions *in vivo* [111].

Tabatt *et al.* found that a SLN and liposome composed of the same cationic lipids showed similar *in vitro* transfection efficiencies [97]. An advantage of SLN over liposome formulations is that the former can be produced on a large scale and under favourable technology parameters, becoming a competitor of the well-established repertoire of non-viral transfection agents lead by cationic liposomes. The positively charged SLNs would bind to polyanionic DNA via electrostatic force, leading to a SLN-DNA lipoplex that will protect DNA from interaction with small molecules in the environment [69] and will be taken into the cell by endocytosis [98].

Calvo *et al.* reported that submicron SLNs could be taken up more efficiently into cell by endocytosis than microparticles [113]. Therefore, both size and zeta potential values are important physicochemical particle properties, as they determine the physical stability as well as the biopharmaceutical properties of the preparation [114].

In vitro studies on human liver cancer cells demonstrated a very low degree of toxicity of both SLNs and SLN–DNA lipoplexes. Moreover, these lipoplexes were able to promote transfection of liver cancer cells [6]. Lipoplexes composed by glycerol palmitostearate (Precirol®ATO 5), DOTAP and polysorbate 80 were shown not decrease cell viability, depicting good stability properties after storage for a month at 4 °C [10]. There must be some equilibrium between the gene protection degree, the binding forces of DNA to SLNs, and the DNA topology. This equilibrium is determined by cationic lipid/DNA ratio and it must be optimized with every new formulation.

DOTAP/cetyl palmitate-based SLN-DNA lipoplexes showed high tolerance and good transfection efficiency even in the presence of serum [98]. These lipoplexes demonstrated excellent storage stability even after steam sterilization, being thereafter a promising platform for further development of non-viral transfection agents and *in vivo* gene medicines.

Gene therapy is also becoming a promising strategy for cancer therapy. It has been proposed for use alone or in combination with cytotoxic drug treatment. Exciting progress has been obtained in areas such as reconstituting p53 expression in p53-mutant tumour cells, protecting hematopoietic cells from cytotoxic drugs, repressing HER-2/neu expression in breast cancer cells and targeting the neovasculature formed in cancerous tissues [115]. Pre-clinical evidence and early clinical trials strongly suggest also the possibility of gene therapy for liver cancer management. Bondi *et al.* prepared cationic SLNs based on glycerol behenate and dimethyl-dioctadecyl ammonium bromide (DDAB), with the aim to obtain lipoplexes useful as plasmid transfection vector for gene delivery capable of treating hepatocellular carcinoma [6]. The obtained cationic SLNs were able to bind efficiently to DNA to achieve a shift in its electrophoretic mobility. DNA is transfected in HuH-6 cells, and this good transfection efficiency was associated with very low degree of cytotoxicity.

7.6.3 Vaccines

When delivered by parenteral or mucosal routes, the majority of antigens are ineffective due to proteolytic degradation. Antigen incorporation or their adsorption onto colloidal carriers is an established procedure to improve vaccination strategies [2].

It has been reported that liposomes can act as useful immunological adjuvants, inducing both cellular and humoral immunity for a variety of bacterial and viral antigens relevant to human diseases [116–118]. Protein antigens may be covalently linked to triglycerides using a controlled lipidation. The resulting antigens have the ability to self assemble in water, mimicking viral particles that enhance both humoral and cellular immune responses at systemic and mucosal levels.

The application of solid lipid nanoparticles for antigen delivery is still an area poorly explored. Saraf *et al.* produced anionic and cationic lipid particles (1.6 μm) by the double emulsion solvent evaporation technique, to deliver the protein antigens (HBsAg) for intranasal immunization against hepatitis B [52]. After intranasal administration to rats the particles were taken up from the mucosa by the alveolar macrophages, eliciting higher specific mucosal antibody responses when compared with the free and the alum-adsorbed antigen. This response was also higher than that observed upon intramuscular vaccination with the same formulations. The specific anti-HBsAg sIgA levels induced intranasally by the cationic particles were the highest in the nasal, pulmonary and salivary glands, as determined in nasal washes, in bronchoalveolar lavages and in saliva. On the other hand, the intramuscularly administered alum-adsorbed antigen resulted in a superior IgG level in serum.

7.7 Conclusions

Commercial feasibility is the most desired characteristic of any novel delivery system to achieve real clinical applications. A system that has great clinical efficacy but poor commercial feasibility is simply not useful. Obviously, such feasibility is governed by the cost of the material, the ease of manufacturing and the possibility for scaling up the production lines. Considering these facts, in the case of lipid nanoparticles the scenario is promising. As SLNs are based on the easily available triglycerides and similar lipids, the cost of the materials would not be an issue in most of the cases, unlike synthetic or semi-synthetic polymers, or even phospholipids.

Several advantageous features can be pointed out for lipid nanoparticles, such as their small size and versatility to carry very chemically different drug molecules (e.g. peptides, proteins, nucleic acids and antigens), as well as biocompatibility, biodegradability and usefulness for several administration routes. The most recent achievements of such particles to carry the above-mentioned biomacromolecules, in particular for parenteral and mucosal administration, have been reviewed in this chapter.

References

1. Devalapally, H., Chakilam, A. and Amiji, M.M. (2007) Role of nanotechnology in pharmaceutical product development. *J. Pharm. Sci.*, **96**, 2547–2565.

2. Almeida, A.J. and Souto, E. (2007) Solid lipid nanoparticles as a drug delivery system for peptides and proteins. *Adv. Drug Deliv. Rev.*, **59**, 478–490.
3. Caliceti, P., Brossa, A., Salmaso, S. *et al.* (2006) Preparation of protein loaded solid lipid nano-particles by compressed fluid process. *Proc. Intern. Symp. Control. Rel. Bioact. Mater.*, **33**, 383.
4. Asasutjarit, R., Lorenzen, S.I., Sirivichayakul, S. *et al.* (2007) Effect of solid lipid nanoparticles formulation compositions on their size, zeta potential and potential for *in vitro* pHIS-HIV-hugag transfection. *Pharm. Res.*, **24**, 1098–1107.
5. Brioschi, A., Zenga, F., Zara, G.P. *et al.* (2007) Solid lipid nanoparticles: could they help to improve the efficacy of pharmacologic treatments for brain tumors? *Neurol. Res.*, **29**, 324–330.
6. Bondi, M.L., Azzolina, A., Craparo, E.F. *et al.* (2007) Novel cationic solid-lipid nanoparticles as non-viral vectors for gene delivery. *J. Drug Target*, **15**, 295–301.
7. Choi, M.J., Kim, J.H. and Maibach, H.I. (2006) Topical DNA vaccination with DNA/Lipid based complex. *Curr. Drug Deliv.*, **3**, 37–45.
8. Choi, S.H., Jin, S.E., Lee, M.K. *et al.* (2008) Novel cationic solid lipid nanoparticles enhanced p53 gene transfer to lung cancer cells. *Eur. J. Pharm. Biopharm.*, **68**, 545–554.
9. Pedersen, N., Hansen, S., Heydenreich, A.V. *et al.* (2006) Solid lipid nanoparticles can effectively bind DNA, streptavidin and biotinylated ligands. *Eur. J. Pharm. Biopharm.*, **62**, 155–162.
10. del Pozo-Rodriguez, A., Delgado, D., Solinis, M.A. *et al.* (2007) Solid lipid nanoparticles: formulation factors affecting cell transfection capacity. *Int. J. Pharm.*, **339**, 261–268.
11. Montana, G., Bondi, M.L., Carrotta, R. *et al.* (2007) Employment of cationic solid-lipid nanoparticles as RNA carriers. *Bioconjug. Chem.*, **18**, 302–308.
12. Kim, H.R., Kim, I.K., Bae, K.H. *et al.* (2008) Cationic solid lipid nanoparticles reconstituted from low density lipoprotein components for delivery of siRNA. *Mol. Pharm.*, **5** (4), 622–631.
13. Mehnert, W. and Mader, K. (2001) Solid lipid nanoparticles: production, characterization and applications. *Adv. Drug Deliv. Rev.*, **47**, 165–196.
14. Muller, R.H. and Keck, C.M. (2004) Drug delivery to the brain – realization by novel drug carriers. *J. Nanosci. Nanotechnol.*, **4**, 471–483.
15. Muller, R.H. and Keck, C.M. (2004) Challenges and solutions for the delivery of biotech drugs – a review of drug nanocrystal technology and lipid nanoparticles. *J. Biotechnol.*, **113**, 151–170.
16. Muller, R.H., Mader, K. and Gohla, S. (2000) Solid lipid nanoparticles (SLN) for controlled drug delivery – a review of the state of the art. *Eur. J. Pharm. Biopharm.*, **50**, 161–177.
17. Souto, E.B. and Müller, R.H. (2007) Lipid nanoparticles (SLN and NLC) for drug delivery, in *Nanoparticles for Pharmaceutical Applications*, (eds A.J. Domb, Y. Tabata, M.N.V. Ravi Kumar and S. Farber), American Scientific Publishers, Los Angeles, California, pp. 103–122.
18. Uner, M. (2006) Preparation, characterization and physico-chemical properties of solid lipid nanoparticles (SLN) and nanostructured lipid carriers (NLC): their benefits as colloidal drug carrier systems. *Pharmazie*, **61**, 375–386.
19. Wissing, S.A., Kayser, O. and Muller, R.H. (2004) Solid lipid nanoparticles for parenteral drug delivery. *Adv. Drug Deliv. Rev.*, **56**, 1257–1272.
20. Smith, A. and Hunneyball, I.M. (1986) Evaluation of polylactid as a biodegradabledrug delivery system for parenteral administration. *Int. J. Pharm.*, **30**, 215–230.
21. Lasic, D.D. (1998) Novel applications of liposomes. *Tibetech*, **16**, 307–321.
22. Dingler, A. and Gohla, S. (2002) Production of solid lipid nanoparticles (SLN): scaling up feasibilities. *J. Microencapsul.*, **19**, 11–16.
23. Muller, R.H., Ruhl, D., Runge, S. *et al.* (1997) Cytotoxicity of solid lipid nanoparticles as a function of the lipid matrix and the surfactant. *Pharm. Res.*, **14**, 458–462.
24. Muller, R.H., Maassen, S., Weyhers, H. and Mehnert, W. (1996) Phagocytic uptake and cytotoxicity of solid lipid nanoparticles (SLN) sterically stabilized with poloxamine 908 and poloxamer 407. *J. Drug Target*, **4**, 161–170.

25. Müller, R.H. and Lucks, J.S. (1996) Arzneistoffträger aus festen Lipidteilchen – Feste Lipid Nanosphären (SLN). European Patent 0605497: Germany.
26. Souto, E.B., Petersen, R.D. and Müller, R.H. (2008) Lipid nanoparticles with solid matrix for dermal delivery, in *Modified Release Drug Delivery Technology*, (eds M.J. Rathbone, J. Hadgraft, M.S. Roberts and M. Lane), Marcel Dekker Inc., New York, Basel, Hong Kong, pp. 357–379.
27. Schubert, M.A. and Muller-Goymann, C.C. (2005) Characterisation of surface-modified solid lipid nanoparticles (SLN): influence of lecithin and nonionic emulsifier. *Eur. J. Pharm. Biopharm.*, **61**, 77–86.
28. Dubes, A., Parrot-Lopez, H., Abdelwahed, W. *et al.* (2003) Scanning electron microscopy and atomic force microscopy imaging of solid lipid nanoparticles derived from amphiphilic cyclodextrins. *Eur. J. Pharm. Biopharm.*, **55**, 279–282.
29. Shahgaldian, P., Da Silva, E., Coleman, A.W. *et al.* (2003) Para-acyl-calix-arene based solid lipid nanoparticles (SLNs): a detailed study of preparation and stability parameters. *Int. J. Pharm.*, **253**, 23–38.
30. Müller, R.H., Mäder, K., Lippacher, A. and Jenning, V. (1998) Fest-flüssig (halbfeste) Lipidpartikel und Verfahren zur Herstellung hochkonzentrierter Lipidpartikeldispersionen. In PCT application PCT/EP00/04565.
31. Gasco, M.R. (1993) Method for producing solid lipid microspheres having a narrow size distribution. US Patent 5 250 236: Italy.
32. Heurtault, B., Saulnier, P., Pech, B. *et al.* (2002) A novel phase inversion-based process for the preparation of lipid nanocarriers. *Pharm. Res.*, **19**, 875–880.
33. Sjostrom, B.B.B. (1992) Preparation of submicron drug particles in lecithin-stabilized o/w emulsions I. Model studies of the precipitation of cholesteryl acetate. *Int. J. Pharm.*, **88**, 53–62.
34. Charcosset, C., El-Harati, A. and Fessi, H. (2005) Preparation of solid lipid nanoparticles using a membrane contactor. *J. Control. Release*, **108**, 112–120.
35. Charcosset, C. and Fessi, H. (2005) A new process for drug loaded nanocapsules preparation using a membrane contactor. *Drug Dev. Ind. Pharm.*, **31**, 987–992.
36. Sarmento, B., Martins, S., Ferreira, D. and Souto, E.B. (2007) Oral insulin delivery by means of solid lipid nanoparticles. *Int. J. Nanomed.*, **2**, 743–749.
37. Siekmann, B. and Westesen, K. (1996) Investigations on solid lipid nanoparticles prepared by precipitation in o/w emulsions. *Eur. J. Pharm. Biopharm.*, **43**, 104–109.
38. Huang, G., Zhang, N., Bi, X. and Dou, M. (2008) Solid lipid nanoparticles of temozolomide: potential reduction of cardial and nephric toxicity. *Int. J. Pharm.*, **355**, 314–320.
39. Liu, J., Zhu, J., Du, Z. and Qin, B. (2005) Preparation and pharmacokinetic evaluation of Tashinone IIA solid lipid nanoparticles. *Drug Dev. Ind. Pharm.*, **31**, 551–556.
40. Trotta, M., Debernardi, F. and Caputo, O. (2003) Preparation of solid lipid nanoparticles by a solvent emulsification-diffusion technique. *Int. J. Pharm.*, **257**, 153–160.
41. Chattopadhyay, P., Shekunov, B.Y., Yim, D. *et al.* (2007) Production of solid lipid nanoparticle suspensions using supercritical fluid extraction of emulsions (SFEE) for pulmonary delivery using the AERx system. *Adv. Drug Deliv. Rev.*, **59**, 444–453.
42. Jovanovic, N., Bouchard, A., Hofland, G.W. *et al.* (2004) Stabilization of proteins in dry powder formulations using supercritical fluid technology. *Pharm. Res.*, **21**, 1955–1969.
43. Dos Santos, R., Richard, I., Pech, J. *et al.* (2002) Microencapsulation of protein particles within lipids using a novel supercritical fluid process. *Int. J. Pharm.*, **242**, 69–78.
44. Dos Santos, R., Richard, I., Thies, J. *et al.* (2003) A supercritical fluid-based coating technology. 3: preparation and characterization of bovine serum albumin particles coated with lipids. *J. Microencapsul.*, **20**, 110–128.
45. Olbrich, C., Gessner, A., Kayser, O. and Muller, R.H. (2002) Lipid-drug-conjugate (LDC) nanoparticles as novel carrier system for the hydrophilic antitrypanosomal drug diminazenediaceturate. *J. Drug Target*, **10**, 387–396.
46. Olbrich, C., Gessner, A., Schroder, W. *et al.* (2004) Lipid-drug conjugate nanoparticles of the hydrophilic drug diminazene-cytotoxicity testing and mouse serum adsorption. *J. Control. Release*, **96**, 425–435.

47. Florence, A.T., Hillery, A.M., Hussain, N. and Jani, P.U. (1995) Factors affecting the oral uptake and translocation of polystyrene nanoparticles: histological and analytical evidence. *J. Drug Target*, **3**, 65–70.
48. Bargoni, A., Cavalli, R., Caputo, O. *et al.* (1998) Solid lipid nanoparticles in lymph and plasma after duodenal administration to rats. *Pharm. Res.*, **15**, 745–750.
49. Garcia-Fuentes, M., Prego, C., Torres, D. and Alonso, M.J. (2005) A comparative study of the potential of solid triglyceride nanostructures coated with chitosan or poly(ethylene glycol) as carriers for oral calcitonin delivery. *Eur. J. Pharm. Sci.*, **25**, 133–143.
50. Zhang, N., Ping, Q., Huang, G. *et al.* (2006) Transport characteristics of wheat germ agglutinin-modified insulin-liposomes and solid lipid nanoparticles in a perfused rat intestinal model. *J. Nanosci. Nanotechnol.*, **6**, 2959–2966.
51. Zhang, N., Ping, Q., Huang, G. *et al.* (2006) Lectin-modified solid lipid nanoparticles as carriers for oral administration of insulin. *Int. J. Pharm.*, **327**, 153–159.
52. Saraf, S., Mishra, D., Asthana, A. *et al.* (2006) Lipid microparticles for mucosal immunization against hepatitis B. *Vaccine*, **24**, 45–56.
53. Videira, M.A., Botelho, M.F., Santos, A.C. *et al.* (2002) Lymphatic uptake of pulmonary delivered radiolabelled solid lipid nanoparticles. *J. Drug Target*, **10**, 607–613.
54. Videira, M.A., Gano, L., Santos, C. *et al.* (2006) Lymphatic uptake of lipid nanoparticles following endotracheal administration. *J. Microencapsul.*, **23**, 855–862.
55. Cavalli, R., Gasco, M.R., Chetoni, P. *et al.* (2002) Solid lipid nanoparticles (SLN) as ocular delivery system for tobramycin. *Int. J. Pharm.*, **238**, 241–245.
56. Sznitowska, M., Gajewska, M., Janicki, S. *et al.* (2001) Bioavailability of diazepam from aqueous-organic solution, submicron emulsion and solid lipid nanoparticles after rectal administration in rabbits. *Eur. J. Pharm. Biopharm.*, **52**, 159–163.
57. Wong, H.L., Bendayan, R., Rauth, A.M. *et al.* (2007) Chemotherapy with anticancer drugs encapsulated in solid lipid nanoparticles. *Adv. Drug Deliv. Rev.*, **59**, 491–504.
58. Wong, H.L., Bendayan, R., Rauth, A.M. and Wu, X.Y. (2004) Development of solid lipid nanoparticles containing ionically complexed chemotherapeutic drugs and chemosensitizers. *J. Pharm. Sci.*, **93**, 1993–2008.
59. Muller, R.H., Runge, S., Ravelli, V. *et al.* (2006) Oral bioavailability of cyclosporine: solid lipid nanoparticles (SLN) versus drug nanocrystals. *Int. J. Pharm.*, **317**, 82–89.
60. Muller, R.H., Runge, S.A., Ravelli, V. *et al.* (2008) Cyclosporine-loaded solid lipid nanoparticles (SLN): drug-lipid physicochemical interactions and characterization of drug incorporation. *Eur. J. Pharm. Biopharm.*, **68**, 535–544.
61. Varia, J.K., Dodiya, S.S. and Sawant, K.K. (2008) Cyclosporine a loaded solid lipid nanoparticles: optimization of formulation, process variable and characterization. *Curr. Drug Deliv.*, **5**, 64–69.
62. Zhang, Q., Yie, G., Li, Y. *et al.* (2000) Studies on the cyclosporin A loaded stearic acid nanoparticles. *Int. J. Pharm.*, **200**, 153–159.
63. Ugazio, E., Cavalli, R. and Gasco, M.R. (2002) Incorporation of cyclosporin A in solid lipid nanoparticles (SLN). *Int. J. Pharm.*, **241**, 341–344.
64. Ewesuedo, R.B. and Ratain, M.J. (2003) Principles of cancer chemotherapy, in *Oncologic Therapeutics* (eds E.E. Vokes and HMG), Springer, New York, pp. 19–66.
65. Wong, H.L., Bendayan, R., Rauth, A.M. *et al.* (2007) Chemotherapy with anticancer drugs encapsulated in solid lipid nanoparticles. *Adv. Drug Deliv. Rev.*, **59**, 491–504.
66. Date, A.A., Joshi, M.D. and Patravale, V.B. (2007) Parasitic diseases: liposomes and polymeric nanoparticles versus lipid nanoparticles. *Adv. Drug Deliv. Rev.*, **59**, 505–521.
67. Fundaro, A., Cavalli, R., Bargoni, A. *et al.* (2000) Non-stealth and stealth solid lipid nanoparticles (SLN) carrying doxorubicin: pharmacokinetics and tissue distribution after i.v. administration to rats. *Pharmacol. Res.*, **42**, 337–343.
68. Cavalli, R., Bocca, C., Miglietta, A. *et al.* (1999) Albumin adsorption on stealth and non-stealth solid lipid nanoparticles. *STP Pharma. Sci.*, **9**, 183–189.
69. Olbrich, C., Bakowsky, U., Lehr, C.M. *et al.* (2001) Cationic solid-lipid nanoparticles can efficiently bind and transfect plasmid DNA. *J. Control. Release*, **77**, 345–355.

70. Müller, R.H. and Tabatt, K. (2002) ransoPlex-novel transfection agents based on solid lipid nanoparticles (SLN). In Annual Meeting of AAPS, Toronto, Canada, pp. 10–14.
71. Heiati, H., Tawashi, R. and Phillips, N.C. (1998) Solid lipid nanoparticles as drug carriers: II. Plasma stability and biodistribution of solid lipid nanoparticles containing the lipophilic prodrug 3′-azido-3′-deoxythymidine palmitate in mice. *Int. J. Pharm.*, **174**, 71–80.
72. Yang, S., Zhu, J., Lu, Y. *et al.* (1999) Body distribution of camptothecin solid lipid nanoparticles after oral administration. *Pharm. Res.*, **16**, 751–757.
73. Zara, G.P., Bargoni, A., Cavalli, R. *et al.* (2002) Pharmacokinetics and tissue distribution of idarubicin-loaded solid lipid nanoparticles after duodenal administration to rats. *J. Pharm. Sci.*, **91**, 1324–1333.
74. Goppert, T.M. and Muller, R.H. (2005) Polysorbate-stabilized solid lipid nanoparticles as colloidal carriers for intravenous targeting of drugs to the brain: comparison of plasma protein adsorption patterns. *J. Drug Target*, **13**, 179–187.
75. Goppert, T.M. and Muller, R.H. (2005) Adsorption kinetics of plasma proteins on solid lipid nanoparticles for drug targeting. *Int. J. Pharm.*, **302**, 172–186.
76. Scholer, N., Hahn, H., Muller, R.H. and Liesenfeld, O. (2002) Effect of lipid matrix and size of solid lipid nanoparticles (SLN) on the viability and cytokine production of macrophages. *Int. J. Pharm.*, **231**, 167–176.
77. Scholer, N., Olbrich, C., Tabatt, K. *et al.* (2001) Surfactant, but not the size of solid lipid nanoparticles (SLN) influences viability and cytokine production of macrophages. *Int. J. Pharm.*, **221**, 57–67.
78. Scholer, N., Zimmermann, E., Katzfey, U. *et al.* (2000) Preserved solid lipid nanoparticles (SLN) at low concentrations do cause neither direct nor indirect cytotoxic effects in peritoneal macrophages. *Int. J. Pharm.*, **196**, 235–239.
79. Koziara, J., Oh, J., Akers, W. *et al.* (2005) Blood compatibility of cetyl alcohol/polysorbate-based nanoparticles. *Pharm. Res.*, **22**, 1821–1828.
80. Olbrich, C. and Muller, R.H. (1999) Enzymatic degradation of SLN-effect of surfactant and surfactant mixtures. *Int. J. Pharm.*, **180**, 31–39.
81. Wissing, S.A., Kayser, O. and Muller, R.H. (2004) Solid lipid nanoparticles for parenteral drug delivery. *Adv. Drug Deliv. Rev.*, **56**, 1257–1272.
82. Olbrich, C., Scholer, N., Tabatt, K. *et al.* (2004) Cytotoxicity studies of Dynasan 114 solid lipid nanoparticles (SLN) on RAW 264.7 macrophages-impact of phagocytosis on viability and cytokine production. *J. Pharm. Pharmacol.*, **56**, 883–891.
83. Liu, J., Gong, T., Fu, H. *et al.* (2008) Solid lipid nanoparticles for pulmonary delivery of insulin. *Int. J. Pharm.*, **356**, 333–344.
84. Battaglia, L., Trotta, M., Gallarate, M. *et al.* (2007) Solid lipid nanoparticles formed by solvent-in-water emulsion-diffusion technique: development and influence on insulin stability. *J. Microencapsul.*, **24**, 660–672.
85. Liu, J., Gong, T., Wang, C. *et al.* (2007) Solid lipid nanoparticles loaded with insulin by sodium cholate-phosphatidylcholine-based mixed micelles: preparation and characterization. *Int. J. Pharm.*, **340**, 153–162.
86. Garcia-Fuentes, M., Torres, D. and Alonso, M.J. (2005) New surface-modified lipid nanoparticles as delivery vehicles for salmon calcitonin. *Int. J. Pharm.*, **296**, 122–132.
87. Martins, S., Silva, A.C., Ferreira, D. and Souto, E.B. (2009) Improving oral absorption of Samon calcitonin by trimyristin lipid nanoparticles. *J. Biomed. Nanotechnol.*, **5**, 76–83.
88. Garcia-Fuentes, M., Alonso, M.J. and Torres, D. (2005) Design and characterization of a new drug nanocarrier made from solid-liquid lipid mixtures. *J. Colloid Interface Sci.*, **285**, 590–598.
89. Goppert, T.M. and Muller, R.H. (2005) Protein adsorption patterns on poloxamer- and poloxamine-stabilized solid lipid nanoparticles (SLN). *Eur. J. Pharm. Biopharm.*, **60**, 361–372.
90. Scanlon, K.J. (2004) Anti-genes: siRNA, ribozymes and antisense. *Curr. Pharm. Biotechnol.*, **5**, 415–420.
91. Li, B.J., Tang, Q., Cheng, D. *et al.* (2005) Using siRNA in prophylactic and therapeutic regimens against SARS coronavirus in Rhesus macaque. *Nat. Med.*, **11**, 944–951.

92. Alonso, M.J. (2004) Nanomedicines for overcoming biological barriers. *Biomed. Pharmacother.*, **58**, 168–172.
93. Matzke, M.A. and Birchler, J.A. (2005) RNAi-mediated pathways in the nucleus. *Nat. Rev. Genet*, **6**, 24–35.
94. Li, W., Nicol, F. and Szoka, F.C. Jr. (2004) GALA: a designed synthetic pH-responsive amphipathic peptide with applications in drug and gene delivery. *Adv. Drug Deliv. Rev.*, **56**, 967–985.
95. Cavallaro, G., Campisi, M., Licciardi, M. *et al.* (2006) Reversibly stable thiopolyplexes for intracellular delivery of genes. *J. Control. Release*, **115**, 322–334.
96. Licciardi, M., Campisi, M., Cavallaro, G. *et al.* (2006) Synthesis and characterization of polyaminoacidic polycations for gene delivery. *Biomaterials*, **27**, 2066–2075.
97. Tabatt, K., Kneuer, C., Sameti, M. *et al.* (2004) Transfection with different colloidal systems: comparison of solid lipid nanoparticles and liposomes. *J. Control. Release*, **97**, 321–332.
98. Tabatt, K., Sameti, M., Olbrich, C. *et al.* (2004) Effect of cationic lipid and matrix lipid composition on solid lipid nanoparticle-mediated gene transfer. *Eur. J. Pharm. Biopharm.*, **57**, 155–162.
99. Rudolph, C., Schillinger, U., Ortiz, A. *et al.* (2004) Application of novel solid lipid nanoparticle (SLN)-gene vector formulations based on a dimeric HIV-1 TAT-peptide *in vitro* and *in vivo*. *Pharm. Res.*, **21**, 1662–1669.
100. Tokunaga, M., Hazemoto, N. and Yotsuyanagi, T. (2004) Effect of oligopeptides on gene expression: comparison of DNA/peptide and DNA/peptide/liposome complexes. *Int. J. Pharm.*, **269**, 71–80.
101. Hyndman, L., Lemoine, J.L., Huang, L. *et al.* (2004) HIV-1 Tat protein transduction domain peptide facilitates gene transfer in combination with cationic liposomes. *J. Control. Release*, **99**, 435–444.
102. El-Aneed, A. (2004) Current strategies in cancer gene therapy. *Eur. J. Pharmacol.*, **498**, 1–8.
103. El-Aneed, A. (2004) An overview of current delivery systems in cancer gene therapy. *J. Control. Release*, **94**, 1–14.
104. Freitas, C. and Muller, R.H. (1999) Correlation between long-term stability of solid lipid nanoparticles (SLN) and crystallinity of the lipid phase. *Eur. J. Pharm. Biopharm.*, **47**, 125–132.
105. Faneca, H., Simoes, S. and de Lima, M.C. (2002) Evaluation of lipid-based reagents to mediate intracellular gene delivery. *Biochim. Biophys. Acta*, **1567**, 23–33.
106. Sakurai, F., Inoue, R., Nishino, Y. *et al.* (2000) Effect of DNA/liposome mixing ratio on the physicochemical characteristics, cellular uptake and intracellular trafficking of plasmid DNA/ cationic liposome complexes and subsequent gene expression. *J. Control. Release*, **66**, 255–269.
107. Remaut, K., Sanders, N.N., Fayazpour, F. *et al.* (2006) Influence of plasmid DNA topology on the transfection properties of DOTAP/DOPE lipoplexes. *J. Control. Release*, **115**, 335–343.
108. Elouahabi, A. and Ruysschaert, J.M. (2005) Formation and intracellular trafficking of lipoplexes and polyplexes. *Mol. Ther.*, **11**, 336–347.
109. Rejman, J., Bragonzi, A. and Conese, M. (2005) Role of clathrin- and caveolae-mediated endocytosis in gene transfer mediated by lipo- and polyplexes. *Mol. Ther.*, **12**, 468–474.
110. Elouahabi, A., Thiry, M., Pector, V. *et al.* (2003) Calorimetry of cationic liposome-DNA complex and intracellular visualization of the complexes. *Methods Enzymol.*, **373**, 313–332.
111. Han, S., Mahato, R.I., Sung, Y.K. and Kim, S.W. (2000) Development of biomaterials for gene therapy. *Mol. Ther.*, **2**, 302–317.
112. Edelstein, M.L., Abedi, M.R. and Wixon, J. (2007) Gene therapy clinical trials worldwide to 2007 – an update. *J. Gene Med.*, **9**, 833–842.
113. Calvo, P., Alonso, M.J., Vila-Jato, J.L. and Robinson, J.R. (1996) Improved ocular bioavailability of indomethacin by novel ocular drug carriers. *J. Pharm. Pharmacol.*, **48**, 1147–1152.
114. Weyenberg, W., Filev, P., Van den Plas, D. *et al.* (2007) Cytotoxicity of submicron emulsions and solid lipid nanoparticles for dermal application. *Int. J. Pharm.*, **337**, 291–298.

115. Parker, L.P., Wolf, J.K. and Price, J.E. (2000) Adenoviral-mediated gene therapy with Ad5CMVp53 and Ad5CMVp21 in combination with standard therapies in human breast cancer cell lines. *Ann. Clin. Lab. Sci.*, **30**, 395–405.
116. Agger, E.M., Cassidy, J.P., Brady, J. *et al.* (2008) Adjuvant modulation of the cytokine balance in Mycobacterium tuberculosis subunit vaccines; immunity, pathology and protection. *Immunology*, **124**, 175–185.
117. Bergers, J.J., ten Hagen, T.L., van Etten, E.W. and Bakker-Woudenberg, I.A. (1995) Liposomes as delivery systems in the prevention and treatment of infectious diseases. *Pharm. World. Sci.*, **17**, 1–11.
118. Gregoriadis, G. (1999) DNA vaccines: a role for liposomes. *Curr. Opin. Mol. Ther.*, **1**, 39–42.

8

Dendrimers in Delivery of Biopharmaceuticals

Chandan Thomas and Fakhrul Ahsan
Texas Tech University Health Sciences Center, Department of Pharmaceutical Sciences, School of Pharmacy, Amarillo, TX, USA

8.1 Introduction

The polymer industry has achieved tremendous advances since the Nobel Laureate Hermann Staudinger first explicated the molecular basis of high molecular weight natural and synthetic polymeric materials in the 1920s. In a seminal paper, Staudinger showed that macromolecules can be produced by joining a large number of monomers to each other via covalent bonding. In fact, the modern era of structured and controlled macromolecular design evolved from the principles that Staudinger laid out nearly ninety years ago [1]. Polymer-based drug delivery provides an array of possibilities by which drug solubility and release, and thus pharmacokinetics and pharmacodynamics, can be controlled. Furthermore, polymer-based drug delivery is a very active area of research in today's pharmaceutical industry because of the fact that many polymers are reported to be biochemically inert and non-toxic. Such polymers are often referred to as 'biocompatible' or 'biodegradable' polymers.

Dendrimers, which are three-dimensional hyperbranched macromolecules, have been found to have characteristics of an ideal drug delivery vehicle, and therefore they have been widely investigated as potential carriers of drugs, genes and vaccines [2–4]. Two separate groups of investigators carried out pioneering work on dendrimers in the early 1980s. Synthesis of a dendrimer was first reported by Tomalia and his coworkers. During

Delivery Technologies for Biopharmaceuticals: Peptides, Proteins, Nucleic Acids and Vaccines Edited by Lene Jorgensen and Hanne Mørck Nielsen

the same time frame, Newkome and colleagues published a report on the synthesis of another type of dendrimer. It is interesting to point out that although the two groups gave different names to their dendrimeric structures, the meaning of two terms is the same. Donald Tomalia called the new family of molecules 'dendrimers', which comes from the Greek word 'dendron', meaning tree; whereas Newkome and his group coined the term 'arborol' from the Latin word 'arbor', which also means tree [5–7].

Typically a dendrimer consists of a multifunctional core, branched units and surface groups. The branched units that exist as layers are called a 'generation', and these are the repeating monomeric units of dendrimers [8]. In addition to being hyperbranched, they are monodipersed, three-dimensional molecules that have well defined molecular weights as well as host–guest entrapment properties. Tomalia's landmark work on a novel three-dimensional polymer, poly(amido amine) (PAMAM) dendrimer, in the mid 1980s spurred interest in the design and synthesis of large and precisely defined dendrimeric molecules. It is now possible to control not only the shape and size but also the dimension, density, polarity and solubility of dendrimers; and this can be achieved either by attaching specialized functional groups to the dendrimer surface or by building branching units with functional groups that can modify the physicochemical properties of the macromolecules [6, 9–11]. PAMAM dendrimers were the first complete family of dendrimers to be synthesized, characterized and made commercially available.

Whereas the synthesis of traditional macromolecules results in polydispersed structures with varying molecular weights, the process of synthesis of dendrimers produces monodispersed and structure-controlled macromolecules similar to those that exist in biological systems. Novel macromolecular structures such as dendrimers are synthesized by the 'divergent' method as reported initially by Tomalia [6], which involves a two-step iterative reaction sequence that produces concentric shells (generations) of dendritic β-alanine units around a central initiator core. Another method of dendrimer synthesis is called the 'convergent' method, which also is based on the principle of building layer after layer; however, in this synthetic approach, addition of the new layers progresses inwardly, meaning that growth begins at the terminal groups and ends at the core. The details of dendrimer synthesis have been reviewed extensively elsewhere [9] and will not be covered in this chapter.

8.1.1 Properties of Dendrimers as a Carrier System

Dendrimers are an ideal carrier system for drug delivery because they offer flexibility to modify chemistry, size, molecular weight, surface characteristics and internal structure, which modern drug delivery and targeting strategies demand. Furthermore, some dendrimers have been shown to be relatively safe and to have low immunogenicity [12]. Depending on the application in drug delivery, it is possible to synthesize highly predictable and ordered dendrimeric structures of generation 0 to 5 or more (Figure 8.1). When relatively large nanocarriers are to be used in drug delivery, a higher generation dendrimer may be the preferred option. The diameter of PAMAM dendrimer with an ethylenediamine core, for example, increases from 1.1 to 12.4 nm [13–15] when the generation grows from 1 to 10. The lower generation dendrimers (0–3) show an ellipsoidal shape whereas the higher generation ones [4–10] show roughly spherical shapes [15].

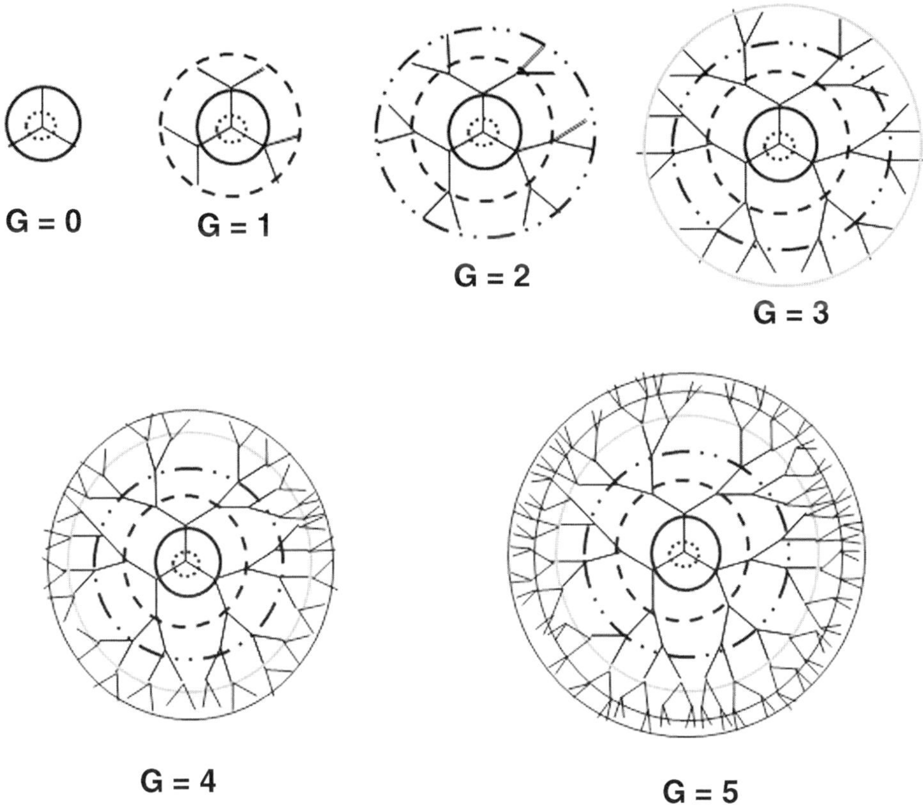

Figure 8.1 *Configuration of dendrimers with different generations (Reprinted from Bai, S., Thomas, C., Rawat, A., and Ahsan, F. (2006), Recent progress in dendrimer-based nanocarriers,* Critical Reviews in Therapeutic Drug Carrier Systems, ***23****, 437–495, with permission from Begell House, Inc.)*

The ability of dendrimers to interact with a variety of drug molecules has attracted a great deal of interest in dendrimer-based drug delivery. Indeed, the presence of a number of highly reactive functional groups like amide (NH_2), carboxyl (COOH) and hydroxyl (OH) in the dendrimers makes them highly reactive. A wide range of drug and other molecules can be stored or encapsulated in the intrinsic cavities of dendrimers. These cavities are hydrophobic in nature and are, therefore, suitable for interacting with drugs that are poorly water soluble. Furthermore, the presence of reactive groups such as nitrogen and oxygen in the internal cavities also facilitates interactions with various drug molecules by formation of hydrogen bonds [12, 16, 17]. The addition of appropriate capping groups on the dendrimer surface can be used to modify the charge and solubility of dendrimers or to enhance their affinity for specific cellular targets [18, 19]. The presence of functional groups with a high density of surface charge promotes both electrostatic

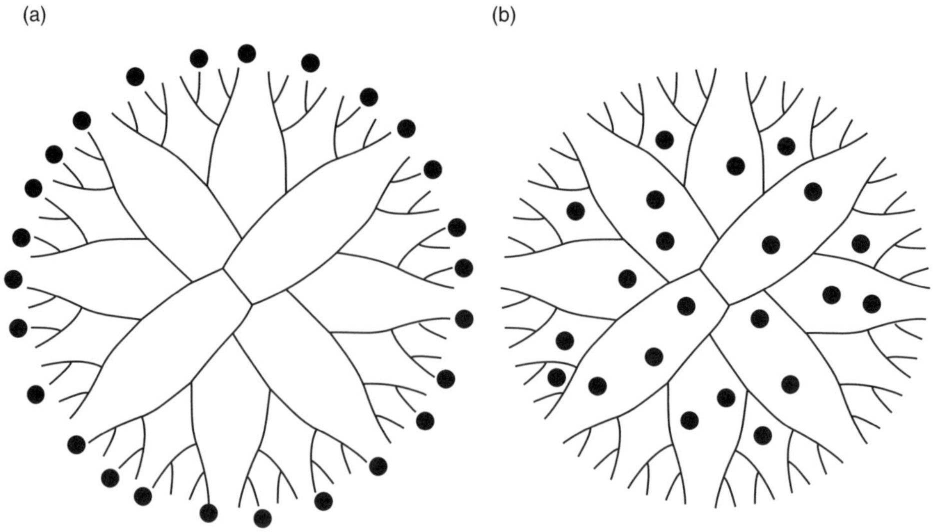

Figure 8.2 *Potential interactions between dendrimers and drug molecules: (a) electrostatic or covalent conjugation, (b) encapsulation (Reprinted from Cheng, Y., Xu, Z., Ma, M., and Xu, T. (2008) Dendrimers as drug carriers: applications in different routes of drug administration,* Journal of Pharmaceutical Sciences, ***97****, 123–143, with permission from Wiley-Liss, Inc., a subsidiary of John Wiley and Sons, Inc.)*

interactions and the ability of dendrimers to conjugate drug molecules via covalent bonding. Potential interactions between dendrimers and drug molecules are illustrated in Figure 8.2 [8, 20, 21].

Furthermore, recently developed surface-modified dendrimers are reported to address some of the problems associated with conventional dendrimers, including toxicity and poor solubility. Modification of surface amino groups with either neutral or anionic moieties reduces the toxicity and propensity to accumulate in the liver that usually occurs with dendrimers containing polycationic surfaces [12]. Pegylation of dendrimers can also address the biocompatibility and solubility issues of dendrimers. Ihre and coworkers [22] reported the design and synthesis of a polyester dendritic scaffold with a aliphatic polyester backbone based on 2,2-bis(hydroxymethyl) propanoic acid. This delivery system was water soluble and reduced the drug's toxicity during its delivery to the target tissue. Such a property would be important in cases of tumour targeting, for example. Furthermore, poly(ethylene glycol) (PEG) moieties conjugated to the dendrimer surface provide a hydrophilic shell around the hydrophobic core of the dendrimer, thus giving rise to a unimolecular micelle. Conjugation with PEG can also improve the pharmacokinetic properties of the attached drug by increasing its circulation time in the blood. In fact, pegylated dendrimer has been used to increase the circulation time of drugs with short half-lives because of the ability of pegylated carriers to avoid recognition by cells of reticuloendothelial system (RES) [9, 23].

8.2 Case I – Application of Dendrimers in Delivery of Large Molecular Weight Drugs

Dendrimers have been used as a carrier for delivery of a large molecular weight non-peptide drug, low molecular weight heparin [24]. Low molecular weight heparins (LMWHs) are negatively charged oligosaccharides used in the treatment of deep vein thrombosis (DVT) and pulmonary embolism. The use of LMWH in an outpatient basis, however, has been limited because of the requirement of daily subcutaneous injections. Attempts have been made to deliver LMWH via non-invasive routes, including the nasal and pulmonary routes [25–27]. However, the presence of carboxylic acid and sulfate groups in the glycosaminoglycan units of the LMWH renders the molecule highly anionic and an unlikely candidate for absorption via the mucosa, including the nasal and pulmonary routes.

Based on published reports on the complexation between DNA and dendrimer, we hypothesized that negatively charged LMWH can form a complex with positively charged dendrimer to facilitate absorption of the drug across biological membranes. Generation 2 (G2), 2.5 (G2.5) and 3 (G3) PAMAM dendrimers have been used in these studies. G2 and G3 are full-generation dendrimers whereas G2.5 is a half-generation dendrimer. *In vitro* characterization studies were performed to assess the differences in the interaction/complexation between LMWHs and dendrimers of different generations. Enoxaparin, a commercially available LMWH, was used for all the experiments in this study. Fourier Transformed Infrared Spectroscopy (FTIR) studies suggest that an electrostatic interaction occurs between cationic amino groups of the dendrimer and anionic groups such as sulfate or carboxylic acid of the LMWH. Similar interactions also occurred between DNA and dendrimers [28]. Further evidence of interactions between dendrimer and LMWH was observed in the azure A assay, which is a colorimetric assay based on the principle that a positively charged phenothiazine dye binds with the negatively charged sulfate groups in a heparin molecule. This method has previously been used to study the interaction between neutrophil elastase and heparin [29, 30]. The results of both the FTIR analysis and azure assay demonstrated that an electrostatic interaction occurs between LMWH and dendrimer.

The ability of PAMAM dendrimers to enhance pulmonary absorption of negatively charged LMWH was evaluated in a rodent model. LMWH formulated with only saline failed to show an increase in plasma anti-factor Xa level required for an anti-thrombotic effect in rats. A plasma anti-factor Xa level of 0.2 U/ml or higher is considered to produce a therapeutic effect in rodent models. Similarly, when LMWH was formulated with 0.25% or 0.5% G2 PAMAM dendrimer, no appreciable increase in plasma anti-factor Xa level was observed compared to plain LMWH. However, when G2 dendrimer was used at a concentration of 1%, a significant increase in plasma anti-factor Xa level was observed compared to the control group (Figure 8.3a). Furthermore, when the concentration of G2 dendrimer was increased to 2%, an additional increase in plasma anti-factor Xa level was observed. In the case of G3 dendrimers, plasma anti-factor Xa levels were above therapeutic levels only when the concentration of dendrimer was 0.5%. Interestingly, when G3 dendrimer was used at concentrations of 0.25%, 1%, and 2%, the anti-factor Xa levels were sub-therapeutic (Figure 8.3b). Half-generation (G2.5) dendrimer failed to show an increase in anti-factor Xa levels at any of the concentrations employed in the

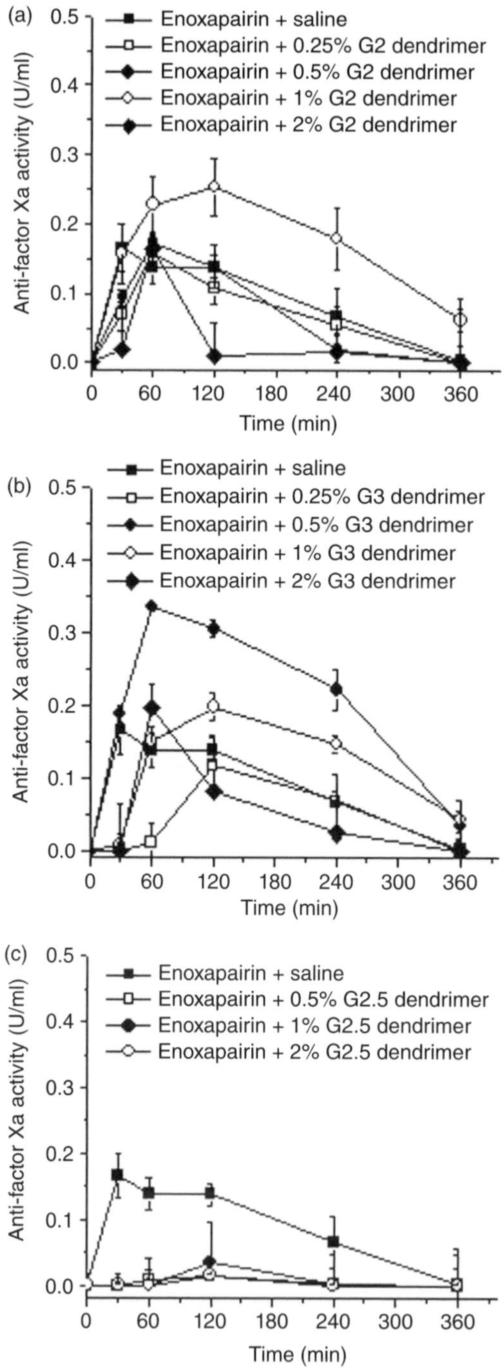

Figure 8.3 *Changes in anti-factor Xa activity after pulmonary administration of enoxaparin formulated in saline or in the presence of (a) G2 PAMAM dendrimer, (b) G3 PAMAM dendrimer and (c) G2.5 PAMAM dendrimer (Reprinted from Bai, S., Thomas, C., and Ahsan, F. (2007) Dendrimers as a carrier for pulmonary delivery of enoxaparin, a low molecular weight heparin,* Journal of Pharmaceutical Sciences, ***96****, 2090–2106, with permission from Wiley-Liss, Inc., a subsidiary of John Wiley and Sons, Inc.)*

study (Figure 8.3c). The relative bioavailability of a 1% G2 or 0.5% G3 dendrimer-based formulation was twofold higher compared to the saline control. However, other formulations used in the studies failed to show a statistically significant change in the relative bioavailability.

It is proposed that the likely reason for the G2 and G3 dendrimers to have shown increased pulmonary absorption was due to the positive surface charge of the dendrimers, which formed a complex with the negatively charged LMWH. The complexation led to either reduction in the net negative charge of LMWH or made the complex electrostatically neutral. It is argued that the reduction in negative surface charge aids in the absorption of the negatively charged LMWH to a greater extent compared to LMWH formulation containing G2.5 dendrimer.

The efficacy of PAMAM dendrimers and LMWH formulations in the treatment of DVT in a rat jugular vein thrombosis model was also tested. Two formulations that showed the highest bioavailability (1% G2 PAMAM dendrimer plus LMWH and 0.5% G3 PAMAM dendrimer plus LMWH) were tested and thrombus weight was used as an indicator of anti-thrombotic effect. The efficacy of the formulations was compared against a saline control and plain LMWH administered by both the pulmonary and subcutaneous routes. A significant reduction in thrombus weight was observed when LMWH plus 1% G2 or 0.5% G3 dendrimer was administered via the pulmonary route. The thrombus weights were 1.3 ± 0.2 or 0.8 ± 0.5 mg, respectively, versus 5.3 ± 0.7 mg in rats that received saline. When LMWH plus saline was administered by the subcutaneous route, the reduction in thrombus weights observed was statistically significant compared to the pulmonary formulations of LMWH plus dendrimer. The *in vivo* data suggest that the dendrimer-based LMWH formulation was effective in preventing DVT in a rodent model.

Evaluated, too, was the effect of the formulations on the mucociliary transport rate (MTR) – the primary defence mechanism of the human body against inhaled pathogens – in a frog palate model. Also, biochemical changes that may occur in the lungs following administration of the formulations were studied by measuring injury markers in broncho-alveolar lavage fluid. The data show that the formulations did not adversely affect the MTR or produce extensive damage to the lungs. It is proposed that dendrimers could potentially be used as carriers for pulmonary delivery of large molecular weight drugs.

8.3 Case II – Application of Dendrimers in Gene Delivery

The major goal of gene therapy is to correct genetic disorders by either incorporating a therapeutic gene or silencing a defective gene. Progress towards gene delivery over the past decade has been disappointing. The major limitations to gene therapy are safety and efficacy of the carrier used to deliver the gene. Of the two currently studied carriers for gene delivery, non-viral vectors are preferred over viral vectors because of their reduced toxicity, non-immunogenicity and ease of preparation. In fact, there has been a decline in the use of viral vectors since the approval in China of adenoviral based therapy for direct injection into tumours of the neck and head [31]. Non-viral vectors, including liposomes, nanoparticles and microparticles, have been extensively studied in recent years because of complications associated with viral vector-based delivery. Very recently, dendrimers have been used as non-viral vectors for gene delivery.

8.3.1 Case IIA – Application of PAMAM-PLGA Microparticle Conjugates in Gene Delivery

PAMAM dendrimers have been successfully used to enhance the transfection efficiency of poly(D,L-lactide-*co*-glycolide) (PLGA) microparticles. PAMAM dendrimers of different generations were conjugated to the surface of PLGA microparticles to increase the transfection efficiency of PLGA-based particles [32]. PLGA microparticles were prepared by the water-oil-water (w/o/w) double emulsion method and loaded with VR1255 plasmid DNA (pDNA), which is a 6.4 kb cDNA encoding firefly luciferase driven by the cytomegalovirus (CMV) promoter/enhancer. Cationic microspheres were prepared by conjugating activated PLGA microspheres with PAMAM dendrimers. PLGA particles were activated by a modified 1-ethyl-3-[3-dimethylaminopropyl]carbodiimide-*N*-hydroxysuccinimide (EDC/NHS) chemistry and then combined with different generations of PAMAM dendrimer (G3, G4, G5 and G6). The microspheres thus prepared were denoted as PG3, PG4, PG5, and PG6. The amount of dendrimer conjugated to PLGA microparticles was determined by a colorimetric method [33]. Each milligram of particles contained 5.0, 5.5, 4.6 and 5.4 μg of G3, G4, G5 and G6 dendrimer, respectively. pDNA was loaded onto the surface of the cationic microparticles according to a published procedure [34]. pDNA-loaded PLGA microparticles or PAMAM conjugated-PLGA microparticles were of uniform size with a smooth surface.

The zeta potential of PAMAM-modified particles was between +20 and +40 mV and it was reduced to +0.5 to +13.5 mV upon binding with pDNA. The authors argued that the reduction in zeta potential was because of the binding of negatively charged pDNA with positively charged PAMAM-conjugated particles. However, it is important to note that the zeta potential of blank PLGA microparticles was around −50 mV. As a result of conjugation of PAMAM dendrimer to PLGA particles the zeta potential became positive, which aided the binding of the pDNA. In fact, the amount of pDNA loaded in PAMAM-modified microspheres was fivefold greater than that in unmodified PLGA particles ($P < 0.001$). Of the particles tested, PG6 microparticles showed the highest pDNA loading compared to dendrimers of other generations ($P < 0.05$).

The ability of PAMAM-PLGA microparticles to resist acidification was tested by an acid titration assay as reported by Tang *et al.* [33]. In fact, the ability to buffer against pH changes, known as the proton sponge hypothesis, is essential for endophagosomal escape into the cytoplasm [33, 35]. The buffering capacity of the particles increased with increasing generations of dendrimers, and the presence of secondary and tertiary amine structures are believed to confer buffering capacity to PAMAM dendrimers. The cytotoxicity of the conjugates was evaluated by (3-(4,5-Dimethylthiazol-2-yl)-2,5-diphenyltetrazolium bromide) assay (MTT assay) in two types of cell lines: the African green monkey kidney fibroblast-like cell line (COS7) and human embryonic kidney (HEK293) cells. Dose-dependent toxicity was also evaluated. Although the toxicity of PAMAM dendrimers has been reported to vary with generation, PAMAM-PLGA microparticles showed no generation-dependent toxicity. Indeed, the data in this study suggest that the differences in toxicity between different generations of dendrimers were not statistically significant. Cytotoxicity produced by the particles varied depending on the cell type used. The viability of COS7 cells exposed to PAMAM-PLGA microparticles was similar to that of cells exposed to unmodified PLGA microparticles. However, in the

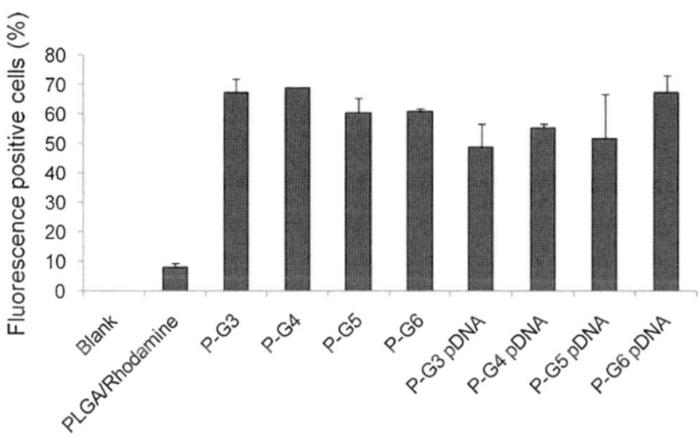

Figure 8.4 *Flow cytometry data of fluorescence-labelled PLGA microparticles that have been incubated with HEK293 cells. The cellular uptake of the PLGA formulations is presented as a percentage of fluorescence-labelled cells (*n *= 3 per group). PLGA microparticle formulations were added at doses of 0.5 mg/well in HEK293 cells. The concentrations of PLGA particles were chosen from an estimated pDNA loading of 5 µg/mg particles and a target pDNA dose of 2.5 µg/well (Reprinted from Zhang, X.Q., Intra, J., Salem, A.K. (2007) Conjugation of polyamidoamine dendrimers on biodegradable microparticles for nonviral gene delivery,* Bioconjugate Chemistry, ***18****, 2068–2076, with permission from American Chemical Society)*

case of HEK293 cells, PAMAM-PLGA microparticles showed only a moderately higher toxicity compared to the unmodified microparticles. Based on the cell-type-dependent toxicity data, the authors argued that PAMAM-PLGA particles could possibly be used for targeting drugs to certain cell types with little or no toxicity.

The cellular uptake of PLGA-pDNA was compared with that of PAMAM-PLGA-pDNA microparticles by conjugating fluorescein isothiocyanate (FITC)-labelled dendrimers with the microparticles. FITC-labelled microparticles loaded with and without pDNA were incubated in HEK293 cells for 16 hours. Microparticle uptake and internalization were evaluated using flow cytometry and confocal microscopy. Orthogonal sections from stacked images confirmed that the microparticles were more likely located inside the cells rather than on the surface. The uptake (Figure 8.4) of PAMAM-PLGA-pDNA microparticles was six to sevenfold higher compared to the uptake of unmodified microparticles ($P < 0.001$). The size of the microparticles did not play a major role in cellular uptake. However, compared to unmodified PLGA particles, an increase in the uptake was observed with PAMAM-PLGA particles and this increase was attributed to the positive surface charge of the particles. Interestingly, the uptake of modified particles was independent of the generation of dendrimers.

Expression of luciferase in COS7 and HEK293 cells was also evaluated. PAMAM-PLGA microparticles showed a significantly higher transgene expression level compared to the unmodified microparticles. Similar to the cellular uptake results, the transfection efficiencies induced by the microparticles were generation-independent. PG3 pDNA microparticles showed the highest luciferase expression in both cell lines. Compared to unmodified microparticles, the transfection efficiency produced by PG3 particles was 11-fold and 7-fold greater in COS7 and HEK293 cells, respectively. Similar to earlier studies, although there was an increase in zeta potential and buffering capacity with increase in generation, no generation-dependent increase in transfection efficiency was observed. However, the authors argued that because G3 dendrimers are easy to synthesize, less expensive and have reduced toxicity once released from the microparticles, this particular generation of dendrimer (G3) would be advantageous over other dendrimer generations. Overall, the increase in transfection efficiency observed was mainly due to the increased pDNA loading and enhanced ability of modified dendrimers to interact with the cell surface. The authors concluded that because delivery of pDNA entrapped in PLGA microparticles has been effective in animal models and phase II clinical trials [36–39], such cationic surface-modified PLGA polymers hold immense promise as gene delivery vehicles.

8.3.2 Case IIB – Application of Dendrimers in siRNA Delivery

The use of short double-stranded RNAs (siRNA) has recently gained in popularity as a means to specifically downregulate expression of target genes. This approach of regulating the expression of target genes is termed RNA interference (RNAi). In fact, RNAi has emerged as a widely accepted and important tool for the identification and silencing of target genes. Dendrimers, especially PAMAM dendrimers, have been used successfully for siRNA delivery [40, 41]. In addition to PAMAM dendrimers, monodispersed dendritic poly(L-lysine)s (DPKs) have also shown promise as a gene delivery carrier both *in vitro* and *in vivo* [42–46].

Ohsaki and coworkers [43] showed that KG6, a generation 6 DPK with 128 surface amino groups, can be used for transfection of genes into a number of cultivated cell lines without significant toxicity. In a separate study, Inoue *et al.* [47] used KG6 in combination with a peptide to knockdown target genes that are thought to be involved in the pharmacological action of metformin, a drug commonly used in the treatment of hyperglycemia associated with type 2 diabetes. However, the precise mechanism of action of metformin is not entirely understood. The drug is believed to lower blood sugar levels via serine-threonine kinase 11 (STK11/LKB1), an upstream AMPK kinase. As this oral anti-diabetic is a cationic organic compound, organic cationic transporter 1 (OCT1) has been reported to be involved in the facilitated transport of this drug. OCT1 is also known to play a role in the hepatic uptake of metformin [48–52]. The authors used KG6 as a carrier for siRNA in order to identify the role of three important genes – GAPDH (glyceraldehyde-3-phosphate dehydrogenase), PEPCK (phosphoenolpyruvate carboxykinase) and OCT1 – involved in gluconeogenesis in rat hepatoma H4IIEC3 cells. Two types of transfection, forward and reverse, were used in this study. For forward transfection, the cells were plated the day before the transfection and transfection reagents with siRNA were added to the wells the following day. In reverse transfection, the complexes were first prepared by mixing the reagents with the siRNA inside the wells and then the cells were added. Forward transfection used Opti-MEM (Opti-MEM® I Reduced-SerumMedium, Invitrogen) as a diluent for the transfection reagents. The siRNA delivery efficiency was assessed by using fluorescence microscopy, flow cytometry and GAPDH gene-silencing experiments. Opti-MEM was used as the diluent in the reverse transfection experiment, and transfection efficiency was estimated by PEPCK and OCT1 gene knockdown.

Fluorescence microscopy revealed that KG6 was efficacious in delivering the fluorescein-labelled oligonucleotide into the cells, although a significant amount of the fluorescence was localized in the endosomal compartment. It was hypothesized that the complexes were unable to be released from the endosomal compartment after endocytic uptake. To enhance the release of the complex, the authors combined KG6 with Endo Porter (KG6/EP), a weakly basic amphiphilic peptide, which has been to shown to promote the release of oligonucleotide complexes from the endosomal compartment [53]. The KG6/EP combination led to widespread release of fluorescence in the cytosol, similar to that seen with the commercially available transfection reagent Lipofectamine 2000 (LF2000).

Flow cytometry was performed to compare the transfection efficiency of the KG6/EP complex with that of a number of commercial transfection reagents, including HiPerFect, DoFect-GT1 and LF2000, as well as EP and KG6 alone. The uptake efficiency of each of the reagents was quantified from fluorescence-activated cell sorting (FACS). KG6 showed an uptake of 90% whereas EP showed no uptake. KG6 showed a bell shaped histogram, whereas DoFect-GT1 and LF2000 showed a broad distribution pattern without any peaks. It was hypothesized that the high transfection efficiency of KG6 was likely because dendrimers are monodiperse symmetrical macromolecules. The failure of EP to enhance the uptake efficiency of KG6 suggests that EP's role is limited to the release of KG6 from the endosome into the cytosol.

The ability of the KG6/EP complex in carrying out siRNA-mediated gene knockdown was investigated by using the GADPH gene as a model target. The efficiency of EP, LF2000, KG6 and KG6/EP in silencing GADPH was also studied and compared. GADPH was efficiently knocked down at the protein expression level when either KG6/EP or LF2000 was used. Similar results were obtained in the dose-dependency studies at the mRNA expression level. No change in expression level was observed when KG6 or EP was used alone. LF2000 was slightly superior to KG6/EP at low concentrations of siRNA, as indicated by real time quantitative Reverse Transcription Polymerase Chain Reaction (qRT-PCR) results. However, LF2000 caused detachment of cells from the surface of the wells. The 3-(4,5-dimethylthiazol-2-yl)-5-(3-carboxymethoxyphenyl)-2-(4-sulfophenyl)-2H-tetrazolium inner salt (MTS) cytotoxicity assay showed that neither EP nor KG6 was cytotoxic, although LF2000 was significantly cytotoxic. KG6/EP did not affect cell viability at low siRNA concentrations, although a slight decrease in cell viability was observed at high siRNA concentration.

As mentioned above, KG6 based delivery of siRNA was used to inhibit gluconeogenesis in H4IIEC3 cells. In a preliminary study, the authors showed that glucose production decreased with increasing doses of metformin (Figure 8.5) and it reached a plateau at 1 mM metformin. The authors then used phosphoenolpyruvate carboxykinase (PEPCK), a well-known marker for gluconeogenesis, in the knockdown experiments to analyse RNAi-mediated loss of function in H4IIEC3 cells. siRNA-induced changes in glucose production were used as a marker for functional validation of PEPCK siRNA. A reduction in PEPCK expression was observed with increasing siRNA concentration. In agreement with the reduction in PEPCK expression, a reduction in glucose production was also observed. The data suggest that siRNA-mediated knockdown of PEPCK led to a reduction in gluconeogenesis in H4IIEC3 cells.

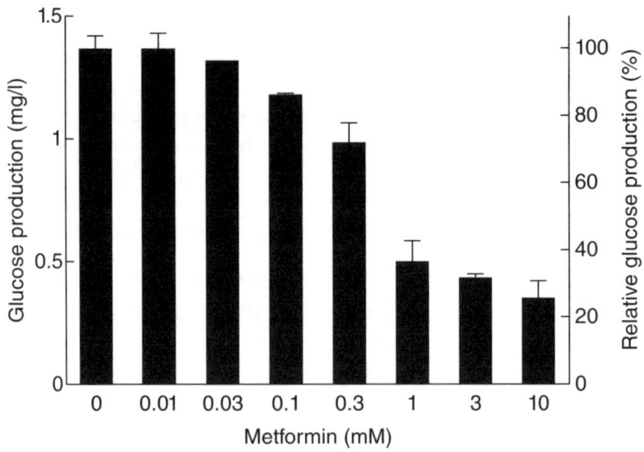

Figure 8.5 *The effect of metformin on the inhibition of gluconeogenesis. Starved H4IIEC3 cells were treated with various concentrations of metformin in glucose-free medium. The medium was collected after 3 h and glucose output was determined (Reprinted from Inoue, Y., Kurihara, R., Tsuchida, A.,* et al. *(2008) Efficient delivery of siRNA using dendritic poly(L-lysine) for loss-of-function analysis,* Journal of Controlled Release, ***126****, 59–66, with permission from Elsevier)*

The authors also investigated the expression of OCT1 and OCT2 in H4IIEC3 cells by measuring mRNA expression levels using qRT-PCR. OCT1 was found to be expressed in H4IIEC3 cells although OCT2 showed no expression (Figure 8.6a). Once expression of OCT1 in rat hepatoma H4IIEC3 cells was confirmed, OCT1 was used as a target gene to investigate the siRNA-mediated knockdown of OCT1. The authors used KG6/EP to transfect siRNA into H4IIEC3 cells both in the presence and absence of metformin. Upon transfection, expression of OCT1 mRNA and glucose production were measured. The qRT-PCR data showed that the expression of OCT1 mRNA declined in

Figure 8.6 *Knockdown of OCT1 attenuates the ability of metformin to inhibit gluconeogenesis. (a) mRNA expression levels of OCT1 and OCT2 in H4IIEC3 cells. Relative mRNA expression levels of OCT1 and OCT2 were determined by qRT-PCR. Data represent the mean ± S.D. (*n *= 2), N.D.: not detected. H4IIEC3 cells were transfected with different concentrations of OCT1 siRNA (10, 30 and 100 nM) using KG6/EP. (b) The relative expression levels of OCT1 mRNA were determined using qRT-PCR. Data represent the mean ± S.D. (*n *= 2). (c) Measurement of glucose production. Metformin (1 mM) was added 3 h before glucose measurement. Data represent the mean ± S.D. (*n *= 3). (d) Differences in glucose production between metformin treated and untreated cells. The data are expressed as metformin action compared to the value in mock-transfected cells (Reprinted from Inoue, Y., Kurihara, R., Tsuchida, A.,* et al. *(2008) Efficient delivery of siRNA using dendritic poly(L-lysine) for loss-of-function analysis,* Journal of Controlled Release, ***126****, 59–66, with permission from Elsevier)*

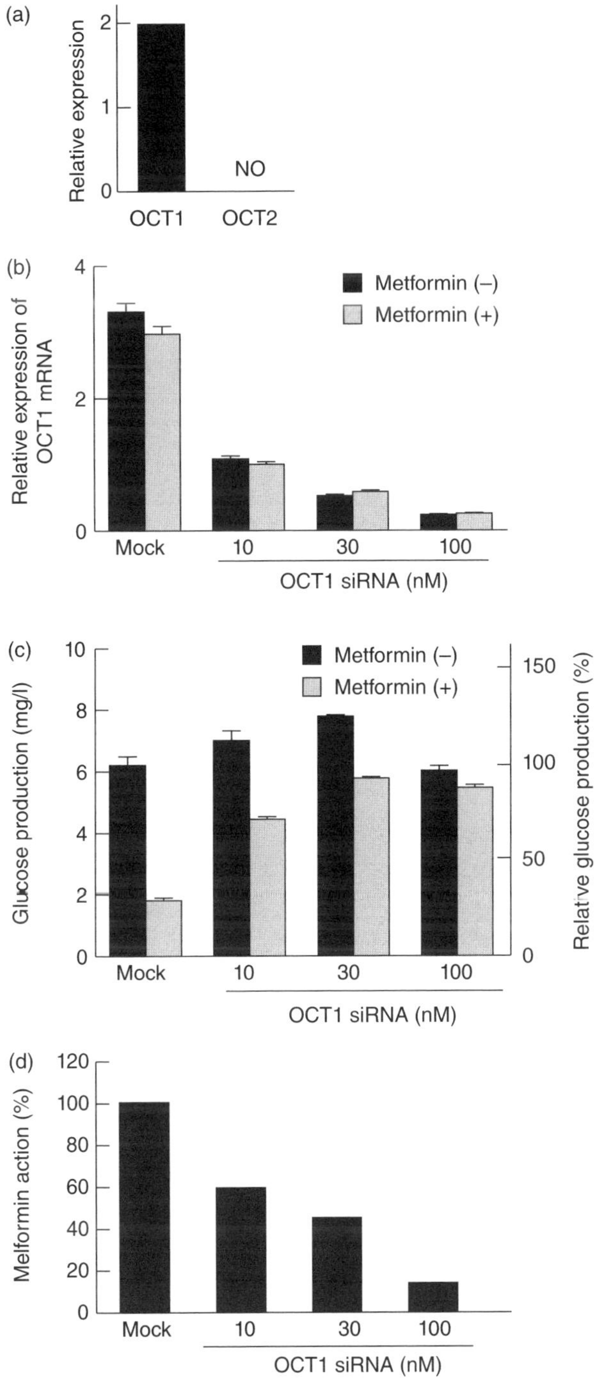
(a)
Relative expression
2
1
0
NO
OCT1
OCT2
(b)
Relative expression of OCT1 mRNA
4
2
0
Metformin (–)
Metformin (+)
Mock
10
30
100
OCT1 siRNA (nM)
(c)
Glucose production (mg/l)
10
8
6
4
2
0
Relative glucose production (%)
150
100
50
0
Metformin (–)
Metformin (+)
Mock
10
30
100
OCT1 siRNA (nM)
(d)
Melformin action (%)
120
100
80
60
40
20
0
Mock
10
30
100
OCT1 siRNA (nM)

a concentration-dependent manner (Figure 8.6b). Upon siRNA-mediated knockdown of OCT1, glucose production was significantly increased in the presence of metformin, but in the absence of metformin only a small increase in glucose production was observed (Figure 8.6c). The authors also estimated metformin action and showed that it was reduced upon OCT1 knockdown. Overall, this study showed that the combination of KG6, a dendrimer, and EP, an amphiphilic peptide, can be used as a carrier to deliver siRNA that could effectively knockdown genes involved in the pharmacological action of meformin.

8.4 Case III – Application of Dendrimers in Vaccine Delivery

Over the past several decades, immunologists and pharmaceutical scientists have explored a variety of approaches to amplify the immunogenicity of antigens that are non- or poorly immunogenic when administered alone [54, 55]. Poly(D,L-lactide-*co*-glycolide) (PLGA) and poly(lactic acid) (PLA based particles, for example, have been extensively studied for their ability to increase the immune response to many recombinant antigens, including hepatitis B surface antigen. Other particulate carriers that have been tested as immuno-adjuvants include chitosan and liposomes [56, 57]. It has been shown that the immune response to hepatitis B vaccine can be enhanced by using tetradecylmaltoside as an adjuvant [58]. Dendrimers have also shown promise in vaccine delivery. In this section, the use of PAMAM dendrimer in enhancing the antigenicity of recombinant ovalbumin (OVA) antigen is summarized[59].

Mannosylation of antigens is reported to increase the immune responses to many antigens. However, the method for antigen mannosylation has a number of limitations. The current method of mannosylation of nonglycosylated antigens requires expression of recombination proteins in yeast. This is a highly complex technique that requires the 'know how' of recombinant DNA technology; the process of characterization and optimization is also tedious. Furthermore, as this method involves direct attachment of mannose residues to the antigen, there is the risk of irreversibly modifying the lysine residues in the main antigenic epitopes. Irreversible modification could significantly alter the epitope and thus compromise the immunogenicity of the antigen [60–65].

In a recently published paper, Sheng *et al.* [64] used PAMAM dendrimer to circumvent the problems associated mannosylation of antigens. In their study, G4 PAMAM dendrimer was used to synthesize a highly branched dendritic structure containing a known number of mannose resides, which the authors called mannosylated dendrimer (MD). MD was chemically conjugated to OVA and with little or no changes to the lysine residues; MD conjugated to OVA is denoted as MDO. Details of the method of conjugation and characterization of conjugated antigen can be found the original paper by Sheng and coworkers [64].

The authors studied the efficacy of MDO and OVA in stimulating the T-cell response *in vitro* in bone marrow dendritic cells (BMDCs) collected from femurs and tibias of C57BL/6 mice. In these studies, BMDCs were pre-pulsed with MDO or OVA and co-incubated with OTI (CD8+) or OTII T (CD4+) cells. Of the various MDO test samples (containing 2–200 μg mannose residues per ml), MDO containing 40 μg/ml mannose produced maximal T-cell proliferation. Compared to OVA-pulsed BMDCs, MDO pre-pulsed BMDCs showed higher levels of OTI and OTII T-cell proliferation. This *in vitro* study suggests that MDO was more effective than OVA in presentation of major histocompatibility complex (MHC) class I and MHC class II in BMDCs.

To determine the significance of the extent of mannosylation, the survival and T-cell stimulation capacity of BMDCs were studied after treatment with OVA, MDO and DO (plain dendrimer linked to OVA). With increasing doses of OVA, an increase in cell death was observed. Similarly, with increasing mannose content in MDO, an increase in cell death was observed. The T-cell proliferative response to OVA was also increased with increasing OVA concentration, and the maximum increase was observed at 320 μg/ml. However, $CD8^+$ T-cell proliferation was independent of mannose content in MDO. Nevertheless, mannose content-dependent proliferation was observed in OTII T-cells. In other words, the capacity of BMDCs to stimulate OTI cells was not influenced by the mannose content of mannosylated-dendrimer-OVA, but the mannose content of MDO aided in increasing OTII T-cell proliferation. The data suggest that the amount or extent of mannosylation can influence MHC class II presentation.

The *in vivo* immune response was evaluated in mice after intradermal injection of plain OVA and MDO. The dosing was performed on days 0, 11 and 18 and the antibody levels were detected in the blood after 28 days; animals were euthanized on day 29 for ELISPOT assay. An additional group of mice received a booster injection on day 38 and was challenged with B16-OVA melanoma cells on day 45. The data presented in Figure 8.7 show that mice immunized with 25 μg MDO had a statistically significant increase in interferon-γ (IFN-γ) levels. The SIINFEKL-specific T-cell response (SIINFEKL is a peptide used as control; H-2Kb-restricted SIINFEKL and I-Ab-restricted OVA323–339 ISQAVHAAHAEINEAGR) was enhanced only in mice that were injected with MDO (Figure 8.7a). OVA-specific IgG levels were higher for MDO-treated animals than those treated with OVA (Figure 8.7b). Because MDO induced a strong SIINFEKL-specific CD8+ response, the authors performed additional experiments to determine MDO's ability to induce cross-presentation of SIINFEKL in the lymph environment following immunization. For the cross-presentation assay, a group of mice was injected with MDO, OVA and phosphate buffered saline (PBS) as control into the hind footpads. Popliteal lymph node (LN) cells were isolated 24 hours after the injection. The LN cells isolated from PBS-treated group were pulsed with SIINFEKL peptide. The LN cells immunized with MDO produced a higher level of SIINFEKL-specific OTI T-cell proliferation in comparison to the OVA-treated group (Figure 8.7c). This set of experiments suggests that MDO enhances cross-presentation of OVA in the lymph compartment.

In the challenge studies, OVA-treated mice failed to prevent B16-OVA tumour growth (Figure 8.8a). However, the MDO-treated group showed a much delayed or no tumour growth for the first 11 days of treatment. With the increase in dose from 12.5 μg to 25 μg, better tumour protection was observed (Figure 8.8b). At the end of 11 days of tumour challenge, one mouse from the 12.5 μg dose group and two mice from the 25 μg MDO-treated group showed no tumour growth. These three animals survived for a period of six months post challenge. It has been reported that the generation of a tumour antigen-specific CD8+ T-cell response is essential for tumour protection in mice. The data from the challenge study agree with the ELISPOT data presented in Figure 8.7 [66, 67]. Overall, this paper documents the fact that dendrimers can be used for mannosylation

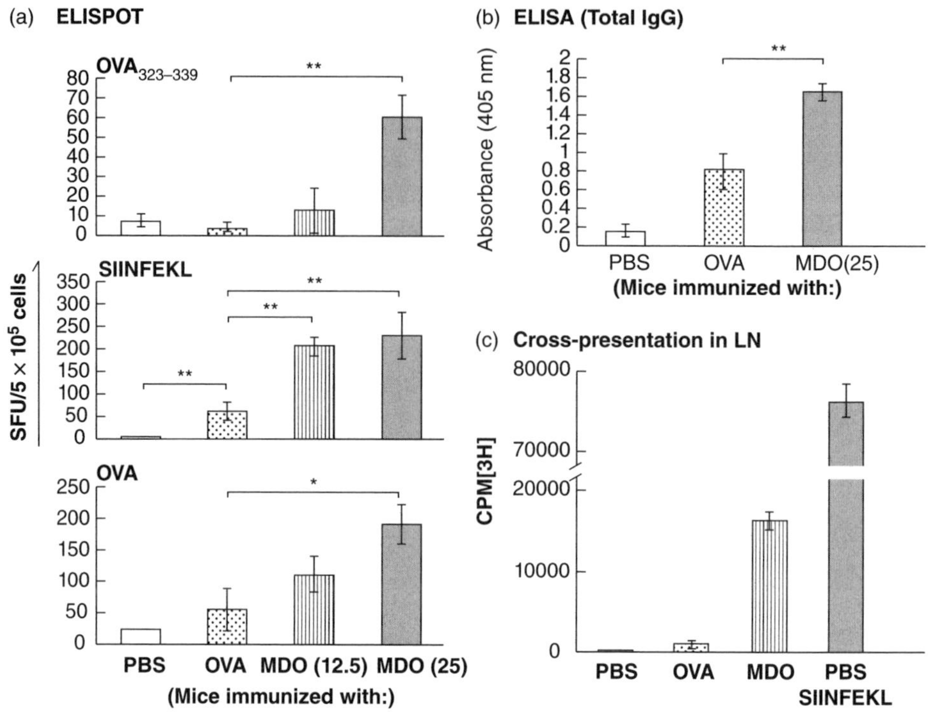

Figure 8.7 *MDO induces high levels of cellular and humoral immunity, as well as cross-presentation, in immunized mice. (a, b) Mice (n = 4) immunized with PBS, OVA (25 μg), MDO (12.5 or 25 μg) were euthanized after three immunizations, and levels of $OVA_{323-339}$ (CD4 epitope)-, SIINFEKL (CD8 epitope)-, and OVA-specific T-cell IFN-γ responses (evaluated by the ELISPOT assay in triplicate) and the total IgG level to OVA (determined at 1 : 400 serum dilution in the ELISA) were compared. (c) Popliteal LN cells isolated from groups of four mice injected separately with PBS, OVA (25 μg) or MDO (25 μg) into footpads were evaluated for their capacities in OTI T-cell stimulation* in vitro. *LN cells (2×10^4) isolated from injected mice were co-cultured with 5×10^4 OTI T cells in quadruplicate. Peak T-cell proliferation on day 2 was compared. LN cells isolated from PBS-immunized mice were used as the background control, and a portion of these cells was pulsed with SIINFEKL (1 μg/mL) as a positive control. Data shown are representative of two experiments. SFU: spot-forming unit; *$P < 0.05$, **$P < 0.01$. (Reproduced from Sheng, K.C., Kalkanidis, M., Pouniotis, D.S.,* et al. *(2008) Delivery of antigen using a novel mannosylated dendrimer potentiates immunogenicity* in vitro *and* in vivo, *European Journal of Immunology,* ***38****, 424–436, with permission from Wiley-VCH Verlag GmbH & Co. KGaA)*

of antigens. Dendrimer based mannosylation amplifies immunogenicity by facilitating binding and recognition of antigens by DCs, increasing antigen uptake, processing and presentation, and enhancing T-cell stimulation. In composite, this study suggests that dendrimer can be successfully used to circumvent the problems of mannosylation and amplify the immune response to antigens.

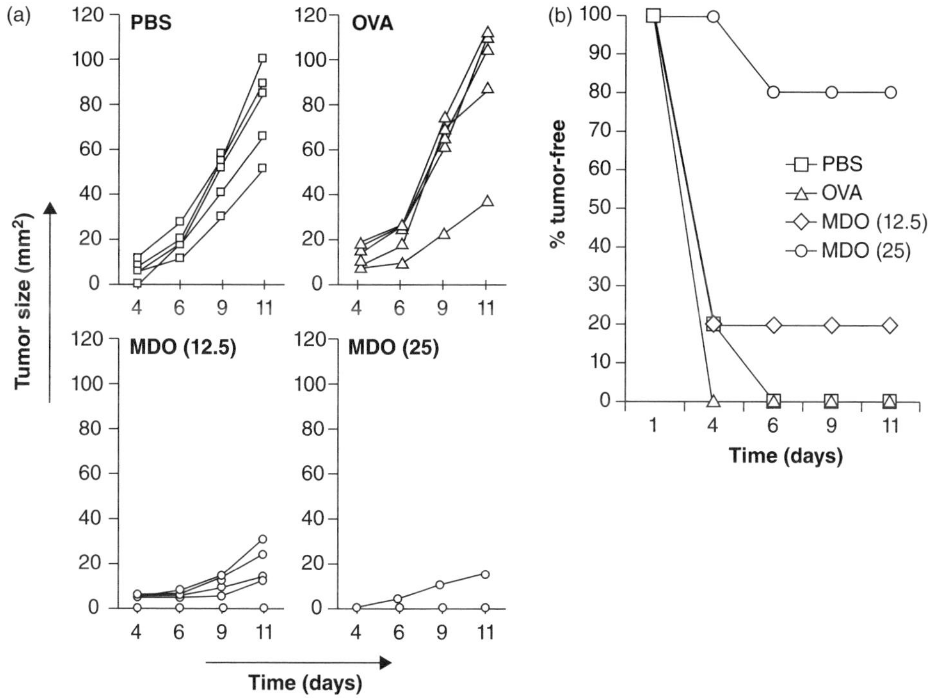

Figure 8.8 *Mice pre-immunized with MDO significantly delayed or showed no B16-OVA melanoma development. (a) Mice (n = 5) pre-immunized with PBS, OVA or MDO of specific doses were inoculated subcutaneously with 10^6 B16-OVA tumour cells into the abdomen. Tumour sizes were measured every 2–3 days over 11 days after tumour inoculation. (b) The percentages of tumour-free mice were determined for each immunized group. Data shown are representative of two experiments. (Reproduced from Sheng, K.C., Kalkanidis, M., Pouniotis, D.S.,* et al. *(2008) Delivery of antigen using a novel mannosylated dendrimer potentiates immunogenicity* in vitro *and* in vivo, European Journal of Immunology, ***38****, 424–436, with permission from Wiley-VCH Verlag GmbH & Co. KGaA)*

8.5 Concluding Remarks

The application of dendrimers has been broadened significantly since the discovery of these macromolecules in the 1980s. The pharmaceutical and biomedical use of dendrimers ranges from drug and gene delivery to medical diagnostics. Dendrimers can address many drug delivery problems, including solubility, absorption and site specificity. In the future, dendrimers may also find applications in vaccine delivery, especially in DNA-vaccine delivery. Based on the huge number of manuscripts published on dendrimer based drug delivery, it is very likely that dendrimer based pharmaceutical formulations or drug delivery systems will soon be available in the market.

References

1. Mulhaupt, R. (2004) Hermann Staudinger and the origin of macromolecular chemistry. *Angew. Chem. Int. Ed.*, **43**, 1054–1063.
2. Patri, A.K., Majoros, I.J. and Baker, J.R. (2002) Dendritic polymer macromolecular carriers for drug delivery. *Curr. Opin. Chem. Biol.*, **6**, 466–471.
3. Qiuand, L.Y. and Bae, Y.H. (2006) Polymer architecture and drug delivery. *Pharm. Res.*, **23**, 1–30.
4. Al-Jamal, K.T., Sakthivel, T. and Florence, A.T. (2005) Dendrisomes: vesicular structures derived from a cationic lipidic dendron. *J. Pharm. Sci.*, **94**, 102–113.
5. Newkome, G.R., Yao, Z., Baker, G.R. and Gupta, V.K. (1985) Cascade molecules: a new approach to micelles. A [27]-Arborol. *J. Org. Chem.*, **50**, 2003–2004.
6. Tomalia, D.A., Baker, H., Dewald, J.R. *et al.* (1985) A new class of polymers: starburst-dendritic macromolecules. *Polym. J. (Tokyo)*, **17**, 117–132.
7. Klajnertand, B. and Bryszewska, M. (2001) Dendrimers: properties and applications. *Acta Biochim. Pol.*, **48**, 199–208.
8. Aulenta, F., Hayes, W. and Rannard, S. (2003) Dendrimers: a new class of nanoscopic containers and delivery devices. *Eur. Polym. J.*, **39**, 1741–1771.
9. Bai, S., Thomas, C., Rawat, A. and Ahsan, F. (2006) Recent progress in dendrimer-based nanocarriers. *Crit. Rev. Ther. Drug Carrier Syst.*, **23**, 437–495.
10. Tomalia, D.A. (1995) Dendrimer molecules. *Sci. Am.*, **272**, 62–66.
11. Tomalia, D.A., Naylor, A.M. and Goddard, W.A. (1990) Starburst dendrimers: molecular-level control of size, shape, surface chemistry, topology and flexibility from atoms to macroscopic matter. *Angew. Chem. Int. Ed.*, **29**, 138.
12. Gilliesand, E.R. and Frechet, J.M. (2005) Dendrimers and dendritic polymers in drug delivery. *Drug Discov. Today*, **10**, 35–43.
13. Esfandand, R. and Tomalia, D.A. (2001) Poly(amidoamine) (PAMAM) dendrimers: from biomimicry to drug delivery and biomedical applications. *Drug Discov. Today*, **6**, 427–436.
14. Svensonand, S. and Tomalia, D.A. (2005) Dendrimers in biomedical applications – reflections on the field. *Adv. Drug Deliv. Rev.*, **57**, 2106–2129.
15. Tomalia, D.A. (2005) Birth of a new macromolecular architecture: dendrimers as quantized building blocks for nanoscale synthetic polymer chemistry. *Prog. Polym. Sci.*, **30**, 294–324.
16. D'Emanueleand, A. and Attwood, D. (2005) Dendrimer–drug interactions. *Adv. Drug Deliv. Rev.*, **57**, 2147–2162.
17. Devarakonda, B., Hill, R.A. and de Villiers, M.M. (2004) The effect of PAMAM dendrimer generation size and surface functional group on the aqueous solubility of nifedipine. *Int. J. Pharm.*, **284**, 133–140.
18. Gupta, U., Agashe, H.B., Asthana, A. and Jain, N.K. (2006) Dendrimers: novel polymeric nanoarchitectures for solubility enhancement. *Biomacromolecules*, **7**, 649–658.
19. Jensen, A.W., Maru, B.S., Zhang, X. *et al.* (2005) Preparation of fullerene-shell dendrimer-core nanoconjugates. *Nano. Lett.*, **5**, 1171–1173.
20. Frechetand, J.M.J. and Hawker, C.J. (1996) Synthesis and properties of dendrimers and hyperbranched polymers. *Compr. Polym. Sci.*, **S2**, 71–132.
21. Cheng, Y., Qu, H., Ma, M. *et al.* (2007) Polyamidoamine (PAMAM) dendrimers as biocompatible carriers of quinolone antimicrobials: an *in vitro* study. *Eur. J. Med. Chem.*, **42**, 1032–1038.
22. Ihre, H.R., Padilla De Jesus, O.L., Szoka, F.C. Jr. and Frechet, J.M. (2002) Polyester dendritic systems for drug delivery applications: design, synthesis, and characterization. *Bioconjug. Chem.*, **13**, 443–452.
23. Jevprasesphant, R., Penny, J., Jalal, R. *et al.* (2003) The influence of surface modification on the cytotoxicity of PAMAM dendrimers. *Int. J. Pharm.*, **252**, 263–266.
24. Bai, S., Thomas, C. and Ahsan, F. (2007) Dendrimers as a carrier for pulmonary delivery of enoxaparin, a low-molecular weight heparin. *J. Pharm. Sci.*, **96**, 2090–2106.
25. Mustafa, F., Yang, T., Khan, M.A. and Ahsan, F. (2004) Chain length-dependent effects of alkylmaltosides on nasal absorption of enoxaparin. *J. Pharm. Sci.*, **93**, 675–683.

26. Yang, T., Mustafa, F. and Ahsan, F. (2004) Alkanoylsucroses in nasal delivery of low molecular weight heparins: *in vivo* absorption and reversibility studies in rats. *J. Pharm. Pharmacol.*, **56**, 53–60.
27. Yang, T., Mustafa, F., Bai, S. and Ahsan, F. (2004) Pulmonary delivery of low molecular weight heparins. *Pharm. Res.*, **21**, 2009–2016.
28. Braun, C.S., Vetro, J.A., Tomalia, D.A. *et al.* (2005) Structure/function relationships of polyamidoamine/DNA dendrimers as gene delivery vehicles. *J. Pharm. Sci.*, **94**, 423–436.
29. Cadene, M., Boudier, C., de Marcillac, G.D. and Bieth, J.G. (1995) Influence of low molecular mass heparin on the kinetics of neutrophil elastase inhibition by mucus proteinase inhibitor. *J. Biol. Chem.*, **270**, 13204–13209.
30. Gundry, S.R., Klein, M.D., Drongowski, R.A. and Kirsh, M.M. (1984) Clinical evaluation of a new rapid heparin assay using the dye azure A. *Am. J. Surg.*, **148**, 191–194.
31. Pearson, S., Jia, H. and Kandachi, K. (2004) China approves first gene therapy. *Nat. Biotechnol.*, **22**, 3–4.
32. Zhang, X.Q., Intra, J. and Salem, A.K. (2007) Conjugation of polyamidoamine dendrimers on biodegradable microparticles for nonviral gene delivery. *Bioconjug. Chem.*, **18**, 2068–2076.
33. Tangand, M.X. and Szoka, F.C. (1997) The influence of polymer structure on the interactions of cationic polymers with DNA and morphology of the resulting complexes. *Gene Ther.*, **4**, 823–832.
34. Singh, M., Briones, M., Ott, G. and O'Hagan, D. (2000) Cationic microparticles: a potent delivery system for DNA vaccines. *Proc. Natl. Acad. Sci. USA*, **97**, 811–816.
35. Lee, C.C., MacKay, J.A., Frechet, J.M. and Szoka, F.C. (2005) Designing dendrimers for biological applications. *Nat. Biotechnol.*, **23**, 1517–1526.
36. Garcia, F., Petry, K.U., Muderspach, L. *et al.* (2004) ZYC101a for treatment of high-grade cervical intraepithelial neoplasia: a randomized controlled trial. *Obstet. Gynecol.*, **103**, 317–326.
37. Klencke, B., Matijevic, M., Urban, R.G. *et al.* (2002) Encapsulated plasmid DNA treatment for human papillomavirus 16-associated anal dysplasia: a Phase I study of ZYC101. *Clin. Cancer Res.*, **8**, 1028–1037.
38. McKeever, U., Barman, S., Hao, T. *et al.* (2002) Protective immune responses elicited in mice by immunization with formulations of poly(lactide-co-glycolide) microparticles. *Vaccine*, **20**, 1524–1531.
39. Sheets, E.E., Urban, R.G., Crum, C.P. *et al.* (2003) Immunotherapy of human cervical high-grade cervical intraepithelial neoplasia with microparticle-delivered human papillomavirus 16 E7 plasmid DNA. *Am. J. Obstet. Gynecol.*, **188**, 916–926.
40. Tsutsumi, T., Hirayama, F., Uekama, K. and Arima, H. (2007) Evaluation of polyamidoamine dendrimer/alpha-cyclodextrin conjugate (generation 3, G3) as a novel carrier for small interfering RNA (siRNA). *J. Control. Release*, **119**, 349–359.
41. Zhou, J., Wu, J., Hafdi, N. *et al.* (2006) PAMAM dendrimers for efficient siRNA delivery and potent gene silencing. *Chem. Commun.*, 2362–2364.
42. Kawano, T., Okuda, T., Aoyagi, H. and Niidome, T. (2004) Long circulation of intravenously administered plasmid DNA delivered with dendritic poly(L-lysine) in the blood flow. *J. Control. Release*, **99**, 329–337.
43. Ohsaki, M., Okuda, T., Wada, A. *et al.* (2002) *In vitro* gene transfection using dendritic poly(L-lysine). *Bioconjug. Chem.*, **13**, 510–517.
44. Okuda, T., Kawakami, S., Akimoto, N. *et al.* (2006) Gylated lysine dendrimers for tumor-selective targeting after intravenous injection in tumor-bearing mice. *J. Control. Release*, **116**, 330–336.
45. Okuda, T., Kawakami, S., Maeie, T. *et al.* (2006) Biodistribution characteristics of amino acid dendrimers and their PEGylated derivatives after intravenous administration. *J. Control. Release*, **114**, 69–77.
46. Yamagata, M., Kawano, T., Shiba, K. *et al.* (2007) Structural advantage of dendritic poly(L-lysine) for gene delivery into cells. *Bioorg. Med. Chem.*, **15**, 526–532.
47. Inoue, Y., Kurihara, R., Tsuchida, A. *et al.* (2008) Efficient delivery of siRNA using dendritic poly(L-lysine) for loss-of-function analysis. *J. Control. Release*, **126**, 59–66.

48. Shaw, R.J., Kosmatka, M., Bardeesy, N. *et al.* (2004) The tumor suppressor LKB1 kinase directly activates AMP-activated kinase and regulates apoptosis in response to energy stress. *Proc. Natl. Acad. Sci. USA*, **101**, 3329–3335.
49. Shaw, R.J., Lamia, K.A., Vasquez, D. *et al.* (2005) The kinase LKB1 mediates glucose homeostasis in liver and therapeutic effects of metformin. *Science*, **310**, 1642–1646.
50. Woods, A., Johnstone, S.R., Dickerson, K. *et al.* (2003) LKB1 is the upstream kinase in the AMP-activated protein kinase cascade. *Curr. Biol.*, **13**, 2004–2008.
51. Zang, M., Zuccollo, A., Hou, X. *et al.* (2004) AMP-activated protein kinase is required for the lipid-lowering effect of metformin in insulin-resistant human HepG2 cells. *J. Biol. Chem.*, **279**, 47898–47905.
52. Zhou, G., Myers, R., Li, Y. *et al.* (2001) Role of AMP-activated protein kinase in mechanism of metformin action. *J. Clin. Invest.*, **108**, 1167–1174.
53. Summerton, J.E. (2005) Endo-Porter: a novel reagent for safe, effective delivery of substances into cells. *Ann. NY Acad. Sci.*, **1058**, 62–75.
54. Jackson, D.C., O'Brien-Simpson, N., Ede, N.J. and Brown, L.E. (1997) Free radical induced polymerization of synthetic peptides into polymeric immunogens. *Vaccine*, **15**, 1697–1705.
55. Patterson, L.J., Robey, F., Muck, A. *et al.* (2001) A conformational C4 peptide polymer vaccine coupled with live recombinant vector priming is immunogenic but does not protect against rectal SIV challenge. *AIDS Res. Hum. Retroviruses*, **17**, 837–849.
56. Bivas-Benita, M., van Meijgaarden, K.E., Franken, K.L. *et al.* (2004) Pulmonary delivery of chitosan-DNA nanoparticles enhances the immunogenicity of a DNA vaccine encoding HLA-A*0201-restricted T-cell epitopes of Mycobacterium tuberculosis. *Vaccine*, **22**, 1609–1615.
57. Feng, L., Qi, X.R., Zhou, X.J. *et al.* (2006) Pharmaceutical and immunological evaluation of a single-dose hepatitis B vaccine using PLGA microspheres. *J. Control. Release*, **112**, 35–42.
58. Thomas, C., Rawat, A., Bai, S. and Ahsan, F. (2008) Feasibility study of inhaled hepatitis B vaccine formulated with tetradecylmaltoside. *J. Pharm. Sci.*, **97**, 1213–1223.
59. Boasand, U. and Heegaard, P.M. (2004) Dendrimers in drug research. *Chem. Soc. Rev.*, **33**, 43–63.
60. Agnes, M.C., Tan, A., Jordens, R. *et al.* (1998) Strongly increased efficiency of altered peptide ligands by mannosylation. *Int. Immunol.*, **10**, 1299–1304.
61. Lam, J.S., Mansour, M.K., Specht, C.A. and Levitz, S.M. (2005) A model vaccine exploiting fungal mannosylation to increase antigen immunogenicity. *J. Immunol.*, **175**, 7496–7503.
62. Levitzand, S.M. and Specht, C.A. (2006) The molecular basis for the immunogenicity of Cryptococcus neoformans mannoproteins. *FEMS Yeast Res.*, **6**, 513–524.
63. Luong, M., Lam, J.S., Chen, J. and Levitz, S.M. (2007) Effects of fungal N- and O-linked mannosylation on the immunogenicity of model vaccines. *Vaccine*, **25**, 4340–4344.
64. Sheng, K.C., Kalkanidis, M., Pouniotis, D.S. *et al.* (2008) Delivery of antigen using a novel mannosylated dendrimer potentiates immunogenicity *in vitro* and *in vivo*. *Eur. J. Immunol.*, **38**, 424–436.
65. Zhong, G., Wang, J., Xu, M. *et al.* (2005) Enhanced maturation and functional capacity of dendritic cells induced by mannosylated L2 domain of ErbB2 receptor. *Scan. J. Immunol.*, **62**, 108–116.
66. Ikeda, H., Chamoto, K., Tsuji, T. *et al.* (2004) The critical role of type-1 innate and acquired immunity in tumor immunotherapy. *Cancer Sci.*, **95**, 697–703.
67. Pietersz, G.A., Li, W., Popovski, V. *et al.* (1998) Parameters for using mannan-MUC1 fusion protein to induce cellular immunity. *Cancer Immunol. Immunother.*, **45**, 321–326.

Peptides and Proteins

9

Modification of Peptides and Proteins

Susanne Hostrup,[1] *Kasper Huus*[2] *and Henrik Parshad*[1]
[1]Novo Nordisk A/S, Bagsværd, Denmark
[2]Novo Nordisk A/S, Måløv, Denmark

9.1 Introduction

In the early drug development phases covering preformulation and formulation development, the physicochemical properties of the protein drug are assessed. Efforts are made to optimize these, as the physical and chemical properties of protein drugs significantly determine their performance during development, manufacturing, shelf-life and clinical use. Endogenous peptides and proteins may serve as interesting candidates for protein drugs. They are, however, limited by inherent properties, such as expression at low concentration, low exposure due to fast clearance and rapid degradation. By contrast, development of protein drugs requires sufficiently high levels of exposure as well as retention of activity throughout the stages of purification, formulation, storage and administration. Various technologies have proved successful in modulating the exposure and, thereby, obtaining more desirable pharmacokinetics. These may be based on: molecular engineering, such as glycoengineering and classical amino acid substitution; chemical modification, such as protein conjugation, PEGylation and lipidization of the protein; or protein fusion. In this chapter PEGylation, lipidization and amino acid substitution are discussed, with emphasis on the pharmacokinetic and pharmaceutical implications of these technologies.

9.2 PEGylated Peptides and Proteins

Peptides and proteins are known to be susceptible to fast clearance and degradation after administration. The attachment of one or more poly(ethylene glycol) (PEG) chains to the

Delivery Technologies for Biopharmaceuticals: Peptides, Proteins, Nucleic Acids and Vaccines Edited by Lene Jorgensen and Hanne Mørck Nielsen

peptide or protein is one of the technologies used to overcome this challenge. The polymer is non-toxic, non-immunogenic, non-antigenic and highly soluble in water and a number of products containing PEG-conjugated proteins have already been marketed or are undergoing clinical trials.

The PEG chain can adopt various conformations dependent on solution conditions and may consequently be located as a random-coil domain adjacent to the protein [1] or exhibit non-specific interactions with protein to perform screening of the protein surface from the solvent [2]. PEGylation provides increased stability against attack from proteases and reduced clearance from the body. This increases the residence time in the body and, thereby, compensates the activity decrease often associated with protein conjugation. An increased physical stability of the protein after PEGylation may also be observed, which is highly favourable for the formulated drug product.

9.2.1 Methods

The aim of this section is to describe the advantages and challenges of PEGylation with regard to drug delivery of proteins. For more extensive information on PEGylation chemistries and process optimization some excellent reviews are available [3–6].

9.2.1.1 PEGylation Chemistries

A large number of PEG reagents are available and they are produced in different configurations (Figure 9.1) and with different molecular weights.

The coupling of PEG most often relies on nucleophilic attack of a polypeptide amino acid side chain functional group at the electrophilic centre of an activated PEG chain. The type of functional group added to PEG is dependent on the target for PEGylation at the protein molecule. For monofunctional PEGs, the hydroxy group at the other terminal end of the PEG chain is often substituted by a methoxy group to create monomethoxy PEGs (mPEGs) [7].

Primary amines are good nucleophiles and are often used as the first target for PEGylation, since they are highly abundant in proteins at the N-terminal of the protein chain (α-NH_2) and in the side chain of lysine amino acid residues (ε-NH_2). The primary amines can be reacted with PEGs activated with eletrophilic groups by acylation or alkylation reactions.

The limitations of such a general PEGylation strategy are the high number of positional isomers and polyPEGylated products formed, and that the purification of these mixtures

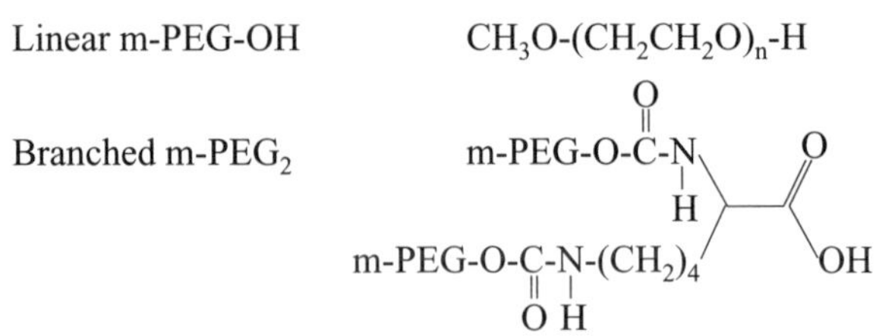

Figure 9.1 *Structure of linear and branched mPEGs*

is difficult. Some marketed products have been approved by the United States Food and Drug Administration (FDA) as mixtures of different isomers (PEG-asparaginase (Oncaspar®) and PEG adenosine deaminase (Adagen®)), where evidence for reproducibility of the PEGylation process could be obtained. However, characterization of each isomer is compulsory and it involves extensive analytical work [8]. Thus, more selective PEGylation strategies are very attractive.

Several approaches for site-specific targeting have been described [4, 5, 9, 10]. These approaches may for example involve the exploitation of the difference in pK_a values of α-NH_2 and ε-NH_2, the rare occurrence of cysteines in proteins, or the attachment of PEG to O-glycans [4, 11–13].

9.2.2 Conceptual Considerations on PEGylation

9.2.2.1 Choice of PEGylation Site and Linker

The choice of PEGylation site and linker is highly dependent on the properties of the individual protein and must be determined empirically in each case. Knowledge of protein structure, stability and the amino acids essential for biological activity may help to perform more informed choices. The use of active site protection during PEGylation may lead to better biological activity [14]; however, site-specific PEGylation at, for example, thiol groups appears to be a more general approach that reduces the risk of interfering with binding sites or active sites of the protein.

An elegant way of overcoming the challenge of decreased bioactivity is to direct the PEGylation to the natural glycosylation site of a protein, since these sites are often not essential for activity. This was done with Human Interleukin-2 (IL-2), where the glycosylated threonine in the molecule was substituted by cysteine by protein engineering, after which the protein was PEGylated by a thioether linkage. The PEGylated species had full activity relative to the unmodified molecule [15].

It has also been suggested to use releasable PEG (rPEG) conjugates in order to maintain bioactivity [16, 17]. In these cases, the classical prodrug ideas was used by conjugating a non-stable PEG linker to the protein, which undergoes 1,6-benzyl elimination in plasma. As an example, full activity of lysozyme was regained after release of PEG from the molecule [17]. The hurdle using releasable PEGs is the instability of the conjugates in aqueous solutions, and thus rPEG conjugates must be formulated as a lyophilized product.

9.2.2.2 Choice of PEG Size and Structure

The first generation of PEG-conjugated proteins were often modified with multiple PEG chains of low molecular mass that were randomly attached to the molecule. The second generation of PEGs included larger PEGs (20–40 kDa) and branched PEGs (Figure 9.1) that offer better protection of the protein even when monoPEGylated.

The bioactivity of conjugated proteins will often be reduced because of steric hindrance in the binding site region of the molecule. In the case of antibodies and enzymes it has been reported that the binding activity decreases with the number of PEGs attached [18]. Thus, the use of fewer PEG chains of higher molecular mass as, for example, branched PEGs is typically related with less loss of activity, since interference with amino acids essential for receptor binding can be avoided. Furthermore, the decrease in binding activity is often accounted for by the increased circulation time of the conjugated protein. It is also important to note that synthetic polymers such as PEG are polydisperse with a Gaussian distribution of molecular weights. This will inevitably result in differences in residence time of conjugated proteins, which demands a low variation in the raw material and extensive characterization of the end product.

9.2.3 Pharmacokinetic Aspects

Most often the decision of development of a PEGylated variant of a protein is based on the promising effects on the pharmacokinetic and immunogenic properties [18]. The physicochemical properties of a PEG-conjugated protein are more similar to the polymer than the protein and, thus, the *in vivo* fate of the conjugate is severely changed relative to the native protein, as can be observed for example by an increased circulation time of PEGylated proteins and a significantly lower volume of distribution [7].

Parenterally administered proteins are rapidly cleared by the reticuloendothelial system (RES), kidney, spleen and liver. Clearance is dependent on hydrodynamic size and charge as well as specific receptor-binding capabilities [7]. The attachment of PEG to proteins may alter these properties, for example the high water coordination of PEG increases the hydrodynamic volume of PEG up to 3–5 times that of a globular protein having the same molecular weight [4, 8]. The cut-off for glomerular filtration of PEG is approximately 30 kDa [19] and, thus, renal clearance is circumvented by attachment of PEG of this size or larger. These large size conjugates are cleared more slowly, mainly trough hepatic clearance [5].

The increases in circulation time of PEGylated proteins as compared to their non-conjugated counterparts often exhibit a significant longer action profile and larger exposure [20]. The longer circulation time may result in a more constant concentration of drug in the blood, which may reduce the toxicity of the protein, since a lower amount of drug may induce the same response and sharp plasma concentration peaks are avoided [18, 21].

A further advantage of PEGylation is the reduction of immunogenicity. Proteins and peptides are known to elicit immune responses in humans to varying degrees. The perceived risk becomes larger with non-native proteins than for native human proteins [22]; the structural stability of the protein is also important [23]. Decreased immunogenicity and antigenicity after PEGylation has been observed for several proteins, for example PEGylated insulin [9]. The mechanism by which PEG-conjugation reduces immune responses is probably due to shielding of immunogenic determinants on the protein. However, it is also generally known that protein aggregates are far more immunogenic than soluble proteins [22]. Consequently, the increased solubility observed for PEGylated proteins may also reduce the immunogenicity.

It is important to note that PEGylation is a trade-off between increased circulation, decreased immunogenicity and decreased biological activity, and the effects are not easily predicted.

The biological activity of Pegasys® (PEGylated Interferon α-2a) only retains 7% of the antiviral activity of the native protein, but superior *in vivo* performance is still observed because of improved pharmakokinetics [24]. Furthermore, PEGylation of INFα with different PEGs resulted in completely different pharmacokinetic profiles [24, 25].

Accordingly, care should be taken when evaluating the effect of PEGylation by *in vitro* assays. Many PEGylated proteins would have failed their development if choices had solely been based on *in vitro* activity assays.

9.2.4 Pharmaceutical Aspects

Even though PEGylation is primarily used to improve pharmacokinetic and immunogenic properties of proteins, some beneficial pharmaceutical effects should also be considered. These are largely associated with increased solution stability and more favourable formulation properties.

Many proteins are poorly soluble in water, which represents a significant challenge in protein formulation. The poly(ethylene glycol) chain is highly soluble in water with solubility 5–10 times larger than a soluble protein of comparable molecular weight [4]. These solution properties are transferred to the respective protein upon conjugation, which leaves the protein more soluble and less prone to aggregation. The high solubility of PEGylated proteins may also be favourable when high protein doses are needed [18].

Beneficial effects of PEGylation have also been observed after lyophilization. Freeze-drying of PEGylated human growth hormone (hGH) resulted in a shorter dissolution time and lower moisture content of the freeze-dried cake compared to the native protein [26].

A drawback of PEGylation in regard to formulation properties is the increase in viscosity of the solution containing PEGylated proteins that may be observed relative to the unmodified protein. The viscosity has been observed to increase exponentially with the PEG chain length [27] and this may have consequences for the handling and syringeability of solutions aimed for pharmaceutical purposes.

9.2.4.1 Solution Stability

The improved *in vivo* chemical stability is associated with enhanced stability towards proteases due to steric shielding of protein domains that are susceptible to proteolytic attack, which accordingly leads to increased circulation time as described above. In addition, PEGylation might change the charge distribution and the isoelectric point of the protein, which may alter the pH-related stability.

The improved physical stability, however, is mainly due to enhanced solubility of the protein and may be the single motive for PEGylation, since development of protein drugs in solution is likely to fail due to poor physical stability.

9.2.4.2 Stability Towards Solution Aggregation

The positive effects of PEGylation on the physical stability of proteins are believed to be both specific and non-specific. PEG-conjugation may specifically block residues that participate in the hydrophobic interactions that drive aggregation or the contribution to increased physical stability may be non-specific steric shielding to decrease hydrophobic interactions [9].

PEGylation of granulocyte colony-stimulating factor (G-CSF) was found to reduce the formation of insoluble aggregates significantly [28] and, thus, the physical stability of a drug product can be greatly improved by PEGylation. The authors found that soluble aggregates were formed instead through the same pathway as the aggregation of native

G-CSF. The results indicated that the differences in solubility of the aggregates were due to poly(ethylene glycol) chains facilitating solvation of the aggregate by water. Furthermore, the stability against formation of insoluble aggregates was only slightly reduced by decreasing the size of the PEG chain from 20 kDa to 5 kDa, and thus conjugation of low molecular weight PEGs can be considered if the desired goal is improved stability [28]. For glucagons, however, an increase in fibrillation tendency was observed when the size of the attached PEG chain was reduced from 5 kDa to 2 kDa [29].

It is important to note that the conjugation method chosen for PEGylation may also influence the aggregation properties of the PEGylated protein, since the charge distribution of end product may be different. By conjugation through acylation the charge at the target amino acid is neutralized, whereas this is not the case for conjugates prepared by alkylation [5]. In fact, for G-CSF a faster aggregation was observed after acylation of the N-terminal residue compared to aggregation after conjugation through alkylation [12]. The authors suggested that this could be due to neutralization of charge at the N-terminal α-amino group of G-CSF.

9.2.4.3 *Stability Towards Interface-Induced Aggregation*

As mentioned above, increased solubility of PEGylated proteins may lead to improved properties against solution aggregation. Interface-induced aggregation of proteins is recognized as a major stability challenge in the commercial production of pharmaceutical proteins [30, 31]. Many proteins tend to adsorb to interfaces experienced during manufacture, storage and delivery, and may cause unfolding and subsequent aggregation of the protein.

Adsorption studies by Daly *et al.* showed that PEGylation of lysozyme reduced the number of molecules per unit area after adsorption to a silica/water interface, probably due to steric exclusion by PEG, and increases the protein-to-protein distance [32]. Furthermore, the adsorption was shown to be less irreversible, which is an indication of the protein being less unfolded in the adsorbed state. In another study, the grafting of one 20 kDa PEG to lysozyme did not decrease the extent of aggregation, whereas two PEG chains significantly reduced the extent of aggregation [33]. It can be speculated that PEGylation may be even more efficient in protecting proteins from adsorption to hydrophobic interfaces, since the PEG chain may provide shielding of the hydrophobic areas on the exterior of the protein from the interface. However, PEG itself is surface active on hydrophobic materials and adsorption of PEGylated proteins may therefore proceed by direct interaction of the PEG chain and the corresponding interface [1].

9.2.4.4 *Effect on Self-Association Behaviour*

Proteins are commonly observed to self-associate non-covalently in solution, for example insulin, glucagon [34, 35]. The self-association is dependent on many factors such as temperature, pH, ionic strength and concentration, and the physical stability of the protein is often related to the association state of the protein in solution. PEG-conjugated proteins may exhibit differences in self-association pattern compared to the native protein and, thus, the physical stability of PEGylated species should be taken into account.

Human insulin in solution consists of a mixture of monomeric, dimeric and hexameric states, with hexamers being the predominant association state at concentrations ≥ 2 mM in the absence of zinc [36]. Hexameric insulin is observed to be more physical stable than the other association states [37]. In the PEGylated form, human insulin molecules are predominantly observed to be in their dimeric association state, as they are probably impaired from forming hexamers due to steric hindrance [9]. However, the physical stability measured as the resistance to form fibrils during shaking of PEGylated human insulin was found to be 36–40 times higher than the native protein for the most successful candidates [9]. Thus, even though PEG-conjugation may lead to impaired or decreased self-association compared to the native protein, this does not necessarily indicate a decreased physical stability of the PEGylated form.

9.2.5 PEGylated Proteins in Advanced Delivery Systems

PEGylation may also increase the success of advanced delivery systems for proteins, such as microspheres. As an example, PEGylation of ribonuclease A appeared to be a useful strategy to enhance the release of the protein from poly(lactic-*co*-glycolic acid) (PLG) microspheres [1]. Whereas only 50% of the unmodified protein was released from the microspheres over three months, 65% and 95% of monoPEGylated and diPEGylated RNase A was released, respectively. The authors suggests that the protein release is reduced by protein adsorption to the PLG matrix, since this causes a decrease in concentration of proteins that are free to diffuse through the matrix pores. PEGylation of RNase A was shown to reduce the adsorption to PLG, and thus the release rate and extent are increased [1]. For other proteins a reduction in encapsulation-induced aggregation and significant increase in residual activity has been observed after encapsulation in poly(D,L-lactide-*co*-glycolide) (PLGA) microspheres [38, 39].

9.3 Lipidization of Peptides and Proteins

Pharmaceutically relevant peptides and proteins are in general hydrophilic. For various reasons, they may be made more hydrophobic by conjugation with lipophilic groups. The reasons for lipidization of pharmaceutical peptides and proteins are, typically, improvement of delivery and prolongation of the half-life, either via sustained release from the administration site or via prolonged circulation time. Various types of lipophilic groups, linkers and chemical means of conjugation have been examined depending on the purpose of lipidization.

Lipidization, particularly acylation, is also a naturally occurring post-transcriptional modification involving hundreds of proteins. The purposes include addressing a particular subcellular localisation, enhancing membrane association and, possibly, impact on structure and stability of the protein. The fatty acyl groups generally contain 14–20 carbons [40, 41].

9.3.1 Methods of Lipidization

Lipidization may be obtained by acylation of the ε-amino group of lysine residues. Other targets for acylation are N-terminal α-amino groups in general [42–44]. Acylation of N-terminal glycine residues is a common *in vivo* post-transcriptional modification process.

Lipidization has also been carried out at cysteine residues following cleavage of intra-disulfide bonds with a reducing agent. This type of lipidization is typically reversible [45–49], although non-reversible lipidization via intra-chain disulfide bonds has also been reported [50]. Similarly, lipidization can also be done at the free sulfhydryl group of a cysteine residue [51].

9.3.2 Conceptual Considerations on Lipidization

The choice of lipophilic groups and site of conjugation is highly dependent on the purpose of lipidization and the nature of the drug molecule. Generally speaking, perturbation of the structure and of the receptor-interacting site of the peptide or protein should be avoided.

One way of increasing the circulation half-life of peptides and proteins is to attach a lipophilic group that binds to albumin. Albumin is a transport protein found at high concentrations both in plasma and in the interstitial fluid. Free fatty acids bind to albumin, and therefore lipidization of a protein with long chain fatty acids and a negative charge may give a conjugate with good affinity for albumin. Since albumin is also present subcutaneously, affinity for albumin may slow the absorption following subcutaneous injection [42, 52]. Other concepts of lipidization may apply if the required effect is primarily, for example, the formation of slowly absorbed complexes at the injection site, increased permeability across absorption barriers or interaction with advanced delivery systems.

9.3.3 Improved Pharmacokinetics

Several insulin molecules with lipophilic groups attached have been described in the literature. The long-acting insulin detemir (Levemir®), which has been marketed since 2004, and the recently approved (2009) long-acting GLP-1 analogue liraglutide (Victoza®) are, to our best knowledge, the only marketed lipophilically modified protein drugs. Since insulin detemir is the most thoroughly investigated lipidated pharmaceutical peptide, it will be the focus of this section.

The aim of the development of insulin detemir has been to obtain a long-acting insulin with a smoother and more reproducible effect profile than the traditional insulin suspensions. Saturated fatty acids with chain lengths from 10 to 16 carbon atoms were attached by acylation to the ε-amino group of Lys^{B29} with the purpose of increasing the affinity for human serum albumin (HSA), and thereby increasing the circulation time of the peptide. The affinity gradually increased when the carbon chain length was increased from C_{10} to C_{14} but there was a decrease when going from C_{14} to C_{16}. Studies in rabbits showed that the initial fall in blood glucose correlated to the affinity of the insulin derivate for albumin [52]. In a pig study with radioactively labelled insulin derivates, it was shown that the disappearance from the injection site correlated with the albumin affinity [42]. A typical drawback of lipidization is reduced receptor affinity. Insulin receptor studies showed that the derivates with the highest affinity for albumin exhibited reduced receptor affinity (around 50% relative to native human insulin) [42, 52]. Earlier studies with B1 and B29 lipidated insulin also showed reduced biological activity [44, 53].

The early publications describing the concept of insulin detemir only related to the effect of the albumin binding. However, it has later been found that albumin binding is not the primary effect of prolonged action. A study, which combined several physicochemical and pharmacokinetic experiments, has been conducted in an attempt to elucidate the mechanism of protraction of insulin detemir following subcutaneous injection [54]. It was found that the protracted action was primarily caused by prolonged retention at the injection site. The retention is mainly ascribed to the formation of dihexamers, which are formed through hydrophobic interaction of the attached fatty acids. Further retention is caused by albumin binding at the injection site. Following absorption into the blood stream, the half-life of monomeric insulin detemir is further prolonged compared to native human insulin due to albumin binding [54]. The concept of slowing the insulin absorption from the injection site has been actively exploited in newer generations of lipidated insulin derivates. For example, a *N*-litocholic acid acylated insulin derivate (Lys^{B29}-litocholyl-γ-Glu-des-(B30) human insulin) has lower albumin affinity than insulin detemir, but due to the formation of large macromolecular complexes (>5000 kDa) the absorption from the subcutaneous injection site was even slower than that of insulin detemir [55, 56].

The binding of lipidated insulin to albumin in the blood circulation may provide a buffer against fluctuations in free insulin concentration. Accordingly, it has been found in clinical trials that albumin binding insulin derivates provide lower within-subject variability than other long-acting basal insulin products. The lower variability is most likely also a consequence of the albumin binding insulin derivates being soluble. They were compared to the long-acting neutral protamine hagedorn (NPH) and insulin glargine, which exist as crystals or amorphous precipitates at the injection site, respectively. Dissolution of the crystals or precipitates is associated with larger variability in the absorption rate compared to the soluble lipidated derivates [57–59].

Another example of improved pharmacokinetic profile obtained by fatty acid acylation is the glucagon-like Peptide-1 (GLP-1) derivate, liraglutide. Native GLP-1 is a 30-amino acid peptide hormone with important metabolic properties of use in type 2 diabetes treatment. However, the potential of using GLP-1 in therapy is limited by a very short biological half-life (1–2 minutes) for the circulating peptide. This very short half-life is caused by specific enzymatic cleavage by dipeptidyl peptidase IV (DDP-IV) which completely inactivates the peptide [60]. Liraglutide is conjugated with a C_{16} fatty acid via a γ-Glu spacer. The effects of this acylation are slowed release from the injection site, increased circulation due to albumin binding and protection against DDP-IV degradation. The change in plasma half-life is dramatic. Following subcutaneous injection in pigs, the plasma half-life was increased to 14 hours compared to 1.2 hours observed for native GLP-1 [43]. The increased half-life of liraglutide means that it can be used as aonce daily subcutaneous injection in the treatment of type 2 diabetes [61, 62]. Liraglutide was recently approved and marketed (trade name Victoza®) in Europe.

9.3.4 Improved Delivery via Non-Parenteral Delivery Routes

Improved non-parenteral delivery due to the increased hydrophobic character of lipidated peptides and proteins has been demonstrated in numerous studies; most of them concern oral delivery of smaller peptides. Peptides are generally too hydrophilic to cross the gastrointestinal (GI) tract mucosa. In addition, they are degraded enzymatically in the GI tract. A number of studies have found increased absorption properties of lipidated peptides and, in many cases, also decreased enzymatic degradation. Several studies have combined lipidization with other modifications of the peptides in order to optimize the oral absorption [63–65]. Some studies have shown significantly increased permeability across Caco-2 cell monolayers; however, this does not necessarily correlate with increased *in vivo* absorption [50, 66]. It should be noted that there are several examples of increased liver retention of lipidated peptides [45, 66]. This effect may be seen as an unwanted increased first-pass metabolism, but it also provides opportunities regarding liver-specific targeting.

Dermal and transdermal absorption may be increased by lipidated peptides. For example, a C_{16} derivate of interferon α (19.3 kDa) was absorbed 5–6 times more than the native protein, but at the same time the bioactivity was reduced by 50% [67]. The absorption increased with the fatty acid chain length from C_{12} to C_{16} but it decreased from C_{16} to C_{18}. A further decrease was seen with a mono-unsaturated C_{18} [68].

Lipidization generally increases cellular uptake of peptides and proteins, since the lipophilic groups interact with the membranes. Also, it has been found that lipidization may be a useful method for increasing passage across the blood–brain barrier [69]. Improved absorption following nasal administration has also been demonstrated [70].

9.3.5 Pharmaceutical Aspects

The consequence of lipidization is an influence on the physicochemical properties of the peptide or protein molecule. Obviously, the molecule becomes overall more hydrophobic and, in general, attains amphiphilic properties. These changes are likely to influence properties such as solubility, self-association and surface activity. The extent of these effects is particularly dependent on the nature of the lipophilic group and the size relative to the size of the native peptide, as well as possible linkers and other modifications. Furthermore, the modification may directly influence the structure and stability of the protein. In all cases, the site of attachment also plays an important role.

9.3.5.1 Solubility

Obviously, attachment of lipophilic groups will render a peptide or a protein less soluble in aqueous solutions. This may pose a problem in the formulation of the resulting derivate. The increase in overall lipophilicity of the derivate depends on the size of the peptide/protein and on the size and lipophilicity of the attached group. Acylation of lysine groups with simple fatty acids increases the net negative charge of the resulting derivate by one at physiological pH. This extra negative charge will increase the solubility [43]. Thus, decreased solubility due to lipidization may be counteracted by other polar and/or charge modifications to the conjugated molecule.

9.3.5.2 Intermolecular Associations and Surface Activity

The attachment of lipophilic groups to peptides and proteins gives the resulting derivates amphiphilic properties. This makes the derivates more surface active. The increased surface activity has been actively exploited in the food and cosmetics industry where it is desirable, for example, to improve foaming properties. Due to the increased surface activity, lipidated proteins adsorb more easily at air–water interfaces [71]. This may decrease the stability of the proteins by causing increased aggregation of (partially) unfolded protein.

The amphiphilic properties of the lipidated derivates may also give them other surfactant-like properties. It is expected that the lipophilic groups of the derivates will interact in order to minimize the contact with water. Due to the size and bulkiness of most of the protein derivates, they will rarely form micelles but rather associate into some less well-defined structures of varying shape and size. For example, dynamic light scattering studies of salmon calcitonin showed that while the non-derivated protein (sCT) mainly existed as monomers, the lipophilic conjugate, Mal-sCT (conjugated with a ε-maleimido lysine derivate of palmitic acid), formed macromolecular complexes consisting of 10–15 monomers. The complexes were reversible upon dilution [50].

9.3.5.3 Structural and Stability Effects

The effect of attaching lipophilic groups on the secondary and tertiary structure of peptides and proteins is highly dependent on where on the molecule the group is attached. For example, a large difference in the content of secondary structural elements was observed when comparing sCT and Mal-sCT. In this case, fatty acid conjugation was targeted at an intra-disulfide bond [50]. On the other hand, X-ray studies of insulin detemir have shown that the structure and self-association properties are not altered compared to native human insulin, since the conjugation is targeted a terminal residue, which is not critical for the structural properties of the molecule [72].

Several studies have found a higher free energy of unfolding of lipidated proteins compared to their non-lipidated counterparts. The results can be explained in terms of entropic effects. The lipophilic groups probably stabilize hydrophobic areas near the surface of the protein and shield these from the aqueous environment [73–76]. Lipidization has also been shown to protect peptides against enzymatic degradation [77, 78]. This effect, which can be explained with steric hindrance, can both increase the absorption and prolong the circulation half-life of lipidated peptides.

9.3.6 Lipidization in Advanced Delivery Systems

Lipidization can be applied to improve the incorporation of peptides and proteins with different types of delivery vehicles. For example, lipidization has been shown to influence protein incorporation in emulsions, liposomes and PLGA particles. Lipidated proteins also have possible uses in systems for biosensors, targeted delivery, immunoassay and recognition abilities [71, 79, 80].

9.3.6.1 Prodrug Concept: Reversible Lipidization

The reduced bioactivity often observed for lipidated peptides can be avoided by the use of reversible lipidization, where the original peptide is regenerated *in vivo*. Reversible lipidization is typically done by reducing intramolecular disulfide bonds and conjugating lipophilic groups to the resulting cysteine residues. *In vivo*, the process is reversed and the original peptide is gradually regenerated. For example, the activity of lipidated desmopressin with two C_{16} fatty acid chains was prolonged from one day to more than four days following subcutaneous injection in rats. The prolonged activity was mainly attributed to slow elimination and a prolonged tissue retention [49].

9.4 Modification of the Primary Structure of Peptides and Proteins

A widely applied mechanism for modulating stability and pharmacokinetic properties is manipulation of the primary structure by application of molecular engineering. Several protein design and engineering strategies have been applied to design novel proteins with improved properties, such as stability, while maintaining the biological activity. The techniques may roughly be divided into three categories: (i) rational and (ii) semi-rational based approaches, and (iii) combinatorial design, where random mutations at random protein sites are introduced by, for example, error-prone polymerase chain reaction (PCR), DNA shuffling [81] and phage display [82]. Extensive reviews of the application of the techniques in improving protein stability have been published [83–85]. The two first approaches are described more in detail.

9.4.1 Strategies for Designing Protein Analogues

Typically, rational based protein engineering uses high resolution three-dimensional structure information, and knowledge of sequences controlling the stability and activity is required. Optimally, the tertiary structure and sequences of the protein have been identified that are critical for the properties: stability, biological activity and solubility (Figure 9.2), so that specific residues or regions may be targeted for optimising specific properties.

Thus, a rational approach in this sense requires high quality data on the protein structure and function for site-directed mutagenesis.

An example of a semi-rational based protein design is the sequence homology approach, also termed the consensus concept, which has been applied to design stabilizing mutations [89]. The approach is based on the assumption that conserved amino acids at particular

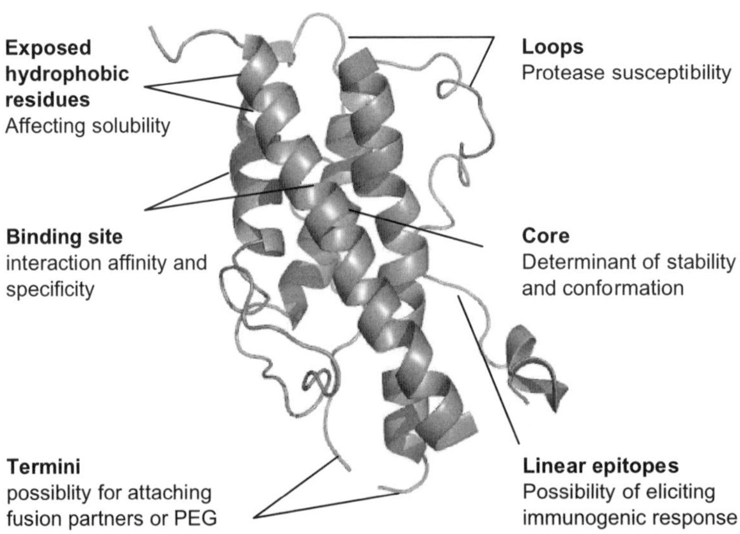

Figure 9.2 *Example of a model protein (human growth hormone) where the roles of specific amino acid regions have been fictively assigned. (modified from [86]). While rationally designing proteins, change of the primary structure in a given region may affect a wide range of properties at the same time. This figure was made using PyMOL [87] and the growth hormone structure file 1HGU [88]*

positions of homologous sequence alignment contribute more to the protein stability than residues which occur less frequently. Thus, a sequence database screening of the particular protein family is mandatory for performing sequence alignment. Using the protein alignment, potential stabilizing residues may be identified and tested individually and in combination.

More recently, computational screening techniques have been developed to find the optimal sequence for a defined three-dimensional structure allowing all or part of the sequence to change simultaneously. This method, termed protein design automation (PDA), ranks the fit of sequences to the three-dimensional structure using physical chemical functions that model the energetic interactions of the sequence of protein atoms, including steric, van der Waals, solvation and electrostatic interactions. A more thorough description of this method can be found elsewhere [90–93].

9.4.2 Advantages and Disadvantages

Whereas the pure rational approach as described has the force of relying on sound chemical and thermodynamic rationals, the method has the drawback that only a limited amount of analogues may be screened as compared to the *in silico* method. It has been reported that simultaneous consideration and modification of more than 60 residues may be done by computational methods whereas only 10 residues are estimated to be considered by the aforementioned approach [94]. This improves the possibility of finding novel protein sequences conferring improved stability to the protein. Moreover, Protein Design Automation (PDA) has the advantage of being able to control the location and type of mutation even though it is an ultra high-throughput screening tool. Thus, mutations may be directed towards the core of the protein, for instance, while the surface structure of the protein is kept intact, which may have the positive effect that biological activity and antigenicity properties are kept similar to the native protein. This is a vital advantage, as modification of amino acids in one region may potentially affect a wide range of protein properties, and thereby emphasizes the need for assessing multiple thermodynamic parameters simultaneously for successful protein design.

9.4.3 Pharmaceutical Aspects – General Principles for Improving Protein Stability

Thermostable proteins have been studied to elucidate factors controlling protein stability [95–98]. The studies suggested that thermophiles do not contain unique structural characteristics. However, these studies, along with other empirical findings given below, recognized some general traits that form the basis for improved protein stability by site-directed mutation. Figure 9.2 illustrates some of these. These should not be regarded as general rules but rather as guiding principles to be considered in the process of optimising the protein stability. These include:

- Replacement of free cysteine residues due to formation of unwanted intra- or intermolecular disulfide bonds. Cysteine to serine mutations have been successfully applied for therapeutically relevant proteins such as granulocyte colony-stimulating factor (G-CSF) and interferon (IFN) [99, 100].
- Introduction of mutations that decrease flexibility since proteolytic sites are often located in flexible loops [101].

- Optimization of interactions such as salt bridges, hydrogen bonding networks, disulfide bonds, hydrophobic interaction and hydrophobic burial in the core packing [93, 102].
- Reduction of the entropy of the unfolded state, and thereby an increase in the free energy difference between the folded and unfolded state. Introduction of disulfide bonds has been suggested to enhance the stability by this effect [103].

9.4.3.1 Design of a More Stable hGH Protein Analog–an Example

Recombinant human growth hormone (hGH, ~22 kDa) is a marketed protein therapeutic for treatment of dwarfism and used for children suffering low levels of hGH. hGH is a protein and the bioavailability after oral administration is very poor. Thus, hGH has been administered subcutaneously or intramuscularly several times a week. As hGH inherently has poor stability in solution, evidently, there has been an incentive to develop stable or sustained release formulations so as to increase the bioavailability and convenience of use for the patients. To this end, stable aqueous formulations of hGH have been successfully developed [104, 105] and marketed in easy to inject and convenient pen systems.

Another approach to overcome the inherent instability of hGH is to develop a more thermostable variant of hGH resulting in a longer shelf-life. In order to develop a new hGH analogue that has the desired attributes, PDA was used to computationally develop new hGH analogues. Only the core residues of the protein were targeted for redesigning the protein, since inclusion of surface proteins may increase the probability of an immunogenic response to the mutated protein. Besides, focusing on the core residues simplifies the thermodynamic considerations used for screening of new analogues. 45 amino acid residues in the core protein were chosen for optimization. An attempt was made to identify amino acid substitutions that could lower the energy of the wild-type protein, and thereby make the protein thermodynamically more stable. Two mutants (CORE1 and CORE2) were identified; these had six and eight amino acid substitutions, respectively, as compared to the wild type. The mutants along with the wild-type hGH were expressed, isolated and the thermal melting point was assessed, as this value is generally accepted to be indicative of the thermostability of the protein. Full bioactivity of the mutants as compared to the wild type was also confirmed. The designed hGH analogues showed enhancement of 13 °C and 16 °C, respectively (CORE1 and CORE2). The increased stability is suggested to stem from improved van der Waals packing interactions and increased burial of hydrophobic groups in the core, and replacement of unfavourable hydrogen bond acceptors and donors with hydrophobic residues. It is noteworthy that in the CORE2 protein, which exhibited highest thermal stability, eight mutations had been introduced of which four were serine → alanine substitutions. It would have been interesting to study this mutation more in-depth individually, but apparently these mutations contribute to enhance the thermal stability of hGH protein to a considerable extent and highlight amino acid mutation as a viable method for increasing the stability of a therapeutic protein [93].

A similar strategy (PDA) has been successfully applied to optimize the physical and chemical stability of granulocyte colony-stimulating factor (G-CSF, ~19 kDa). A 10-fold improvement in storage stability was observed for the mutant G-CSF as compared to the wild type under accelerated conditions. The increased stability of the mutant was suggested to be due to improved packing interactions, significant improvement in helical propensity triggered by Gly28 → Ala28 substitution and reduced aggregation due to elimination of a free cysteine. The mutant was tested to be as biological active as the wild type [94].

9.4.4 Pharmacokinetic Aspects

Molecular engineering has also been successfully applied with the primary aim of altering the pharmacokinetic profile of a therapeutic protein in order to achieve superior clinical treatment. For instance, modern insulin analogues have been designed (and marketed) with the rationale of achieving more rapid acting insulin analogues compared to native human insulin. Human insulin has been reported to act within 30–60 minutes of injection whereas modern insulin analogues (insulin lispro (Humalog®), insulin aspart (NovoRapid®) and insulin glusine (Apidra®)) act within 5–15 minutes of injection [106, 107]. Basically, the mode for achieving the rapid onset of action for the above analogues has been to reduce the tendency of self-association of insulin, so that monomers are formed faster as these are readily absorbed into the systemic circulation upon injection and dilution in the subcutaneous layer. For insulin aspart this has been achieved by exchange of proline for aspartic acid at position B28, which results in charge repulsion between monomers and steric hindrance due to local conformation change at the C- terminus of the B-chain and, thereby, reduces self-association of insulin into dimers and hexamers [108]. Insulin lispro differs from human insulin by switching the two amino acids at B28 (Pro → Lys) and B29 (Lys → Pro). The modification reduces the non-polar and β-sheet interactions between insulin monomers, resulting in reduced self-association and thereby increased rate of absorption of the insulin analogue [108]. For insulin glusine, association of monomers to dimers has been reduced by mutations of amino acid residues: B3 (Asp → Lys) and B29 (Lys → Glu). The exchange of the neutral and hydrophilic asparagine at B3 for the basic and hydrophilic lysine has been reported to affect the monomer–dimer equilibrium by introducing slight steric and electrostatic repulsion. Similarly, the negative charge and smaller size of glutamic acid at B29 reduces dimerization of the insulin analogue [109]. So for all analogues faster onset of action compared to native human insulin has been confirmed, and although the mutations are different for each analogue, the overall target of reducing self-association of insulin is evident for all three analogues.

However, altering the primary structure of proteins may alter the stability profile, as described in the previous section. Insulin at higher concentration exists in a stabilized self-associated (hexameric) state in the presence of zinc ions while it is less stable in the dimeric and monomeric state [110]. As the insulin analogues augment the equilibrium towards the monomeric state, the stability of insulin becomes a challenge which may be overcome by selection of an appropriate formulation. Insulin aspart and lispro both contain zinc and phenol in their respective formulations as stabilizers to shift the equilibrium towards the hexameric state. However, insulin glusine does not contain these stabilizers but is formulated with polysorbate 20 instead to overcome the stability issue.

9.4.5 Manipulating the Isoelectrical Point (pI) by Molecular Engineering

Molecular engineering has also successfully been applied to manipulate the isoelectrical point (pI) of a protein. The rationale may be to increase or decrease solubility at a given pH value. Design of the long-acting insulin analogue, insulin glargine (Lantus®), a once daily basal insulin product, is an example of how pI-shift has resulted in a blockbuster product. Two basic residues (arginine) have been added to the C-terminus of the B chain of human insulin, and at the same time the acidic asparagine at A21 has been exchanged for glycine, resulting in a rise of the isoelectrotic point from 5.5 to about 7.0 [109]. The drug product is formulated and injected as an acidic solution (pH 4.0) and the reduced solubility at physiological pH facilitates prolonged absorption from the injection site as the insulin analogue precipitates in the subcutis after injection. Thus, molecular engineering is a useful tool to manipulate pI and, thereby, the solubility-pH profile.

In conclusion, amino acid substitution is an interesting pathway for manipulating protein solubility, stability and the pharmacokinetic profile. However, a keen understanding of the protein structure, activity and chemistry should be established for a rational approach to be applied so that either of the two parameters are not compromised by optimising the third.

9.5 General Considerations on Processing and Characterization

Even though solid improvements in pharmacokinetics may be obtained by attachment of lipophilic groups or PEG chains to proteins, the challenges and costs of processing must be considered.

In general, only pure protein preparations are modified in order to reduce validation and separation challenges after modification. Thus, the production of functionalized therapeutic proteins involves additional processing steps when compared to their non-modified counterparts, and the high production cost of the purified non-modified protein used as raw material in the acylation or PEGylation reaction demands focus on maximizing the reaction extent and specificity of the process [6].

In addition, it must be taken into account that the methods of analysis used for characterization of the starting material will not always be feasible for the modified proteins, since protein modification may lead to altered size, charge, hydrophobicity, metal chelation, hydrogen bonding and affinity of the proteins. Especially, preparative and analytical separation methods are affected by the conjugation of PEG to the protein [6]. In order to separate modified proteins from other reaction product mixture components, differences in size, charge and hydrophobicity can be exploited.

9.6 Conclusions

The fast development within protein engineering and modification in the last decades is of great potential for turning peptides and proteins into blockbuster pharmaceuticals. Proteins that may have been thought of no use in drug delivery due to poor stability, fast clearance or extensive degradation may suddenly be given another chance after modification.

Conducting modifications is not an easy task and there is a risk that modification in one part of a protein may affect another part in a suboptimal way. Thus, a thorough knowledge of the protein, especially with regard to amino acids essential for bioactivity and stability, is a prerequisite for a successful approach.

Proteins that are lipidated or PEGylated often display a decreased bioactivity compared to the native protein. It must be borne in mind that the structure and function of the native protein has been optimized through many thousand years of evolution, and modification of the protein close to areas involved in receptor binding will therefore inherently lead to decreased binding affinity. However, the improvements in stability and residence time in the body often compensate for the loss of activity, and the overall value of protein modification can, therefore, often only be predicted by *in vivo* studies.

References

1. Daly, S.M., Przybycien, T.M. and Tilton, R.D. (2005) Adsorption of poly(ethylene glycol)-modified ribonuclease a to a poly(lactide-co-glycolide) surface. *Biotechnol. Bioeng.*, **90**, 856–868.
2. Morar, A.S., Schrimsher, J.L. and Chavez, M.D. (2006) PEGylation of proteins: a structural approach. *Biopharm. Int.*, **19**, 34–49.
3. Morpurgo, M. and Veronese, F.M. (2004) Conjugates of peptides and proteins to polyethylene glycols. *Methods Mol. Biol.*, **283**, 45–70.
4. Roberts, M.J., Bentley, M.D. and Harris, J.M. (2002) Chemistry for peptide and protein PEGylation. *Adv. Drug Deliv. Rev.*, **54**, 459–476.
5. Pasut, G. and Veronese, F.M. (2006) PEGylation of proteins as tailored chemistry for optimized bioconjugates. *Adv. Polym. Sci.*, **192**, 95–134.
6. Fee, C.J. and Van Alstine, J.A. (2006) PEG-proteins: reaction engineering and separation issues. *Chem. Eng. Sci.*, **61**, 924–939.
7. Harris, J.M., Martin, N.E. and Modi, M. (2001) Pegylation: a novel process for modifying pharmacokinetics. *Clin. Pharmacokinet.*, **40**, 539–551.
8. Veronese, F.M. and Pasut, G. (2005) PEGylation, successful approach to drug delivery. *Drug Discov. Today*, **10**, 1451–1458.
9. Hinds, K.D. and Kim, S.W. (2002) Effects of PEG conjugation on insulin properties. *Adv. Drug Deliv. Rev.*, **54**, 505–530.
10. Sato, H. (2002) Enzymatic procedure for site-specific pegylation of proteins. *Adv. Drug Deliv. Rev.*, **54**, 487–504.
11. Kinstler, O., Molineux, G., Treuheit, M. *et al.* (2002) Mono-N-terminal poly(ethylene glycol)-protein conjugates. *Adv. Drug Deliv. Rev.*, **54**, 477–485.
12. Kinstler, O.B., Brems, D.N., Lauren, S.L. *et al.* (1996) Characterization and stability of N-terminally PEGylated rhG-CSF. *Pharm. Res.*, **13**, 996–1002.
13. DeFrees, S., Wang, Z.G., Xing, R. *et al.* (2006) GlycoPEGylation of recombinant therapeutic proteins produced in *Escherichia coli*. *Glycobiology*, **16**, 833–843.
14. Caliceti, P., Schiavon, O., Sartore, L. *et al.* (1993) Active site protection of proteolytic enzymes by poly(ethylene glycol) surface modification. *J. Bioact. Compat. Polym.*, **8**, 41–50.
15. Goodson, R.J. and Katre, N.V. (1990) Site-directed pegylation of recombinant interleukin-2 at its glycosylation site. *Biotechnology*, **8**, 343–346.
16. Greenwald, R.B., Yang, K., Zhao, H. *et al.* (2003) Controlled release of proteins from their poly(ethylene glycol) conjugates: drug delivery systems employing 1,6-elimination. *Bioconjug. Chem.*, **14**, 395–403.
17. Lee, S., Greenwald, R.B., McGuire, J. *et al.* (2001) Drug delivery systems employing 1,6-elimination: releasable poly(ethylene glycol) conjugates of proteins. *Bioconjug. Chem.*, **12**, 163–169.
18. Chapman, A.P. (2002) PEGylated antibodies and antibody fragments for improved therapy: a review. *Adv. Drug Deliv. Rev.*, **54**, 531–545.
19. Yamaoka, T., Tabata, Y. and Ikada, Y. (1994) Distribution and tissue uptake of poly(ethylene glycol) with different molecular weights after intravenous administration to mice. *J. Pharm. Sci.*, **83**, 601–606.

20. Basu, A., Yang, K., Wang, M.L. *et al.* (2006) Structure-function engineering of interferon-beta-1b for improving stability, solubility, potency, immunogenicity, and pharmacokinetic properties by site-selective mono-PEGylation. *Bioconjug. Chem.*, **17**, 618–630.
21. Harris, J.M. and Chess, R.B. (2003) Effect of pegylation on pharmaceuticals. *Nat. Rev. Drug Discov.*, **2**, 214–221.
22. Chirino, A.J., Ary, M.L. and Marshall, S.A. (2004) Minimizing the immunogenicity of protein therapeutics. *Drug Discov. Today*, **9**, 82–90.
23. Hermeling, S., Crommelin, D.J., Schellekens, H. and Jiskoot, W. (2004) Structure-immunogenicity relationships of therapeutic proteins. *Pharm. Res.*, **21**, 897–903.
24. Bailon, P., Palleroni, A., Schaffer, C.A. *et al.* (2001) Rational design of a potent, long-lasting form of interferon: a 40 kDa branched polyethylene glycol-conjugated interferon alpha-2a for the treatment of hepatitis C. *Bioconjug. Chem.*, **12**, 195–202.
25. Wang, Y.S., Youngster, S., Grace, M. *et al.* (2002) Structural and biological characterization of pegylated recombinant interferon alpha-2b and its therapeutic implications. *Adv. Drug Deliv. Rev.*, **54**, 547–570.
26. Mosharraf, M., Malmberg, M. and Fransson, J. (2007) Formulation, lyophilization and solid-state properties of a pegylated protein. *Int. J. Pharm.*, **336**, 215–232.
27. Manjula, B.N., Tsai, A., Upadhya, R. *et al.* (2003) Site-specific PEGylation of hemoglobin at Cys-93(beta): correlation between the colligative properties of the PEGylated protein and the length of the conjugated PEG chain. *Bioconjug. Chem.*, **14**, 464–472.
28. Rajan, R.S., Li, T.S., Aras, M. *et al.* (2006) Modulation of protein aggregation by polyethylene glycol conjugation: G-CSF as a case study. *Protein Sci.*, **15**, 1063–1075.
29. Stigsnaes, P., Frokjaer, S., Bjerregaard, S. *et al.* (2007) Characterisation and physical stability of PEGylated glucagon. *Int. J. Pharm.*, **330**, 89–98.
30. Wang, W. (1999) Instability, stabilization, and formulation of liquid protein pharmaceuticals. *Int. J. Pharm.*, **185**, 129–188.
31. Wang, W. (2005) Protein aggregation and its inhibition in biopharmaceutics. *Int. J. Pharm.*, **289**, 1–30.
32. Daly, S.M., Przybycien, T.M. and Tilton, R.D. (2005) Adsorption of poly(ethylene glycol)-modified lysozyme to silica. *Langmuir*, **21**, 1328–1337.
33. Daly, S.M., Przybycien, T.M. and Tilton, R.D. (2007) Aggregation of lysozyme and of poly(ethylene glycol)-modified lysozyme after adsorption to silica. *Colloid. Surf. B*, **57**, 81–88.
34. Blundell, T., Dodson, G., Hodgkin, D. and Mercola, D. (1972) Insulin: the structure in the crystal and its reflection in chemistry and biology. *Adv. Prot. Chem.*, **26**, 279–402.
35. Blundell, T. (1983) The conformation of glucagon, in *Glucagon I* (ed. P.J. Lefébre), Springer Verlag, Germany, pp. 37–56.
36. Hansen, J.F. (1991) The self-association of zinc-free human insulin and insulin analogue B13-glutamine. *Biophys. Chem.*, **39**, 107–110.
37. Brange, J. and Langkjaer, L. (1993) Insulin structure and stability. *Pharm. Biotechnol.*, **5**, 315–350.
38. Al-Azzam, W., Pastrana, E.A., King, B. *et al.* (2005) Effect of the covalent modification of horseradish peroxidase with poly(ethylene glycol) on the activity and stability upon encapsulation in polyester microspheres. *J. Pharm. Sci.*, **94**, 1808–1819.
39. Castellanos, I.J., Al-Azzam, W. and Griebenow, K. (2005) Effect of the covalent modification with poly(ethylene glycol) on alpha-chymotrypsin stability upon encapsulation in poly(lactic-co-glycolic) microspheres. *J. Pharm. Sci.*, **94**, 327–340.
40. Boutin, J.A. (1997) Myristoylation. *Cell Signal.*, **9**, 15–35.
41. Roy, M.O., Uppenberg, J., Robert, S.P. *et al.* (1997) Crystallization of monoacylated proteins: influence of acyl chain length. *Eur. Biophys. J.*, **26**, 155–162.
42. Markussen, J. (1996) Soluble, fatty acid acylated insulins bind to albumin and show protracted action in pigs. *Diabetologia*, **39**, 281–288.
43. Knudsen, L.B., Nielsen, P.F., Huusfeldt, P.O. *et al.* (2000) Potent derivatives of glucagon-like peptide-1 with pharmacokinetic properties suitable for once daily administration. *J. Med. Chem.*, **43**, 1664–1669.

44. Hashimoto, M., Takada, K., Kiso, Y. and Muranishi, S. (1989) Synthesis of palmitoyl derivatives of insulin and their biological activities. *Pharm. Res.*, **6**, 171–176.
45. Yuan, L., Wang, J. and Shen, W.C. (2005) Reversible lipidization prolongs the pharmacological effect, plasma duration, and liver retention of octreotide. *Pharm. Res.*, **22**, 220–227.
46. Wang, J., Chow, D., Heiati, H. and Shen, W.C. (2003) Reversible lipidization for the oral delivery of salmon calcitonin. *J. Control. Release*, **88**, 369–380.
47. Wang, J., Wu, D. and Shen, W.C. (2002) Structure–activity relationship of reversibly lipidized peptides: studies of fatty acid-desmopressin conjugates. *Pharm. Res.*, **19**, 609–614.
48. Ekrami, H.M., Kennedy, A.R. and Shen, W.C. (1995) Water-soluble fatty acid derivatives as acylating agents for reversible lipidization of polypeptides. *FEBS Lett.*, **371**, 283–286.
49. Wang, J., Shen, D. and Shen, W.C. (1999) Preparation, purification, and characterization of a reversibly lipidized desmopressin with potentiated anti-diuretic activity. *Pharm. Res.*, **16**, 1674–1679.
50. Cheng, W., Satyanarayanajois, S. and Lim, L.Y. (2007) Aqueous-soluble, non-reversible lipid conjugate of salmon calcitonin: Synthesis, characterization and *in vivo* activity. *Pharm. Res.*, **24**, 99–110.
51. Wan, L., Chen, Y.H. and Chang, T.W. (2003) Improving pharmacokinetic properties of adrenocorticotropin by site-specific lipid modification. *J. Pharm. Sci.*, **92**, 1882–1892.
52. Kurtzhals, P., Havelund, S., Jonassen, I. *et al.* (1995) Albumin binding of insulins acylated with fatty acids: characterization of the ligand-protein interaction and correlation between binding affinity and timing of the insulin effect *in vivo*. *Biochem. J.*, **312**, 725–731.
53. Geiger, R., Obermeier, R., Teetz, V. *et al.* (1980) Biological activity of insulin analogues substituted at the amino group of B1-phenylalanine, in *Insulin Chemistry, Structure and Function of Insulin and Related Hormones* (eds D. Brandenburg and A. Wollmer), Walter de Gruyter & Co, Berlin, pp. 409–415.
54. Havelund, S., Plum, A., Ribel, U. *et al.* (2004) The mechanism of protraction of insulin detemir, a long-acting, acylated analog of human insulin. *Pharm. Res.*, **21**, 1498–1504.
55. Whittingham, J.L., Jonassen, I., Havelund, S. *et al.* (2004) Crystallographic and solution studies of N-lithocholyl insulin: a new generation of prolonged-acting human insulins. *Biochemistry*, **43**, 5987–5995.
56. Jonassen, I., Havelund, S., Ribel, U. *et al.* (2006) Biochemical and physiological properties of a novel series of long-acting insulin analogs obtained by acylation with cholic acid derivatives. *Pharm. Res.*, **23**, 49–55.
57. Klein, O., Lynge, J., Endahl, L. *et al.* (2007) Albumin-bound basal insulin analogues (insulin detemir and NN344): comparable time-action profiles but less variability than insulin glargine in type 2 diabetes. *Diabetes Obes. Metab.*, **9**, 290–299.
58. Heise, T., Nosek, L., Ronn, B.B. *et al.* (2004) Variability of insulin detemir in comparison to NPH insulin and insulin glargine in people with type 1 diabetes. *Diabetes*, **53**, 1614–1620.
59. Kurtzhals, P. (2005) How to achieve a predictable basal insulin? *Diabetes Metab.*, **31**, 4S25–4S33.
60. Gutniak, M.K., Linde, B., Holst, J.J. and Efendic, S. (1994) Subcutaneous injection of the incretin hormone glucagon-like peptide 1 abolishes postprandial glycemia in NIDDM. *Diabetes Care*, **17**, 1039–1044.
61. Arulmozhi, D.K. (2006) GLP-1 based therapy for type 2 diabetes. *Eur. J. Pharm. Sci.*, **28**, 96–108.
62. Ribel, U., Larsen, M.O., Rolin, B. *et al.* (2002) NN2211: a long-acting glucagon-like peptide-1 derivative with anti-diabetic effects in glucose-intolerant pigs. *Eur. J. Pharmacol.*, **451**, 217–225.
63. Toth, I., Malkinson, J.P., Flinn, N.S. *et al.* (1999) Novel lipoamino acid- and liposaccharide-based system for peptide delivery: application for oral administration of tumor-selective somatostatin analogues. *J. Med. Chem.*, **42**, 4010–4013.
64. Toth, I., Flinn, N., Hillery, A. *et al.* (1994) Lipidic conjugates of luteinizing hormone releasing hormone (LHRH)+ and thyrotropin releasing hormone (TRH)+ that release and protect the native hormones in homogenates of human intestinal epithelial (Caco-2) cells. *Int. J. Pharm.*, **105**, 241–247.

65. Flinn, N., Hussain, I., Shaw, A. *et al.* (1996) Oral absorption studies of lipid-polylysine conjugates of thyrotropin releasing hormone (TRH1) and luteinizing hormone releasing hormone (LHRH1). *Int. J. Pharm.*, **138**, 167–174.
66. Blanchfield, J.T., Gallagher, O.P., Cros, C. *et al.* (2007) Oral absorption and *in vivo* biodistribution of alfa-conotoxin MII and a lipidic analogue. *Biochem. Bioph. Res. Co.*, **361**, 97–102.
67. Foldvari, M., Attah-Poku, S., Hu, J. *et al.* (1998) Palmitoyl derivatives of interferon alpha: potential for cutaneous delivery. *J. Pharm. Sci.*, **87**, 1203–1208.
68. Foldvari, M., Baca-Estrada, M.E., He, Z. *et al.* (1999) Dermal and transdermal delivery of protein pharmaceuticals: lipid-based delivery systems for interferon alpha. *Biotechnol. Appl. Biochem.*, **30** (Pt 2), 129–137.
69. Kellam, B., Drouillat, B., Dekany, G. *et al.* (1998) Synthesis and *in vitro* evaluation of lipoamino acid and carbohydrate-modified enkephalins as potential antinociceptive agents. *Int. J. Pharm.*, **161**, 55–64.
70. Gozes, I., Bardea, A., Reshef, A. *et al.* (1996) Neuroprotective strategy for Alzheimer disease: intranasal administration of a fatty neuropeptide. *Proc. Natl. Acad. Sci. USA*, **93**, 427–432.
71. Magdassi, S., Kamyshny, A. and Baszkin, A. (2001) Interfacial properties of hydrophobically modified biomolecules: fundamental aspects and applications. *J. Disper. Sci. Technol.*, **22**, 313–322.
72. Whittingham, J.L., Havelund, S. and Jonassen, I. (1997) Crystal structure of a prolonged-acting insulin with albumin-binding properties. *Biochemistry*, **36**, 2826–2831.
73. Olsen, H.B. and Kaarsholm, N.C. (2000) Structural effects of protein lipidation as revealed by LysB29-myristoyl, des(B30) insulin. *Biochemistry*, **39**, 11893–11900.
74. Yonemoto, W., McGlone, M.L. and Taylor, S.S. (1993) N-myristylation of the catalytic subunit of cAMP-dependent protein kinase conveys structural stability. *J. Biol. Chem.*, **268**, 2348–2352.
75. Traxler, K.W. and Dewey, T.G. (1994) Effects of depalmitoylation on physicochemical properties of rhodopsin. *Biochemistry*, **33**, 1718–1723.
76. Kennedy, M.T., Brockman, H. and Rusnak, F. (1996) Contributions of myristoylation to calcineurin structure/function. *J. Biol. Chem.*, **271**, 26517–26521.
77. Wang, J. and Shen, W.C. (2000) Gastric retention and stability of lipidized Bowman-Birk protease inhibitor in mice. *Int. J. Pharm.*, **204**, 111–116.
78. Delie, F., Couvreur, P., Nisato, D. *et al.* (1994) Synthesis and *in vitro* study of a diglyceride prodrug of a peptide. *Pharm. Res.*, **11**, 1082–1087.
79. Pignatello, R. and Pecora, T.M.G. (2007) Conjugation of thymopentin (TP5) with lipoamino acid residues increases the hydrolytic stability and preserves the biological activity. *Pharmazie*, **62**, 663–667.
80. Fahmy, T.M., Samstein, R.M., Harness, C.C. and Mark Saltzman, W. (2005) Surface modification of biodegradable polyesters with fatty acid conjugates for improved drug targeting. *Biomaterials*, **26**, 5727–5736.
81. Hecky, J. and Muller, K.M. (2005) Structural perturbation and compensation by directed evolution at physiological temperature leads to thermostabilization of beta-lactamase. *Biochemistry*, **44**, 12640–12654.
82. Kotz, J.D., Bond, C.J. and Cochran, A.G. (2004) Phage-display as a tool for quantifying protein stability determinants. *Eur. J. Biochem.*, **271**, 1623–1629.
83. Bommarius, A.S., Broering, J.M., Chaparro-Riggers, J.F. and Polizzi, K.M. (2006) High-throughput screening for enhanced protein stability. *Curr. Opin. Biotech.*, **17**, 606–610.
84. Eijsink, V.G.H., Bjørk, A., Seidnes, S. *et al.* (2004) Rational engineering of enzyme stability. *J. Biotechnol.*, **113**, 105–120.
85. Eijsink, V.G.H., Seidnes, S., Borchert, T.V. and Van Den Burg, B. (2005) Directed evolution of enzyme stability. *Biomol. Eng.*, **22**, 21–30.
86. Marshall, S.A., Lazar, G.A., Chirino, A.J. and Desjarlais, J.R. (2003) Rational design and engineering of therapeutic proteins. *Drug Discov. Today*, **8**, 221.
87. DeLano, W.L. (2002) *The PyMOL Molecular Graphics System*, DeLano Scientific, Palo Alto, CA, USA.

88. Chantalat, L., Jones, N.D., Korber, F. *et al.* (1995) The crystal structure of wild-type growth hormone at 2.5 Angstrom resolution. *Protein Pept. Lett.*, **2**, 333–340.
89. Zakrzewska, M., Krowarsch, D., Wiedlocha, A. and Otlewski, J. (2004) Design of fully active FGF-I variants with increased stability. *Protein Eng. Des. Sel.*, **17**, 603–611.
90. Dahiyat, B.I. and Mayo, S.L. (1996) Protein design automation. *Protein Sci.*, **5**, 895–903.
91. Dahiyat, B.I. and Mayo, S.L. (1997) De novo protein design: fully automated sequence selection. *Science*, **278**, 82–87.
92. Dahiyat, B.I. and Mayo, S.L. (1997) Probing the role of packing specificity in protein design. *Proc. Natl. Acad. Sci. USA*, **94**, 10172–10177.
93. Filikov, A.V., Hayes, R.J., Luo, P. *et al.* (2002) Computational stabilisation of human growth hormone. *Protein Sci.*, **11**, 1452–1461.
94. Luo, P., Hayes, R.J., Chan, C. *et al.* (2002) Development of a cytokine analog with enhanced stability using computational ultrahigh throughput screening. *Protein Sci.*, **11**, 1218–1226.
95. Eidness, M.K., Richie, K.A., Burden, A.E. *et al.* (1997) Dissecting contributions to the thermostability of pyrococcus furiosus rubredoxin: b-sheet chimeras. *Biochemistry*, **36**, 10406–10413.
96. Jiang, X., Bishop, E.J. and Farid, R.S. (1997) A de novo designed protein with properties that characterize natural hyperthermophilic proteins. *J. Am. Chem. Soc.*, **119**, 838–839.
97. Lee, B. and Vasumatzis, G. (1997) Stabilization of protein structures. *Curr. Opin. Biotech.*, **8**, 423–428.
98. Rees, D.C. and Adams, M.W.W. (1995) Hyperthermophiles: taking the heat and loving it. *Structure*, **3**, 251–254.
99. Arakawa, T., Prestrelski, S.J., Narhi, L.O. *et al.* (1993) Cysteine 17 of recombinant human granulocyte-colony stimulating factor is partially solvent-exposed. *J. Protein Chem.*, **12**, 525–531.
100. Lin, L. (1998) Betaseron, in *Characterisation of Biotechnology Pharmaceutical Products* (eds F. Brown, A. Lubiniecki and G. Murano), Karger, Basel, pp. 97–104.
101. Nishi, N., Itoh, A., Fujiyama, A. *et al.* (2005) Development of highly stable galectins: truncation of the linker peptide confers protease-resistance on tandem-repeat type galectins. *FEBS Lett.*, **579**, 2058–2064.
102. Malakauskas, S.M. and Mayo, S.L. (1998) Design, structure and stability of a hyperthermophilic protein variant. *Nat. Struct. Biol.*, **5**, 470–475.
103. Pantoliano, M.W., Ladner, R.C., Bryan, P.N. *et al.* (1987) Protein engineering of subtilisin BPN': enhanced stabilization through the introduction of two cysteines to form a disulfide bond. *Biochemistry*, **26**, 2077–2082.
104. Kappelgaard, A.-M., Bojesen, A., Skydsgaard, K. *et al.* (2004) Hormone: preservatives and buffers. *Horm. Res.*, **62**, 98–103.
105. O'Conner, B.H. and Oeswein, J.Q. (1998) Human growth hormone aqueous formulation. Genentech Inc. US Patent 5,763,394.
106. Bethel, M.A. and Feinglos, M.N. (2005) Basal insulin therapy, in type 2 diabetes. *J. Am. Board Fam. Pract.*, **18**, 199–204.
107. Garg, S.K., Rosenstock, J. and Ways, K. (2005) Optimised basal-bolus insulin regimens in type 1 diabetes: insulin glusine versus regular human insulin in combination with basal insulin glargine. *Endocr. Pract.*, **11**, 11–17.
108. Brange, J. and Volund, A. (1999) Insulin analogs with improved pharmacokinetic profiles. *Adv. Drug Deliver. Rev.*, **35**, 307–335.
109. Becker, R.H.A. (2007) Insulin glusine complementing basal insulins: a review of structure and activity. *Diabet. Technol. Ther.*, **9**, 109–121.
110. Brange, J., Andersen, L., Laursen, E.D. and Meyn, G.R.E. (1997) Towards understanding insulin fibrillation. *J. Pharm. Sci.*, **86**, 517–525.

10

Nanocarriers for the Delivery of Peptides and Proteins

Kenneth Lundstrom
Pan Therapeutics, Lutry, Switzerland

10.1 Introduction

Modern drug discovery and development have recently experienced much attention for being inefficient and expensive. Much of the criticism relates, obviously, to the ever increasing requirements for more efficient and safer drugs. One of the major issues in drug development has been and still is the question of drug delivery. Intravenous, subcutaneous and intramuscular injections are routes for parenteral drug delivery [1]. Local delivery by epidural and intra-articulate injection represents other modes for parenteral delivery. Although some advances have been made in non-invasive drug delivery by applying transdermal, pulmonary and nasal routes they have been hampered by the poor mucosal membrane permeability and hostile delivery environment (gastrointestinal tract).

Attempts to provide well-defined concentrations and extended mode of action, especially of protein and peptide-based drugs, have particularly focused on sustained-release systems. Indeed, this approach has resulted in a number of drug delivery systems, which certainly have provided improvement in drug efficacy, reduction in pain due to a less frequent injection regimen, decreased drug doses and less adverse side effects.

Recent development has seen new innovations applying the popular nanoparticle approach. The definition of nanoparticles is quite broad and mainly refers to the size of the particles, which are employed for the delivery of the drug. Obviously, the size is therefore smaller than the micrometer range, although in certain applications engineered

Delivery Technologies for Biopharmaceuticals: Peptides, Proteins, Nucleic Acids and Vaccines Edited by Lene Jorgensen and Hanne Mørck Nielsen

nanoparticles can provide properties of assembly into larger structures, that is microparticles.

The types of drugs that nanoparticles can harbour are almost limitless. In this context, small molecule drugs, protein and peptide-based drugs as well as nucleic acids have successfully been introduced into nanoparticles. Moreover, they have demonstrated superior properties in both preclinical and clinical studies in comparison to the naked drug. Likewise, the composition of nanoparticles varies to a great extent. For instance, they consist of different polymers and liposomes, as well as combinations thereof. Additionally, a number of systems have been developed that are based on *in situ* depot formation, precipitation, thermal gelling, cross-linking and thermoplastic semisolids. This chapter provides a review of recent applications of nanoparticles in drug delivery with a special emphasis on the delivery of protein and peptide-based drugs. Examples of drug entities other than peptides and proteins are included to illustrate the full potential of the delivery systems. Additionally, comments are added to the present view on the possibility of using the delivery systems to deliver peptides and proteins.

10.2 Polymeric Nanoparticles

A number of microsphere formulations have been applied to generate micro- and nanoparticulate systems for both subcutaneous and intramuscular injections. Because microspheres are non-retrievable after administration, they are generally prepared synthetically as biodegradable formulations. As described below, the chemical structure and composition of the microspheres represent a large variation, which also has enabled the formulation of a wide range of delivery systems with major differences in respect to drug loading properties, release rate and duration of action.

10.2.1 Lactic/Glycotide Polymers

Poly(lactic acid) (PLA), polyglycolic and poly(lactic-*co*-glycolic acid) (PLGA) are commonly used polymers for which biodegradable microsphere formulations have been obtained. Because of their hydrophobicity and relatively high gas transition temperature, PLGA polymers are dissolved in organic solvents before further processing. Application of different encapsulation methods include single or double (water/oil/water) emulsion methods. Also, spray-drying methods have been developed, although ink jet-based technology has allowed better control of production and a more uniform particle size distribution [2]. As proteins in contact with solvents are sensitive to degradation during the encapsulation procedure, an anhydrous spray freeze-drying procedure has been developed by Alkermes [3]. In this context, PLGA formulations of the recombinant human growth hormone (rhGH) showed improved total insulin-like growth factor-1 (IGF-1) responses in juvenile rhesus monkeys. Furthermore, the rhGH PLGA formulations demonstrated good safety profiles in growth hormone-deficient adults. Unfortunately, the high production costs led to discontinuation of the commercialization. In another approach, a coaxial ultrasonic atomizer was applied to generate microcapsules of proteins with a thin layer of PLGA [4]. This method allowed efficient encapsulation of lysozyme, which demonstrated no loss of functional integrity and a near zero-order kinetics over 50 days. Additionally, the use of supercritical carbon dioxide ($scCO_2$) for the encapsulation of insulin in PLA microspheres resulted in the retention of more than 80% of the hyperglycemic activity of insulin [5].

The success of peptide-PLGA polymer formulations has resulted in the approval and commercialization of a number of products. Commonly, a sustained-release profile of 1–4 months has been obtained for such peptides as leuprolide (Lupron Depot® by TAP), octreotide (Sandostatin LAR® by Novartis) and triptorelin (Trelstar® by Debiopharm). Moreover, the first rhGH PLGA formulation (Nutropin Depot® by Genentech) was launched in 2000 (discontinued in 2004). A sustained-release PLGA formulation for the small molecule risperidone (Risperdal Consta® by Johnson & Johnson) has been developed for the treatment of schizophrenia. Furthermore, PLGA microspheres for sustained release of other small molecules [6–8], peptides [9–11], vaccines [12] and proteins [13] have been evaluated.

10.2.2 Polycaprolactones

In attempts to achieve extended sustained release of drug molecules, poly-ε-caprolactones (PCLs) have been evaluated [14]. PLCs are biodegradable semicrystalline polyesters that have demonstrated a very slow degradation and sustained drug release has been obtained for more than a year. A number of different delivery systems based on caprolactones have been developed in the form of microspheres, nanospheres and implants. Various types of molecules, such as antigens, anti-hypersensitive drugs, chemotherapeutic agents and antibiotics, have been encapsulated in PCLs. In attempts to obtain appropriate drug release characteristics microspheres of PCLs, either alone or as PCL copolymers, have been formulated. One characteristic of PLCs is that their degradation will not result in an acidic microenvironment, which in the case of PLAs and PLGAs has a negative impact on the stability of therapeutic proteins.

An *in vivo* study in streptozotocin-induced hypoglycemic rats demonstrated that subcutaneous injection of insulin encapsulated in the biodegradable depot-forming PLC microspheres controlled the drug release and plasma sugar levels more efficiently than injection of naked insulin [15]. According to preliminary pharmacokinetic evaluation, the plasma insulin concentration remained steady during the study period of 60 days.

Novel synthetic biodegradable amphiphilic copolymers based on PCLs and chitosan were synthesized by coupling the hydroxyl end groups on preformed PCL chains and the amino groups present on 6-O-triphenylmethyl chitosan and by removing the protective 6-O-triphenylmethyl groups in acidic aqueous solution [16]. This approach resulted in spherical or elliptic nanoparticles in water.

10.2.3 Polyphosphoesters

Another approach has been to formulate biodegradable polyphosphoesters as carriers for sustained release of small molecule drugs [17], proteins [18] and nucleic acids [19]. In this context, the nerve growth factor (NGF) delivery from polyphosphoester microspheres was evaluated for the repair of nerve injury [20]. This approach demonstrated peripheral nerve regeneration and could provide the potential for improved functional recovery in patients whose injured nerves are reconstructed by entubulation.

Recently, novel biodegradable hydrogels were obtained by photo-cross-linking macromers based on polyphosphoesters and poly(ethylene glycol) (PEG) [21]. These macromers were compatible to osteoblast cells and did not exhibit any significant cytotoxicity up to 0.5 mg/ml. Viability cell staining demonstrated that the majority of the osteoblast cells remained viable after encapsulation, showing their potential as vehicles for tissue engineering. Furthermore, a series of polymers with a phosphoester backbone have been synthesized [22]. Introducing different charge groups in the side chain connected to the backbone resulted in safe and effective non-viral gene carriers with different properties related to DNA binding capacity, cytotoxicity, DNA protection, biodegradability, DNA release kinetics and transfection efficiency.

A poly(lactide-*co*-ethylphosphate) microsphere formulation of paclitaxel (Paclimer®, Guilford Pharmaceuticals) has been applied for the treatment of ovarian cancer. Intraperitoneal injection of Paclimer® in a phase I trial on 12 ovarian cancer patients demonstrated paclitaxel delivery at a therapeutic level (0.1–0.2 ng/ml) of sustained release for eight weeks [23]. Low but persistent levels of paclitaxel in the plasma suggested that paclitaxel continued to be released during the study period. Moreover, significant peritoneal abnormality findings, which included the presence of residual polymer filaments months after Paclimer® treatment, indicated a slow degradation of the polymer preparation.

10.2.4 Polyanhydrides

Another type of biodegradable copolymer is represented by polyanhydrides, which are based on the condensation reaction of two fatty acids to form a hydrophobic backbone with hydrolytically labile anhydride linkages [24, 25]. Variation in fatty acid ratio and type used (hydrophobic or hydrophilic) has allowed control of degradation rates, which therefore can define the release rate from weeks to months. Different methods such as spray-drying, hot-melt encapsulation and emulsion methods have been used for polyanhydride formulations. Polyanhydrides have been applied for a variety of therapeutic agents, such as local anaesthetic agents, anticancer drugs, antibiotics, anticoagulants, anti-inflammatory agents, growth hormones and neuroactive drugs [24, 26]. For instance, the prevention of *Pseudomonas aeruginosa* colonization by poly(anhydride-esters) composed of non-steroidal anti-inflammatory drugs that biodegrade to salicylic acid (SA) and adipic acid have been studied *in vitro* and *in vivo*, respectively [27]. Soluble SA at concentrations up to 300 mg/l did not affect the growth rate or viability of *P. aeruginosa*, which suggested that SA is not directly toxic to bacteria. When long-term (three-day) biofilm accumulation was studied for salicylate-based polymer in comparison to controls a dramatic reduction in biofilm formation was observed. Furthermore, it was demonstrated that salicylate-based polymer prevents biofilm accumulation in a recombinant *P. aeruginosa* strain (pMHLAS) with a fluorescent reporter gene prior to the *las* regulon. In another example, nifedipine was encapsulated in poly(sebacic anhydride-*co*-Pluronic F68/F127) microspheres, which allowed controlled and sustained drug release [28].

10.2.5 Polyorthoesters

The condensation of diols and a diketene acetal is the basis for polyorthoesters (POEs). Glycolic acid and lactic acid have been used to control the hydrolytic degradation of the ortho ester linkages, resulting in the proprietary class of POEs named Biochronomer® developed by AP Pharma [29]. Despite irradiation at 24 kGy and storage at room temperature the POEs showed good stability profiles [30]. Due to the thermoplasticity of the Biochronomer® polymers they can be manufactured as microparticles and microspheres. For instance, POE microspheres of bupivacaine have been developed to achieve prolonged anaesthetic efficacy [31]. Additionally, block polymers of POE and PEG have been demonstrated to form microspheres with high encapsulation efficiency of water-soluble compounds such as peptides and proteins [32].

Various POE formulations have been applied for ocular delivery of 5-fluorouracil (5-FU) as an adjunct to glaucoma filtering surgery [33]. Excellent biocompatibility was obtained after subconjunctival, intravitreal and suprachoroidal injections. The polymer lifetime varied in the various sites investigated from five weeks for subconjunctival injections to three months for intravitreal injections to more than six months for intracameral and suprachoroidal injections.

10.2.6 Block Copolymers with a Specific Emphasis on Medusa

A number of block copolymers have been formulated for sustained drug delivery. For instance, polyether ester multiblock copolymers based on hydrophilic PEG and hydrophobic poly(buthylene terephalate) (PBT) have shown to be biocompatible and biodegradable [34]. Polyactive(R) (poly(ethylene oxide)-poly(buthylene terephtalate) (PEO-PBT)) copolymers have been evaluated in orthopedic surgery and dentistry [35]. In goats, histological examination showed good implant–bone contact at different follow-up times in the distal femora. In another study PEG-PBT copolymers were applied to seven patients with different causes of burn injury [36]. After three months normal generation of dermal tissue occurred although the collagen bundles were organized as in scar tissue. A mild inflammatory response was observed due to the presence of fragments of the dermal substitute in the deeper layers of the neodermis. After one year, the extent of wound contraction after successful overgrafting ranged from 30 to 57%, and some fragments of the copolymer were still observed.

Poly(amino acid) polymers have been frequently applied for subcutaneous drug delivery. The Medusa® technology developed by Flamel Technologies exists as two different formulations [37]. Medusa I is based on an amphiphilic block polymer consisting of L-leucine and L-glutamate (pBLE) [38]. Nanoparticles are self-assembled in an aqueous solution with poly-leucine chains packed inside the structure and poly-glutamate chains exposed to the water. The nanoparticles are composed of 95% water and 5% polymer with an approximate size of 200 nm in diameter. Non-covalent interaction associates proteins with nanoparticles and it is believed that the slow release of protein occurs by replacement of therapeutic protein by endogenous protein (Figure 10.1). This procedure leads to protein release over two days. Basulin® has been developed as a long-acting formulation of native insulin and has shown a favourable profile in Phase I and Phase IIa pharmacokinetic and pharmacodynamic studies in healthy volunteers and in Type 1 diabetes mellitus patients in comparison to Lantus® (Sanofi Aventis), a registered drug for diabetes treatment.

The second generation Medusa II polymer consists of a backbone of poly-L-glutamate with hydrophobic molecules of α-tocopherol (vitamin E) randomly grafted to some of the glutamate units through a hydrolysable ester bond [39, 40]. In water the lateral vitamin E groups self-assemble into hydrophilic nanodomains leading to aggregation of the hydrophilic glutamate chains. Generally, Medusa nanoparticles are aggregates of 10–15

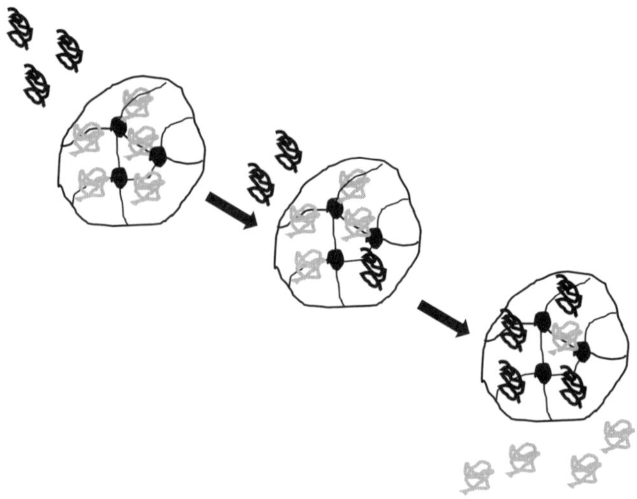

Figure 10.1 Hypothetical replacement of therapeutic protein by endogenous protein (Endogenous protein: black; therapeutic protein: light gray.)

polymer chains with a mean diameter of 20 to 50 nm and an average molecular mass of 400–500 kDa. Single and repeated injections in animal models have demonstrated the low toxicity of Medusa polymers. Only low-grade inflammation was observed at the injection site. Also, after repeated injections in rats and monkeys no undesirable immune responses were induced. The Medusa technology has been applied for extended release of interleukin-2 (IL-2) (IL-2-XL) [37]. In comparison to the registered IL-2 drug Proleukin the peak concentration of IL-2-XL was reduced 100-fold. In beagle dogs and Cynomolgus monkeys the corresponding decrease was thirteen and two times, respectively. In a Phase I/II clinical trial on renal carcinoma patients, the Cmax values for IL-2-XL were twofold lower than for Proleukin after a single dose injection. Moreover, extended release (seven days) was only observed after IL-2-XL administration.

The Medusa technology has also been applied for sustained delivery of interferon-α2b (IFN-α2b) [37]. A dose-escalating Phase I/II study in 53 hepatitis C virus (HCV) positive patients showed a ninefold IFN-α2b peak concentration after Medusa-based administration compared with Viraferon® (Schering-Plough). In contrast to Viraferon, IFN-α2b-XL was still detected in serum seven days post-injection. Furthermore, reduction in mean viral load was higher after administration of IFN-α2b-XL (1.27 log) than Viraferon (0.97 log). Most importantly, a robust antiviral response (0.98 log versus 0.57 log) was observed in patients previously resistant to IFN-α therapy.

10.2.7 Cross-Linked Dextran

The advantage of applying cross-linked dextrans for protein delivery constitutes its microsphere formation in a completely aqueous medium [41]. Chemical or physical cross-linking is the basis of network formation for the design of hydrogels based on a variety of both natural and synthetic polymers [42]. Cross-linking of dextran and the formation of microspheres has been obtained for an aqueous solution of dextran derivative with hydroxyethyl methacrylate (dex-HEMA) in a PEG solution [43, 44]. It has been demonstrated that it is possible to adjust the protein release from days to months [45]. The sustained release profile was demonstrated for IL-2 in a mouse tumour model [46]. In rats cross-linked dextran microspheres have demonstrated excellent biocompatibility [47]. Furthermore, a single subcutaneous injection of microspheres with human growth hormone (hGH) resulted in prolonged drug release for two weeks without any initial burst release (www.octoplus.nl). The presence of elevated levels of insulin-like growth factor-1 and IGF-binding protein-3 suggested that hGH activity was retained *in vivo*.

10.3 *In Situ* Depot Forming Systems

The popularity of *in situ* depot forming systems is due to the far less invasive and costly administration procedure compared to implantation systems [48]. The composition of these systems can be either liquid or semisolid [49, 50]. Generally, the carrier is dissolved or dispersed in a solvent system and drug is present either in a dispersed or dissolved form. The solid depots are formed at the injection site after subcutaneous or intramuscular injection. The *in situ* depot forming systems can be divided into different classes, such as systems based on precipitation, thermal gelling, cross-linking and thermoplastic solids.

10.3.1 Precipitation Systems

In these systems the liquid vehicle consists of a water-resistant biodegradable carrier, which is dissolved in an organic solvent. The drug component is present in a liquid form. Upon administration the organic solvent dissipates, resulting in water diffusion, phase separation and carrier precipitation. This leads to depot formation from where the drug then can be released. The release kinetics can be controlled by the degradation/erosion rates of the carrier.

Precipitation systems, such as the Atrigel® technology based on PLGA polymers, have resulted in an approved product for long-term delivery of leuprolide acetate (www.gltinc.com) [51]. Furthermore, application of more lipophilic solvents, for instance in Alzamer® (benzyl benzoate), demonstrated a decrease in irritation and reduction in the initial drug burst [52]. Another approach has been to dissolve sucrose acetate isobutyrate in ethanol or benzyl alcohol, which facilitates the administration with small gauge needles [53]. In this context, a long-acting formulation for bupivacaine using the SABER® system (Durect) has been applied for post surgical pain management (www.durect.com). The SABER system has also been used for peptide and protein delivery and has demonstrated sustained release of intact recombinant hGH *in vivo* for at least seven days [54].

10.3.2 Thermal Gelling Systems

In attempts to develop thermal gelling systems, hydrophobic PLGA polymers have been combined with hydrophilic PEG. For instance, ReGel® is a free-flowing liquid at room temperature and only transforms to a gel at body temperature. Modifications in hydrophobic/hydrophilic ratio and the concentration, polydispersity and molecular weight of the polymer(s) allow for variation in drug release rate [55]. Incorporation of paclitaxel, commonly used for the treatment of solid tumours, into ReGel generated OncoGel® (www.macromed.com). Because paclitaxel is entrapped within the hydrophobic domain of the gel the drug release is sustained for six weeks. Using this approach it was demonstrated that ReGel-based paclitaxel delivery could efficiently inhibit neointimal hyperplasia in canine vascular grafts [56]. Furthermore, OngoGel has been used in a phase II clinical trial for the treatment of oesophageal cancer (www.macromed.com). Constant release of insulin for two weeks was also obtained for ReGel depot formulation in rats [57].

Alkaline deacetylation of chitin, the natural component of shrimp and crab shells, results in the biocompatible and biodegradable cationic polymer chitosan. The combination of chitosan and glycerophosphate (GP), an anionic polyphosphate salt, has provided an interesting injectable *in situ* gelling system [58]. A formulation containing homogeneously dispersed camptothecin was implanted intratumorally into a subcutaneous mouse tumour model (RIF-l) [59]. Animals treated with the chitosan-GP camptothecin implants showed significantly longer tumour growth delay (TGD) compared to untreated animals or mice treated systemically with camptothecin by intraperitoneal injection. Moreover, no toxicity in terms of loss of body weight was observed.

Another thermal gelling system is based on polyoxyethylene and polyoxypropylene units (Poloxamer®), a water-soluble non-ionic surfactant, which shows low viscosity at room temperature but rapidly gels at body temperature [60]. Despite the better compatibility of the Poloxamer system with protein delivery due to no need for applying organic solvents, there has been some concern related to the lack of biodegradation and cytotoxicity issues. In a study in rats it was demonstrated that the Poloxamer 407-induced hyperlipidemia was associated with alterations in plasma lecithin cholesterol acyl transferase (LCAT), cholesteryl ester transferase protein (CETP), hepatic lipase (HL) and lipoprotein lipase (LPL) activities [61].

10.3.3 Cross-Linked Systems

In situ cross-linking of PEG-based copolymers containing multiple thio (–SH) groups with $\alpha\omega$-divinylsulfone-PEG resulted in hydrogel formation [62]. Water-soluble drugs including peptides and proteins are physically entrapped in the hydrogel. The system is really suitable for protein delivery as no organic solvents are applied in the preparation. Fluorescein-labelled bovine serum albumin demonstrated excellent release *in vitro*. Likewise, release of such model proteins as erythropoietin and RANTES showed sustained release for 2–4 weeks *in vivo*. Furthermore, prolonged biological activity of the released proteins was demonstrated in animal models [62].

Recently, hydrogel cross-linking has been applied for sustained delivery of plasmid DNA [63]. Local and sustained delivery of a plasmid DNA encoding vascular endothelial growth factor (VEGF) was evaluated in the ischaemic hind limbs of mice. A significant improvement in the recovery of blood perfusion was observed after hydrogel-based DNA delivery compared with injection of naked VEGF encoding plasmid DNA. These hydrogels still need to be verified for peptides and proteins. In another study, superporous hydrogels of poly(vinyl alcohol) and poly(vinyl pyrrolidone) were prepared by a double emulsion process and physically cross-linked by freeze–thaw cycling [64]. An increase in the volume of organic solvent resulted in increased porosity due to cross-linking of the hydrogel solution around the droplets in the emulsion. These hydrogels are potentially good candidates for tissue engineering of cartilage because of their capacity to mimic mature cartilage mechanically and simultaneously presenting a porous matrix for cell migration and proliferation. Furthermore, these hydrogels may also perform well for the delivery of peptides and proteins.

A natural acidic polysaccharide polymer has been purified from the aloe plant by DelSite Biotechnologies (GelSite®). The presence of calcium induces gel formation of the polymer, which allows entrapment of water-soluble drugs after subcutaneous or intramuscular injection [65]. Additionally, GelSite has a specific binding affinity to heparin-binding proteins such as fibroblast growth factor and VEGF. Furthermore, it has been demonstrated that this binding property enhances the control of drug release without interfering with the biological function of the therapeutic protein.

Additional applications have been for biodegradable comb-branched polymers, which consist of amine-modified poly(vinyl alcohol) backbone grafted with PLGA side chains [66]. Anionic drugs can bind to the highly positive charged surface of the nanoparticles. In the presence of ions in the body fluids drug-loaded nanoparticles will form hydrogels after injection. The degradation of the polymer and the electrostatic interactions between polymer and drug dictate the drug release. It has been demonstrated *in vitro* that insulin was released for at least four days after an initial drug burst [67]. Recently, photo cross-linkable (PCL) systems have been modified with 2-isocyanatoethylmethacrylate to form a macromer, which were cross-linked by UV irradiation using Irgacure 2959 as the photo-initiating agent [68]. The morphology of PCL networks was evaluated by scanning electron microscopy immediately after cross-linking and following biodegradation in human plasma. Furthermore, thrombosis and haemolysis tests confirmed the haemocompatibility of membranes.

10.3.4 Thermoplastic Semisolids

Application of highly flexible diols has presented the opportunity to engineer POEs with a molecular weight <5 kDa, which are semisolid at room temperature. Mixing at low temperature in the absence of organic solvents has provided an attractive delivery method especially for protein drugs. Improved injectability has been obtained after mixing the semisolid formulations with 20% monomethoxy poly(ethylene glycol) (molecular weight 550 kDa). Surface erosion of the polymer is the basis of controlled drug release due to the hydrophobic nature of POEs [69]. This approach has been taken for the delivery of mepivacaine and gansinetron (www.appharma.com). Delivery of 3% mepivacaine was verified in a Phase II clinical trial for post-operative pain management and gansitron was tested for control of chemotherapy-induced nausea. Although no reports so far are available for peptides or proteins, the efficient localized sustained delivery of the drug campothecin by a biodegradable thermoplastic polymer suggests a wide application range [70].

Doxorubicin (DOX) containing thermally sensitive poly(N-isopropylacrylamide-*co*-N, N-dimethylacrylamide)-b-poly(d,l-lactide-*co*-glycolide) [P(NIPAAm-*co*-DMAAm)-b-poly(D,L-lactide-*co*-glycolide) (PLGA)] have been synthesized with an average micelle size of 200 nm [71]. The drug loading capacity varied depending on the PLGA block. The DOX-loaded micelles were stable in PBS containing serum at 37 °C, but degradation above 39.5 °C triggered the release of DOX. Therefore, PLGA micelles may be applicable for anti-cancer drug delivery.

10.4 Conclusions

A number of drug delivery systems based on polymers in different compositions and formulations have been developed. These consist of polymer nanocarriers and *in situ* depot systems. Many of the engineered formulations have demonstrated good long-term stability. Furthermore, low toxicity has been observed in studies in various animal species. Most importantly, efficient and extended drug delivery has been shown for various indications in several clinical trials.

Among the different systems described here, their specific application to peptides and proteins depends on a number of factors. The complexity of the production procedure may prevent the use of an otherwise efficient delivery method. Likewise, a proper control of particle size is of great importance for the feasibility and quality of commercial scale production. In this context, the block copolymers seem to provide a good basis, but other approaches including other types of polymeric nanoparticles should not be excluded. On the other hand, the less invasive approaches and cheaper administration procedure of *in situ* depot forming systems have their own advantages. Naturally, further development is required, but nanocarrier delivery approaches have already now confirmed its potency as a serious player in modern drug delivery.

References

1. Shi, Y. and Li, C. (2005) Current advances in sustained-release systems for parenteral drug delivery. *Expert Opin. Drug Deliv.*, **2**, 1039–1058.
2. Tarcha, P.J., Verlee, D., Hui, H.W. *et al.* (2007) The application of ink-jet technology for the coating and loading of drug-eluting stents, *Ann. Biomed. Eng.*, **35**, 1791–1799.
3. Jones, A.J., Putney, S., Johnson, O.L. and Cleland, J.L. (1997) Recombinant human growth hormone poly(lactic-co-glycolic acid) microsphere formulation development. *Adv. Drug Deliv. Rev.*, **28**, 71–84.
4. Yeo, Y. and Park, K. (2004) A new microencapsulation method using an ultrasonic atomizer based on interfacial solvent exchange. *J. Control. Release*, **100**, 379–388.
5. Elvassore, N., Bertucco, A. and Caliceti, P. (2001) Production of protein-loaded polymeric microcapsules by compressed CO_2 in a mixed solvent. *Ind. Eng. Chem. Res.*, **40**, 795–800.
6. Gupte, A. and Ciftci, K. (2004) Formulation and characterization of paclitaxel, 5-FU and paclitaxel + 5-FU microspheres. *Int. J. Pharm.*, **276**, 93–106.
7. Lagarce, F., Renaud, P., Faisant, N. *et al.* (2005) Baclofen-loaded microspheres: preparation and efficacy testing in a new rabbit model. *Eur. J. Pharm. Biopharm.*, **59**, 449–459.
8. Naraharisetti, P.K., Lew, M.D., Fu, Y.-C. *et al.* (2005) Gentamicin-loaded discs and microspheres and their modifications: characterization and *in vitro* release. *J. Control. Release*, **102**, 345–359.
9. Cui, F., Cun, D., Tao, A. *et al.* (2005) Preparation and characterization of melittin-loaded poly(D, L-lactic acid) or poly(D, L-lactic-co-glycolic acid) microspheres made by the double emulsion method. *J. Control. Release*, **107**, 310–319.
10. Kostanski, J.W., Dani, B.A., Reynolds, G.A. *et al.* (2000) Evaluation of Orntide microspheres in a rat animal model and correlation to *in vitro* release profiles. *AAPS Pharm. Sci. Tech.*, **1**, E27.

11. Blanco-Prieto, M.J., Campanero, M.A., Besseghir, K. *et al.* (2004) Importance of single or blended polymer types for controlled *in vitro* release and plasma levels of a somastatin analogue entrapped in PLA/PLGA microspheres. *J. Control. Release*, **96**, 437–448.
12. Jiang, W., Gupta, R.K., Deshpande, M.C. and Schwendeman, S.P. (2005) Biodegradable poly(lactic-co-glycolic acid) microparticles for injectable delvery of vaccine antigens. *Adv. Drug Deliv. Rev.*, **57**, 391–410.
13. Sinha, V.R. and Trehan, A. (2003) Biodegradable microspheres for protein delivery. *J. Control. Release*, **90**, 261–280.
14. Sinha, V.R., Bansal, K., Kaushik, R. *et al.* (2004) Poly-ε-caprolactone microspheres and nanospheres: an overview. *Int. J. Pharm.*, **278**, 1–23.
15. Shenoy, D.B., D'Souza, R.J., Tiwari, S.B. and Udupa, N. (2003) Potential applications of polymeric microsphere suspension as subcutaneous depot for insulin. *Drug Dev. Ind. Pharm.*, **29**, 555–563.
16. Yu, H., Wang, W., Chang, X. *et al.* (2006) Synthesis and characterization of the biodegradable polycaprolactone-graft-chitosan amphiphilic copolymers. *Biopolymers*, **15**, 233–242.
17. Harper, E., Dang, W., Lapidus, R.G. *et al.* (1999) Enhanced efficacy of a novel controlled release paclitaxel formulation (PACLIMER delivery system) for local-regional therapy of lung cancer tumor nodules in mice. *Clin. Cancer Res.*, **5**, 4242–4248.
18. Xu, X., Yu, H., Gao, S. *et al.* (2002) Polyphosphoester microspheres for sustained release of biologically active nerve growth factor. *Biomaterials*, **23**, 3765–3772.
19. Wang, J., Mao, H.-Q. and Leong, K.W. (2001) A novel biodegradable gene carrier based on polyphosphoester. *J. Am. Chem. Soc.*, **123**, 9480–9481.
20. Xu, X., Yee, W.C., Hwang, P.Y. *et al.* (2003) Peripheral nerve regeneration with sustained release of poly(phosphoester) microencapsulated nerve growth factor within nerve guide conduits. *Biomaterials*, **24**, 2405–2412.
21. Du, J.Z., Sun, T.M., Weng, S.Q. *et al.* (2007) Synthesis and characterization of photo-cross linked hydrogels based on biodegradable polyphosphoesters and poly (ethylene glycol) copolymers. *Biomacromolecules*, **8**, 3375–3381.
22. Mao, H.Q. and Leong, K.W. (2005) Design of polyphosphoester-DNA nanoparticles for non-viral gene delivery. *Adv. Genet.*, **53PA**, 275–306.
23. Armstrong, D.K., Fleming, G.F., Markman, M. and Bailey, H.H. (2006) A phase I trial on intraperitoneal sustained-release paclitaxel microspheres (Paclimer) in recurrent ovarian cancer: a Gynecologic Oncology Group study. *Gynecol. Oncol.*, **102**, 391–396.
24. Kumar, N., Langer, R.S. and Domb, A.J. (2002) Polyanhydrides: an overview. *Adv. Drug Deliv. Rev.*, **54**, 889–910.
25. Jain, J.P., Modi, S. and Kumar, N. (2008) Hydroxy fatty acid based polyanhydride as drug delivery system: synthesis, characterization, *in vitro* degradation, drugs release and biocompatibility. *J. Biomed. Mater. Res. A*, **84A**, 740–752.
26. Jain, J.P., Modi, S., Domb, A.J. and Kumar, N. (2005) Role of polyanhydrides as localized drug carrier. *J. Control. Release*, **103**, 541–563.
27. Bryers, J.D., Jarvis, R.A., Lebo, J. *et al.* (2006) Biodegradation of poly(anhydride-esters) into non-steroidal anti-inflammatory drugs and their effect on *Pseudomonas aeruginosa* biofilms *in vitro* and the foreign-body response *in vivo*. *Biomaterials*, **27**, 5039–5048.
28. Shelke, N.B. and Aminabhavi, T.M. (2007) Synthesis and characterization of novel poly(sebacic anhydride-co-Pluronic F68/F127) biopolymeric microspheres for the controlled release of nifepidine. *Int. J. Pharm.*, **345**, 51–58.
29. Heller, J. and Barr, J. (2005) Biochronomer technology. *Expert Opin. Drug Deliv.*, **2**, 169–183.
30. Ng, S.Y., Shen, H.R., Lopez, E. *et al.* (2000) Development of a poly(ortho ester) prototype with a latent acid in the polymer backbone for 5-fluorouracil delivery. *J. Control. Release*, **65**, 367–374.
31. Deng, J.-S., Li, L., Tian, Y. *et al.* (2003) *In vitro* characterization of polyorthoester microparticles containing bupivacaine. *Pharm. Dev. Technol.*, **8**, 31–38.
32. Yang, Y.-Y., Wan, J.-P., Chung, T.-S. *et al.* (2001) POE-PEG-POE triblock copolymeric microspheres containing protein. I. Preparation and characterization. *J. Control. Release*, **75**, 115–128.

33. Heller, J. (2005) Ocular delivery using polyortho esters. *Adv. Drug Deliv. Rev.*, **57**, 2053–2062.
34. Jansen, J.A., Ruijter, J.E., Janssen, P.T. and Paquay, Y.G. (1995) Histological evaluation of a biodegradable polyactive®/hydroxyapatite membrane. *Biomaterials*, **16**, 819–827.
35. Sakkers, R.J., Dalmeyer, R.A., de Wijn, J.R. and van Blitterswijk, C.A. (2000) Use of bone-bonding hydrogen copolymers in bone: an *in vitro* and *in vivo* study of expanding PEO-PBT copolymers in goat femora. *J. Biomed. Mater. Res.*, **49**, 312–318.
36. Mensik, I., Lamme, E.N., Riesle, J. and Brychta, P. (2002) Effectiveness and safety of the PEGT/PBT copolymer scaffold as dermal substitute in scar reconstruction wounds (feasibility trial). *Cell Tissue Bank*, **3**, 245–253.
37. Chan, Y.-P., Meyrueix, R., Kravtzoff, R. *et al.* (2007) Review on Medusa®: a polymer-based sustained release technology for protein and peptide drugs. *Expert Opin. Drug Deliv.*, **4**, 441–451.
38. Constancis, A., Meyrueix, R., Bryson, N. *et al.* (1999) Macromolecular colloids of diblock poly(amino acids) that bind insulin. *J. Colloid Interface Sci.*, **217**, 357–368.
39. Chan, Y.-P., Angot, S. and Breyne, O. (2003) Polyaminoacids functionalized by alpha tocopherol. WO 200310403.
40. Huille, S., Nicolas, F. and Brison, N. (2003) Particles based on polyamino-acid(s) and methods for preparing same. US Patent 6630171.
41. Franssen, O., Vendervennet, L., Roders, P. and Hennink, W.E. (1999) Degradable dextran hydrogels: controlled release of a model protein from cylinders and microspheres. *J. Control. Release*, **60**, 211–221.
42. van Tomme, S.R. and Hennink, W.E. (2007) Biodegradable dextran hydrogels for protein delivery applications. *Expert Rev. Med. Devices*, **4**, 147–164.
43. Stenekes, R.J.H., Franssen, O., Bommel, W.M. *et al.* (1999) The use of aqueous PEG/dextran phase separation for the preparation of dextran microspheres. *Int. J. Pharm.*, **189**, 29–32.
44. Chung, J.T., Vlugt-Wensink, K.D.F., Hennink, W.E. and Zhang, Z. (2005) Effect of polymerization conditions on the network properties of dex-HEMA microspheres and macro-hydrogels. *Int. J. Pharm.*, **288**, 51–61.
45. Cadee, J.A., Degroot, C.J., Jiskoot, W. *et al.* (2002) Release of recombinant human interleukin-2 from dextran-based hydogels. *J. Control. Release*, **778**, 1–13.
46. Koten, J.W., van Luyn, M.J., Cadee, J.A. *et al.* (2003) IL-2 loaded dextran microspheres with attractive histocompatibility properties for local IL-2 cancer therapy. *Cytokine*, **24**, 57–66.
47. de Groot, C.J., van Luyn, M.J., van Dijk-Wolthuis, W.N. *et al.* (2001) *In vitro* biocompatibility of biodegradable dextran-based hydrogels tested with human fibroblasts. *Biomaterials*, **22**, 1197–1203.
48. Packhauser, C., Schnieders, J., Osters, C.G. and Kissel, T. (2004) *In situ* forming parental drug delivery systems: an overview. *Eur. J. Biopharm.*, **58**, 445–455.
49. Martini, A. and Lauria, S. (2003) Sustained release injectable products. *Am. Pharm. Rev.*, **6**, 16–20.
50. Tipton, A.J. and Dunn, R.L. (2000) *In situ* gelling systems, in *Sustained-Release Injectable Products* (eds J.H. Senior, S. Senior and M.L. Radomsky), Interpharm Press, Boca Rota, FL, USA, pp. 241–278.
51. Sartor, O. (2003) Eligard: leuprolide acetate is a novel sustained-release delivery system. *Urology*, **61**, 25–31.
52. Sudmann, B., Anfinsen, O.G., Bang, G. *et al.* (1993) Assessment in rats of a new bioerodible bone-wax-like polymer. *Acta Orthop. Scand.*, **64**, 336–339.
53. Josh, G.P. (2003) Chronogesic (DURECT). *IDrugs*, **6**, 368–371.
54. Okumu, F.W., Dao le, N., Fielder, P.J. *et al.* (2002) Sustained delivery of human growth hormone from a novel gel system: SABER. *Biomaterials*, **23**, 4353–4358.
55. Zentner, G.M., Rathi, R., Shih, C. *et al.* (2001) Biodegradable block copolymers for delivery of proteins and water-insoluble drugs. *J. Control. Release*, **72**, 203–315.
56. Masaki, T., Rathi, R., Zentner, G.M. *et al.* (2004) Inhibition of neointimal hyperplasia in vascular grafts by sustained perivascular delivery of paclitaxel. *Kidney Int.*, **66**, 2061–2069.
57. Choi, S. and Kim, S.W. (2003) Controlled release of insulin from injectable biodegradable triblock copolymer depot in ZDF rats. *Pharm. Res.*, **20**, 2008–2010.

58. Chenite, A., Chaput, C., Wang, D. *et al.* (2000) Novel injectable neutral solutions of chitosan form biodegradable gels *in situ*. *Biomaterials*, **21**, 2155–2161.
59. Berrada, M., Serreqi, A., Dabbarh, F. *et al.* (2005) A novel non-toxic camptothecin formulation for cancer chemotherapy. *Biomaterials*, **261**, 2115–2120.
60. Yong, C.S., Choi, J.S., Quan, Q.Z. *et al.* (2001) Effect of sodium chloride on the gelation temperature, gel strength and bioadhesive forces of poloxamer gels containing diclofenac sodium. *Int. J. Pharm.*, **226**, 195–205.
61. Wasan, K.M., Subramanian, R., Kwong, M. *et al.* (2003) Poloxamer 407-induced alterations in the activities of enzymes regulating lipid metabolism in rats. *J. Pharm. Sci.*, **6**, 189–197.
62. Qiu, Q., Stefanos, S., Ma, J. *et al.* (2002) A hydrogel prepared by *in situ* crosslinking of a thiol-containing poly(ethylene glycol)-based copolymer: a new biomaterial for protein drug delivery. *Biomaterials*, **24**, 11–18.
63. Kong, H.J., Kim, E.S., Huang, Y.C. and Mooney, D.J. (2008) Design of biodegradable hydrogel for the local and sustained delivery of angiogenic plasmid DNA. *Pharm. Res.*, **25**, 1230–1238.
64. Spiller, K.L., Laurencin, S.J., Charlton, D. *et al.* (2008) Superporous hydrogels for cartilage repair: evaluation of the morphological and mechanical properties. *Acta Biomater.*, **4**, 17–25.
65. Ni, Y. and Yates, K.M. (2001) Pectic substance as a growth factor stabilizer. US Patent 6313103.
66. Oster, C.G., Wittmar, M., Unger, F. *et al.* (2004) Design of amine-modified graft polyesters for effective gene delivery using DNA-loaded nanoparticles. *Pharm. Res.*, **21**, 927–931.
67. Packhauser, C., Wittmar, M. and Kissel, T. (2005) Novel *in situ* forming implants for bioactive macromolecules based on positively charged nanoparticles, characterization of depots and investigation of insulin release. Controlled Release Society Annual Meeting, Miami, FL, USA.
68. Ferreira, P., Coelho, J.F. and Gil, M.H. (2008) Development of a new photocrosslinkable biodegradable bioadhesive. *Int. J. Pharm.*, **352**, 172–181.
69. Heller, J., Barr, J., Ng, S., *et al.* (2002) Poly(ortho esters): synthesis, characterization, properties and uses. *Adv. Drug Deliv. Rev.*, **54**, 1015–1039.
70. Hatefi, A., Knight, D. and Amsden, B. (2004) A biodegradable injectable thermoplastic for localized camptothecin delivery. *J. Pharm. Sci.*, **93**, 1195–1204.
71. Liu, S.Q., Tong, Y.W. and Yang, Y.Y. (2005) Incorporation and *in vitro* release of doxorubicin in thermally sensitive micelles from poly(N-isopropylacrylamide-co-N,N-dimethylacrylamide)-b-ply(D,L-lactide-co-glycolide) with varying compositions. *Biomaterials*, **26**, 5064–5074.

11

Polymer-Based Delivery Systems for Oral Delivery of Peptides and Proteins

Bruno Sarmento,[1,2] *Domingos Ferreira*[1] *and Teófilo Vasconcelos*[1,3]
[1] University of Porto, Department of Pharmaceutical Technology, Faculty of Pharmacy, Porto, Portugal
[2] Health Sciences Research Center, Department of Pharmaceutical Sciences, Instituto Superior de Ciências da Saúde, Gandra, Portugal
[3] Laboratory of Pharmaceutical Development, BIAL, S. Mamede do Coronado, Portugal

11.1 Introduction

Peptidic drugs represent a significant part of the new pharmaceuticals coming on the market every year [1, 2]. At present, they hold a strong position in the novel area of nanomedicine, using nanotechnology for medical applications for both academic and industrial fields [3].

However, parenteral administration of some therapeutic proteins has not been efficient because the drug is eliminated too rapidly from the body. Thus, research continues on non-parenteral delivery routes but the high molecular weight of proteins, combined with their hydrophilic and charged nature, renders transport through membranes very difficult. Moreover, various proteases are present at potential sites of administration, namely in the gastrointestinal tract, as well as in the lungs, buccal epithelium and skin. Finally, cells of the immune system located at the point of entry may absorb and eliminate the protein, thus reducing bioavailability and increasing the potential for an immune response [4]. Although protein delivery via non-parenteral routes remains under investigation, no major progress has until now been reported [1].

Delivery Technologies for Biopharmaceuticals: Peptides, Proteins, Nucleic Acids and Vaccines Edited by Lene Jorgensen and Hanne Mørck Nielsen

The success of a protein formulation depends not only on the stability of the delivery system but also on the ability of the protein to maintain the native structure and activity during preparation and the delivery, as well as during long-term storage of the formulation [5]. Polymer-based delivery systems have been proposed as valid approaches to provide such protective conditions. Marketed products based on polymeric-based delivery systems are also useful tools for diagnosing and treating prominent diseases [6]. For controlled release in targeting particular organs/tissues, such polymeric-based systems have attracted considerable attention as carriers of not only proteins or peptides but also DNA in gene therapy [6]. They can effectively deliver the proteins to a target site and thus increase the therapeutic benefit while minimizing side effects [7]. A variety of natural and synthetic polymers has been explored for the preparation of delivery systems because of their biocompatibility and biodegradability. Furthermore, they are versatile for formulation and promote sustained release properties. Over the past few decades, there has been considerable interest in developing biodegradable polymeric systems as effective protein delivery devices. However, there is still a lack of protein formulations in the market, and even less for oral application.

This chapter presents the outstanding contributions in the field of polymeric systems as carriers for oral protein delivery. Although an extensive list of research manuscripts is found throughout the literature revealing potential use of polymeric systems for oral protein delivery, only those that demonstrate effective and well-established oral bioactivity *in vivo* are considered in this review.

11.2 Advances in Oral Protein Delivery

Oral administration is the preferable route for drug delivery because it is non-invasive, avoids pain and discomfort associated with injections, decreases risk of contamination and is compatible with the molecular weight and chemical structure of most drugs. For protein drugs, it is also physiologically desirable, since the exogenous protein imitates the physiological pathway undergoing first hepatic bypass. Moreover, among the non-invasive routes of administration that have been evaluated for the delivery of proteins, the oral route remains the most convenient, although it is not the most efficient for peptides and proteins due to their low absorption rate. The main reasons for the low oral bioavailability are related to chemical and conformational stabilities, cellular and luminal enzymatic degradation in the gastrointestinal tract, poor intrinsic permeation across the intestinal membrane [8–10] and clearance mechanisms such as first-pass and excretion in the bile [11]. However, there are peptide drugs, such as cyclosporine A and desmopressin, that are available in oral dosage forms, indicating that peptide or even polypeptide structure molecules should not be an absolute limitation for oral delivery [12–14].

Advances on developing an effective oral delivery system for proteins requires a comprehensive perception of their specific physicochemical properties, such as molecular weight, hydrophobicity, ionization coefficient and pH stability, as well as of the biological barriers that limit protein absorption through the gastrointestinal tract [15]. The important therapeutic proteins and peptides being explored for oral delivery include insulin, calcitonin, interferons, human growth hormone, glucagons, gonadotropin-releasing hormones, enkephalin and, in general, vaccines, enzymes, hormone analogues and enzyme inhibitors

[16]. These are outstanding proteins used in the pharmaceutical development, mostly due to their well-established physicochemical properties and social impact of their therapeutic applications.

Strategies to improve the oral bioavailability of proteins have ranged from changing their physicochemical properties by modification of their lipophilicity and enzyme susceptibility, to adding novel functionality using transport-carrier molecules that are recognized by endogenous transport-carrier systems in the gastrointestinal tract and/or to their inclusion in specially adapted drug carrier systems [8]. Nonetheless, protein association with polymer-based carriers, such as polymeric microparticles, nanoparticles, hydrogels or patches, is one of the most promising approaches proposed to improve oral protein bioavailability. Polymer-based carriers can protect proteins from the gastrointestinal environment and allow the modulation of physicochemical and protein release properties and consequently the biological behaviour [17]. Also, from the perspective of improving oral absorption, the major effect of carriers is to increase epithelial membrane permeability, thereby leading to higher bioavailability [18].

There are many other criteria that must be satisfied to bring an oral protein formulation to the market. For example, the low bioavailability implies a large variation in absorption and a high manufacturing cost, which is unacceptable for the development of most peptide and protein drugs. For proteins like insulin that have a relatively narrow therapeutic window, the effects on intestinal absorption of age, genomic factors, physiological conditions and other individual variations must be carefully investigated. Finally, most peptide and protein drugs require chronic administration, and hence the effects of long-term oral administration of absorption carriers on both the intestinal and systemic physiology must also be carefully evaluated [12].

11.3 Intestinal Considerations for Protein Absorption

Human intestinal mucosa extends to a surface area of more than 200 m^2 in adults [19], thus being very appealing to an extensive absorption target. Morphologically, different aspects contribute to its selective absorption barrier, namely the anatomic structure of the gastrointestinal tract with a single layer of columnar epithelial cells covering the luminal surface combined with a variety of more specialized cells. The epithelial cell layer covers a region of loose connective tissue, the lamina propria, containing blood and lymph vessels [20]. Transit along the gastrointestinal tract and mixing within intestinal segments is accomplished by contractions of separate groups of muscular layers found beneath the lamina propria, the muscularis mucosa, the circular and the longitudinal muscle layers. Gastric glands, villi, microvilli and crypts are found along the gastrointestinal tract. Differences in permeability to peptidic drugs and drugs in general may be related to the phospholipid composition of the plasma membranes of enterocytes [21].

The main challenges usually reported involve the fate of proteins in the gastrointestinal tract, mostly overcoming the enzymatic degradation and the insufficient permeation of proteins through the intestinal walls, achieving epithelial permeability and conserving the bioactivity of the proteins during the formulation process. Thus, the inactivation of intestinal proteolytic enzymes located in the gut lumen (e.g. trypsin) or at the brush border (e.g. aminopeptidases) contributes to improved protein absorption.

Even though the proteolytic activity is higher in a few segments of the gastrointestinal tract, this route has certain segments that have lower proteolytic activity, for example in the colon. The colon has captured attention as a site for the delivery of these molecules because of its greater responsiveness to absorption enhancers, protease inhibitors and novel bioadhesive and biodegradable polymers. Although the success rate of these approaches, when applied separately is pretty low, when used in combinations these agents have demonstrated wonders in increasing the drug bioavailability [22].

Although the passive transcellular route is, in general, quantitatively dominating for drug transport across the intestinal epithelium, both paracellular and transcellular routes have been explored to examine intestinal absorption of proteins formulated in polymer-based systems. The paracellular pathway is commonly identified as limiting to protein absorption, due to low surface area and tightness of the junctions of the intercellular spaces [23]. Still, the paracellular route of protein transport has been in focus, not only because its deficient in proteolytic activity, but also due to the aqueous-filled channel through which polypeptide drugs prefer to diffuse because of their hydrophilic nature [23]. Strategies to modify the physicochemical properties of the intestinal wall and to modulate the tight junctions associated with the paracellular pathway have been mentioned; the paracellular transport enhancement of proteins has been achieved by binding to epithelia by charged polymers like chitosan or polyacrylates, resulting in a structural rearrangement of tight junction-associated proteins or by a reduction of free extracellular calcium ion concentration. Some examples of peptide drugs that have been shown to permeate the intestinal mucosa via the paracellular route include octreotide, salmon calcitonin or insulin [23].

The transcellular route of protein absorption mainly comprises the membranous epithelial cells (M-cells) of the Peyer's patches in the gut-associated lymphoid tissue (GALT), but it also involves the much more numerous gut mucus enterocytes by diffusion and endocytosis [24]. During endocytotic processes various types of membrane vesicles have been observed, ranging from the actin-dependent formation of phagosomes, involved in particle uptake, to smaller clathrin-coated endocytosis vesicles, responsible for the internalization of extracellular fluid and receptor-bound ligands [25]. By contrast, passive adsorption does not require specific ligand-receptor interactions at the cell's surface and, therefore, this mechanism may have potential for oral delivery of macromolecules and colloidal carriers, although its efficiency remains to be demonstrated [26].

Peyer's patches are specific structures dispersed through the lymphoid nodules of the intestinal mucosa called O-MALT (Organized Mucosa-Associated Lymphoid Tissue). They are rich in M-cells, which are specialized in antigen sampling and are also a potential portal for oral delivery of peptides and proteins, since they possess a high transcytotic capacity and are able to transport a broad range of materials, including nanoparticles [27, 28]. Uptake by M-cells has been demonstrated to result in physiologic action of proteins after oral administration of nanoparticles [29, 30]. Adsorptive endocytosis occurs through clathrin-coated pits and vesicles, fluid phase endocytosis and phagocytosis [31]. In general, mucoadhesive polymeric systems are able to increase the relative pharmacological bioavailability of proteins. Different approaches, namely those composed of nanoparticles, seem to promote the transcellular absorption through the membranous epithelial cells (M-cells) of the Peyer's patches in the gut-associated lymphoid tissue.

11.4 Polymer-Based Delivery Systems for Oral Delivery of Proteins

11.4.1 Hydrogels

Hydrogels, are polymeric three-dimensional networks able to swell in the presence of an aqueous medium [32]. Hydrogels can be classified as neutral or ionic, based on the nature of the side groups, and as amorphous, semi-crystalline, hydrogen-bonded, supermolecular structures. Additionally, they can be homopolymer or copolymer networks, based on the method of preparation [33]. Hydrogels can be superporous [34–36] or comb-type grafted hydrogels [37, 38] with fast response times, self-assembling hydrogels from hybrid graft copolymers with property controlling protein domains [38], or genetically engineered triblock copolymers [39] with advanced properties.

Hydrogels have been pointed out as outstanding systems to administer proteins orally due to the exhibition of dramatic changes in network structure or swelling behaviour in response to body temperature, gastrointestinal variations in pH, or even in electric field or ionic strength [40, 41]. Hydrogel responsive properties may allow a targeted delivery of protein to its absorption site. This versatility can be exploited to target delivery and to protect proteins from gastrointestinal decomposition. Moreover, physiologic (pH, enzymes etc.) or non-physiologic (magnetic field, light etc.) set offs can be used for temporal or spatial control delivery of biomolecules in the host [33, 41]. These delivery systems can also be designed to release macromolecules in response to increased concentration of a specific compound like glucose [42, 43]. Nevertheless, the oral application of hydrogels for protein delivery is still limited, and only few studies have been performed so far.

In oral protein delivery, methacrylic acid (MAA) and methoxy-terminated poly(ethylene glycol) (PEG) mono methacrylate hydrogels have been described as ideal for the oral delivery of peptides and proteins due to their large change in network structure over a small pH range [44, 45]. It is mainly used in the form of microparticles (see below), however, poly(ethylene glycol) grafted poly(methacrylic acid) hydrogel film has also been prepared to load insulin. It was observed that the total insulin absorption in rats was dependent on both the amount of the polymer present and the concentration of insulin, with a maximum relative oral bioavailability of 8% [46].

Another promising advance for effective intestinal absorption of protein drug molecules can be achieved by delivery systems based on superporous hydrogels (SPHs) [47]. Superporous hydrogels are a new generation of hydrogels, which are able to swell very quickly due to their highly porous structure [34]. SPH composite or the so-called shuttle systems maintain the dosage form at the site of drug absorption by mechanical interaction with the intestinal membranes. These systems reach a 'double phase time-controlled release profile' and target the dosage form to the desired site(s) of the intestinal tract [47]. Release studies from SPH revealed that buserelin, octreotide and insulin were almost completely released and that SPH were able to open tight junctions and enhance the protein intestinal transport [48] Moreover, these polymers were proved to be cytocompatible [49] and *in vivo* studies indicate that the absorption was increased for both insulin [50] and octreotide [51] using SPH delivery systems.

Hydrogels are, thus, still in the early pipeline of protein delivery. The hydrophilic moieties combined with adaptive physical structure appear to be synergic factors to the protein formulation and release.

11.4.2 Patches and Tablets

Approximately 40% of all dosage forms are tablets. In addition to their long lasting shelf-life and due to their convenient route of administration patient compliance is very high. Depending on the drug carrier matrix and the excipients used, peptide liberation can be adjusted to delay or prolong release. If polymers are used as drug carrier matrices for tablets, the polymer might form a gel after contact with the liquids of mucosal membranes [52].

Mucoadhesive polymers in the form of tablets have been established as efficient carriers for poorly absorbed orally proteins [53]. Besides not being absorbed from the intestinal tract and guaranteeing, in that way, desirable absence of systemic toxic side effects, polymeric tablets may offer intimate contact to the intestinal mucosa due to mucoadhesive properties. Further, it potentially provides prolonged residence time in the intestine, higher concentration of proteins on the mucosa and sustained release of the active drug [53].

It has been reported that the mucoadhesive polyacrylates like carbopol and polycarbophil are capable of enhancing the intestinal absorption of peptides by reducing the metabolic activity of both luminal and membrane bound proteolytic enzymes and by opening of the intestinal intercellular junctions [54]. Mucoadhesive intestinal patches of carbopol to orally deliver therapeutic doses of insulin into systemic circulation have been developed, showing that intestinal patches localize insulin near the mucosa and protect it from proteolytic degradation [55]. These patches were fabricated using a mucoadhesive pressed matrix of carbopol, pectin and sodium carboxymethylcellulose and sliced into disks, which were coated with a solution of ethylcellulose in acetone. *In vivo* experiments showed that intestinal insulin patches with doses in the range of 1–10 U/kg induced dose-dependent hypoglycemia in normal rats with a maximum drop in blood glucose level of 75% [55].

Thiolated polymers, or so-called thiomers, represent a new class of efficient mucoadhesive polymers with improved mucoadhesive and permeation enhancing properties that are believed to form disulfide bonds with cysteine-rich subdomains of mucus glycoproteins [56]. A patch system consisting of a multilayered oral delivery system for insulin comprising thiolated polycarbophil as a polymeric matrix layer and a water-insoluble backing layer of ethylcellulose, was able to prevent attack of intestinal luminal enzymes against insulin and provided enhancement of intestinal absorption [57]. It is well established that for inducing greater levels of absorption and stability at the intestinal epithelium, the use of a multilayered patch system is recommended. From a technological standpoint, these protective, rate controlling and adhesive properties are also ideal for oral dosage forms intended for delivery to the intestinal mucosa [58]. By using enteric coating of the commercial acrylate and methacrylic acid copolymer Eudragit, the intact transport of the dosage form through the stomach was guaranteed. Six hours after administration of the patch system, the mean maximum decrease in blood glucose level at 31.6% of the initial value could be observed [57]. In another study, a 2-iminothiolane thiolated polymer was covalently linked to chitosan to provide an improved efficacy of orally administered insulin. The resulting chitosan-4-thiobutylamidine (chitosan-TBA) conjugate was compressed into insulin tablets, showing controlled release of insulin and *in vitro* mucoadhesive/cohesive properties of chitosan at least 60-fold improved by the immobilization of thiol groups on the polymer [59]. After oral administration of chitosan-TBA-insulin tablets to non-diabetic conscious rats, the blood glucose level decreased significantly for 24 hours. A similar formulation composed of TBA homogenized with salmon calcitonin and chitosan led to a significant decrease in the plasma calcium level of the animals. The addition of glutathione to the tablets led to a further improvement in the oral bioavailability of salmon calcitonin with an earlier onset of action and a decrease in the calcium level of about 10% [60]. Calcitonin was also formulated into smaller minitablets of chitosan-TBA conjugate, resulting in similar hypocalcemia effect as after oral administration [61].

In addition, other mucoadhesive agents have been used to induce site-specific contact of peptides to the intestinal wall. This allows for local delivery of peptides to sites in the gastrointestinal tract, generating greater levels of absorption and stability. The use of permeation enhancers to augment uptake and transport through the intestinal wall, as well as protease inhibitors to protect peptides from enzymatic degradation, is also being explored. A comprehensive example was achieved by a concomitant oral administration of a capsule formulation containing insulin coated with Eudragit and sodium cholate which improved relative hypoglycemia compared with Eudragit coated capsules with insulin alone [62]. A caprylic acid derivative as a specific calcitonin carrier, which is thought to facilitate the transmembrane passage of calcitonin by inducing conformational changes in the calcitonin molecules and forming non-covalent bonds with the carrier, showed that oral delivery of calcitonin is feasible with reproducible absorption and systemic biological efficacy [63]. Such an oral formulation could facilitate the use of calcitonin in the treatment of osteoporosis and other bone diseases.

Although all these approaches improve the oral bioavailability of large molecules, none of them offers a complete solution for adequate and safe oral administration [58].

11.4.3 Microparticles

Microparticles of both biodegradable and non-biodegradable polymers have been investigated for sustained release of proteins. Since non-biodegradable polymers may cause problems of toxicity and clearance, the interest has been focused on biodegradable polymers [64]. Biodegradable polymers can be used to prepare microparticles by several methods, each with advantages and disadvantages. It is essential to select an encapsulation process fulfilling the requirements of high encapsulation efficiency, optimal protein loading and high yield of microparticles of the desired size. Also, protein:polymer ratio must be as large as possible to reduce the mass of the material to be administered. Stability of the encapsulated protein, batch uniformity and inter-batch reproducibility, adjustable release profiles, low burst effect and free-flowing or non-aggregating conditions are also crucial parameters of the encapsulation [65].

Mucoadhesive natural polymers like chitosan, alginate or pectin in the form of microparticles have been extensively explored to encapsulate proteins. It has been taking advantage of their ability to intensify the contact with the mucous layer of the intestine and to extend the residence time at the possible site of protein absorption without dilution or degradation of proteins in the intestinal fluid [66–69].

Insulin-loaded chitosan microparticles showed a significant hypoglycemic effect on the blood glucose level of streptozocin-diabetic rats after oral administration compared to administration of insulin in solution [70]. The same improvement of the hypoglycemic effect was observed when insulin was encapsulated into pectin microparticles and administered orally [71]. The ability of chitosan phthalate derivatives and succinate-insulin microparticles to enhance intestinal absorption and thus improve the relative pharmacological availability of insulin was confirmed by monitoring the plasma glucose and insulin level of streptozotocin-induced diabetic rats after oral administration [72, 73]. Because chitosan suffers from high solubility in gastric pH 2.0 and low solubility at physiological pH 7.4, thereby limiting its use as a carrier in oral protein delivery, chitosan phthalate [72] and chitosan succinate [73] were proposed because of their low solubility at acidic pH and high solubility at basic pH. Furthermore, chitosan phthalate and chitosan succinate microparticles protect insulin from acid and enzymatic degradation in the gastrointestinal tract.

Alginate [74] and alternative alginate-based microparticles by introducing additional copolymers on the alginate matrix like dextran sulfate [75] or by using alginate derivatives like alginate-cysteine [76], have been proposed to administer proteins orally, revealing most appropriate *in vitro* release profiles for intestinal purposes. One approach to make the mucoadhesion more efficient is to exploit a receptor-mediated process within the gastrointestinal tract with lectins, a process which occurs widely in nature and has the unique ability to recognize and bind to specific carbohydrate residues [77]. *In vivo* studies showed that the blood glucose levels of diabetic rats after oral administration of insulin were even lower when insulin was encapsulated into microparticles of alginate conjugated to lectin (wheat germ agglutinin, WGA) compared to simple alginate microparticles, thus taking advantage of the mucoadhesive properties of WGA [78].

Synthetic biodegradable polymers, such as the United States Food and Drug Administration (FDA)-approved poly(lactic acid) (PLA) and poly(lactide-*co*-glycolide acid) (PLGA), may be potentially useful for oral protein delivery vehicles [79]. Such microparticles can be easily prepared with high protein encapsulation efficiencies, as well as with the capability to effectively protect encapsulated proteins from degradation in the gastrointestinal tract, making this technology attractive for delivering labile drugs like proteins. Moreover, PLGA and its derivatives are able to provide sustained delivery of proteins [79]. Oral efficiency of insulin incorporated into PLGA microparticles was determined by measuring reduction of plasma glucose levels after feeding diabetic rats, which were observed over a period of six hours [80]. PLGA microparticles were able to adhere to the intestinal epithelium and traverse the absorptive cells since they were observed in spleen and liver [80]. Increasing the time that the proteins spend in the intestine after dosing would allow for greater absorption and increased bioavailability. PLGA microparticles complexed with positively charged micromagnets were effectively immobilized *in vitro* and gavaged to rats fitted with a magnetic belt [81]. This strategy retained higher amounts of insulin in the intestine and exhibited longer-term glucose reduction compared with regular PLGA microparticles [82].

Another polymer that has been used for microparticle oral protein delivery is the pH-sensitive poly(acrylic acid) derivate Eudragit, demonstrating successful incorporation of calcitonin and proving a controlled protein release *in vivo* [83]. The microparticle delivery system exhibited a distinct increase of the pharmacological effect compared to free calcitonin. Eudragit microparticles containing insulin also demonstrated prominent hypoglycemic effect after oral administration, suggesting not only that the polymer could protect insulin against proteolytic degradation in the gastrointestinal tract but also that it promotes the absorption of the protein through the intestinal epithelium [84]. It was observed that the co-administration of aprotinin and bile salts, as protease inhibitor and absorption enhancer, respectively, incorporated into insulin-loaded Eudragit microparticles produced prolonged and significant reduction of blood glucose level as compared with the formulation of insulin alone [85].

Microparticulate systems based on cross-linked poly(methacrylic acid) grafted with poly(ethylene glycol) (P(MAA-g-PEG)) as well as cross-linked poly(acrylic acid) grafted with poly(ethylene glycol) (P(AA-g-PEG)) have been developed and widely tested to elucidate the insulin intestinal absorption [86] and its functional activity in diabetic rats [87]. These microparticles are multifunctional carriers showing high insulin incorporation efficiency, a rapid insulin release in the intestine based on their pH-dependent complexation properties, enzyme-inhibiting effects and mucoadhesive characteristics [88]. PEG dimethacrylates of molecular weight ranging from 400 to 4000 and methacrylic acid were further copolymerized by suspension polymerization to obtain pH-sensitive hydrogel microparticles and insulin was loaded by partitioning from concentrated insulin solution [87]. *In vitro* release studies of insulin-loaded microparticles showed the minimal insulin leakage (18–25%) at acidic pH (2.5) and significantly higher release at more basic pH (7.4). In studies with diabetic rats, the blood glucose level was reduced for animals that received the insulin-loaded microparticles and the effect lasted for 8–10 h. It was also observed that two capsules per day of microparticles containing 80 IU/kg of insulin dose were sufficient to control the blood glucose level of fed diabetic rats to between 100 and 300 mg/dl.

Similar insulin-loaded polymer microparticles also composed of cross-linked poly(methacrylic acid) and poly(ethylene glycol) were administered orally to both healthy and type 1 and type 2 diabetic rats and further used for *in vivo* multiple oral administration studies using diabetic rats [88]. The cytotoxicity of these copolymers was investigated in contact with Caco-2 cell cultures and was found to exhibit no cytotoxicity on the cell cultures in the concentration range tested [86]. Furthermore, they were shown to open the tight junctions between cells, increasing the available area for diffusion across the cell monolayer, and thus increasing the permeability of insulin across the monolayer. In the multiple administration study, insulin-loaded microparticles significantly suppressed the postprandial rise in blood glucose and showed continuous hypoglycemic effects following three times/day oral administration to both diabetic rat groups in the presence of foods [88]. Polymer composition and the particle size of insulin-loaded polymer microparticles using these polymers strongly influenced insulin bioavailability. Microparticles with diameters of $<53\,\mu$m composed of a 1 : 1 molar ratio of methacrylic acid/ethylene glycol units showed the most pronounced hypoglycemic effects following oral administration to healthy rats, achieving a 9.5% pharmacological availability compared to subcutaneous insulin injection [88].

11.4.4 Nanoparticles

The use of nanoparticles for protein delivery has also been the focus in the field of long-term systemic delivery, intended to last for several days or weeks, depending on the degradability and erosion of the nanomeric matrix. Nanoparticulate drug delivery systems show a promising approach to obtain desirable drug pharmacodynamic properties by altering the biopharmaceutic and pharmacokinetic properties of the molecule [89].

It is nowadays generally assumed that nanoparticles can enhance the oral bioavailability of encapsulated therapeutic peptides and proteins. Moreover, their submicron size and their large specific surface area favour the absorption of the whole particle compared to larger carriers like microparticles. Consequently, it has already been extensively shown that nanoencapsulation of proteins, similarly to microencapsulation, protects them against the harsh environment of the gastrointestinal tract and increases their transmucosal diffusion [10]. Apart from the enhancing potential for systemic administration, nanoparticulate drug delivery systems can also be used for site-specific delivery, thus avoiding toxicity due to non-specific distribution, improving patient compliance and providing favourable clinical outcomes [89].

One of the advantages of nanoparticles, when administered orally, is that they can be absorbed transcellularly, although in small quantities, through the M-cells of the Peyer's patches and also through enterocytes [24]. Nanoparticle size, electrical charge and surface hydrophobicity are the main factors that influence the uptake by Peyer's patches. It is well accepted that hydrophobic, negatively charged protein-loaded nanoparticles smaller that 1 μm potentially show the best absorption rate [26, 90]. However, no generalities must be taken because other conditions can regulate the final absorption. The uptake of nanoparticles carrying proteins directly by enterocytes was also demonstrated [10].

To improve the uptake of nanoparticles, surface modifications [91, 92] and enhancing mucoadhesion properties [93, 94] are usually explored with efficiency to promote the contact of proteins with the intestinal epithelium, increasing the concentration at the site of absorption. The nanoparticle surface can easily be modified by adsorption or chemical grafting of certain molecules such as poly(ethylene glycol), poloxamers, and lectins [95]. In the gastrointestinal tract, the cationic nanoparticles are favoured to bind to the mucous layer, which is negatively charged, thus cationic polymers are selected as preferential mucoadhesive coating.

Chitosan has been by far the most employed natural polymer for developing nanoparticles intended for oral protein delivery, not only because it shows the special feature of adhering to mucous surface but also because it transiently opens or expands the tight junctions between epithelial cells [96, 97]. Apart from its ability to complex negatively charged polymers, an important property of chitosan is that it gels in the presence of specific polyanions such as tripolyphosphate. Adhesive chitosan nanoparticles were useful in increasing the relative pharmacological bioavailability of insulin [98, 99] and calcitonin [100] for at least 24 hours. This is associated with the ability to weaken the intestinal barrier in a reversible way, still maintaining the cellular viability in Caco-2 cells. A distinct advantage of proper particle size helped to increase the drug absorption and consequent physiological effect [96]. Similarly, chitosan nanoparticles were responsible for the improvement of cyclosporine oral absorption rate [101].

However, since chitosan is insoluble at physiological pH, water-soluble derivatives of chitosan have been used to produce protein-loaded nanoparticles. N-trimethyl chitosan (TMC), a partially quaternized chitosan derivative [102], n-(2-hydroxyl) propyl-3-trimethyl ammonium chitosan chloride [103] and chitosan-graft methyl methacrylate monomers [104] have replaced chitosan in the encapsulation of proteins, providing an improved stability due to enhanced electrostatic interactions, which increased the loading efficiency. Furthermore, various *in vivo* studies in different animal and cell models confirmed the ability of water-soluble derivatives of chitosan-based nanoparticles to increase the oral availability of the peptide octreotide [105, 106] and the intestinal absorption of insulin and buserelin [102, 104, 107]. N-trimethyl chitosan has been PEGylated in an attempt to increase the solubility of chitosan in water and to improve its biocompatibility [108, 109]. Thus, insulin has been entrapped into these nanoparticles composed of PEGylated trimethyl chitosan copolymers, which significantly enhanced cellular uptake by Caco-2 cells.

Chitosan has also been used as a hydrophilic polymeric coating on the surface of protein-loaded nanoparticles [110–112], increasing transport through the intestinal membrane. The surface composition of the nanoparticles is an important factor for the improvement of the efficiency of nanoparticle formulations. Due to the mucoadhesive properties of chitosan, the potential of chitosan-coated nanoparticles as carriers for peptide delivery is improved [110].

Alginate has also been applied to deliver insulin orally [111, 113]. When administered orally to diabetic rats, basal serum glucose levels were lowered by more than 40% by sustaining hypoglycemia for over 18 hours. Recently, insulin-loaded dextran-alginate nanoparticles were coated with chitosan, PEG4000 and albumin and induced a marked reduction in blood glucose level from four hours up to 48 hours after oral administration [114].

A different attempt to deliver proteins orally has been to encapsulate them into dextran-based nanoparticles. A main benefit in the use of dextran sulfate is its ability to reach a high level of protein entrapment and to maintain its bioactive structure [115]. Insulin-loaded dextran nanoparticles coated with chitosan showed a release profile that was suitable for oral application and elicited a significant hypoglycemic effect after administration to diabetic animals [116]. *In vivo* results also indicate that oral delivery of insulin-loaded vitamin B12-coated dextran nanoparticles have considerable promise in complementing the therapy of diabetes [117, 118]. These particles take advantage of the use of vitamin B12–dextran nanoparticle conjugates, which target the systemic circulation through vitamin B12-IF-IFR (intrinsic factor receptor) ligand-mediated endocytosis via ileocytes of thc intcstine [118].

The introduction of the use of synthetic polymers in protein delivery was based on the advantage of sustaining release over a period of days to several weeks compared to natural polymers, which have a relatively short duration of drug release. Aliphatic polyesters such as PLA, PLGA and poly (ε-caprolactone) have attracted significant interest as protein carriers [7, 10, 119] due to their biocompatibility and biodegradability. This class of polymers degrades via the hydrolytic cleavage of the ester bonds in their backbone [120]; the role of enzymatic involvement in biodegradation is unclear. As a majority of proteins and peptides are hydrophilic, any attempt to load them into hydrophobic polyester nanoparticles becomes a challenging task. To minimize the problems related to protein stability and improve protein association efficiency, the w/o/w double emulsion method is mostly preferred forproducing nanoparticles [79].

Proteins and peptides like calcitonin [90], insulin [121, 122] or cyclosporine [123] were encapsulated into PLGA nanoparticles and *in vivo* oral uptake was demonstrated. Additionally, PLGA nanoparticles showed significantly improved intestinal uptake and oral bioavailability when compared to commercial cyclosporine formulations and lower nephrotoxicity upon the chronic administration in animals [123]. In addition, modifications of PLGA nanoparticles have been proposed to improve the properties of PLGA nanoparticles. These include a blended matrix formed by the PLGA copolymer and polyoxyethylene derivatives (PLGA:poloxamer and PLGA:poloxamine compositions) [124] and PLGA-mannosamine nanoparticles [125] and PLGA and HPMCP (hydroxypropyl methoxycellulose phthalate), a pH sensitive cellulose coating designed for enterocoating of oral dosage forms [126]; they result in good insulin entrapment ability and reduced significantly the serum glucose level over 24 hours, compared with simple PLGA nanoparticles. Coating calcitonin-loaded PLGA nanoparticles with chitosan reduced significantly the blood calcium level compared with uncoated nanoparticles administered intragastrically, and the reduced calcium level was sustained for a period of 48 hours [127]. Also, bile salts could be used to improve the bioavailability of PLGA nanoparticles, when administered in a bile salt emulsion, by protecting them during their transport through the gastrointestinal tract and enhancing their absorption by the intestinal epithelia [128]. This attempt is hypothesized to improve protein oral delivery. The conjugation of lectins onto PLGA nanoparticles has been demonstrated to effectively improve the intestinal absorption of thymopentin, a pentapeptide, which is clinically used in the treatments of autoimmune diseases [129]. Moreover, *ex vivo* bioadhesion on rat intestinal mucosa confirmed an important increase of interaction between lectin-conjugated nanoparticles and the intestinal segments [130].

Nanoparticles in the form of vesicles made by amphiphilic copolymers of PLA and a triblock of ethylene oxide and propylene oxide (PEO-PPO-PEO) demonstrated superior properties to deliver insulin orally, decreasing the blood glucose concentration in a diabetic animal model and maintaining a significant hypoglycemic effect for 24 hours [131]. Insulin was orally delivered encapsulated in nanoparticles prepared with poly(-ε-caprolactone) blended with the acrylic polymer Eudragit, decreasing fasted glycemia in a dose-dependant manner [132]. In addition, the anti-diabetic effect was improved by the mucoadhesive properties of Eudragit allowing the intestinal uptake of insulin.

A novel poly(acrylic acid)-cysteine conjugate poly(vinyl pyrrolidone) (PAA-Cys/PVP) nanoparticulate system was recently proposed to encapsulate insulin and provided a stabilized system within a gastric environment, resulting in a significant serum insulin concentration and a blood sugar reduction after oral administration, suggesting that the mucoadhesive properties of PAA-Cys played an essential role in the absorption of insulin from the gastric environment [133].

Poly(alkylcyanoacrylates) (PACA) are another group of polymers that have also been used to deliver proteins orally because of their excellent adhesive properties resulting from the strong bonds formed with most polar substrates, including mucosa, tissues and skin, and absence of toxicological effects [134]. PACA nanoparticles can be prepared either by polymerization of alkylcyanoacrylate or directly from the polymers and these nanoparticles have been shown for more than two decades to be able to encapsulate insulin and to increase its activity as assessed by a reduction of glycemia after oral administration [135, 136]. Later, it was possible to develop similar insulin PACA nanoparticles without the use of the oily core maintaining the hypoglycemic ability [137]. Then, octreotide [138] and calcitonin [139] were also demonstrated to posses a better oral availability after encapsulation into PACA nanoparticles.

Recently, novel pH-sensitive poly(methacrylic acid)-chitosan-poly(ethylene glycol) (PCP) nanoparticles were prepared. Nanoparticles were obtained spontaneously during polymerization without using organic solvents or surfactants/steric stabilizers and insulin was incorporated into the nanoparticles by diffusion filling method. PCP nanoparticles exhibited good protein encapsulation efficiency and pH responsive release profile under *in vitro* conditions, suggesting good ability for oral peptide delivery [140], but additional *in vivo* assays are required to confirm this trend. However, similar networks of methacrylic acid grafted with poly(ethylene glycol), and acrylic acid grafted with poly(ethylene glycol) nanospheres caused a significant reduction of serum glucose with respect to that of a control animal [141].

More complex polymers have also been used to administer proteins orally in the form of nanoparticles. Styrene nanoparticles having poly-N-isopropylacrylamide (PNIPAAm) chains on their surfaces were tested as carrier for calcitonin in rats [142]. After oral administration, the absorption of calcitonin was significantly enhanced and this enhancement effect was considerably increased by introducing cationic poly(vinyl amine) (PVAm) groups to the surface of PNIPAAm nanoparticles, mainly due to increased mucoadhesion effect [143].

11.5 Conclusions

The development of new technologies to produce novel drug delivery systems has been one of the most attractive directions to overcome the peptide and protein instability both *in vitro* and *in vivo*. Increased focus on materials to be used as matrices, such as natural polymers (e.g. chitosan, alginate, cellulose sulfate, cyclodextrins, starch and dextrans) and synthetic polymers (e.g. PLA, PLGA, PACA, PMMA and other amphiphilic polymers) combined with suitable methodologies has provided optimum physical conditions to formulate proteins. The selection of the matrix material to be used as delivery system for a particular peptide or protein is related to several optimized parameters, that is the drug molecule must be active upon administration of the developed system and no toxic residues must be produced by metabolism. In addition, formulations should be able to demonstrate maintenance of peptide and protein stability, and to release the bioactive in such a quantity to act according to its therapeutic purposes. Moreover, conjugated complexes between the bioactive molecule and the matrix material should be considered as a novel chemical entity, which must be evaluated in terms of safety as well. Peptide and protein stability is the major concern when developing delivery systems based on nanoparticles and the integrity is often affected by the production procedure applied. Thus, the choice of a particular methodology should be performed keeping in mind that the processing parameters might interfere with peptide and protein stability. Therefore, the major challenge of producing biodegradable polymeric systems containing stable peptides and proteins is circumvention of technical obstacles related to both production procedures and administration routes.

In the present review, a significant number of relevant polymer-based carriers for oral administration of peptides and proteins have been summarized, portraying optimistic perspectives to overcome the classical protein absorption drawbacks in the oral tract.

Significant advances have been made in the field of nanoencapsulation of therapeutic proteins. The development of improved formulation methods with the ability to overcome the physical and chemical intestinal barriers resulting in higher oral bioavailability is being explored with different degrees of success.

49. Dorkoosh, F.A., Setyaningsih, D., Borchard, G. *et al.* (2002) Effects of superporous hydrogels on paracellular drug permeability and cytotoxicity studies in Caco-2 cell monolayers. *Int. J. Pharm.*, **241**, 35–45.
50. Dorkoosh, F.A., Verhoef, J.C., Borchard, G. *et al.* (2002) Intestinal absorption of human insulin in pigs using delivery systems based on superporous hydrogel polymers. *Int. J. Pharm.*, **247**, 47–55.
51. Dorkoosh, F.A., Verhoef, J.C., Verheijden, J.H. *et al.* (2002) Peroral absorption of octreotide in pigs formulated in delivery systems on the basis of superporous hydrogel polymers. *Pharm. Res.*, **19**, 1532–1536.
52. Bernkop-Schnurch, A., Krauland, A.H., Leitner, V.M. and Palmberger, T. (2004) Thiomers: potential excipients for non-invasive peptide delivery systems. *Eur. J. Pharm. Biopharm.*, **58**, 253–263.
53. Bernkop-Schnurch, A., Schwarz, V. and Steininger, S. (1999) Polymers with thiol groups: a new generation of mucoadhesive polymers? *Pharm. Res.*, **16**, 876–881.
54. Luessen, H.L., de Leeuw, B.J., Langemeyer, M.W. *et al.* (1996) Mucoadhesive polymers in peroral peptide drug delivery. VI. Carbomer and chitosan improve the intestinal absorption of the peptide drug buserelin *in vivo*. *Pharm. Res.*, **13**, 1668–1672.
55. Whitehead, K., Shen, Z. and Mitragotri, S. (2004) Oral delivery of macromolecules using intestinal patches: applications for insulin delivery. *J. Control. Release*, **23**, 37–45.
56. Schmitz, T., Leitner, V.M. and Bernkop-Schnurch, A. (2005) Oral heparin delivery: design and *in vivo* evaluation of a stomach-targeted mucoadhesive delivery system. *J. Pharm. Sci.*, **94**, 966–973.
57. Grabovac, V., Foger, F. and Bernkop-Schnurch, A. (2008) Design and *in vivo* evaluation of a patch delivery system for insulin based on thiolated polymers. *Int. J. Pharm.*, **348**, 169–174.
58. Taoand, S.L. and Desai, T.A. (2005) Gastrointestinal patch systems for oral drug delivery. *Drug Discov. Today*, **10**, 909–915.
59. Krauland, A.H., Guggi, D. and Bernkop-Schnurch, A. (2004) Oral insulin delivery: the potential of thiolated chitosan-insulin tablets on non-diabetic rats. *J. Control. Release*, **95**, 547–555.
60. Guggi, D., Kast, C.E. and Bernkop-Schnurch, A. (2003) *In vivo* evaluation of an oral salmon calcitonin-delivery system based on a thiolated chitosan carrier matrix. *Pharm. Res.*, **20**, 1989–1994.
61. Guggi, D., Krauland, A.H. and Bernkop-Schnurch, A. (2003) Systemic peptide delivery via the stomach: *in vivo* evaluation of an oral dosage form for salmon calcitonin. *J. Control. Release*, **92**, 125–135.
62. Hosny, E.A., Ghilzai, N.M., al-Najar, T.A. and Elmazar, M.M. (1998) Hypoglycemic effect of oral insulin in diabetic rabbits using pH-dependent coated capsules containing sodium salicylate without and with sodium cholate. *Drug Dev. Ind. Pharm.*, **24**, 307–311.
63. Buclin, T., Cosma Rochat, M., Burckhardt, P. *et al.* (2002) Bioavailability and biological efficacy of a new oral formulation of salmon calcitonin in healthy volunteers. *J. Bone Miner. Res.*, **17**, 1478–1485.
64. Sinhaand, V.R. and Trehan, A. (2003) Biodegradable microspheres for protein delivery. *J. Control. Release*, **90**, 261–280.
65. Singh, R., Singh, S. and Lillard, J.W. (2008) Past, present, and future technologies for oral delivery of therapeutic proteins. *J. Pharm. Sci.*, **97**, 2497–2523.
66. Reis, C.P., Ribeiro, A.J., Neufeld, R.J. and Veiga, F. (2007) Alginate microparticles as novel carrier for oral insulin delivery. *Biotechnol. Bioeng.*, **96**, 977–989.
67. Silva, C.M., Ribeiro, A.J., Figueiredo, I.V. *et al.* (2006) Alginate microspheres prepared by internal gelation: development and effect on insulin stability. *Int. J. Pharm.*, **311**, 1–10.
68. Georgeand, M. and Abraham, T.E. (2006) Polyionic hydrocolloids for the intestinal delivery of protein drugs: alginate and chitosan – a review. *J. Control. Release*, **114**, 1–14.
69. Shimoda, J., Onishi, H. and Machida, Y. (2001) Bioadhesive characteristics of chitosan microspheres to the mucosa of rat small intestine. *Drug Dev. Ind. Pharm.*, **27**, 567–576.

70. Huang, H., Tian, H., Li, X. and Zhao, G. (2001) Hypoglycemic effect of chitosan-microcapsulated insulin on the blood glucose level of streptozotocin-diabetic Wistar rats. *Sheng Wu Yi Xue Gong Cheng Xue Za Zhi*, **18**, 425–427, 460.
71. Musabayane, C.T., Munjeri, O., Bwititi, P. and Osim, E.E. (2000) Orally administered, insulin-loaded amidated pectin hydrogel beads sustain plasma concentrations of insulin in streptozotocin-diabetic rats. *J. Endocrinol.*, **164**, 1–6.
72. Ubaidulla, U., Khar, R.K., Ahmed, F.J. and Panda, A.K. (2007) Development and *in vivo* evaluation of insulin-loaded chitosan phthalate microspheres for oral delivery. *J. Pharm. Pharmacol.*, **59**, 1345–1351.
73. Ubaidulla, U., Kishan, R., Fahan, K. *et al.* (2007) Development and characterization of chitosan succinate microspheres for the improved oral bioavailability of insulin. *J. Pharm. Sci.*, **96**, 3010–3023.
74. Jerry, N., Anitha, Y., Sharma, C.P. and Sony, P. (2001) *In vivo* absorption studies of insulin from an oral delivery system. *Drug Deliv.*, **8**, 19–23.
75. Silva, C., Ribeiro, A., Veiga, F. and Sousa, A. (2006) Insulin release from alginate microspheres reinforced with dextran sulfate. *Chem. Ind. Chem. Eng. Q.*, **12**, 40–46.
76. Greimel, A., Werle, M. and Bernkop-Schnurch, A. (2007) Oral peptide delivery: *in vitro* evaluation of thiolated alginate/poly(acrylic acid) microparticles. *J. Pharm. Pharmacol.*, **59**, 1191–1198.
77. Ponchel, G. and Irache, J.-M. (1998) Specific and non-specific bioadhesive particulate systems for oral delivery to the gastrointestinal tract. *Adv. Drug Deliv. Rev.*, **34**, 191–219.
78. Kim, B.-Y., Jeong, J.H., Park, K. and Kim, J.-D. (2005) Bioadhesive interaction and hypoglycemic effect of insulin-loaded lectin-microparticle conjugates in oral insulin delivery system. *J. Control. Release*, **102**, 525–538.
79. Mundargi, R.C., Babu, V.R., Rangaswamy, V. *et al.* (2008) Nano/micro technologies for delivering macromolecular therapeutics using poly(D,L-lactide-co-glycolide) and its derivatives. *J. Control. Release*, **125**, 193–209.
80. Li, Y., Jiang, H.L., Jin, J.F. and Zhu, K.J. (2004) Bioadhesive fluorescent microspheres as visible carriers for local delivery of drugs. II: uptake of insulin-loaded PCEFB/PLGA microspheres by the gastrointestinal tract. *Drug Deliv.*, **11**, 335–340.
81. Cheng, J., Teply, B., Jeong, S. *et al.* (2006) Magnetically responsive polymeric microparticles for oral delivery of protein drugs. *Pharm. Res.*, **23**, 557–564.
82. Teply, B.A., Tong, R., Jeong, S.Y. *et al.* (2008) The use of charge-coupled polymeric microparticles and micromagnets for modulating the bioavailability of orally delivered macromolecules. *Biomaterials*, **29**, 1216–1223.
83. Lamprecht, A., Yamamoto, H., Takeuchi, H. and Kawashima, Y. (2004) pH-sensitive microsphere delivery increases oral bioavailability of calcitonin. *J. Control. Release*, **98**, 1–9.
84. Jain, D., Panda, A.K. and Majumdar, D.K. (2005) Eudragit S100 entrapped insulin microspheres for oral delivery. *AAPS Pharm. Sci. Tech.*, **6**, E100–E107.
85. Rajan, D.S., Mandal, U.K., Gowda, K.V. *et al.* (2004) Oral delivery system of insulin microspheres: effect on relative hypoglycemia of diabetic albino rats. *Boll. Chim. Farm.*, **143**, 315–318.
86. Fossand, A.C. and Peppas, N.A. (2004) Investigation of the cytotoxicity and insulin transport of acrylic-based copolymer protein delivery systems in contact with Caco-2 cultures. *Eur. J. Pharm. Biopharm.*, **57**, 447–455.
87. Kumar, A., Lahiri, S.S. and Singh, H. (2006) Development of PEGDMA: MAA based hydrogel microparticles for oral insulin delivery. *Int. J. Pharm.*, **323**, 117–124.
88. Morishita, M., Goto, T., Nakamura, K. *et al.* (2006) Novel oral insulin delivery systems based on complexation polymer hydrogels: single and multiple administration studies in type 1 and 2 diabetic rats. *J. Control. Release*, **110**, 587–594.
89. Devalapally, H., Chakilam, A. and Amiji, M.M. (2007) Role of nanotechnology in pharmaceutical product development. *J. Pharm. Sci.*, **96**, 2547–2565.
90. Yoo, H.S. and Park, T.G. (2004) Biodegradable nanoparticles containing protein-fatty acid complexes for oral delivery of salmon calcitonin. *J. Pharm. Sci.*, **93**, 488–495.

91. Bhattarai, N., Ramay, H.R., Chou, S.-H. and Zhang, M. (2006) Chitosan and lactic acid-grafted chitosan nanoparticles as carriers for prolonged drug delivery. *Int. J. Nanomed.*, **1**, 181–187.
92. Vila, A., Gill, H., McCallion, O. and Alonso, M.J. (2004) Transport of PLA-PEG particles across the nasal mucosa: effect of particle size and PEG coating density. *J. Control. Release*, **98**, 231–244.
93. Garcia-Fuentes, M., Torres, D. and Alonso, M.J. (2005) New surface-modified lipid nanoparticles as delivery vehicles for salmon calcitonin. *Int. J. Pharm.*, **296**, 122–132.
94. Tobio, M., Sanchez, A., Vila, A. *et al.* (2000) The role of PEG on the stability in digestive fluids and *in vivo* fate of PEG-PLA nanoparticles following oral administration. *Colloids. Surf. B*, **18**, 315–323.
95. Rodrigues, J.S., Santos-Magalhaes, N.S., Couvreur, L.C.B.B. *et al.* (2003) Novel core(polyester)-shell(polysaccharide) nanoparticles: protein loading and surface modification with lectins. *J. Control. Release*, **92**, 103–112.
96. Pan, Y., Zheng, J.-M., Zhao, H.-Y. *et al.* (2002) Relationship between drug effects and particle size of insulin-loaded bioadhesive microspheres. *Acta Pharm. Sinic.*, **23**, 1051–1056.
97. Janes, K.A., Calvo, P. and Alonso, M.J. (2001) Polysaccharide colloidal particles as delivery systems for macromolecules. *Adv. Drug Deliv. Rev.*, **47**, 83–97.
98. Pan, Y., Li, Y.J., Zhao, H.Y. *et al.* (2002) Bioadhesive polysaccharide in protein delivery system: chitosan nanoparticles improve the intestinal absorption of insulin *in vivo*. *Int. J. Pharm.*, **249**, 139–147.
99. Ma, Z., Lim, T.M. and Lim, L.-Y. (2005) Pharmacological activity of peroral chitosan-insulin nanoparticles in diabetic rats. *Int. J. Pharm.*, **293**, 271–280.
100. Prego, C., Torres, D. and Alonso, M.J. (2006) Chitosan nanocapsules as carriers for oral peptide delivery: effect of chitosan molecular weight and type of salt on the *in vitro* behaviour and *in vivo* effectiveness. *J. Nanosci. Nanotechnol.*, **6**, 1–8.
101. El-Shabouri, M.H. (2002) Positively charged nanoparticles for improving the oral bioavailability of cyclosporin-A. *Int. J. Pharm.*, **249**, 101–108.
102. van der Merwe, S.M., Verhoef, J.C., Verheijden, J.H.M. *et al.* (2004) Trimethylated chitosan as polymeric absorption enhancer for improved peroral delivery of peptide drugs. *Eur. J. Pharm. Biopharm.*, **58**, 225–235.
103. Xu, Y., Du, Y., Huang, R. and Gao, L. (2003) Preparation and modification of N-(2-hydroxyl) propyl-3-trimethyl ammonium chitosan chloride nanoparticle as a protein carrier. *Biomaterials*, **24**, 5015–5022.
104. Qian, F., Cui, F., Ding, J. *et al.* (2006) Chitosan graft copolymer nanoparticles for oral protein drug delivery: preparation and characterization. *Biomacromolecules*, **7**, 2722–2727.
105. Thanou, M., Verhoef, J.C., Marbach, P. and Junginger, H.E. (2000) Intestinal absorption of octreotide: N-trimethyl chitosan chloride (TMC) ameliorates the permeability and absorption properties of the somatostatin analogue *in vitro* and *in vivo*. *J. Pharm. Sci.*, **89**, 951–957.
106. Thanou, M., Verhoef, J.C., Verheijden, J.H.M. and Junginger, H.E. (2001) Intestinal absorption of octreotide using trimethyl chitosan chloride: studies in pigs. *Pharm. Res.*, **18**, 823–828.
107. Kotze, A.F., Luessen, H.L., de Leeuw, B.J. *et al.* (1997) N-trimethyl chitosan chloride as a potential absorption enhancer across mucosal surfaces: *in vitro* evaluation in intestinal epithelial cells (Caco-2). *Pharm. Res.*, **14**, 1197–1202.
108. Mao, S., Shuai, X., Unger, F. *et al.* (2005) Synthesis, characterization and cytotoxicity of poly(ethylene glycol)-graft-trimethyl chitosan block copolymers. *Biomaterials*, **26**, 6343–6356.
109. Mao, S., Germershaus, O., Fischer, D. *et al.* (2005) Uptake and transport of PEG-graft-trimethyl-chitosan copolymers as insulin nanocomplexes by epithelial cells. *Pharm. Res.*, **22**, 2058–2068.
110. Garcia-Fuentes, M., Prego, C., Torres, D. and Alonso, M.J. (2005) A comparative study of the potential of solid triglyceride nanostructures coated with chitosan or poly(ethylene glycol) as carriers for oral calcitonin delivery. *Eur. J. Pharm. Sci.*, **25**, 123–133.

111. Sarmento, B., Ribeiro, A., Veiga, F. *et al.* (2007) Alginate/chitosan nanoparticles are effective for oral insulin delivery. *Pharm. Res.*, **24**, 2198–2206.
112. Lin, Y.H., Mi, F.L., Chen, C.T. *et al.* (2007) Preparation and characterization of nanoparticles shelled with chitosan for oral insulin delivery. *Biomacromolecules*, **8**, 146–152.
113. Reis, C.P., Ribeiro, A.J., Houng, S. *et al.* (2007) Nanoparticulate delivery system for insulin: design, characterization and *in vitro/in vivo* bioactivity. *Eur. J. Pharm. Sci.*, **30**, 392–397.
114. Reis, C.P., Veiga, F.J., Ribeiro, A.J. *et al.* (2008) Nanoparticulate biopolymers deliver insulin orally eliciting pharmacological response. *J. Pharm. Sci.*, **97**, 5290–5305.
115. Sarmento, B., Ribeiro, A., Veiga, F. and Ferreira, D. (2006) Development and characterization of new insulin containing polysaccharide nanoparticles. *Colloids Surf. B*, **53**, 193–202.
116. Sarmento, B., Ribeiro, A., Veiga, F. *et al.* (2007) Oral bioavailability of insulin contained in polysaccharide nanoparticles. *Biomacromolecules*, **8**, 3054–3060.
117. Chalasani, K.B., Russell-Jones, G.J., Yandrapu, S.K. *et al.* (2007) A novel vitamin B12-nanosphere conjugate carrier system for peroral delivery of insulin. *J. Control. Release*, **117**, 421–429.
118. Chalasani, K.B., Russell-Jones, G.J., Jain, A.K. *et al.* (2007) Effective oral delivery of insulin in animal models using vitamin B12-coated dextran nanoparticles. *J. Control. Release*, **122**, 141–150.
119. Panyam, J. and Labhasetwar, V. (2003) Biodegradable nanoparticles for drug and gene delivery to cells and tissue. *Adv. Drug Deliv. Rev.*, **55**, 329–347.
120. Houchin, M.L. and Topp, E.M. (2007) Chemical degradation of peptides and proteins in PLGA: a review of reactions and mechanisms. *J. Pharm. Sci.*, **97**, 2395–2404.
121. Kumar, P.S., Ramakrishna, S., Saini, T.R. and Diwan, P.V. (2006) Influence of microencapsulation method and peptide loading on formulation of poly(lactide-co-glycolide) insulin nanoparticles. *Pharmazie*, **61**, 613–617.
122. Barichello, J.M., Morishita, M., Takayama, K. and Nagai, T. (1999) Encapsulation of hydrophilic and lipophilic drugs in PLGA nanoparticles by the nanoprecipitation method. *Drug Dev. Ind. Pharm.*, **25**, 471–476.
123. Italia, J.L., Bhatt, D.K., Bhardwaj, V. *et al.* (2007) PLGA nanoparticles for oral delivery of cyclosporine: nephrotoxicity and pharmacokinetic studies in comparison to Sandimmune Neoral. *J. Control. Release*, **119**, 197–206.
124. Csaba, N., Gonzalez, L., Sanchez, A. and Alonso, M.J. (2004) Design and characterisation of new nanoparticulate polymer blends for drug delivery. *J. Biomat. Sci. Polym. Ed.*, **15**, 1137–1151.
125. Alonso-Sande, M. (2007) Nanopartículas poliméricas modificadas con manosa: nuevos sistemas destinados a la administración oral de macromoléculas terapéuticas. PhD Thesis, University of Santiago de Compostela, ISBN: 978-84-9750-769-1.
126. Cui, F., Tao, A., Cun, D. *et al.* (2007) Preparation of insulin loaded PLGA-Hp55 nanoparticles for oral delivery. *J. Pharm. Sci.*, **96**, 421–427.
127. Kawashima, Y., Yamamoto, H., Takeuchi, H. *et al.* (2000) Mucoadhesive DL-lactide/glycolide copolymer nanospheres coated with chitosan to improve oral delivery of elcatonin. *Pharm. Dev. Technol.*, **5**, 77–85.
128. Samstein, R.M., Perica, K., Balderrama, F. *et al.* (2008) The use of deoxycholic acid to enhance the oral bioavailability of biodegradable nanoparticles. *Biomaterials*, **29**, 703–708.
129. Yin, Y., Chen, D., Qiao, M. *et al.* (2006) Preparation and evaluation of lectin-conjugated PLGA nanoparticles for oral delivery of thymopentin. *J. Control. Release*, **116**, 337–345.
130. Yin, Y., Chen, D., Qiao, M. *et al.* (2007) PLGA nanoparticles loaded with thymopentin: *ex vivo* bioadhesion and *in vivo* biodistribution. *J. Control. Release*, **123**, 27–38.
131. Xiong, X.Y., Li, Y.P., Li, Z.L. *et al.* (2007) Vesicles from Pluronic/poly(lactic acid) block copolymers as new carriers for oral insulin delivery. *J. Control. Release*, **120**, 11–17.
132. Damge, C., Maincent, P. and Ubrich, N. (2007) Oral delivery of insulin associated to polymeric nanoparticles in diabetic rats. *J. Control. Release*, **117**, 163–170.
133. Deutel, B., Greindl, M., Thaurer, M. and Bernkop-Schnurch, A. (2008) Novel insulin thiomer nanoparticles: *in vivo* evaluation of an oral drug delivery system. *Biomacromolecules*, **9**, 278–285.

134. Vauthier, C., Dubernet, C., Fattal, E. *et al.* (2003) Poly(alkylcyanoacrylates) as biodegradable materials for biomedical applications. *Adv. Drug Deliv. Rev.*, **55**, 519–548.
135. Damge, C., Michel, C., Aprahamian, M. and Couvreur, P. (1988) New approach for oral administration of insulin with polyalkylcyanoacrylate nanocapsules as drug carrier. *Diabetes*, **37**, 246–251.
136. Damge, C., Vrancks, H., Balschmidt, P. and Couvreur, P. (1997) Poly(alkyl cyanoacrylate) nanospheres for oral administration of insulin. *J. Pharm. Sci.*, **86**, 1407–1500.
137. Mesiha, M.S., Sidhom, M.B. and Fasipe, B. (2005) Oral and subcutaneous absorption of insulin poly(isobutylcyanoacrylate) nanoparticles. *Int. J. Pharm.*, **288**, 289–293.
138. Damge, C., Vonderscher, J., Marbach, P. and Pinget, M. (1997) Poly(alkyl cyanoacrylate) nanocapsules as a delivery system in the rat for octreotide, a long-acting somatostatin analogue. *J. Pharm. Pharmacol.*, **49**, 949–954.
139. Lowe, P.J. and Temple, C.S. (1994) Calcitonin and insulin in isobutylcyanoacrylate nanocapsules: protection against proteases and effect on intestinal absorption in rats. *J. Pharm. Pharmacol.*, **46**, 547–552.
140. Sajeesh, S. and Sharma, C.P. (2006) Novel pH responsive polymethacrylic acid-chitosan-polyethylene glycol nanoparticles for oral peptide delivery. *J. Biomed. Mater. Res. B. Appl. Biomater.*, **76**, 298–305.
141. Foss, A.C., Goto, T., Morishita, M. and Peppas, N.A. (2004) Development of acrylic-based copolymers for oral insulin delivery. *Eur. J. Pharm. Biopharm.*, **57**, 163–169.
142. Sakuma, S., Suzuki, N., Sudo, R. *et al.* (2002) Optimized chemical structure of nanoparticles as carriers for oral delivery of salmon calcitonin. *Int. J. Pharm.*, **239**, 185–195.
143. Sakuma, S., Sudo, R., Suzuki, N. *et al.* (2002) Behavior of mucoadhesive nanoparticles having hydrophilic polymeric chains in the intestine. *J. Control. Release*, **81**, 281–290.

12

Advanced Pulmonary Delivery of Peptides or Proteins Using Polymeric Particles

Yu Seok Youn,[1] *Kang Choon Lee,*[2] *You Han Bae,*[3] *Kun Na*[4] *and Eun Seong Lee*[4]

[1]Pusan National University, College of Pharmacy, Busan, Republic of Korea
[2]SungKyunKwan University, College of Pharmacy, Suwon City, Republic of Korea
[3]University of Utah, Department of Pharmaceutics and Pharmaceutical Chemistry, Salt Lake City, UT, USA
[4]The Catholic University of Korea, Division of Biotechnology, Bucheon-si, Republic of Korea

12.1 Introduction

Biotechnological advancements of the recombinant DNA and solid-phase peptide synthesis techniques have generated a number of peptides and proteins with therapeutic potential. Some peptides and proteins are known to effectively treat a wide variety of specific diseases, which are sometimes untreatable using chemically-based drugs due to their specific biological actions. Hundreds of therapeutic peptides and proteins are now either commercially available or are in the process of development [1].

The development of these peptide and protein agents also presents new challenges in terms of the delivery methods because they have intrinsic disadvantages, that is large molecular size, hydrophilicity and enzyme susceptibility, meaning that many cannot be delivered via mucosal routes, particularly the oral route [1]. Therefore, their administration is still dependent on parenteral routes, which cause the injection pain [2]. Moreover, metabolic disorders treated with peptides and proteins necessitate long-term therapy [3], and thus non-invasive drug delivery systems (DDSs) are considered mandatory for these drugs.

Delivery Technologies for Biopharmaceuticals: Peptides, Proteins, Nucleic Acids and Vaccines Edited by Lene Jorgensen and Hanne Mørck Nielsen

Numerous approaches involving non-parenteral routes have been examined, for example intestinal, nasal, buccal, transdermal and rectal, but most are inadequate in terms of inducing a satisfactory therapeutic response [4]. On the other hand, the pulmonary route shows great promise in the systemic delivery and bioavailability of peptides and proteins, since the lungs are highly permeable and accessible by normal inhalation methods. Actually, the pulmonary route typically has 10 to 200 times greater bioavailability than other non-invasive routes [5, 6]. Thus, various pharmaceutical techniques have been introduced to take advantage of this route. These include formulations using excipients, chemical modification of drugs, polymeric DDSs, particle engineering, inhalation devices and others.

In particular, over the past decade, the uses of polymeric DDSs involving micro or nanocarriers and particle engineering techniques have developed remarkably, and the resulting pharmaceutical techniques have been eagerly applied to the pulmonary delivery of peptides and proteins. This chapter focuses on recent advances in the polymeric approach to the pulmonary delivery of therapeutic peptides and proteins in the context of particulate drug delivery systems, and also briefly describes the physiology of the lungs, the various factors that affect pulmonary delivery, issues associated with sustained/controlled release, stability and toxicity, and *in vivo* efficacy results (Figure 12.1).

12.2 Practical Issues in the Pulmonary Delivery of Peptides and Proteins

12.2.1 Physiological Features of the Lungs

The lungs have unique physiological features and provide many conditions that favour the absorption of peptides and proteins. The lungs consist of the upper (trachea, bronchi and terminal bronchioles) and lower airways (respiratory bronchioles and alveolar sacs) [6]. Furthermore, the lungs offer a spacious surface area (especially alveoli) of ~$100\,m^2$, and alveoli have a highly vascularized structure that facilitates access to the systemic circulation. In addition, the pulmonary epithelium thins from 50–60 μm in the trachea to 0.1–0.5 μm in alveoli, which further aids drug permeation [7]. Therefore, the lungs are viewed to be far more amenable to macromolecules than any other mucosal route. In addition, drugs that enter the body via the lungs can avoid first-pass hepatic elimination, which is inevitable via the oral route [8]. The bioavailabilities of some therapeutic peptides and proteins are listed in Table 12.1 [8–12].

12.2.2 Barriers to the Pulmonary Delivery of Peptides and Proteins

Metabolism in the lungs is much milder than that associated with the oral route because peptides and proteins are heavily degraded by enzymatic hydrolysis in the gut and liver [5, 13]. Nonetheless, peptides and proteins are exposed to substantial proteolytic attack in the lungs because they also contain many peptidases, for example trypsin, chymotrypsin, cathepsin-B, leucine aminopeptidase and dipeptidyl peptidase IV [14–17]. In particular, unlike proteins with relatively large molecular size (>6000 Da), small peptides of less than 30 amino acids are easily broken down by peptidases in the lungs. Thus, contrary to initial expectations, some peptides delivered via the pulmonary route are hardly detectable in the systemic circulation [5]. To address these problems, increasing

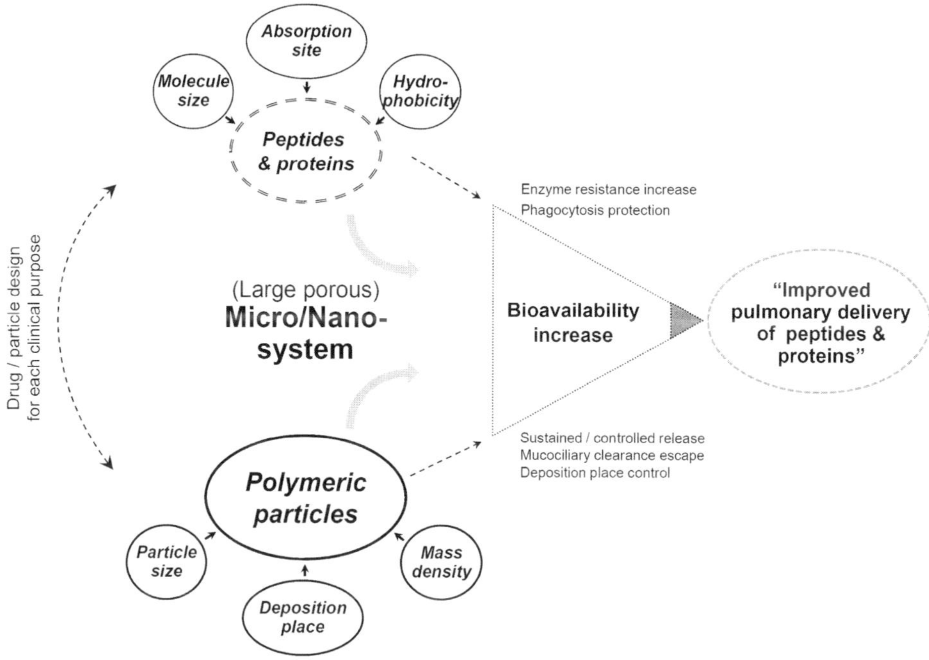

Figure 12.1 *Overview of pulmonary delivery of peptides and proteins using polymeric particles based on practical influence factors*

Table 12.1 *Pulmonary bioavailabilities of various peptides and proteins*

Peptide or protein	Molecular weight (kDa)	Peak concentration time (T_{max}, hour)	Bioavailability for subcutaneous route (%)	Ref.
Arginine-glycine-aspartic acid (RGD peptide)	0.65	0.68	22	[8]
Leuprolide	1.21	0.37	50	[8]
Calcitonin	3.4	0.25	17	[8]
Insulin	5.8	0.33	14	[9, 10, 11]
Granulocyte colony-stimulating factor (G-CSF)	18.6	2.5	12	[12]
Interferon-α (IFN-α)	19	4.5	37.5	[8]
Human growth hormone (hGH)	22	6.0	19.5	[8]
Albumin	68	20	4.5	[8]
Immunoglobulin G (IgG)	150	16	1.7	[8]

peptide resistance to lung peptidases and decreasing peptidase activity have been adopted as strategies for enhancing the therapeutic effects of peptides. It is in this context that the utilities of blocked/PEGylated peptides or lung peptidase inhibitors have been examined [14, 16, 18, 19].

Pulmonary macrophages have been shown to secrete peroxidases, inflammatory/immunomodulatory mediators and other host defence molecules that are able to degrade peptides and proteins. Moreover, proteins are likely to be subjected to precipitation and aggregation in alveolar lining fluid, and these aggregates are then taken up and destroyed by macrophages and phagocytic cells [17]. Furthermore, considerable amounts can be removed by mucociliary clearance [20]. Hence, in order to increase the bioavailability of peptides and proteins, methods that overcome these defensive mechanisms should be found.

12.2.3 Factors Affecting the Pulmonary Delivery of Peptide and Protein Particles

12.2.3.1 Drug-Associated Factors

Molecular Sizes of Peptides and Proteins. The absorption rates of macromolecules are primarily dependent on molecular size and, in general, larger molecular species are taken up at lower rates in the lungs. As molecular size increases and linear peptides become spherical proteins, the absorption tends to be delayed. Proteins such as insulin, growth hormone and cytokines have been shown to reach peak concentrations within several hours, whereas the concentrations of larger proteins (e.g. albumin, IgG) may peak at a day after administration [6, 8]. On the other hand, smaller peptides (chemically blocked forms with good solubility, e.g. RGD peptide and leuprolide) achieve systemic peak concentrations only 10–30 minutes after administration [8].

Absorption Place and Mode. Small hydrophobic molecules are absorbed throughout the lungs by simple passive diffusion, whereas large hydrophilic molecules are thought to be absorbed by specific transporters or via tight junctions [7]. However, the absorption of peptides and proteins in the lungs is not well understood and, in particular, it has not been determined whether peptides and proteins are absorbed paracellularly or transcellularly [6, 7]. Electrical resistance, which reflects the tightness of cell junctions across the lung epithelium, has been reported to diminish from the trachea to the distal airways because the density of tight junctions in airways is about five times greater than it is in the alveoli. However, evidence obtained during a study of insulin absorption suggests that the distal airways immediately before terminal alveoli are the optimal absorption sites for most macromolecules [21]. On the other hand, high molecular mass Immunoglobulin G (IgG) is absorbed optimally in the larger airways, rather than in the alveoli, by receptor-mediated transcytosis [22, 23].

Hydrophilicity and Hydrophobicity. Unless the administered species is insoluble due to severe hydrophobicity, hydrophobicity helps drug molecules traverse through lung epithelial membranes. Unfortunately, however, most peptides and proteins are hydrophilic, which can be a real obstacle to absorption in the lungs [6]. Thus, strategies are needed to increase hydrophobicity and to conceal severe hydrophilicity.

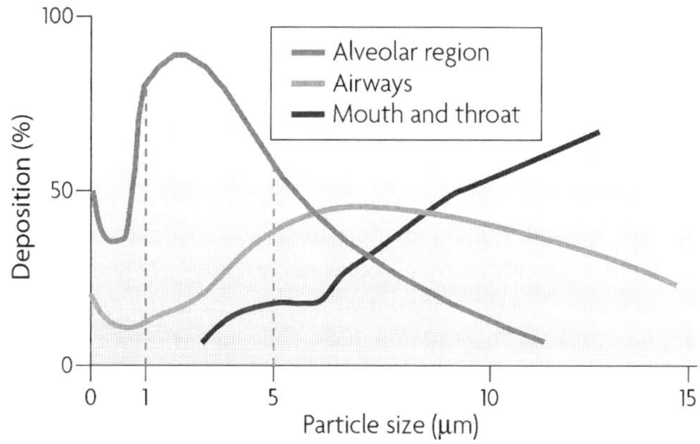

Figure 12.2 *Effect of particle size on the deposition of aerosol particles in the human respiratory tract (Reproduced with permission [6])*

12.2.3.2 Formulation Aspects

Particle Size and Pulmonary Deposition Location. In general, smaller particles can reach deeper into the lungs, that is larger particles tend to deposit in the upper airways or in the mouth and throat whereas smaller particles can move deep into the alveolar region [6]. Specifically, peptide and protein particles with diameters of >5–6 μm are deposited in the upper airways and are subjected to a vigorous mucociliary clearance, whereas small particles of <1 μm are rapidly exhaled from the airways during normal tidal breathing [1, 6] (Figure 12.2). Thus, pharmaceutical aerosols must be in the respirable size range 1–3 μm, to bypass deposition in the upper airways and to deposit deep in the lungs [1]. However, particles of 1–2 μm are easily removed by macrophages [24], which raises the issue of mass density.

Mass Density and Aerodynamic Diameter. Thus far, therapeutic dry powder aerosols have been made with mass densities (ρ) of ~1 ± 0.5 g/cm^3 and mean geometric diameters (d) of <5 μm [25]. However, particles with lower mass density can be made larger in size, which may reduce phagocytic clearance by macrophages and increase bioavailability. Mass median aerodynamic diameter (MMAD), d_a, is defined as the diameter of a unit density sphere having the same gravitational settling velocity as the particle in question, and is defined by the following relationship based on geometric diameter and mass density: $d_a = d \times \rho^{1/2}$ [26, 27]. Recently, particles with a low mass density of <~0.4 g/cm^3, an aerodynamic diameter of 1–5 μm and a large actual geometric diameter of >5–10 μm have been proposed for advanced delivery systems, as exemplified by porous microparticles [25, 28–30].

12.3 Polymeric Microparticles for Delivering Peptides and Proteins

Microparticles have gained much attention as non-invasive carrier systems for protein delivery to the lungs [31–33]. These particles containing peptide or proteins may be successfully delivered to the lungs by various inhalers, for example pressured metered-dose inhalers (pMDIs), nebulizers and dry power inhalers (DPIs) [34]. However, the pulmonary deliveries of peptides and proteins are complicated by the anatomic structure of the human respiratory system and the disposition effect by the respiration process [35]. In this respect, the availabilities of inhaled particles can be enhanced or reduced by altering the characteristics of particles, for example particle sizes, microviscosities, surface energy, porosity and degradation characteristics [36]. As mentioned above, particles measuring 1–5 μm can be removed by alveolar macrophages that recognize small foreign particles, and particles of ~0.5–1 μm are not well deposited in the lungs [37, 38]. Moreover, particles over 5 μm are likely to be deposited in the upper airways, which reduce their effectiveness [38], and deposition also depends on the aerodynamic diameters of particles [38]. In addition, the release behaviour of drugs from particles also affects concentrations in blood [29]. Furthermore, the toxicities of particles delivered to lung tissues require serious study [39–41].

In one study, as an alternative to large-sized particles, a microparticle agglomerate of nano-sized liposomal particles with a small aerodynamic diameter and a large geometric diameter was constructed by forming cross-links that are cleaved by glucose [42]. When glucose-sensitive microparticles are constructed with embedded insulin, hyperglycemic events can trigger the release of insulin due to the disintegration of microparticles by glucose. Moreover, these microparticle agglomerates accumulate deep in the lungs due to their small aerodynamic diameters. Furthermore, this particle type reduces alveolar macrophage recognition due to the comparatively large geometric diameter of the agglomerated structure.

It is interesting to note that surface-modified poly(lactide-*co*-glycolide) (PLGA) particles with mucoadhesive chitosan, which were produced using the emulsion solvent diffusion method, were more slowly eliminated from the lungs than unmodified PLGA particles [43]. This may be attributed to the long retention time of these particles due to adhesion to bronchial mucus and lung tissue. Moreover, this interaction between chitosan on particle surfaces and lung tissues seemed to contribute to drug absorption and increase in bioavailability.

Supercritical fluid (SCF) drying processes offer a promising, protein-friendly technique for the preparation of inhalable protein-loaded particles [44]. This process led to the fabrication of 6–10 μm microparticles with an aerodynamic diameter of ~4 μm (Figure 12.3a) [44], and produces microparticles with a heterogeneous morphology. Microparticles consisting of N-trimethylchitosan and dextran were used to encapsulate insulin during the SCF drying process and these insulin-loaded microparticles effectively deliver insulin to the lung via an inhaler and rapidly release insulin. The characteristics of the different polymer microparticle carriers are summarized in Table 12.2.

12.4 Porous Microparticles

Porous microparticles (PMs) have usually been used as a biodegradable scaffolds for tissue engineering [45], but have also been applied to encapsulate proteins by closing the open pores of PLGA microparticles after protein entrapment [46]. In order to improve protein loading efficiencies, a fluidized bed reactor in an ethanol vapour phase was used

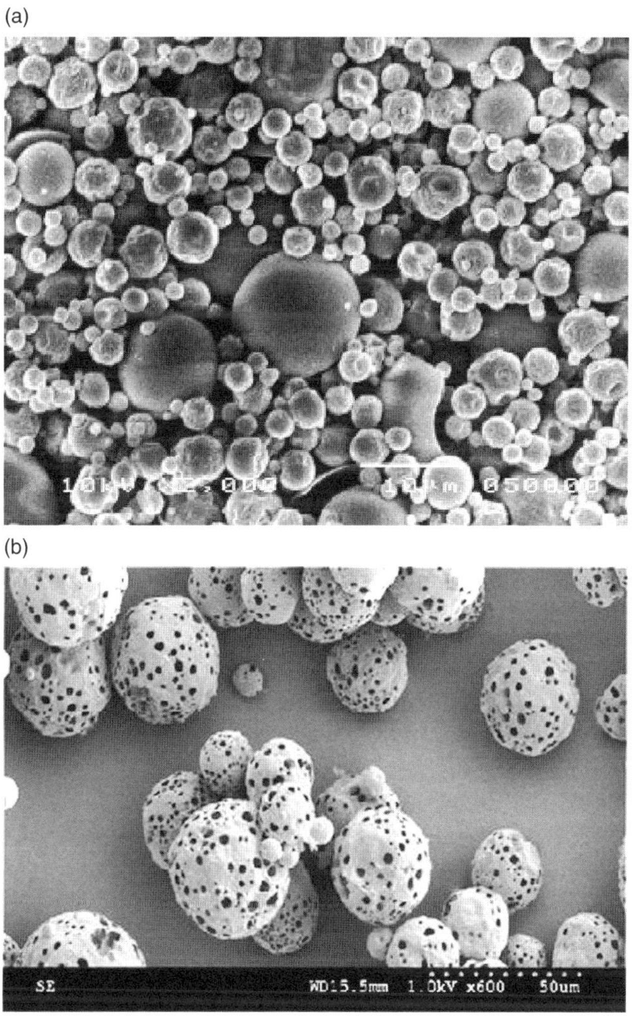

Figure 12.3 (a) Scanning electron microscopy (SEM) images of dextran–insulin microparticles. (Reproduced with permission from Elsevier [44]) (b) The SEM image of porous microparticles (PMs) (Reproduced with permission from Elsevier [29])

for the pore closing process [46]. Electrostatic interaction between protein and heparin-immobilized PMs was found to improve the loading efficiencies of basic fibroblast growth factor (bFGF) and allowed the sustained release of bFGF *in vitro*. Furthermore, it was demonstrated that reversible and specific binding of bFGF to immobilized heparin leads to effective angiogenesis in an animal model [47]. These PMs have been well established using sodium chloride, surfactants, sodium bicarbonate, extractable oils and gas forming agents as porogens [25, 29, 46–48].

Table 12.2 *Various examples of polymeric matrices for particle carriers of peptides and proteins*

Polymers used for vehicles	Encapsulated or entrapped drugs	Carrier formulations	Preparation method	Delivery methods to the lungs	Ref.
Poly (D,L-lactide-*co*-glycolide) (PLGA)	Calcitonin	Nanospheres	Emulsion solvent diffusion method in oil	Nebulization	[43]
PLGA	Insulin		Emulsion solvent diffusion method in water		[52]
Avidin-grafted gelatine	Biotinylated Epidermal growth factor (EGF)	Nanoparticles	Desolvation method	Nebulization	[54]
Poly (n-butyl cyanoacrylate) (PBCA)	Insulin		Emulsion polymerization	Intratracheal instillation	[53]
Polystyrene and silica	–	Porous nanoparticle aggregate particles	Spray-drying method	–	[56]
Heparin immobilized poly (D,L-lactide-*co*-glycolide) [Pluronic F-127]	Basic fibroblast growth factor (bFGF)	Large porous microparticles (microspheres)	Oil-in-water (O/W) single emulsion method	–	[47]
PLGA [SBE-CD][a]	Lysozyme/bovine serum albumin		Double emulsion solvent evaporation method	–	[29, 30]
PLGA or poly(lactic acid-*co* - lysine-graft-lysine) (PLAL-Lys)	Insulin		Single/double emulsion solvent evaporation method	Inhalation with DPIs	[25]
PLGA [HPβCD][a]	Deslorelin		Supercritical fluids	Intratracheal instillation	[37]

[a] Depicts porogens; HPβCD, Hydroxypropyl-β-cyclodextrin; SBE-CD, Sulfobutyl ether β-cyclodextrin.

Recently, PMs (Figure 12.3b) [29] were developed for the delivery of proteins to lungs. PMs accumulate at high levels in deep lung epithelium due to their low densities. For example, porous particles of diameter <5 μm and low density (<0.4 g/cm^3) showed high protein accumulations in lung peripheries, and thus are resistant to phagocytic clearance [29].

In one study, increased porous microparticle accumulation provided a 24-fold higher therapeutic effect than that observed using non-porous microparticles [25]. Moreover, PMs rapidly released encapsulated proteins due to their porous morphology [29]. Proteins are administered frequently due to their short circulation times and instabilities in plasma, which is uncomfortable and inconvenient for patients, particularly those requiring long-term treatment [30]. These requirements raise the topic of long-acting protein formulations delivered via the pulmonary delivery route [29, 30].

In order to further improve drug release kinetics, PMs are needed with delayed drug release characteristics. Lee *et al.* investigated a PM with a low mass density and large size for pulmonary protein delivery. PMs were prepared using the water-in-oil-in-water ($W_1/O/W_2$) double emulsion method using a cyclodextrin derivative (e.g. sulfobutyl ether β-cyclodextrin sodium salt (SBE-CD)) as a porogen [30]. As the amount of SBE-CD was increased, the MMAD decreased to approximately 3 μm (Figure 12.4a) [30]. On the other hand, the ionic complexation of positively charged protein and negatively charged hyaluronate in the W_1 phase was used to prevent protein denaturation in organic solvents (e.g. dichloromethane) and to sustain protein release. In addition, the interaction between protein and hyaluronate increased protein encapsulation efficacy and blocked protein leakage from PMs. Furthermore, the formation of this complex within PMs led to sustained lysozyme (Lys) release for up to seven days (Figure 12.4b) [30]. Similarly, a PM with sucrose ethylacetate butyrate (SAIB) suppressed the rapid release rate of proteins due to the high viscosity of the added SAIB, which allowed bovine serum albumin (BSA) release over seven days [29]. These sustained releases compare with a protein release time of four hours from PMs without protein/hyaluronate complexes or SAIB (Figure 12.4c) [29]. Moreover, PMs containing protein/hyaluronate complexes preserved protein (BSA) bioactivity to more than 98% over the release period. Overall, these preliminary results suggest that PMs with sustained release characteristics might be successfully applied to long-term pulmonary administration.

On the other hand, lipid-based hollow-porous microparticles (PulmoSphere) were evaluated as a potential vehicle for human immunoglobulin (hIgG) delivery to the lung [49]. During inhalation, PulmoSpheres deposited on upper and lower respiratory tracts in mice but were not readily phagocytosed by alveolar macrophages. Moreover, hIgG-formulated PulmoSpheres induced immune responses that resembled those triggered by local and systemic hIgG delivery, which suggests that they can be used to trigger or modulate immune responses. However, as yet no long-acting immunoglobulin delivery system has been devised using this system.

Arnold *et al.* recently reported that a pulmonary drug delivery system based on PMs can provide an effective adjunctive approach to orally administered antibiotics for lung infections [48]. Large porous PLGA microspheres (10–15 μm diameter) were prepared using precision particle fabrication technology, and it was found that PM encapsulated nano-sized drugs were released steadily in the lung. This technology is anticipated to develop further into a novel long-acting protein delivery system. The characteristics of porous particles are summarized in Table 12.2.

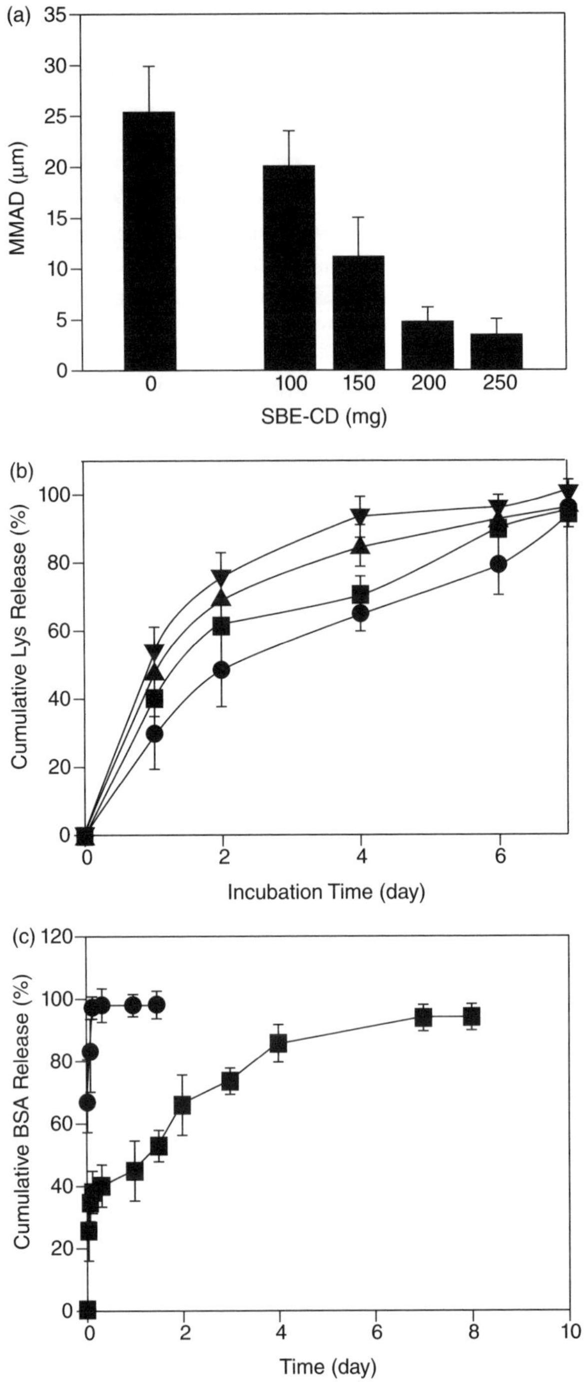

Figure 12.4 *(a) The mass median aerodynamic diameter (MMAD) change of microparticles with the amount of sulfobutyl ether β-cyclodextrin sodium salt (SBE-CD) (n = 3). (Reproduced with permission from Elsevier [30]) (b) The cumulative protein (lysozyme: Lys) release (%) from the porous microparticles (PMs) with ionic strength ionic strength (0.15 (●), 0.30 (■), 0.45 (▲), 0.60 (▼)) (n = 3). (Reproduced with permission from Elsevier [30]) (c) Cumulative protein release (%) from porous microparticles (PMs) with sulfobutyl ether b-cyclodextrin sodium salt (SAIB) (■) and PMs without SAIB (●) (n = 3) (Reproduced with permission from Elsevier [29])*

12.5 Polymeric Nanoparticles

Nanoparticles with geometric diameters of a few hundred nanometers have shown great promise for pulmonary drug delivery. Here, nanoparticles are referred as solid colloidal particles in the size range 1–1000 nm (typically 1–200 nm) [50]. Compared with microparticles, nanoparticles can be deposited in higher concentrations and thus show delayed phagocytic clearance. Nanoparticles have been enthusiastically used as drug carriers in the pharmaceutical area because of their abilities to entrap and encapsulate peptides and proteins [51]. A number of methods of preparing nanoparticles and of improving their pharmacokinetic dispositions *in vivo* have been extensively studied. As is the case for microparticles, the main advantage of polymeric nanoparticles for the pulmonary delivery of peptide and protein drugs is that they offer sustained release in the lung, and thus prolonged therapeutic duration [51, 52].

Although pulmonary drug delivery based on nanoparticles is gaining interest, relatively few studies about peptides and proteins have been published. In a study by Kawashima *et al.*, insulin-loaded PLGA nanospheres with mean diameters of ~400 nm were prepared using a modified emulsion solvent diffusion method in water [52]. The nanosphere dispersions so obtained were nebulized to droplets of 5–7 μm using an ultrasonic nebulizer and were then administered into the tracheas of guinea pigs. This preparation displayed extended hypoglycemia (>48 h) and a significant maximum reduction in blood glucose. Another insulin nanoparticle made of poly (*n*-butyl cyanoacrylate) (PBCA) and dextran as a stabilizing agent with a particle size of ~255 nm was prepared using an emulsion polymerization process by Zhang *et al.* [53]. The nanoparticle colloidal solution was directly instilled into rat tracheas. Animals so treated showed significantly prolonged hypoglycemia as compared with insulin solution controls. In addition, gelatine nanoparticles surface-grafted with avidin were prepared and further conjugated with biotinylated epidermal growth factor (EGF) by Tseng *et al.* [54]. These core-shell nanoparticles were found to be ~220 nm and to have a zeta potential of –9.3 mV. This unique nanoparticle structure was designed to seek out EGF receptors in lung adenocarcinoma and was proved to work *in vitro* (in A549 cells) and *in vivo* (in A549-grafted SCID mice).

In order to endow nanoparticles with a mucoadhesion property, some researchers have used chitosan (CS), a polysaccharide obtained by chitin deacetylation that is biocompatible and biodegradable. Chitosan was used as either a simple coating agent or to provide a nanoparticle structure. Yamamoto *et al.* prepared a calcitonin-encapsulating nanosphere surface-modified with chitosan using an emulsion solvent diffusion method in oil [43]. The nanospheres produced were of ~660 nm and aerosolized well (Figure 12.5A) [43]. They were delivered into the lungs of guinea pigs and treated animals were found to show significant plasma calcium reductions for up to 24 hours. This finding suggests that chitosan-modified nanoparticles are eliminated slowly, indicating protracted nanoparticle retention in the lungs. In another study, Grenha *et al.* used microencapsulated chitosan nanoparticles for lung protein delivery (Figure 12.5B) [53]. This group prepared chitosan/tripolyphosphate (CS/TPP) nanoparticles of ~390 nm and further microencapsulated the nanoparticles with aerosol excipients of sugars (mannitol and lactose) using a spray-drying technique, which increased microsphere formation and improved aerosolization and flow properties [55].

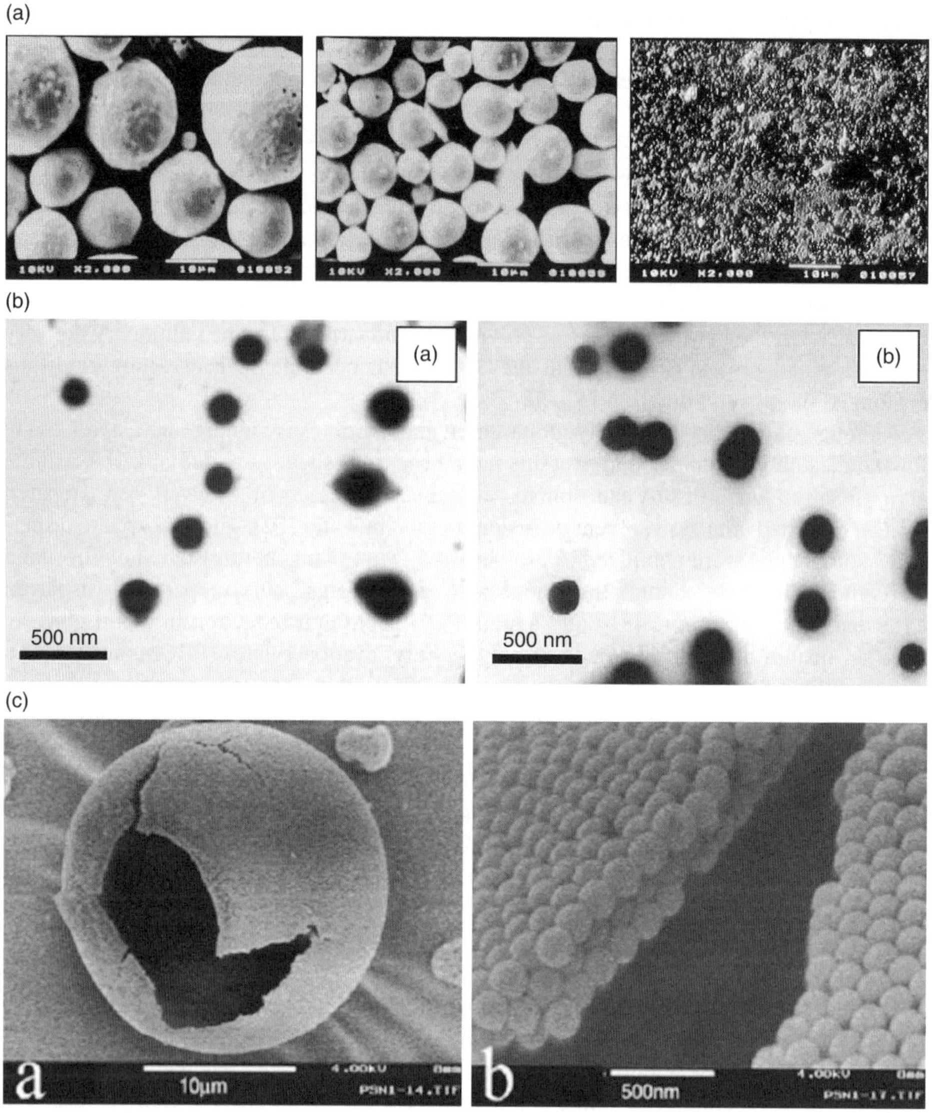

Figure 12.5 (A) SEM images of PLGA nanospheres or microspheres containing calcitonin. D_{50} = (a) 12.5; (b) 7.07; (c) 0.66 µm. (Reproduced with permission from Elsevier [43]) (B) Transmission electron microscopy (TEM) images of chitosan/tripolyphosphate (CS/TPP) nanoparticles (a) freshly prepared based on the ionotropic gelation of CS/TPP and (b) recovered from the microspheres after incorporation of nanoparticles into the microspheres. To recover the nanoparticles from the microspheres, the spray-dried powders (microspheres) were incubated in PBS pH 7.4 for 90 min. (Reproduced with permission from Elsevier [55]) (C) SEM images of (a) a typical hollow spherical large porous microparticles observed from the spray-drying of a solution of polystyrene nanoparticles, (b) a magnified view of the polystyrene nanoparticles in a hollow microparticle (a) (Reproduced with permission. Copyright (2002) National Academy of Sciences, USA [56])

However, the practical utilities of nanoparticles are restricted by their low impaction properties (due to their low inertias) in lung airways, which results in their exhalations from the lungs. Furthermore, their small sizes result in particle aggregation, which hampers the use of liquid and solid nanoparticles as pharmaceuticals. In this regard, a strategy is required to improve their pulmonary practicability and bioavailability. Tsapis *et al.* developed large porous nanoparticle-aggregate particles (PNAPs), also called Trojan particles, by combining the drug release potentials of nanoparticles with the aerosolization potentials of large porous particles (LPPs) (Figure 12.5C) [56]. This novel hybrid particulate system, which has a micron-size collective structure, was designed to efficiently deliver nanoparticles to the lungs. Actually, the resulting dry powders can be applied to the lungs more easily by formulating nanoparticles into larger hollow porous structures. Consequently, the microparticle aggregates consist of only nanoparticles with or without some inert pharmaceutical excipients. Furthermore, Tsapis *et al.* fabricated LPPs by mixing polystyrene and silica nanoparticles with the excipients 1, 2-dipalmitoyl-*sn*-glcero-3-phophocholin (DPPC), lactose, BSA and hydroxypropylcellulose (HPC), and spray-drying. It was found that the LPPs had a solid deformable shell consisting of several NPs and showed a wrinkled structure, depending on molecular component type as well as the size or chemical nature of the NPs. The LPPs also present a low relative density and low aerodynamic diameter.

Using this PNAP-like technology, particles composed of CS/TPP nanoparticles were produced by Grenha *et al.* and were found to have optimal inhalation characteristics, that is (i) appropriate aerodynamic diameters between 2 and 3 μm, (ii) low mass densities of <0.45 g/cm^3, (iii) high insulin loading capacities of 65–80%, and (iv) good insulin release profiles [55]. Also, PNAPs comprised of (i) gelatine and PBCA or (ii) polyacrylate (PMMA-MeOPEGMa) have been intensively studied [57, 58]. The characteristics of polymeric nanoparticle carriers for pulmonary delivery are summarized in Table 12.2.

12.6 Sustained or Controlled Release Issues in Polymeric Particle Formulations

The lungs can rapidly absorb peptides and proteins. As with small molecules, such drugs are reviewed to be absorbed faster after pulmonary dosing than after subcutaneous injection [5]. Although this rapid absorption allows therapeutic concentrations to be readily achieved, it also results in the rapid elimination of drugs from the body. Therefore, sustained or controlled drug release is required to maintain therapeutic efficacy. Most of the polymeric particles mentioned above that involve peptide/protein encapsulation or entrapment allow such sustained release. Several synthetic and natural polymers have been examined in this context, although PLGA has been extensively used as a material of first choice for the sustained release of micro and nanospheres [25, 29, 30, 37, 47, 52]. Copolymers comprised of sebacic acid (SA) and poly(ethylene glycol) (PEG), which impart water insolubility and macrophage resistance, respectively, can be used for this purpose [59, 60]. Various natural polymers, such as chitosan, hyaluronic acid, soluble starches and their synthetic derivatives, such as hydroxypropyl cellulose (HPMC) and trehalose oligosaccharide ester derivatives (OEDs), can also be used as additives or matrices to achieve sustained particle release [61–64].

12.7 Stability Issues of Peptides and Proteins in Particles

It is known that polymeric microparticles, especially PLGA, present serious problems, such as protein denaturation during microparticle fabrication and protein instability within microparticles, and that these problems have continuously interrupted commercialization schedules [28, 65]. Similarly, the microparticles used for pulmonary delivery also require protein stabilization technology [29, 30]. During microparticle fabrication, the loss of structural integrity (such as covalent or non-covalent aggregation, hydrolysis, deamidation, acylation, etc.) may be held responsible for loss of biological or enzymatic activity of a protein [2, 65]. Non-covalent aggregates of proteins are spontaneously formed during the incubation period of protein-encapsulated microparticles [28–30]. In the case of BSA, covalent aggregation is often related to thiol-disulfide exchange [2, 65]. In addition, the moisture penetrated into the polymer matrix during incubation facilitates the degradation mechanisms of the polymer matrix [37]. Deamidation or acylation of proteins continuously occurs after degradates of polymer matrix lead to chemical reactions with proteins [2, 65]. The uses of protein-stabilizing surfactants [66] or non-organic solvents, incorporating protein-stabilizing components [29, 30], and the selection of safe fabrication methods (e.g. SCF drying process) [44], therefore provide means of stably encapsulating proteins.

12.8 Toxicity Issues of Inhaled Particles

Recently, the toxicities of particles intended for pulmonary route delivery have been carefully examined [67–71]. Baker *et al.* reported that the toxicity of C60 fullerence microparticles is due to a longer duration (half-lives ~29 days) in the lung [67]. Comparisons of broncoalveolar (BAL) lavage fluid parameters identified a significant increase in BAL protein concentrations after exposure to C60 fullerene microparticles, despite the absence of gross and microscopic lesions at necropsy [68]. Chitosan microparticles have been reported to have a dose-dependent pro-inflammatory effect in rat lung [69], which was related to increased BAL protein levels, lactate dehydrogenase activity and lung tissue myeloperoxidase activity. Consequently, inflammatory response due to chitosan microparticles was 1.8-fold greater than in controls not containing chitosan microparticles. Furthermore, whereas Brzoska *et al.* concluded that polyalkylcyanoacrylate particles showed little or no cytotoxicity and no inflammatory effect on bronchial epithelial cells [70], Wegesser *et al.* found that particulate matter can invoke inflammatory and/or toxic responses in the lung, and that these vary by deposition region and time, the surface properties of the particles, particle size and materials from which they are constructed [71].

Although the pulmonary delivery of particles raises concerns about toxicity, it also offers the possibility of enhanced therapeutic effects and reduced dosage. Thus, additional investigations are required to resolve toxicity concerns.

12.9 Conclusions

The lungs are viewed as the most effective non-invasive route of introducing peptides and proteins to the systemic circulation. New inhalation devices and formulations have also made the pulmonary administration of these drugs highly feasible. Nevertheless, this pharmaceutical area has many substantial problems, and more efforts should be made in order to produce the inhalation drug products. Recently, Pfizer's Exubera® (pulmonary insulin powder) was withdrawn from the market. Pfizer's failure on Exubera® was not of safety issues, or complicated devices, or even consumer marketing problems; Exubera® was expensive and the diabetic smokers would rather select smoking than using it. Exubera® presenting a short-term therapeutic effect was, also, not attractive; insulin needles used in the conventional insulin administration are so small, they are much painless. In this regard, polymeric particle carriers, such as micro or nanospheres/particles, attract much attention because of the practical advantages they offer. For example, as compared with drug particles of respirable size, they have significant merits in effective dosing and bioavailability, that is they can be used (i) to cause sustained/controlled release and (ii) to protect drugs from lung enzymes. In addition, large porous microparticles with low mass densities and large sizes can be used to control deposition sites or to prevent phagocytosis and mucociliary clearance. Novel large porous nanoparticle aggregates also offer new possibilities in terms of tailoring inhalation techniques. Many challenges still remain in this area, but the polymeric particle formulations reviewed in this chapter offer new practicable possibilities for the effective pulmonary delivery of peptides and proteins.

References

1. Agu, R.U., Ugwoke, M.I., Armand, M. *et al.* (2001) The lung as a route for systemic delivery of therapeutic proteins and peptides. *Respir. Res.*, **2**, 198–209.
2. Taluja, A., Youn, Y.S. and Bae, Y.H. (2007) Novel approach in microparticulate PLGA delivery systems encapsulating proteins. *J. Mater. Chem.*, **17**, 4002–4014.
3. Carino, G.P. and Mathiowitz, E. (1999) Oral insulin delivery. *Adv. Drug Deliv. Rev.*, **35**, 249–257.
4. Lee, V.H.L. and Yamamoto, A. (1990) Penetration and enzymatic barriers to peptide and protein absorption. *Adv. Drug Deliv. Rev.*, **4**, 171–207.
5. Patton, J.S., Fishburn, C.S. and Weers, J.G. (2004) The lungs as a portal of entry for systemic drug delivery. *Proc. Am. Thorac. Soc.*, **1**, 338–344.
6. Patton, J.S. and Byron, P.R. (2007) Inhaling medicines: delivering drugs to the body through the lungs. *Nat. Rev. Drug Discov.*, **6**, 67–74.
7. Patton, J.S. (1996) Mechanisms of macromolecular absorption by the lungs. *Adv. Drug Deliv. Rev.*, **19**, 3–36.
8. Adjei, A.L. and Gupta, P.K. (1997) *Inhalation Delivery of Therapeutic Peptides and Proteins*, Marcel Dekker, Inc, New York, p. 107.
9. Komada, F.S., Iwakawa, N., Yamamoto, H. *et al.* (1994) Intratracheal delivery of peptide and protein agents: absorption from solution and dry powder by rat lung. *J. Pharm. Sci.*, **83**, 863–867.
10. Okumura, K., Iwakawa, S., Yoshida, T. and Komada, F. (1992) Intratracheal delivery of insulin and absorption from solution and aerosol by rat lung. *Int. J. Pharm.*, **88**, 63–73.
11. Yamamoto, A., Umemori, S. and Muranishi, S. (1994) Absorption enhancement of intrapulmonary administered insulin by various absorption enhancers and protease inhibitors in rats. *J. Pharm. Pharmacol.*, **46**, 14–18.
12. Niven, R.W., Whitcomb, L., Kinstler, O. *et al.* (1994) The pulmonary absorption of aerosolized and intratracheally instilled rhG-CSF and monoPEGylated rhG-CSF. *Pharm. Res.*, **12**, 1343–1349.

13. Tronde, A., Norden, B., Marchner, H. *et al.* (2003) Pulmonary absorption rate and bioavailability of drugs *in vivo* in rats: structure–absorption relationships and physicochemical profiling of inhaled drugs. *J. Pharm. Sci.*, **92**, 1216–1233.
14. Kobayashi, S., Kondo, S. and Juni, K. (1994) Study on pulmonary delivery of salmon calcitonin in rats: effects of protease inhibitors and absorption enhancer. *Pharm. Res.*, **11**, 1239–1243.
15. Morita, T., Yamamoto, A., Takakura, Y. *et al.* (1994) Improvement of the pulmonary absorption of ($Asu^{1,7}$)-eel calcitonin by various protease inhibitors in rats. *Pharm. Res.*, **11**, 909–913.
16. Hussain, A., Arnold, J.J., Khan, M.A. and Ahsan, F. (2004) Absorption enhancers in pulmonary protein delivery. *J. Control. Release*, **94**, 15–24.
17. Douglas, G.P., Wisniowski, J., Daugherty, G.L. *et al.* (1997) Nonimmune phagocytosis of liposomes by rat alveolar macrophages is enhanced by vitronectin and is vitronectin-receptor mediated. *Am. J. Respir. Cell. Mol. Biol.*, **17**, 462–470.
18. Patton, J.S., Nagarajan, S. and Clark, A. (1998) Pulmonary absorption and metabolism of peptides and proteins. *Respir. Drug Deliv.*, **7**, 17–24.
19. Youn, Y.S., Kwon, M.J., Na, D.H. *et al.* (2008) Improved intrapulmonary delivery of site-specific PEGylated salmon calcitonin: optimization by PEG size selection. *J. Control. Release*, **125**, 68–75.
20. Niven, R.W. (1995) Delivery of biotherapeutics by inhalation aerosol. *Crit. Rev. Ther. Drug Carrier Syst.*, **12**, 151–231.
21. Boucher, R.C., Stutts, M.J. and Gatzy, J.T. (1981) Regional differences in bioelectric properties and ion flow in excised canine airways. *J. Appl. Physiol.*, **51**, 706–714.
22. Bitonti, A.J., Dumont, J.A., Low, S.C. *et al.* (2004) Pulmonary delivery of an erythropoietin Fc fusion protein in non-human primates through an immunoglobulin transport pathway. *Proc. Natl. Acad. Sci. USA*, **101**, 9763–9768.
23. Low, S.C., Nunes, S.L., Bitonti, A.J. and Dumont, J.A. (2005) Oral and pulmonary delivery of FSH-Fc fusion proteins via neonatal Fc receptor-mediated transcytosis. *Hum. Reprod.*, **20**, 1805–1813.
24. Chono, S., Tanino, T., Seki, T. and Morimoto, K. (2006) Influence of particle size on drug delivery to rat alveolar macrophages following pulmonary administration of ciprofloxacin incorporated into liposomes. *J. Drug Target.*, **14**, 557–566.
25. Edwards, D.A., Hanes, J., Caponetti, G. *et al.* (1997) Large porous particles for pulmonary drug delivery. *Science*, **276**, 1868–1871.
26. Chow, A.H.L., Tong, H.H.Y., Chattopadhyay, P. and Shekunov, B.Y. (2007) Particle engineering for pulmonary drug delivery. *Pharm. Res.*, **24**, 411–437.
27. Hinds, W.C. (ed.) (1999) *Aerosol Technology: Properties, Behavior, and Measurement of Airbone Particles*, 2nd edn, John Wiley & Sons, Inc.
28. Lee, E.S., Kwon, M.J., Lee, H. *et al.* (2006) *In vitro* study of lysozyme in poly(lactide-co-glycolide) microspheres with sucrose acetate isobutyrate. *Eur. J. Pharm. Sci.*, **29**, 435–441.
29. Kwon, M.J., Bae, J.H., Kim, J.J. *et al.* (2007) Long acting porous microparticles for pulmonary drug delivery. *Int. J. Pharm.*, **333**, 5–9.
30. Lee, E.S., Kwon, M.J., Na, K. and Bae, J.H. (2007) Protein release behavior from porous microparticle with lysozyme/hyaluronate ionic complex. *Colloids Surfs. B.*, **55**, 125–130.
31. Huang, Y.Y. and Wang, C.H. (2006) Pulmonary delivery of insulin by liposomal carriers. *J. Control. Release*, **113**, 9–14.
32. Oyabu, T., Ogami, A., Morimoto, Y. *et al.* (2007) Biopersistence of inhaled nickel oxide nanoparticles in rat lung. *Inhal. Toxicol.*, **19** (Suppl. 1), 55–58.
33. Seville, P.C., Li, H.Y. and Learoyd, T.P. (2007) Spray-dried powders for pulmonary drug delivery. *Crit. Rev. Ther. Drug Carrier Syst.*, **24**, 307–360.
34. Bell, J. and Newman, S. (2007) The rejuvenated pressurized metered dose inhaler. *Expert Opin. Drug Deliv.*, **4**, 215–234.
35. Kleinstreuer, C., Zhang, Z. and Donohue, J.F. (2008) Targeted drug-aerosol delivery in the human respiratory system. *Annu. Rev. Biomed. Eng.*, **10**, 195–220.
36. Shoyele, S.A. (2008) Engineering protein particles for pulmonary drug delivery. *Methods Mol. Biol.*, **437**, 149–160.

37. Koushik, K., Dhanda, D.S., Cheruvu, N.P.S. and Kompella, U.B. (2004) Pulmonary delivery of deslorelin: Large porous PLGA particles and HPβCD complexes. *Pharm. Res.*, **21**, 1119–1126.
38. Jones, J. (1984) Clearance of inhaled particles from the alveoli, in *Aerosol and the Lung: Clinical and Experimental Aspects* (ed. S.W. Clarke), Butterworth, London, pp. 49–70.
39. Donaldson, K., Borm, P.J., Oberdorster, G. *et al.* (2008) Concordance between *in vitro* and *in vivo* dosimetry in the proinflammatory effects of low- toxicity, low-solubility particles: the key role of the proximal alveolar region. *Inhal. Toxicol.*, **20**, 53–62.
40. Rogueda, P.G. and Traini, D. (2007) The nanoscale in pulmonary delivery. Part 1: deposition, fate, toxicology and effects. *Expert Opin. Drug Deliv.*, **4**, 595–606.
41. Inoue, K., Takano, H., Yanagisawa, R. *et al.* (2007) Effect of nanoparticles on lung physiology in the presence or absence of antigen. *Int. J. Immunopathol. Pharmacol.*, **20**, 737–744.
42. Karathanasis, E., Bhavane, R. and Annapragada, A.V. (2007) Glucose-sensing pulmonary delivery of human insulin to the systemic circulation of rats. *Int. J. Nanomedicine*, **2**, 501–513.
43. Yamamoto, H., Kuno, Y., Sugimoto, S. *et al.* (2005) PLGA nanosphere with chitosan improved pulmonary delivery of calcitonin by mucoadhesion and opening of the intercellular tight junctions. *J. Control. Release*, **102**, 373–381.
44. Amidi, M., Pellikaan, H.C., de Boer, A.H. *et al.* (2008) Preparation and physicochemical characterization of supercritically dried insulin-loaded microparticles for pulmonary delivery. *Eur. J. Pharm. Biopharm.*, **68**, 191–200.
45. Chung, H.J., Kim, I.K., Kim, T.G. and Park, T.G. (2008) Highly open porous biodegradable microcarriers: *in vitro* cultivation of chondrocytes for injectable delivery. *Tissue Eng. Part A*, **14** (5), 607–615.
46. Kim, H.K., Chung, H.J. and Park, T.G. (2006) Biodegradable polymeric microspheres with 'open/closed' pores for sustained release of human growth hormone. *J. Control. Release*, **112**, 164–174.
47. Chung, H.J., Kim, H.K., Yoon, J.J. and Park, T.G. (2006) Heparin immobilized porous PLGA microspheres for angiogenic growth factor delivery. *Pharm. Res.*, **23**, 1835–1841.
48. Arnold, M.M., Gorman, E.M., Schieber, L.J. *et al.* (2007) Cipro encapsulation in monodisperse large porous PLGA microparticles. *J. Control. Release*, **121**, 100–109.
49. Bot, A.I., Tarara, T.E., Smith, D.J. *et al.* (2000) Novel lipid-based hollow-porous microparticles as a platform for immunoglobulin delivery to the respiratory tract. *Pharm. Res.*, **17**, 275–283.
50. Kreuter, J. (1991) Nanoparticle-based drug delivery systems. *J. Control. Release*, **16**, 169–176.
51. Sung, J.C., Pulliam, B.L. and Edwards, D.A. (2007) Nanoparticles for drug delivery to the lungs. *Trends Biotechnol.*, **25**, 563–570.
52. Kawashima, Y., Yamamoto, H., Takeuchi, H. *et al.* (1999) Pulmonary delivery of insulin with nebulized -lactide/glycolide copolymer (PLGA) nanospheres to prolong hypoglycemic effect. *J. Control. Release*, **62**, 279–287.
53. Zhang, Q., Shen, Z. and Nagai, T. (2001) Prolonged hypoglycemic effect of insulin-loaded polybutylcyanoacrylate nanoparticles after pulmonary administration to normal rats. *Int. J. Pharm.*, **218**, 75–80.
54. Tseng, C.-L., Wang, T.-W., Dong, G.-C. *et al.* (2007) Development of gelatin nanoparticles with biotinylated EGF conjugation for lung cancer targeting. *Biomaterials*, **28**, 3996–4005.
55. Grenha, A., Seijo, B. and Remunan-Lopez, C. (2005) Microencapsulated chitosan nanoparticles for lung protein delivery. *Eur. J. Pharm. Sci.*, **25**, 427–437.
56. Tsapis, N., Bennett, D., Jackson, B. *et al.* (2002) Trojan particles: large porous carriers of nanoparticles for drug delivery. *Proc. Natl. Acad. Sci. USA*, **99**, 12001–12005.
57. Sham, J.O.H., Zhang, Y., Finlay, W.H. *et al.* (2004) Formulation and characterization of spray-dried powders containing nanoparticles for aerosol delivery to the lung. *Int. J. Pharm.*, **269**, 457–467.
58. Hadinoto, K., Zhu, K. and Tan, R.B.H. (2007) Drug release study of large hollow nanoparticulate aggregates carrier particles for pulmonary delivery. *Int. J. Pharm.*, **341**, 195–206.

59. Fiegel, J., Fu, J. and Hanes, J. (2004) Poly(ether-anhydride) dry powder aerosols for sustained drug delivery in the lungs. *J. Control. Release*, **96**, 411–423.
60. Fu, J., Fiegel, J., Krauland, E. and Hanes, J. (2002) New polymeric carriers for controlled drug delivery following inhalation or injection. *Biomaterials*, **23**, 4425–4433.
61. Huang, Y.C., Yeh, M.K. and Chiang, C.H. (2002) Formulation factors in preparing BTM-chitosan microspheres by spray drying method. *Int. J. Pharm.*, **242**, 239–242.
62. Surendrakumar, K., Martyn, G.P., Hodgers, E.C.M. *et al.* (2003) Sustained release of insulin from sodium hyaluronate based dry powder formulations after pulmonary delivery to beagle dogs. *J. Control. Release*, **91**, 385–394.
63. Hamdi, G., Ponchel, G., Besnard, M. and Duchene, D. (1999) Loading and *in vitro* release of salbutamol from degradable cross-linked starch microspheres intended for pulmonary administration. *STP Pharma Sci.*, **9**, 573–577.
64. Davidson, I.G., Langner, E.J., Plowman, S.V. and Blair, J.A. (2003) Release mechanism of insulin encapsulated in trehalose ester derivative microparticles delivered via inhalation. *Int. J. Pharm.*, **254**, 211–222.
65. Shinha, V.R. and Trehan, A. (2003) Biodegradable microspheres for protein delivery. *J. Control. Release*, **90**, 261–280.
66. De Rosa, G., Iommelli, R., La Rotonda, M.I. *et al.* (2000) Influence of the co-encapsulation of different non-ionic surfactants on the properties of PLGA insulin-loaded microspheres. *J. Control. Release*, **69**, 283–295.
67. Baker, G.L., Gupta, A., Clark, M.L. *et al.* (2008) Inhalation toxicity and lung toxicokinetics of C60 fullerene nanoparticles and microparticles. *Toxicol. Sci.*, **101**, 122–131.
68. Deguchi, S., Yamazaki, T., Mukai, S.A. *et al.* (2007) Stabilization of C60 nanoparticles by protein adsorption and its implications for toxicity studies. *Chem. Res. Toxicol.*, **20**, 854–858.
69. Huang, Y.C., Vieira, A., Huang, K.L. *et al.* (2005) Pulmonary inflammation caused by chitosan microparticles. *J. Biomed. Mater. Res. A*, **75**, 283–287.
70. Brzoska, M., Langer, K., Coester, C. *et al.* (2004) Incorporation of biodegradable nanoparticles into human airway epithelium cells-*in vitro* study of the suitability as a vehicle for drug or gene delivery in pulmonary diseases. *Biochem. Biophys. Res. Commun.*, **318**, 562–570.
71. Wegesser, T.C. and Last, J.A. (2008) Lung response to coarse PM: bioassay in mice. *Toxicol. Appl. Pharmacol.*, **230** (2), 159–166.

Nucleic Acids

13

Polymer Microparticles for Nucleic Acid Delivery

Tim Pearce, Jared Hierman and Chun Wang
University of Minnesota, Department of Biomedical Engineering, Minneapolis, MN, USA

13.1 Introduction

In this case study, the focus is on polymer microparticles as a formulation strategy for delivering nucleic acids, either as therapeutics or vaccines. Microparticle formulation is distinct because of its unique micrometre length-scale, which is small enough to be injected or delivered non-invasively to the human body, internalizable by cells through phagocytosis or endocytosis, yet large enough to carry significant payload and to provide sustained/controlled release of drug content. The materials currently used in such formulations are usually hydrophobic polymers or hydrophilic cross-linked polymers that are biodegradable. In the case of hydrophobic polymers, nucleic acids are either encapsulated using emulsion methods or adsorbed to polymer surface. In the case of hydrophilic polymers, nucleic acids are directly entrapped during formation of cross-linked microgels or macro-aggregates.

The polymer of choice for microparticles is the poly(lactic-*co*-glycolic acid) polymer, or PLGA. Various aspects of PLGA microparticles are discussed in detail, including preparation techniques, design and formulation parameters, optimization of processing and the relationship between the presence of additional polycations and microparticle performance. Next, a number of other biomedical polymers are introduced, many of which are novel materials and are only recently being applied to microparticle-based delivery of nucleic acids. Brief but complete discussions on polymer properties, preparation,

Delivery Technologies for Biopharmaceuticals: Peptides, Proteins, Nucleic Acids and Vaccines Edited by Lene Jorgensen and Hanne Mørck Nielsen

nucleic acid loading, release kinetics, cytotoxicity and gene transfection are given for each of the polymers. While the scope of this chapter is limited to polymer microparticle-based delivery of nucleic acids, several excellent, general review articles on gene delivery may be of interest and can serve as further readings on this broad subject [1–5].

13.2 Microparticles Based on Poly(Lactic-*co*-Glycolic Acids) (PLGA)

13.2.1 Overview

Poly(lactic-*co*-glycolic acid) (PLGA) dominates current research on microparticles for nucleic acid delivery. For many years, PLGA has been a very popular biodegradable material for drug delivery and numerous other biomedical applications. It was first introduced as a biocompatible material for use as one of the earliest biodegradable sutures [6] and there are currently a handful of PLGA microparticle-based drug delivery products approved by the United States Food and Drug Administration (FDA) [7]. The clinical and commercial success of PLGA-based medical products has made PLGA a promising material for nucleic acid delivery applications and may bode well for success in the future in terms of product development and commercialization.

PLGA is a random copolymer of lactic acid and glycolic acid, the homopolymers of which, poly(lactic acid) (PLA) and poly(glycolic acid) (PGA), respectively, are also two of the most extensively studied bioresorbable materials [6]. The chains of these polyesters hydrolyze in physiological aqueous conditions, releasing lactic acid and glycolic acid, which naturally occur in the human body and are safely metabolized with negligible toxicity. PGA is less hydrophobic and has a shorter degradation half-life in water than PLA. Therefore, the half-life of PLGA can be manipulated by controlling the ratio of lactic acid units to glycolic acid units in the copolymer. Commonly used PLGA is typically either a 50:50 or a 75:25 copolymer of lactic and glycolic acid units. The copolymerization does not result in a simple combination of the properties of the individual monomer components, and PLGA 50:50 has a shorter half-life than both PGA and PLA and other ratios of copolymers. In the preparation of microparticles, PLGA is dissolved in organic solvents and is used at various molecular weights, with and without chemically modified functional ends such as free acid or end-capped with esters. PLGA microparticles have been used for encapsulation of ribosomes and antisense oligonucleotides, although the vast majority of research has been on plasmid DNA, specifically in carrying DNA vaccines intended to transfect antigen-presenting cells.

13.2.2 Design Parameters

The performance of PLGA microparticles for nucleic acid delivery can depend on a number of parameters or properties of the microparticles, including size, microstructure, surface morphology, loading capacity, release kinetics, the integrity and biological activity of the DNA [8]. The properties of PLGA microparticles have been studied extensively and strategies have been proposed for optimizing many of the properties that are discussed in the following sections.

The size of microparticle is perhaps the most studied design parameter. The diameters of PLGA microparticles under investigation have ranged from 1 μm to several hundreds of microns. Generally, microparticle delivery systems can be divided arbitrarily into two rough categories based on size, namely 'large' and 'small'. Large microparticles in the range of a few tens to hundreds of microns have a high loading capacity and have good biostability and resistance to aggregation due to small surface area. Larger than most cells, these microparticles are entrapped within tissue and constantly release their payload into the interstitial space. The released molecules then diffuse and enter nearby cells, primarily by endocytosis. Popular for traditional drug delivery applications, this delivery methodology is not ideally suited for nucleic acid delivery, because the half-life of these molecules in extra-cellular space is very short due to digestive activity of nucleases, while the ability of naked DNA to infiltrate a cell without assistance is extremely limited with the exception of certain cell types bearing receptors specific for nucleic acids.

The majority of research involving nucleic acid delivery focuses on microparticles with a diameter smaller than 10 microns. These microparticles are designed to enter individual cells intact through endocytosis. Small microparticles, according to one report by Trombone *et al.*, released DNA into the cytoplasm, while large microparticles remained within the endosomes [9]. The maximum limit of phagocytic endocytosis is believed to be 10 microns [10], so delivery systems that target phagocytic cells must be smaller than this size. Meanwhile, the cut-off size for endocytosis in non-phagocytic cells is $0.5\,\mu m$ [11], so even the smallest microparticles can not be taken up by non-phagocytic cells. The size range of 1–10 microns is suitable for delivery of DNA vaccines to antigen-presenting cells that are phagocytic. Singh *et al.* demonstrated that small microparticles containing plasmid DNA vaccine enhanced antibody production, although no response was elicited by 30 micron particles [12].

A property that receives little attention is the overall microstructure of the microparticle. Usually, microparticles are solid, but Mohamed *et al.* have formulated hollow microparticles, which have potential for aerosol applications [13]. Surface morphology can range from smooth and uniform to uniformly dimpled to irregular. Desirable morphology depends on the application, and both smooth and dimpled microparticles have been investigated. Most often smooth microparticles have been formulated, although there has been suggestion that dimpled morphology may assist in aerosol delivery [13].

The capacity of microparticles to carry their payload is measured in terms of mass-to-mass ratio of nucleic acid-to-polymer. Practically, loading capacity must be balanced with encapsulation efficiency (EE%), the percent of the total DNA that becomes incorporated into microparticles after processing. In different microparticle systems, high encapsulation efficiency has been reported at loading levels from 0.1 to 1 mass % [14]. While loading capacity would seem to be of great importance to the performance of the microparticle, several studies have shown minimal effect of DNA load on transfection efficiency, probably because direct gene transfection of cells by microencapsulated DNA is, in general, quite poor.

Release behaviour of polymer microparticles is often inversely related to loading behaviour, with high loading capacity associated with slow release rate. The release of nucleic acids from PLGA microparticles has several features worthy of consideration. The release profile typically is composed of two phases. Initially, a burst release period results in escape of a large percentage of payload from the microparticles over a short period. This is followed by a period of sustained release during which the rate of release is slower but constant. The burst release period can last from hours to days and can result in significant loss of payload before the carrier is able to reach its target inside the body. Regarding plasmid DNA loaded in PLGA microparticles, reported burst release has ranged from 5 to 80% of total loaded DNA. During the sustained release phase, the remaining of content of a microparticle is released at a relatively slow, constant rate over a period of weeks. Important considerations include the rate of release at certain time points as well as the amount of total cumulative release over a given time. The cumulative release from microparticles is often much less than 100%, indicative of incomplete release of content [8, 15, 16].

The release rate of nucleic acids has been shown to vary with microparticle degradation rate, Diez *et al.* showed that the evidence of degradation observed by scanning electron microscopy (SEM) correlated to release rate [17]. Several studies have suggested that the sustained release of plasmid DNA vaccine from microparticles would remove the necessity of repeated booster shots [9]. Cai *et al.* found that a single dose of plasmid DNA vaccine encapsulated in PLGA using the standard double emulsion method described below resulted in protection against pathogen in animals to an extent that was not achievable by fewer than three injections of naked DNA [18]. Although an optimal release profile is frequently the goal of optimization of formulation parameters, some studies have suggested that release rate does not affect immunogenic response to microencapsulated DNA vaccine [12], and others have been debating over whether faster or slower release is desirable in delivery of DNA vaccines. Trombone *et al.* state that while slow release results in prolonged immune response, fast release may ultimately be more effective [9].

Nucleic acids, including plasmid DNA, are susceptible to various destabilizing forces encountered during formulation, and they have a short half-life *in vivo* due to quick degradation by nucleases. The primary purpose of PLGA microparticles in the delivery of nucleic acids is to provide protection against degradation, allowing intact and functional nucleic acids to reach their target. In order to predict the success of a nucleic acid carrying microparticle, it is essential to measure nucleic acid degradation. Plasmid DNA is loaded into microparticles as a super-coiled double-stranded structure that is fully functional. The DNA can be 'nicked' when a phosphodiester bond within one strand is broken, leaving an open circle conformation. If a bond breaks on both strands, the plasmid will be linearized. These three states are discernible by gel electrophoresis as three distinct bands that represent decreasing functionality and transfectivity of the plasmid. Further cleavages to the linear configuration result in shortening and fragmentation of DNA, evidence of which is given by smearing bands in the gel [19]. The amount of degraded versus intact DNA can be quantified using gel electrophoresis, which is a very useful method for evaluating DNA conformation during the formulation and release processes.

Bioactivity is a downstream indicator of the functionality of nucleic acids carried by microparticles; it is intimately related to the structural integrity of the molecules. In the case of plasmid DNA, bioactivity refers to the ability of encapsulated plasmids to transfect cells, or transfectivity. To test this, encapsulated DNA is removed from the microparticles and added to a culture of cells. The transfection efficiency is then measured against intact plasmids to evaluate the ability of structural and functional integrity of the encapsulated DNA [20]. Similar functional assays exist for other type of nucleic acids.

13.2.3 General Strategies and Fabrication of Microparticles

There are two leading approaches to PLGA microparticle-based delivery systems for nucleic acids. In one approach, referred to as 'microencapsulation', nucleic acids are incorporated into the bulk of the microparticle. Sustained release continues as the PLGA polymer degrades. This strategy typically results in a slow, sustained release, which may depend on the size of the nucleic acids encapsulated [8]. Disadvantages of this approach include low loading capacity and a high degree of nucleic acid degradation during the microencapsulation process and during microparticle degradation. Because the payload is carried within the bulk of the microparticle, the nucleic acid is exposed to a low pH microenvironment as lactic acid and glycolic acid are released from the polymer by hydrolysis [8]. Much research involves minimizing these disadvantages and is discussed in the next section.

Another approach involves adsorption of nucleic acids onto the surface of the microparticles [21]. Because PLGA microparticles characteristically have a net negative surface charge in physiologic buffers, negatively charged nucleic acids have very low adsorption directly to the surface. Therefore, microparticles have been coated with various polycations, with which nucleic acids interact favourably through complementary charges. This strategy has shown considerable promise and has demonstrated increased loading capacity while circumventing the need to expose DNA to high shear stress during microparticle formation [22]. However, this strategy is not without drawbacks and the polycations can damage cellular membrane, resulting in cytotoxicity [23].

PLGA microparticles are most often formed using a double emulsion solvent evaporation method. This method is simple, reproducible and scales well for large batches. In a two-step process for microencapsulating DNA, a water/oil/water (w/o/w) emulsion is created. In the formation of the primary w/o emulsion, a small volume of aqueous DNA solution is added to a larger volume of organic solvent, such as dichloromethane (DCM) solubilizing PLGA in a 1 : 10 volume to volume ratio or less. The mixture is homogenized at high speed, resulting in an emulsion of DNA solution within the PLGA solution. The aqueous solution often contains surfactants that improve the stability of the emulsion. To form the second emulsion, the w/o suspension is added to a larger volume (such as 5 : 1) of aqueous solution and again homogenized at high speed. Vesicles of the primary w/o emulsion are suspended within the aqueous solution. Very often the second aqueous solution contains poly(vinyl alcohol) (PVA) or excipients that add to viscosity in order to stabilize the emulsion and prevent fusion of microparticles [8]. The product is stirred for 3–12 hours, allowing the volatile organic solvent to evaporate while the PLGA microparticles harden with DNA entrapped within the polymer matrix. The microparticles are collected by centrifugation and rinsed to remove PVA and excess solvent, and lyophilized prior to storage.

Variations of the double emulsion method and other methods such as spray-drying have been explored by many research groups over the years. One particular interesting innovation in the method of microencapsulation is the creation of microparticles of defined size, morphology and geometry. One report describes that 95% of a batch of microparticles was produced to be less than 5 μm in diameter, spherical, and with smooth surface [18]. Finally, it is of obvious importance to control other parameters and properties of the microparticles through optimization of the microencapsulation process, as described in the following section.

The main disadvantage of the double emulsion method is the necessity for high-speed homogenization steps, which expose the nucleic acids to high shear stress that may cause significant fragmentation and degradation of DNA with high molecular weight. The damaging effect of homogenization is amplified in the processing of small microparticles, as especially high homogenization speed is required. Promising applications for small diameter PLGA microparticles, notably delivery of DNA vaccines, have provided motivation for exploring alternative methods of preventing degradation of nucleic acids during particle formation. Various strategies have been developed, ranging from protection of DNA by packaging it in complexes with cations to modification of the particle formation process itself.

Cryopreparation is one approach to protect DNA from degradation caused by shear stress during homogenization in microparticle processing [19]. In this method, the aqueous DNA solution is frozen prior to the primary emulsion step. Most of the DNA is within the bulk of the frozen phase, and because there is no shear stress inside a solid only a minor amount of DNA at the surface is exposed to shear stress. However, freezing DNA in solution causes DNA damage due to salt crystal formation. This problem is alleviated by the addition of polysaccharides in the DNA solution, which interfere with salt crystallization. Because the freezing process during lyophilization of the microparticles can also cause DNA damage, polysaccharides are sometimes added to the aqueous solution even if a cryopreparation strategy is not used. An additional benefit of cryopreparation is that the movement of DNA is restricted at low temperature during the second homogenization, resulting in improved loading efficiency. A modification of the cryopreparation strategy is to lyophilize the DNA solution, creating a solid-in-oil primary emulsion [14]. However, uniform microparticles were not achieved in this first attempt using the method and the solid phase protruded from the microparticle visibly as observed by SEM.

Another approach to avoid the high shear stress of double emulsion is to introduce the phases in turbulent flow by circulating the mixture through a small diameter needle or microfluidic channel. An emulsion can be formed without using homogenization, and it has been suggested that the nucleic acids are exposed to considerably lower shear stress. Particles smaller than 10 microns have been formed and shown to release DNA that transfected cells *in vitro* [24, 25].

Single emulsion spray-drying is an alternative method to the standard double emulsion technique. A single w/o emulsion is formed as in the previous method, with DNA dissolved in the aqueous solution and PLGA in the organic solution. Using specialized spray-drying apparatus, the emulsion is expelled through a small diameter nozzle as a mist of droplets. The solvent evaporates from the droplets quickly and solid particles are collected at the bottom of the chamber. While lessening the exposure of DNA to high shear stress, a major disadvantage of this method is loss of product during the particle collection process [26]. One study has directly compared spray-drying and double emulsion techniques for the preparation of DNA encapsulating particles and found that similar particle size and DNA loading can be achieved with either method [14]. Ultrasonic atomization is an alternative form of spray-drying. This has previously been used for encap sulation of pharmaceutical proteins and has recently been applied to plasmid DNA encapsulation [26]. Ultrasonic atomization uses energy shockwaves to spray a fine mist at relatively low velocity. The shear forces at work are small compared to those of traditional spray-drying method.

13.2.4 Optimization of Microparticle Formulations

Much can be done to manipulate the conditions and parameters of the formulation process in order to obtain microparticles with the desired properties. For the popular double emulsion method of microparticle fabrication, the stability of the primary w/o emulsion has a central role to dictating properties of the fully formed microparticle. In particular, encapsulation efficiency is found to be correlated to primary emulsion stability [13]. It is theorized that during the second homogenization step, the applied energy disrupts the primary emulsion so that the interior water phase is able to mix with the exterior water phase, allowing escape of DNA into the exterior water phase. This leads to low loading efficiency, unless the stability of the primary emulsion is sufficiently high to provide an energy barrier to mixing.

The water:oil ratio of formulation has a large impact on the stability of emulsions. A water:oil ratio of 1:10 has been shown to form more stable emulsions than a water:oil ratio of 1:5 [13]. Likewise, a water:oil ratio of 1:30 resulted in smaller particles than a water:oil ratio 1:10 [8]. The water:oil ratio most frequently used is 1:10 or 1:20. Ratios beyond 1:20 result in decreased particle size, but also decreased encapsulation efficiency [8].

Microparticle properties can depend on properties of the PLGA polymer in use, including its molecular weight [17], the ratio of lactic and glycolic acids, the presence of chemically modified chain-ends, and polymer concentration in the primary emulsion. De Rosa *et al.* attempted to optimize microparticles through manipulation of PLGA molecular weight and concentration [27]. However, a different study [14] suggested that the influence of the chosen polymer may also depend on other formulation parameters.

The molecular weight (MW) of PLGA in use can influence a range of properties, including size, of the microparticles. Higher molecular weight PLGA had a higher loading efficiency and slower release rate [17]. Another study shows that release rate is influenced by the molecular weight of the PLGA chosen and can be tuned by mixing polymers of different molecular weights [28]. Aggregation of microparticles has been shown to decrease when the molecular weight of polymer decreases. This is explained by more rapid degradation, which reduces surface hydrophobicity and therefore weakens the tendency for particles to aggregate [29]. Although high molecular weight resulted in slower degradation, this same study reports that high molecular weight resulted in gene expression for only three weeks, while low molecular weight sustained gene expression for six months [29]. In general, hydrophilic PLGA has produced microparticles with highly porous morphology and faster release rates than less hydrophilic PLGA. For example, PLGA 75:25 had low EE% compared to 50:50 polymer, whereas PLGA 50:50 is degraded more quickly than PLGA 75:25 [17].

PLGA polymer concentration used in the double emulsion process has been shown to influence encapsulation efficiency. Low polymer concentration allows for escape of nucleic acids from the interior emulsion during processing [8], due to decreased phase viscosity. PLGA alone does not have any effect on protection of DNA during processing, and encapsulated DNA is often slightly damaged (nicked) regardless of what composition of PLGA is used. However, naked DNA without PLGA is completely degraded by homogenization. Therefore, DNA extracted from microparticles has been found to be bioactive and able to transfect cells when complexed with a polycationic reagent [17].

Emulsion stability can be improved by the use of surfactants. Amphiphilic surfactant molecules assemble at phase boundaries, lowering surface energy and stabilizing the emulsion. The amphiphilic strength of a surfactant is described by its hydrophilic–lipophilic balance (HLB), defined as $20 \times M_h/M$, where M_h is the molecular mass of the hydrophilic part and M is the total mass of the surfactant, thus a completely hydrophobic molecule has a HLB of 0, while a completely hydrophilic molecule has a HLB of 20. Mohamed *et al.* compared the effects of various surfactants on emulsion stability [13]. The ideal choice of surfactant seems to depend on other design parameters. In the formation of 10-micron diameter microparticles, 1:10 w/o emulsions were most stabilized by surfactants with more hydrophilic character (HLB greater than 16), while 1:5 w/o emulsions had overall lower stability which was most improved by more equally balanced surfactants (HLB between 8 and 13). In this study, it was determined that a HLB greater than six would result in emulsions stable enough for the formation of microparticles. This study also concluded that the concentration of surfactant affects emulsion stability, and that the impact of surfactant is not completely explained by the HLB, as different surfactants with the same HLB had a range of effects on emulsion stability as well.

The use of surfactants has been shown to affect various properties of microparticles. The morphology of microparticles depended on the surfactants used in the primary emulsion. Unstable primary emulsion leads to hollow capsules. Plasticizing surfactants may entrap water in the matrix [13], lowering the glass transition temperature (T_g) to below physiological temperature. Therefore, the possibility of collapse of the microparticle when warmed above T_g *in vivo* has been a concern. Such a possibility may be avoided, however, if the microparticles are sufficiently large (~100 micron) and are independent of the use of plasticizing or antiplasticizing surfactants [30]. Finally, surfactant type and concentration also have an effect on particle size [11, 15]. An early study showed particle size had an inverse relationship to PVA concentration, while DNA loading had a direct relationship, thus high a concentration of PVA produces small particles with high DNA content [8].

The size of microparticles can also be controlled by the energy applied during emulsification. Homogenization and sonication are both feasible options for emulsification, resulting in similar EE% and biologic activity of encapsulated DNA [8], although homogenization is the more common method in recent studies. There are concerns about the suitability of sonication, regarding the inability to scale the process up for large batches and the possible release of metals from the sonicator tip [8]. Microparticle size is inversely related to homogenization speed [17]. However, this is not a preferred method for controlling size, because high speed homogenization can result in mixing of the primary and secondary aqueous phases and escape of nucleic acids out of the microparticle, and can be a major source of nucleic acid degradation.

13.2.5 Combining Polycations with PLGA Microparticles

Incorporation of polycations into the PLGA microparticle formulation has become an increasingly promising avenue to overcome the limitations of PLGA encapsulation. Polymer cations may be used in several different ways and have been incorporated in each of the three phases of a double emulsion process. If DNA is combined with polycations in solution prior to microparticle processing, the DNA interacts with the polycations, forming complexes. Chitosan [31], protamine sulfate [32], poly(β-amino esters) [33] and, most frequently, poly(ethylene imine) (PEI) [34] have recently been investigated as DNA complexing agents for use with PLGA microparticles. The DNA in these complexes is stabilized in a compacted form [32] and is less susceptible to damage by shear force as well as nucleases. In addition, the high positive charge density of some polycations such as PEI may have a buffering effect, which facilitates release of DNA into the cytoplasm from an endolysosomal compartment by the putative 'proton sponge' mechanism [34]. However, polycations have also been shown to be destructive to cell membranes and may result in very high cytotoxicity. When applied as a delivery method, DNA/polycation complexes have had very high transfection rates, but also high toxicity. Encapsulation within PLGA microparticles slightly reduces *in vitro* transfection rates, but greatly reduces toxicity [34].

Several studies have added cationic components in the organic phase of the double emulsion. Kusonwiriyawong *et al.* described the effect of the cationic amphiphile, stearylamine, on microparticle properties. Stearylamine has strong hydrophobic character due to an 18-carbon tail and was intended to interact with the cell membrane to assist in cellular uptake of DNA. Formulation with stearylamine did not result in increase in microparticle size and protected plasmid DNA from enzymatic degradation, but it also completely prevented release of DNA from the microparticles [35]. Oster *et al.* also added several polycations in the polymer phase of the double emulsion, loading DNA after complete particle formation [11]. Loading efficiency and particle size varied widely depending on the polycation – cetyl trimethyl ammonium bromide (CTAB) or PEI – and the percentage of polycation in the solution. The presence of CTAB resulted in an average 10-fold increase in particle diameter, ascribed to aggregation resulting from the plasticizing effect of CTAB [21]. Increasing CTAB concentration increased particle size, while increasing PEI did not. Microparticles formulated with 50% PEI in the polymer phase showed high EE%. PEI containing microparticles showed protection against DNAse for 24 hours, while CTAB containing microparticles did not [11].

Treatment of the fully formed PLGA microparticle surface by polycations promises significant improvement over standard encapsulation. In theory, the adsorption of DNA onto the cationic surface of a microparticle addresses three limitations of the standard matrix encapsulation strategy. Release of large nucleic acids is faster from the microparticle surface than from the bulk and is not affected by the rate of microparticle degradation [21, 34]. DNA at the surface of the microparticle does not experience the low pH that results at the centre of the microparticle due to release of the acidic by-products of PLGA hydrolysis [36]. Finally, because DNA is adsorbed onto a fully formed microparticle, the damage caused by shear stress during microparticle formulation is minimized. Surface loading has the additional advantages of higher loading efficiency as well as increased bioavailability [37] and endosomal buffering by some polycations.

Cationic surfaces have been created by coating with cationic polymer such as PEI [37] or surfactant such as CTAB [21, 38] added in the exterior aqueous phase of the double emulsion. The polycation becomes entrapped at the surface of the microparticle and results in a highly positive surface. When DNA is added in solution with fully formed cationic microparticles, the DNA associates with the positively charged surface of the microparticles. It was shown that DNA can be loaded onto the surface of a PLGA microparticle if CTAB is added, but will not associate in its absence [21]. A later study showed a relationship between CTAB concentration in the formulation and DNA loading as well as release from 30 micron microparticles. At 1% w/w CTAB, 1% w/w DNA was loaded at nearly 100% and release was 100% of the DNA in the first 24 hours [38]. However, CTAB concentration did not relate to transfection efficiency, although all concentrations tested were improvements over naked DNA. The reason might be that DNA released from CTAB microparticles was intact, but mostly in the nicked form [39]. DNA vaccines delivered by this system to mice produced a detectible immune response using only 10% of the mass of naked DNA required to produce the same response [22].

In a recent study, Davies *et al.* compared several cationic polymer coatings, including PEI, poly(L-lysine) (PLL), trimethyl chitosan and poly(dimethylamino)ethyl methacrylate, on PLGA microparticles of 2–4 microns in diameter [36]. Loading was in the range of 0.02–0.05 mg DNA/mg polymer. It was found that the molecular weight of PEI used in the surface treatment was important to the properties of the microparticle. High molecular weight PEI resulted in more positive surface charge and increased DNA loading. One challenge is preventing aggregation of these microparticles. The use of PVA in the second emulsion increased colloidal stability, but the combination of small particles and high molecular weight PEI coating results in aggregation that can not be prevented by using PVA.

In response to concern over the toxic effect of polycations to cell membrane, several studies have demonstrated the ability to covalently link the polycation to functionalized PLGA microparticles. Kasturi *et al.* used carbodiimide/active ester chemistry to tether PEI to microparticles [37]. This study demonstrated the ability of cationic coated microparticles to outperform standard PLGA microparticles, producing detectible transgene expression at a lower dose of DNA. Meanwhile, cell viability was low in the presence of PEI, but when PEI was conjugated to the microparticles viability was comparable to the microparticles alone. Another study used EDC chemistry to conjugate poly(amido amine) (PAMAM) dendrimers [40] and showed that these had a higher transfection rate with slightly higher toxicity than PLGA microparticles. Although Singh *et al.* did not see significantly enhanced transfection capability of microparticles with highly branched PAMAM dendrimers, Kasturi *et al.* observed increased loading capacity of microparticles with branched PEI coating versus linear PEI, along with slower release and enhanced transfection [37].

A very recent study compared four DNA loading strategies that have been previously described in 2–3 micron microparticles: encapsulation of DNA within PLGA microspheres, encapsulation of PEI/DNA complexes within PLGA microparticles, encapsulation of DNA within PLGA-PEI microparticles and adsorption of DNA onto fully formed PLGA microparticles with covalently bound PEI coating [34]. The three alternative PEI encapsulation methods showed improved buffering effect and DNA loading and resulted in higher transgene expression compared to PLGA encapsulation, and had lower toxicity than PEI alone. The results of this study suggested that DNA-PEI complexes encapsulated within PLGA microparticles resulted in the highest cellular uptake and transfection efficiency, with slightly higher toxicity, compared to the other techniques.

13.2.6 Current Applications

Recently, the vast majority of advanced studies in PLGA microparticles have been aimed at enhancing immunogenic response against cancer [37] or viruses [41] through nucleic acid delivery targeting antigen-presenting cells. These studies involve delivery of oligonucleotides with cationic coated particles [42] and PLGA encapsulation [43], plasmid DNA in standard emulsion [9, 29, 44–46] or plasmid DNA on cationic PLGA microparticles [12, 23, 39, 47] for the purpose of eliciting a desired immune response. *In vivo* results have generally indicated the feasibility of PLGA microparticles for vaccine applications and established superiority over naked nucleic acid delivery. Some of the unique advantages of the microparticle approaches include increased efficacy over other techniques and decreased amount of agent required to stimulate a detectible response [41, 47]. Dendritic cells are readily transfected by cationic microparticles, but not by naked DNA [23]. When compared to a bacterial vector, PLGA microparticles produced stronger immune response at low vehicle dosing and stronger effect based on a comparable dose of DNA delivered [48]. PLGA particles also have demonstrated intrinsic adjuvant properties [47] and microparticles containing oligonucleotides have caused upregulation of CD80 and CD86, both markers of dendritic cell maturation in several studies [37, 43].

Several studies have used PLGA microparticles for combined delivery of DNA antigen and oligonucleotides targeted to antigen-presenting cells [43, 49]. Because PLGA microparticles are efficiently internalized by these cells, they are effective at delivering antigen and oligonucleotides to the same cell, enhancing cellular immune response in mice [43]. Trombone *et al.* demonstrated specificity of PLGA microparticles by showing preferred internalization by CD11c+ and CD11b+ antigen-presenting cells in the draining lymph nodes *in vivo* [9].

Although attracting less attention, nucleic-acid-carrying PLGA microparticles have potential for applications beyond delivery of immunogenic agents. For example, Choleris *et al.* have used PLGA microparticles to carry antisense oligonucleotides to turn off gene expression in the amygdala of mice to study social behaviour of these animals. PLGA microparticles were chosen for this application because of their sustained releasing ability and compatibility with brain tissue [28].

13.3 Other Types of Polymer Microparticles

13.3.1 Alginate

Alginate (often as sodium salt) is an anionic linear polysaccharide containing blocks of (1-4)-linked β-D-mannuronate (M) and its C-5 epimer α-L-guluronate (G) residues that gelate in the presence of di- or polyvalent cations [50]. A natural material derived from brown algae, alginate is biodegradable, biocompatible and mucoadhesive, making the material potentially useful, especially for oral and nasal delivery [50].

Alginate microparticles can be prepared by water/oil emulsion combined with ionic cross-linking. First, a pre-gel solution is formed by adding an oil solvent to the aqueous solution of alginate. A mixture of the cationic cross-linking agent, typically Ca^{2+}, and nucleic acid is then added and stirred, cross-linking the gel and forming microparticles with nucleic acids both trapped within and coating the microparticles. This fabrication method yields microparticles whose size varies from 1–10 microns for alginate/naked DNA formulations [51] to over 100 microns for alginate-polycation-complexed nucleic acid formulations [50]. Typically, these microparticles have asymmetric shapes with a sponge-like surface.

Loading efficiency and release of nucleic acids from alginate microparticles varies dramatically depending on the procedure used to prepare the microparticles. Loading efficiencies between 40 and 95% are achieved by varying the ratio of alginate to cross-linking agent, as well as by addition of a nucleic acid complexing agent such as poly(L-lysine). Typically, there is between 20 and 100% release of nucleic acids within 24 hours [50].

The relatively gentle preparation of the alginate microparticles is widely reported to have no adverse effect on the integrity of the incorporated nucleic acids. Addition of a cationic complexation agent like poly(L-lysine) or an adenovirus to the microparticles protects the nucleic acids from degradation by enzymes and increases the cell uptake efficiency through non-specific adsorptive endocytosis [50].

Alginate microparticles have been used for delivering DNA-based vaccines and can be administered through oral, intranasal, intramuscular, subcutaneous or intraperitoneal routes, with each route modulating the type and level of immune response [52]. The microparticles are thought to be endocytosed by macrophages and dendritic cells. Addition of an adenovirus to DNA carrying alginate microparticles has been shown to enhance the transgene expression and antibody response to the introduced DNA. Transfection efficiency is reported to be around 10% [53].

13.3.2 Chitosan

Chitosan is a linear polysaccharide composed of randomly distributed β-(1-4)-linked D-glucosamine (deacetylated unit) and N-acetyl-D-glucosamine (acetylated unit), derived from the deacetylation of chitin. It is non-toxic, biodegradable, polycationic and exhibits low immunogenicity, making it an attractive material for a non-viral nucleic acid delivery system.

Chitosan-nucleic acid microparticles can be prepared by a coacervation method. Briefly, chitosan is dissolved in acidic solution while the nucleic acid sample is combined with sodium sulfate solution (~20% w/v). Drop-wise addition of the nucleic acid solution into the chitosan followed by mixing produces the chitosan-nucleic acid microparticles, which are recovered by centrifugation after washing with deionized water. The spontaneous incorporation of the nucleic acids with the chitosan occurs due to the cationic nature of chitosan and the anionic nature of the nucleic acids. This procedure can be modified to include PEI, in addition to chitosan, to complex with nuclei acids. Chitosan microparticles range in size from 1 to 4 microns, depending largely on the molecular weight of the chitosan [54] and the concentration of chitosan solution [55]. Additional polycations such as PEI have no effect on the size or polydispersity of the microparticles [56]. PEI-nucleic acid complexes give the microparticles a strong buffering capacity over a wide pH range and may potentially enhance gene transfection efficiency.

The nucleic acid loading efficiencies for chitosan microparticles are consistently high, around 75–90%, regardless of the nucleic acid used. It has been shown that loading efficiency can be improved by increasing the chitosan solution concentration during microparticle formation but at the cost of decreasing the total loading capacity [55]. *In vitro* release profiles for nucleic acid-loaded microparticles show a small burst effect, generally less than 20% of the total nucleic acid released in the first 24 hours, followed by a slow, sustained release for up to 180 days. The release profiles are dependent on chitosan concentration and molecular weight, nucleic acid concentration and the size of the nucleic acid. At higher chitosan concentrations and higher molecular weight there is less total nucleic acid loaded in the microparticles, leading to less total nucleic acid release. However, the ratio of nucleic acid released to total nucleic acid in the microparticle is higher at high chitosan concentrations. Higher nucleic acid concentration leads to a higher rate of release from the microparticle, but a lower percentage of release. A third factor affecting the release is the size of the nucleic acid. Higher molecular weight nucleic acids, such as plasmid DNA, are released slower than smaller nucleic acids, such as antisense oligonucleotides, due to the lower diffusion constants of large molecules through the chitosan matrix [56]. pH-dependent release of DNA from chitosan microparticles has been shown. Gastric pH of around 2, intestinal pH of 6.5 and acidic buffer media pH of 4.5 have been used to show that the release rate from the microparticles increases as medium pH decreases. This pH dependency is caused by the higher solubility of the chitosan particles at more acidic pH conditions [57].

Numerous studies have shown chitosan microparticles encapsulate and release nucleic acids without detriment to the integrity of the cargoes. Chitosan microparticles have shown the ability to effectively protect the nucleic acids from enzymatic degradation by DNase1 and serum for a week [56].

Chitosan-nucleic acid microparticles transfect cells by absorbing to the cell surface and triggering endocytosis. Subsequent swelling of the microparticles may lead to endosomal rupture and the release of DNA into the cytoplasm [58]. The nucleic acids then move from the microparticles into the cell nucleus by diffusion. In some cases, chitosan microparticles show the ability to induce long-term gene expression compared to naked plasmid injection. However, transfection ability is cell-specific, with some cells immune to transfection. Most experiments using chitosan microparticles have suffered from low transfection efficiency compared to other techniques such as commercial lipid-based transfection reagents. Increased transfection has been reported by encapsulating two plasmids into chitosan microparticles. This dual plasmid encapsulation approach remains capable of high encapsulation efficiency (90%) without compromising the structural and functional integrity of DNA [59]. Cell-specific transfection may be realized by attaching targeting ligands to the surface of chitosan microparticles through amine groups. For example, chitosan modified with mannose has been used to target dentritic cells, which express a high number of mannose receptors on the cell surface [60]. Increased transfection of these targeted microparticles can be achieved by using PEI to condense the DNA prior to encapsulation, in order to enable DNA to escape from cell endosome through the putative 'proton sponge' mechanism [61].

13.3.3 Hyaluronan (HA)

Hyaluronan (HA), also called hyaluronic acid and hyaluronate, is a member of the glycosaminoglycan family and is composed of disaccharide repeating units of D-glucuronic acids and D-N-acetylglucosamines linked together via alternating β-1,4 and β-1,3 glycosidic bonds. It is a natural polymer found in the extracellular matrix and connective tissue in many animals, including humans, making HA non-immunogenic and an attractive material for biomedical applications. To date, HA has been used for hydrogels for viscosurgery and viscosupplementation, as scaffolds for wound healing and tissue engineering applications, hydrophilic coatings for medical devices [62] and others.

A simple and gentle cross-linking chemistry using adipic dihydrazide was modified to create HA microparticles containing nucleic acids [63]. In this method, plasmid DNA was added to a mixture of HA and adipic dihydrazide followed by a cross-linking initiator. A second cross-linking reaction froze the nucleic acid within the microparticle. Robust microparticles with a heterogeneous size distribution were produced, with an average diameter of 6 microns and 60% of the microparticles between 3 and 10 microns.

The HA microparticles exhibited a moderate diffusion controlled initial burst release, likely from DNA bound to the surface of the microparticles but not fully incorporated. A second slow and sustained release followed as the microparticles degrade via enzymatically catalysed hydrolysis. Overall, 60% of the loaded DNA was released after 63 days [63].

The released pDNA was capable of sustained *in vivo* and *in vitro* transfection. Positive DNA transcription was seen in skeletal muscle of rats three weeks post injection of the microparticles. Transfection was relatively high initially, corresponding to the burst release of the DNA, and gradually decreases over time. The plasmid DNA released from the microparticles was intact and bioactive but mostly in an uncoiled state. It was also shown that targeted HA microparticles can be made by conjugating antibodies specific to cell receptors to the HA microparticles via protein A. These receptor-targeted microparticles showed up to a sixfold increase in adhesion to their target cells compared to non-targeting microparticles [63].

13.3.4 Macro-Albumin Aggregates (MAA)

Human serum albumin (HSA) is the most abundant blood protein found in humans. At high temperatures albumin aggregates to form large microparticles called macro-albumin aggregates (MAAs). These aggregates are often radiolabelled and used in nuclear medicine to image pulmonary blood flow through the capillaries. Conjugation of PEI to the MAAs creates a micrometre-sized delivery vehicle capable of targeting genetic vaccines to the lungs [64].

HSA-PEI conjugates were synthesized by slowly filtering a PEI solution into a stirred solution of HSA. Raising the temperature of the mixture while heating induces the conjugates to aggregate into MAAs that range in size around 25–100 microns in diameter. MAA-PEI aggregates avidly incorporated plasmid DNA to their surface because of the strong affinity between PEI and DNA. MAAs without PEI did not bind to DNA. Transfection studies with a mouse model have shown that MAA-PEI-DNA conjugates injected intravenously elicited a response specific to the lungs if the injection site was immediately upstream and adjacent to the lungs. The high first-pass clearance resulted in no transfection in tissues other than the lungs. Upregulation of the transgene was present in the lungs after 24 hours, reached a maximum after 48 hours and continued to be detectable even after five days. The expression of the plasmid encoding for an antigen elicited a strong immune response as well as high-levels of systemic humoral and cytotoxic T-cell responses [64]. This study demonstrates the safe and non-immunogenic nature of MAAs and the possibility of pulmonary targeting transgene expression and gene-based vaccine in humans.

13.3.5 Poly(D,L-Lactide-*co*-4-Hydroxy-L-Proline) (PLHP)

Poly(D,L-lactide-*co*-4-hydroxy-L-proline) (PLHP) is a copolymer made by copolymerization of lactide and hydroxyproline, a non-essential amino acid and a component of collagen, gelatine and other human proteins. PLHP is an attractive material for biomedical applications as it degrades into lactic acid and hydroxyproline, both endogenous metabolites.

Fabrication of PLHP microparticles follows the standard double emulsion solvent evaporation technique. In one study [65], plasmid DNA was added drop-wise to a solution of PLHP and organic solvent while vortexing. After sonication to achieve a homogeneous mixture, the solution was added to an emulsifier and stirred, creating the microparticles. PHLP microparticles were heterogeneous in size with most particles measuring 1–5 microns in diameter and carried a negative surface charge. The cytotoxicity of the PLHP particles was minimal and significantly less than PEI and poly(L-lysine) when assessed by assays quantifying cell metabolism and cell morphology [65]. Plasmid DNA was distributed evenly throughout the microparticle with loading efficiency ranging from 83 to 89% depending on the initial amount of DNA. The microparticles exhibited an initial burst release of about 40% of the loaded DNA in the first two hours. A slow, continuous release of another 40% followed over 18 days. All DNA was released before the polymer was completely degraded [65].

Transfection of cells using PHLP microparticles occurred quickly and the level of gene expression continued to increase until day 5. The sustained transfection ability of the PHLP microparticles resulted in two to threefold higher activity than PEI-DNA and PLL-DNA complexes after one week. The transfection mechanism is still unknown, but it was suggested that the microparticles were endocytosed and slowly released DNA intracellularly [65].

13.3.6 The Pluronics

Pluronic is the trade name for a family of non-ionic triblock copolymers marketed by BASF. These polymers are composed of poly(propylene oxide) (PPO) and poly(ethylene oxide) (PEO) blocks of various numbers and lengths. At near-freezing temperatures these polymers are soluble in water but aggregate into microparticles above their phase transition temperature. Addition of a cationic surfactant, benzalkonium chloride (BAK), to the Pluronic polymer allows DNA to adsorb to the surface of these particles [66]. The particles are spherical with a smooth surface and diameters between 0.3 and 2 microns on average. BAK is both bound to the surface of the microparticles through hydrophobic interactions as well as solubilized within the hydrophobic compartments of the microparticles. Unlike many other delivery vesicles, Pluronic-BAK-DNA microparticles have a negative surface charge, suggesting that they may not have the same mechanism of action as cationic microparticles when it comes to entry into cells and gene transfection. The reason for the net negative surface charge is the presence of excess DNA over the amount of BAK found on the microparticle surface; nearly all the available surface BAK is complexed with DNA. Therefore, the total loading of DNA is only dependent on the available BAK, with particle size having no discernable effect.

It was shown that intramuscular injection of the Pluronic-BAK microparticles containing an HIV DNA vaccine into rhesus monkeys generated an enhanced cellular immune response compared to Pluronic-DNA particles lacking BAK [66]. It was hypothesized that the Pluronic-pDNA-BAK microparticles might facilitate DNA delivery by enhancing the permeability of cell membrane. Regardless of the exact mechanism of action, Pluronic-BAK microparticles represent a very attractive formulation, especially for DNA vaccines, as BAK is approved by the FDA for use as a preservative and for intramuscular injection, and the experimental HIV vaccine using this formulation (CRL1005) developed by Merck has been evaluated in a Phase I clinical trial at doses up to 75 mg intramuscularly without evidence of significant local or systemic toxicity.

13.3.7 Poly(Ortho Esters) (POEs)

Poly(ortho esters) (POEs) are a family of pH-labile, biocompatible and biodegradable polymers. Synthesized through a condensation copolymerization of a diketene acetal and diols using an acid catalyst, POEs hydrolyse into monomeric diols and minor amounts of acid [67]. Material properties of POEs such as rigidity, hydrophilicity and degradation rate can be tailored based upon the specific choice of the diols used for polymerization [68].

A water/oil/water double emulsion procedure has been used to form POE microparticles. The resulting particles are smooth, symmetric spheres with no surface defects with an average diameter of approximately five microns, allowing for passive targeting of antigen-presenting cells. Once formed, the microparticles can be stored for up to one year at ambient temperature without polymer or DNA degradation [67].

Microparticles can be loaded with about five micrograms of nucleic acid per milligram of POE. The mild preparation technique leaves the nucleic acids unharmed and in their natural active state. The ortho ester bonds in the backbone of the polymer are responsible for the pH sensitivity of hydrolysis. While stable at neutral pH, the ortho ester bonds destabilize and hydrolyse rapidly around pH 5.0. Depending on the POE composition, 100% of the nucleic acid is released after only 24–72 hours of exposure to the pH 5 environment. This sensitivity allows the microparticles to release nucleic acids only after being internalized to phagosomes of antigen-presenting cells, increasing the delivery efficiency. Because these polymers undergo surface erosion rather than bulk erosion, potentially harmful acidic degradation products simply diffuse away and do not have the chance to become concentrated within or around the microparticles.

POE microparticles are effective DNA vaccine carriers capable of activating antigen-specific primary and secondary humoral as well as cytotoxic T-cell responses. These microparticles have been shown to be more effective than naked DNA injection and PLGA microencapsulated DNA delivery. This is likely due to the rapid and controllable delivery of the DNA from the POE microparticles. Microparticles created by blending PEI into the POE matrix have been shown to dramatically alter tranfection efficiency *in vitro* and nucleic acid release kinetics. Hydrophilic PEI present at the surface of the microparticles results in a faster surface degradation. However, PEI forms complexes with the nucleic acids, thus delaying the release of free nucleic acid into solution when compared to plain POE microparticles. Overall, the POE microparticle gene transfection efficiency increases 10–50-fold with the addition of small amounts of PEI without sacrificing biocompatibility [68]. Although clinical use of PEI is limited because of the potential toxicity at high concentrations, the slow release of minor amounts of PEI from the microparticles appears to be at subtoxic levels. In addition, both PEI and POEs appear to stimulate dendritic cells into maturation, which is a critical step for immune activation. Therefore, such microparticle formulations may be of particular interest for use in vaccine delivery.

13.3.8 Polyacrylamide

Polyacrylamide is a biocompatible polymer that can be readily cross-linked into hydrogels or microgel particles. The most common uses for polyacrylamide are for gel electrophoresis and soft contact lenses. Material characteristics of polyacrylamide gels are largely dependent on the cross-linking agent, and the use of acid-sensitive acetal cross-linkers makes the hydrogel pH-labile [69]. In this case, microparticles, or microgels, are prepared using an inverse micro-emulsion polymerization technique. The aqueous phase containing the acrylamide monomer and nucleic acids, and an acetal cross-linker, is added to an organic phase. After dispersion by sonication, radical polymerization is induced, effectively trapping the nucleic acids within the cross-linked beads. The resulting microparticles have size ranges from 0.2 to 1 micron in diameter with fairly uniform spherical shape and no large surface defects. While polyacrylamide itself is non-toxic, the toxicity of the aldehyde degradation by-products is yet to be determined.

The micro-emulsion polymerization method yields encapsulation efficiency of around 50% with total DNA loading proportional to initial DNA in the polymerization mixture. The largest loading reported is over 4 μg/mg polymer, slightly higher than PLGA particles of comparable size. Release kinetics at physiologic pH of 7 include a burst release of about 10%, likely from a small amount of DNA adsorbed onto the surface of the microparticles. After this burst release further incubation at neutral pH shows no further release of DNA. Once placed in an acidic environment of pH 5, as found in cell lysosomes, full release of the DNA occurs within two hours. The pH sensitivity of these microparticles allows them to specifically degrade and release their DNA only after entering the lysosomal compartments, thus increasing the transfection efficiency. Analysis of released DNA shows that it is composed mostly of the open circular form. Despite the loss of the super-coiled conformation, circular DNA displays roughly equivalent transfection efficiency as super-coiled DNA [69]. It appears that such cross-linked polyacrylamide microparticles may be useful not only for nucleic acid delivery in general but also for vaccine delivery in particular, given the pH-labile degradability and acid-triggered release of cargo.

13.4 Conclusions

Much progress has been made in the past years, particularly in the last decade, on developing polymer microparticles for delivering nucleic acids as well as understanding the mechanism of delivery. Although much work has focused on PLGA-type polymers due to their proven record of safety in humans and low regulatory hurdle, a number of innovative approaches have emerged that use novel polymers or novel combinations of polymers. There are many microparticle fabrication techniques at our disposal that enable excellent control and optimization of microparticle properties. Microparticles have been successfully used for *in vivo* gene delivery to different tissues/organs in various animal models both through local and systemic delivery routes. As a result of these fundamental studies, promising technologies are being evaluated in different stages of clinical trials. Two of the most prominent examples are the prime-boost HIV vaccine formulation by Novartis using DNA coated on cationic PLGA microparticles and the cancer vaccine formulation by MGI using classic PLGA microencapsulation. It is only a matter of time before the first commercial success is seen of polymer microparticles for nucleic acid delivery, mostly likely DNA vaccine delivery, in humans.

The future of fundamental research in polymer microparticles for nucleic acid delivery is likely to rely heavily on the advances in chemical material engineering and molecular/cell biology. Despite the dominant role of the safe but inert PLGA, other novel materials, especially those with intrinsic activities towards modulating cell behaviour, need to be explored further. Better formulation processes are needed to improve the control over microparticle geometry and uniformity in size, and to increase loading capacity. Multicomponent, multicompartment microparticles, such as PLGA microparticles encapsulating molecular complexes of polycations and DNA, are likely to become more finely tuned in their microstructure and composition, in order to control precisely the release of cargoes temporally and spatially, to enhance cell uptake and targeting, and to improve gene transfection. Furthermore, such sophisticated micro-devices will be particularly important and useful if two or more nucleic acids with different modes of action and intracellular targets – such as plasmid DNA, small interfering RNA or oligonucleotides – are to be delivered from the same microparticles.

References

1. Luo, D. and Saltzman, W.M. (2000) Synthetic DNA delivery systems. *Nat. Biotechnol.*, **18**, 33–37.
2. Pack, D.W., Hoffman, A.S., Pun, S. and Stayton, P.S. (2005) Design and development of polymers for gene delivery. *Nat. Rev. Drug Discov.*, **4**, 581–593.
3. Pannier, A.K. and Shea, L.D. (2004) Controlled release systems for DNA delivery. *Mol. Ther.*, **10**, 19–26.
4. Putnam, D. (2006) Polymers for gene delivery across length scales. *Nat. Mater.*, **5**, 439–451.
5. Schaffert, D. and Wagner, E. (2008) Gene therapy progress and prospects: synthetic polymer-based systems. *Gene Ther.*, **15**, 1131–1138.
6. Heller, J. and Hoffman, A.S. (2004) Drug delivery systems, in *Biomaterials Science: An Introduction to Materials in Medicine*, 2nd edn (eds B.D. Ratner, A.S. Hoffman, F.J. Schoen and J.E. Lemons), Elsevier Academic Press, San Diego, California, pp. 628–648.
7. Kohn, J., Abramson, S. and Langer, R. (2004) Bioresorbable and bioerodible materials, in *Biomaterials Science: An Introduction to Materials in Medicine*, 2nd edn (eds B.D. Ratner, A.S. Hoffman, F.J. Schoen and J.E. Lemons), Elsevier Academic Press, San Diego, California, pp. 115–127.
8. Hsu, Y., Hao, T. and Hedley, M.L. (1999) Comparison of process parameters for microencapsulation of plasmid DNA in poly(D,L-lactic-*co*-glycolic) acid microspheres. *J. Drug Target.*, **7**, 313.
9. Trombone, A.P., Silva, C., Almeida, L. *et al.* (2007) Tissue distribution of DNA-Hsp65/TDM-loaded PLGA microspheres and uptake by phagocytic cells. *Genet. Vaccines Ther.*, **5**, 9.
10. Tabata, I. and Ikada, Y. (1990) Phagocytosis of polymer microspheres by macrophages. *Adv. Polym. Sci.*, **94**, 107–141.
11. Oster, C.G., Kim, N., Grode, L. *et al.* (2005) Cationic microparticles *co*nsisting of poly(lactide-*co*-glycolide) and polyethylenimine as carriers systems for parental DNA vaccination. *J. Control. Release*, **104**, 359–377.
12. Singh, M., Briones, M., Ott, G. and O'Hagan, D. (2000) Cationic microparticles: a potent delivery system for DNA vaccines. *Proc. Natl. Acad. Sci. USA*, **97**, 811–816.
13. Mohamed, F. and van der Walle, C.F. (2006) PLGA microcapsules with novel dimpled surfaces for pulmonary delivery of DNA. *Int. J. Pharm.*, **311**, 97–107.

14. Oster, C.G. and Kissel, T. (2005) Comparative study of DNA encapsulation into PLGA microparticles using modified double emulsion methods and spray drying techniques. *J. Microencapsul.*, **22**, 235–244.
15. Capan, Y., Woo, B.H., Gebrekidan, S. *et al.* (1999) Influence of formulation parameters on the characteristics of poly(D,L-lactide-*co*-glycolide) microspheres containing poly(L-lysine) complexed plasmid DNA. *J. Control. Release*, **60**, 279–286.
16. Walter, E., Dreher, D., Kok, M. *et al.* (2001) Hydrophilic poly(DL-lactide-*co*-glycolide) microspheres for the delivery of DNA to human-derived macrophages and dendritic cells. *J. Control. Release*, **76**, 149–168.
17. Díez, S. and Tros de Ilarduya, C. (2006) Versatility of biodegradable poly(d,l-lactic-*co*-glycolic acid) microspheres for plasmid DNA delivery. *Eur. J. Pharm. Biopharm.*, **63**, 188–197.
18. Cai, H., Hu, X.D., Yu, D.H. *et al.* (2005) Combined DNA vaccine encapsulated in microspheres enhanced protection efficacy against mycobacterium tuberculosis infection of mice. *Vaccine*, **23**, 4167–4174.
19. Ando, S., Putnam, D., Pack, D.W. and Langer, R. (1999) PLGA microspheres containing plasmid DNA: preservation of supercoiled DNA via cryopreparation and carbohydrate stabilization. *J. Pharm. Sci.*, **88**, 126–130.
20. Hao, T., McKeever, U. and Hedley, M.L. (2000) Biological potency of microsphere encapsulated plasmid DNA. *J. Control. Release*, **69**, 249–259.
21. Briones, M., Singh, M., Ugozzoli, M. *et al.* (2001) The preparation, characterization, and evaluation of cationic microparticles for DNA vaccine delivery. *Pharm. Res.*, **18**, 1476–1479.
22. O'Hagan, D.T., Singh, M. and Ulmer, J.B. (2004) Microparticles for the delivery of DNA vaccines. *Immunol. Rev.*, **199**, 191–200.
23. Denis-Mize, K.S., Dupuis, M., MacKichan, M.L. *et al.* (2000) Plasmid DNA adsorbed onto cationic microparticles mediates target gene expression and antigen presentation by dendritic cells. *Gene Ther.*, **7**, 2105–2112.
24. del Barrio, G., Novo, F.J. and Irache, J.M. (2003) Loading of plasmid DNA into PLGA microparticles using TROMS (total recirculation one-machine system): evaluation of its integrity and controlled release properties. *J. Control. Release*, **86**, 123–130.
25. Román, B.S., Irache, J.M., Gómez, S. *et al.* (2008) Co-encapsulation of an antigen and CpG oligonucleotides into PLGA microparticles by TROMS technology. *Eur. J. Pharm. Biopharm.*, **70**, 98–108.
26. Ho, J., Wang, H. and Forde, G.M. (2008) Process considerations related to the microencapsulation of plasmid DNA via ultrasonic atomization. *Biotechnol. Bioeng.*, **101**, 172–181.
27. De Rosa, G., Maiuri, M.C., Ungaro, F. *et al.* (2005) Enhanced intracellular uptake and inhibi tion of NF-Kappab activation by decoy oligonucleotide released from PLGA microspheres. *J. Gene Med.*, **7**, 771–781.
28. Choleris, E., Little, S.R., Mong, J.A. *et al.*(2007) Microparticle-based delivery of oxytocin receptor antisense DNA in the medial amygdala blocks social recognition in female mice. *Proc. Natl. Acad. Sci. USA*, **104**, 4670–4675.
29. Jang, J. and Shea, L.D. (2006) Intramuscular delivery of DNA releasing microspheres: microsphere properties and transgene expression. *J. Control. Release*, **112**, 120–128.
30. Rouse, J.J., Mohamed, F. and van der Walle, C.F. (2007) Physical ageing and thermal analysis of PLGA microspheres encapsulating protein or DNA. *Int. J. Pharm.*, **339**, 112–120.
31. Wischke, C., Borchert, H.H., Zimmermann, J. *et al.* (2006) Stable cationic microparticles for enhanced model antigen delivery to dendritic cells. *J. Control. Release*, **114**, 359–368.
32. Dunne, M., Bibby, D.C., Jones, J.C. and Cudmore, S. (2003) Encapsulation of protamine sulphate compacted DNA in polylactide and polylactide-*co*-glycolide microparticles. *J. Control. Release*, **92**, 209–219.
33. Little, S.R., Lynn, D.M., Ge, Q. *et al.* (2004) Poly-beta amino ester-containing microparticles enhance the activity of nonviral genetic vaccines. *Proc. Natl. Acad. Sci. USA*, **101**, 9534–9539.
34. Zhang, X.Q., Intra, J. and Salem, A.K. (2008) Comparative study of poly (lactic-*co*-glycolic acid)-polyethylenimine-plasmid DNA microparticles prepared using double emulsion methods. *J. Microencapsul.*, **25**, 1–12.

35. Kusonwiriyawong, C., Atuah, K., Alpar, O.H. *et al.* (2004) Cationic stearylamine-containing biodegradable microparticles for DNA delivery. *J. Microencapsul.*, **21**, 25–36.
36. Davies, O.R., Head, L., Armitage, D. *et al.* (2008) Surface modification of microspheres with steric stabilizing and cationic polymers for gene delivery. *Langmuir*, **24**, 7138–7146.
37. Kasturi, S.P., Sachaphibulkij, K. and Roy, K. (2005) Covalent conjugation of polyethyleneimine on biodegradable microparticles for delivery of plasmid DNA vaccines. *Biomaterials*, **26**, 6375–6385.
38. Singh, M., Ugozzoli, M., Briones, M. *et al.* (2003) The effect of CTAB concentration in cationic PLG microparticles on DNA adsorption and *in vivo* performance. *Pharm. Res.*, **20**, 247–251.
39. Kaur, R., Rauthan, M. and Vrati, S. (2004) Immunogenicity in mice of a cationic microparticle-adsorbed plasmid DNA encoding Japanese encephalitis virus envelope protein. *Vaccine*, **22**, 2776–2782.
40. Singh, M., Fang, J.H., Kazzaz, J. *et al.* (2006) A modified process for preparing cationic polylactide-*co*-glycolide microparticles with adsorbed DNA. *Int. J. Pharm.*, **327**, 1–5.
41. Wang, F., He, X., Jiang, L. *et al.* (2006) Enhanced immunogenicity of microencapsulated multiepitope DNA vaccine encoding T and B cell epitopes of foot-and-mouth disease virus in mice. *Vaccine*, **24**, 2017–2026.
42. Xie, H., Gursel, I., Ivins, B.E. *et al.* (2005) CpG oligodeoxynucleotides adsorbed onto polylactide-co-glycolide microparticles improve the immunogenicity and protective activity of the licensed anthrax vaccine. *Infect. Immun.*, **73**, 828–833.
43. Zhang, X., Dahle, C.E., Weiner, G.J. and Salem, A.K. (2007) A comparative study of the antigen-specific immune response induced by co-delivery of CpG ODN and antigen using fusion molecules or biodegradable microparticles. *J. Pharm. Sci.*, **96**, 3283–3292.
44. Hedley, M.L., Curley, J. and Urban, R. (1996) Microspheres containing plasmid-encoded antigens elicit cytotoxic T-cell responses. *Nat. Med.*, **4**, 365–368.
45. Jilek, S., Zurkaulen, H., Pavlovic, J. *et al.* (2004) Transfection of a mouse dendritic cell line by plasmid DNA-loaded PLGA microparticles *in vitro*. *Eur. J. Pharm. Biopharm.*, **58**, 491–499.
46. Wang, D., Robinson, D.R., Kwon, G.S. and Samuel, J. (1999) Encapsulation of plasmid DNA in biodegradable poly(D,L-lactic-*co*-glycolic acid) microspheres as a novel approach for immunogene delivery. *J. Control. Release*, **57**, 9–18.
47. Pan, C., Nair, N., Adams, R.J. *et al.* (2008) Dose-dependent protection against or exacerbation of disease by a polylactide glycolide microparticle-adsorbed, alphavirus-based measles virus DNA vaccine in rhesus macaques. *Clin. Vaccine Immunol.*, **15**, 697–706.
48. Parsa, S., Wang, Y., Fuller, J. *et al.* (2008) A comparison between polymeric microsphere and bacterial vectors for macrophage P388D1 gene delivery. *Pharm. Res.*, **25**, 1202–1208.
49. Kaiser-Schulz, G., Heit, A., Quintanilla-Martinez, L. *et al.* (2007) Polylactide-*co*-glycolide microspheres Coencapsulating recombinant tandem prion protein with CpG-oligonucleotide break self-tolerance to prion protein in wild-type mice and induce CD4 and CD8 T cell responses. *J. Immunol.*, **179**, 2797–2807.
50. Gonzalez Ferreiro, M., Tillman, L.G., Hardee, G. and Bodmeier, R. (2002) Alginate/poly-L-lysine microparticles for the intestinal delivery of antisense oligonucleotides. *Pharm. Res.*, **19**, 755–764.
51. Aggarwal, N., HogenEsch, H., Guo, P. *et al.* (1999) Biodegradable alginate microspheres as a delivery system for naked DNA. *Can. J. Vet. Res.*, **63**, 148–152.
52. Mittal, S.K., Aggarwal, N., Sailaja, G. *et al.* (2001) Immunization with DNA, adenovirus or both in biodegradable alginate microspheres: effect of route of inoculation on immune response. *Vaccine*, **19**, 253–263.
53. Higashi, T., Nagamori, F., Sone, T. *et al.* (2004) A novel transfection method for mammalian cells using calcium alginate microbeads. *J. Biosci. Bioeng.*, **97**, 191–195.
54. Akbuğa, J., Ozbaş-Turan, S. and Erdoğan, N. (2004) Plasmid-DNA loaded chitosan microspheres for *in vitro* IL-2 expression. *Eur. J. Pharm. Biopharm.*, **58**, 501–507.
55. Dastan, T. and Turan, K. (2004) *In vitro* characterization and delivery of chitosan-DNA microparticles into mammalian cells. *J. Pharm. Pharm. Sci*, **7**, 205–214.

56. Aral, C. and Akbuga, J. (2003) Preparation and *in vitro* transfection efficiency of chitosan microspheres containing plasmid DNA:poly(L-lysine) complexes. *J. Pharm. Pharm. Sci.*, **6**, 321–326.
57. Guliyeva, U., Oner, F., Ozsoy, S. and Haziroglu, R. (2006) Chitosan microparticles containing plasmid DNA as potential oral gene delivery system. *Eur. J. Pharm. Biopharm.*, **62**, 17–25.
58. Ishii, T., Okahata, Y. and Sato, T. (2001) Mechanism of cell transfection with plasmid/chitosan complexes. *Biochim. Biophys. Acta*, **1514**, 51–64.
59. Ozbas-Turan, S., Aral, C., Kabasakal, L. *et al.* (2003) Co-encapsulation of two plasmids in chitosan microspheres as a non-viral gene delivery vehicle. *J. Pharm. Pharm. Sci.*, **6**, 27–32.
60. Zhou, X., Lui, B., Yu, X. *et al.* (2007) Controlled release of PEI/DNA complexes from mannose-bearing chitosan microspheres as a potent delivery system to enhance immune response to HBV DNA vaccine. *J. Control. Release*, **121**, 200–207.
61. Zhou, X., Lui, B., Yu, X. *et al.* (2007) Enhance immune response to DNA vaccine based on a novel multicomponent supramolecular assembly. *Biomaterials*, **28**, 4684–4692.
62. Allison, D.D. and Grande-Allen, K.J. (2006) Hyaluronan: a powerful tissue engineering tool. *Tissue Eng.*, **12**, 1–10.
63. Yun, Y.H., Goetz, D.J., Yellen, P. and Chen, W. (2004) Hyaluronan microspheres for sustained gene delivery and site-specific targeting. *Biomaterials*, **25**, 147–157.
64. Orson, F.M., Kinsey, B.M., Hua, P.J. *et al.* (2000) Genetic immunization with lung-targeting macroaggregated polyethyleneimine-albumin conjugates elicits combined systemic and mucosal immune responses. *J. Immunol.*, **164**, 6313–6321.
65. Li, Z. and Huang, L. (2004) Sustained delivery and expression of plasmid DNA based on biodegradable polyester, poly(D,L-lactide-*co*-4-hydroxy-L-proline). *J. Control. Release*, **98**, 437–446.
66. Evans, R.K., Zhu, D.M., Casimiro, D.R. *et al.* (2004) Characterization and biological evaluation of a microparticle adjuvant formulation for plasmid DNA vaccines. *J. Pharm. Sci.*, **93**, 1924–1939.
67. Wang, C., Ge, Q., Ting, D. *et al.* (2004) Molecularly engineered poly(ortho ester) microspheres for enhanced delivery of DNA vaccines. *Nat. Mater.*, **3**, 190–196.
68. Nguyen, D.N., Raghavan, S.S., Tashima, L.M. *et al.* (2008) Enhancement of poly(ortho ester) microspheres for DNA vaccine delivery by blending with poly(ethylenimine). *Biomaterials*, **29**, 2783–2793.
69. Goh, S.L., Murthy, N., Xu, M. and Fréchet, J.M. (2004) Cross-linked microparticles as carriers for the delivery of plasmid DNA for vaccine development. *Bioconjug. Chem.*, **15**, 467–474.

14

Pulmonary Delivery of Small Interfering RNA for Novel Therapeutics

Qing Ge, David Evans, John J. Xu, Harry H. Yang and Patrick Y. Lu
Sirnaomics, Inc., Gaithersburg, MD, USA

14.1 Introduction

RNA interference (RNAi) is a naturally occurring, highly specific mode of gene regulation that has broad potential applications in both research and therapeutic settings. RNAi has been called 'one of the most exciting discoveries in biology in the last couple of decades'. Since it was first recognized by Andrew Firc and Crag Mello in 1998, it has quickly become one of the most powerful and indispensable tools in the molecular biologist's toolkit [1]. However, the true value of RNAi will be reflected through its therapeutic application. At the heart of its promise as a powerful therapeutic drug lies the exquisite selectivity: siRNA can destroy its target without directly affecting other genes. The clinical applications appear endless: any gene whose overexpression contributes to disease is a potential target, from viral genes to oncogenes to genes responsible for heart disease, Alzheimer's disease, diabetes, and more. Even though RNAi represents a potent and specific inhibitory mechanism using an endogenous pathway of the cell, the stability and delivery of small interfering RNA (siRNA), an agent acting as a intermediate of RNAi, are still the major hurdles to successful RNAi therapeutics. Understanding and overcoming the challenges to pulmonary siRNA delivery hold a key to viable RNAi therapeutics for treatment of viral infections in the airway, lung cancer, asthma, COPD and other respiratory diseases.

Delivery Technologies for Biopharmaceuticals: Peptides, Proteins, Nucleic Acids and Vaccines Edited by Lene Jorgensen and Hanne Mørck Nielsen

14.2 RNA Interference: Mechanism of Action

Though the role of the RNAi pathway was initially hypothesized to be associated with innate cellular immunity, the mechanism is now recognized as a key component in miRNA mediated regulation of genes associated with early developmental decisions and cellular homeostasis. The mechanics of RNAi are both exquisite and highly discriminating. Two steps are thought to be involved in the RNAi response [2]. Firstly, a double-stranded RNA (dsRNA) molecule is recognized and processed into ~21-nucleotide (nt) double-stranded RNAs, termed short interfering RNAs (siRNAs), by the RNase III-like enzyme Dicer [3–5]. siRNAs are then shuttled by Dicer and other proteins into the RNA-induced silencing complex (RISC), a cytoplasmically-localized ribonucleoprotein complex comprised of an Argonaute protein (Ago), as a single-stranded guide RNA [6–10]. The selection of the RNA strand to be incorporated is governed by the thermodynamic profile of the siRNA duplex termini [11, 12].

Secondly, RISC uses this single-stranded RNA molecule as a guide to 'search' the resident population of messenger RNAs (mRNAs) for complementary sequences, eventually endonucleolytically cleaving (and/or attenuating translation of) these transcripts and effectively downregulating the expression of the targeted gene [3, 6, 13–16]. In this step, Ago2 is the sole siRNA-guided Argonaute protein able to act as an RNA-guided, Mg^{2+}-dependent RNA endonuclease that cleaves a single phosphodiester bond in the target RNA, triggering its destruction [17–23].

Synthetic siRNAs or small hairpin RNAs (shRNAs) can be introduced (transfected) into cells by a variety of methods to target virtually any gene. Alternatively, si/shRNAs expressed from plasmid or viral vector constructs can efficiently induce gene knockdown for extended periods (Figure 14.1). Scientists have co-opted the endogenous RNAi machinery to advance a wide range of studies involving gene function analysis, pathway mapping, drug target validation and host–pathogen interactions [13, 24, 25]. In addition to being a successful research tool, siRNA could also be used as a therapeutic agent. With many reports on the *in vivo* application of siRNA and the success of several early phase clinical trials, a broad therapeutic application of RNAi is coming. Respiratory diseases have attracted particular attention as targets of siRNA-mediated therapeutics, due to the lethality and prevalence of certain illnesses and the lung's accessibility to therapeutic agents via both local and systemic delivery routes. However, one of the major challenges to realize the RNAi therapeutic potential in lung diseases is to deliver the siRNAs to the lung tissue, in particular, to the target cells with high efficiency and high specificity. In this chapter, the focus is on the developmental challenges associated with pulmonary delivery of siRNA. The delivery of shRNAs expressed from DNA vectors is summarized elsewhere.

14.3 Lung Physiology

The respiratory tract is divided into two main parts: the upper respiratory tract (the nose, nasal cavity and the pharynx) and the lower respiratory tract (the larynx, trachea, bronchi and lungs). The trachea, which begins at the edge of the larynx, divides into two bronchi. The bronchi divide into smaller bronchioles, which branch in the lungs forming

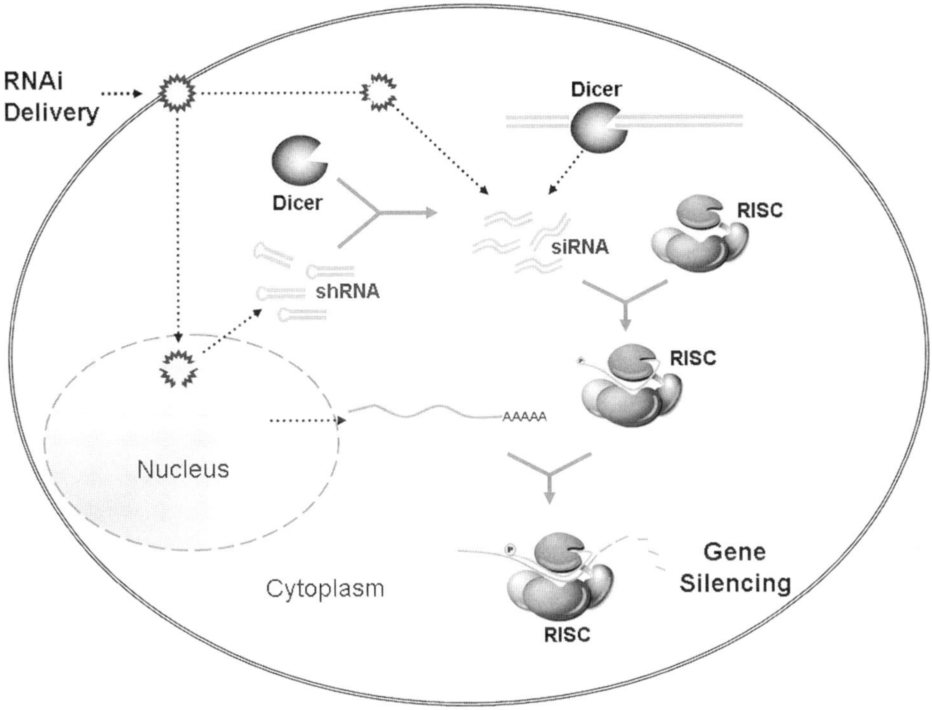

Figure 14.1 RNAi mechanism of action. RNAi therapeutics require efficient RNAi delivery using either direct administration of siRNA molecules or indirect approach of shRNA through a DNA-based expression vector. Both long double-stranded RNA and the shRNA are able to be processed through Dicer into 19–21mer siRNA duplexes with base overhangs. One single-stranded siRNA, either sense or antisense, will be carried by RISC to the targeted single-stranded RNA (mRNA or viral RNA) to facilitate a nuclease-mediated target RNA degradation in a very sequence specific manner. This process results in the target gene expression being silenced

passageways for air. The terminal parts of the bronchi are the alveoli [26]. The pulmonary epithelium is thick (50–60 μm) in the trachea and gradually diminishes to a thickness of 0.2 μm in the alveoli [27, 28]. The alveoli are the functional units and they provide a vast surface area (~100 m^2) and a highly vascularized expanse with access to the entire systemic circulation, a perfect site of gaseous exchange (Figure 14.2). The lungs are equipped with diverse defence mechanisms including surfactant, epithelium, alveolar macrophages (AM) and dendritic cells (DC), the mucociliary escalator and secretory immunolglobulins (sIgs) [29–31]. Surfactant, the underlying watery hypophase, and the epithelial and capillary endothelial cells with the interposed coalesced basal lamina form a thin structural barrier between air and blood [32]. In the process of breathing, some particles are rapidly cleared within the oropharynx, trachea, bronchi and bronchioles by cilia within the mucous layer that lines the epithelia to the throat, where they are swallowed and metabolized.

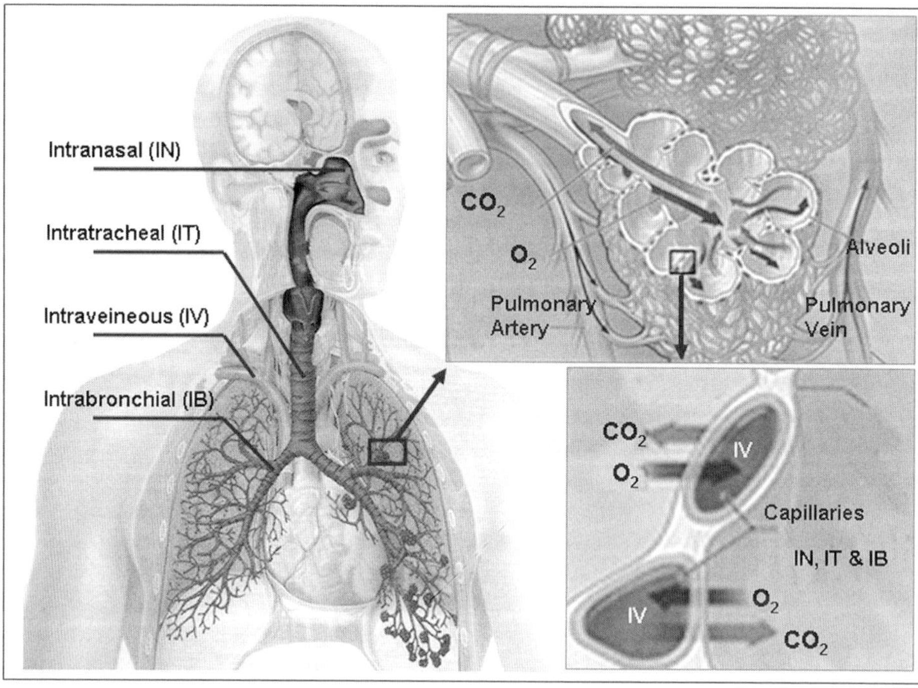

Figure 14.2 *Routes for pulmonary siRNA administration. The siRNA drugs can be administrated into human lung through the airway or blood stream. Airway delivery includes intranasal (IN, upper airway), intratracheal (IT, middle airway) and intrabronchial (IB, lower airway), with aqueous solution or aerosol that will allow siRNA drug to reach various cell types along the entire airway. Intraveneous (i.v., blood stream) delivery will permit siRNA drug to reach the endothelium layer of the lung capillaries on the other side of lung alveoli. Airway delivery is more direct and local, allowing siRNA to reach the epithelium layer of upper, middle and lower respiratory tracts*

Due to its location and physiological function, the respiratory system is susceptible to a number of diseases, and the lungs are prone to a wide range of disorders caused by genetic factors, infection and pollutants in the air. Among them, influenza infection, common cold, asthma, cystic fibrosis and lung cancer are just a few examples. These diseases can be treated via local (intranasal, inhalable delivery) or systemic delivery (oral, intramuscular and intravenous delivery) of therapeutic compounds such as small molecule drugs, antibodies, peptides, DNA and, recently, siRNA [33, 34]. Aerosol therapy is also becoming a popular method to deliver compounds systemically, as shown by the development of inhalable insulin [35].

14.4 Animal Models

Pulmonary delivery studies usually start with tissue cultures of lung epithelial cells. However, the requirements for effective siRNA transfection of the cells in culture are different from those *in vivo*, rendering the predictive power for *in vivo* transfection efficiency questionable. The mouse is a widely accepted model for *in vivo* siRNA studies because of its ready access, ease of handling, statistical validity, availability of various detection methods and, in most cases, availability of appropriate disease models in mice.

However, the interpretation of the results in mice, especially those of delivery efficiency, bioavailability and *in vivo* pharmacokinetics, and the extrapolation of results from the mouse model to a larger animal model or eventually to human studies, needs to take several factors into careful consideration [36–38]. These factors include: (i) anatomical differences between the lungs of different animal models and humans, (ii) physiological differences between species, such as responses of the lung to inflammatory stimuli or pathogens; and (iii) route and method of siRNA administration. For instance, intranasal instillation of siRNAs was used in a number of studies with mouse models [39–44]. However, the delivery results may not be easily extrapolated to human studies because rodents are obligate nose breathers and have a very different lung anatomy.

Some studies used spray instillation (intratracheal instillation or oraltracheal instillation) [45–48] to deliver siRNAs into mouse lungs but the fact that this method avoids oropharyngeal and/or nasopharyngeal deposition could also be a concern when a human study is designed. Other methods such as passive inhalation using exposure chambers may be more clinically relevant in some delivery experiments. An appropriate inhalation device, such as a nebulizer (used in a clinical trial by Alnylam [49]), is also a key factor for efficient delivery and should be considered early to ease the transition from animal to human studies in the development of pulmonary siRNA therapeutics.

14.5 siRNA Delivery to Treat Pulmonary Diseases

14.5.1 Challenges

Like nucleic acid delivery to other organs, the delivery of siRNA to the lung whilst keeping the molecules stable, active and at sufficient concentration in the targeted cells is complicated and involves challenges in every step of the process. Resolving these hurdles may need a physical, chemical or biological approach, and, in some cases, their combination [50]. The extracellular barriers to siRNA delivery depend upon the route of administration. In the case of intravenous (i.v.) delivery, the first hurdle is the size of the siRNA duplex: it is relatively small and thus is rapidly excreted in the urine when administered into the blood stream – even if the siRNA molecules have been chemically modified to improve stability. Secondly, siRNAs are relatively unstable in the serum environment and are rapidly degraded by nucleases within minutes. Thirdly, the non-specific distribution of these siRNA oligos throughout the body and, in particular, the reticuloendothelial system, such as the liver and spleen, will significantly decrease the local concentration in the lung and increase systemic toxicity. Fourthly, the siRNA needs to overcome blockade by endothelial cells in the blood vessels, basal lamina and other tissue barriers to reach the target cells in the bronchi or upper respiratory tract. In the case of delivery to alveoli, only endothelial cells and basal lamina need to be overcome. Finally, when siRNAs reach the target cells, cellular uptake, release from the intracellular compartment to the cytosol and intracellular RNAi activity require efficient endocytosis and endosomal escape of intact siRNA oligos.

Local delivery, including intranasal (IN), intratracheal (IT) and intrabronchial (IB) approaches, through either liquid drops or aerosolation has a number of advantages over systemic delivery. For instance, it is a non-invasive, local and targeted delivery approach that potentially reduces the overall dose and side effects caused by high levels of systemic drug exposure. A rapid onset of action, the avoidance of first-pass metabolism, limited proteolytic activity and possibly limited nuclease activity are among the many benefits of pulmonary delivery [28, 38]. However, both physical and immunologic obstacles in the airway can hinder effective delivery of nucleic acid to epithelial cells or other desired cell types [51]. These include, as mentioned above, removal by beating cilia and mucociliary clearance in the tracheo-bronchial tree, and the interaction of the siRNA/delivery vehicle with the airway surface liquid (ASL, composed of phospholipids, proteins and high molecular weight glycosylated proteins (mucins)) that covers the airway epithelial cells [34, 52]. In patients with cystic fibrosis (characterized by a lung pathology with thick mucus production), airway-directed delivery could be very challenging [53, 54]. Additionally, siRNA cargo targeting epithelial cells may be scavenged by alveolar macrophages and dendritic cells in the airway, not only reducing the dose that reaches its target site but also potentially triggering inflammatory effects that may or may not be beneficial to the treatment of the disease.

Considering the foregoing difficulties, the key to the success of *in vivo* siRNA applications, and delivery to the respiratory tract in particular, is to have delivery vehicles capable of protecting siRNAs from nuclease degradation, efficiently transporting siRNAs through the physical barriers to the target tissue and cell type, and enhancing cellular uptake of the siRNA. For example, to increase the stability in both extracellular and intracellular environments, siRNAs can be chemically modified by a variety of methods, including changing the oligo backbone and replacement of individual nucleotides with nucleotide analogues [55–57]. To increase the delivery efficiency to the desired tissue and cell type, targeting agents such as antibodies and ligands can be conjugated to siRNAs or the delivery vehicles [58]. Cationic components can form complexes with negatively charged siRNAs, protecting the latter from degradation and enhancing cell surface binding and endocytosis. Motifs that have proton sponge or endolytic capabilities can be used to facilitate endosomal escape.

14.5.2 Viral Versus Non-Viral Vectors

Both viral and non-viral vectors are being tested for siRNA delivery to the airway. Genetic material inserted into the vector can encode for si/shRNA that efficiently blocks the production of a specific protein. Retroviral vectors are effective in most cell lines, including primary cells, and provide efficient, uniform delivery and hold promise for long-term silencing of target genes [59, 60]. Adenovirus vectors have been demonstrated to mediate RNAi in a range of animal tissues including an *in vitro* lung model [61, 62]. Though viral based delivery provides a great research tool for gene function analysis and drug target validation, it has a number of safety concerns as a therapeutic approach. Firstly, the immune response to viruses not only impedes gene delivery but also causes complications for the patient. The recent death of Jesse Gelsinger due to severe complications associated with an adenoviral vector highlights this problem [63]. Secondly, some viral vectors, such as lentivirus, may disturb host gene function by inserting its genome at a seemingly random locations in the host chromosome [64]. Thirdly, long-term expression of shRNA from a viral vector may disrupt the processing of endogenous miRNAs [65]. In support of this hypothesis, recent *in vivo* studies that used shRNA viral expression to target hepatitis C virus led to high levels of animal mortality and errors in miRNA processing [66]. Furthermore, such delivery methods have limited ability to regulate the dosing effect *in vivo*. This makes these treatments suitable for chronic rather than acute diseases. Compared to the viral vectors, non-viral delivery usually does not elicit an inflammatory immune response and it is relatively less toxic to the target cells. These advantages enhance the possibility of repeated administrations of siRNA as necessary and make it very practical for siRNA-based therapeutics [67].

Non-viral vectors or delivery carriers can be categorized into lipid- and polycation-based vehicles. Both of them can be positively charged and spontaneously form nanoparticles (complexes) of 50–200 nm in diameter with negatively charged nucleic acids [68, 69]. These nanoparticles protect the integrity of nucleic acid, enhance interactions with the negatively charged cell surface and facilitate the endocytotic process. The fusogenic property of the lipid and the proton sponge effect or endolytic property of the polymer then enhance endosomal release of endocytosed complexes [70–72]. Nucleic acid with or without cationic components can also be protected and delivered by being encapsulated into lipid bilayers or polymer shells to form nano or microparticles [73, 74]. The efficiency of cell-type-specific delivery and cellular uptake can also be greatly enhanced by adding a targeting motif or ligand onto the non-viral carriers. Non-viral carriers for intracellular DNA delivery have been studied for many years. However, effective systems for the delivery of DNA (with a molecular weight of about 1600 kDa) are not always optimal for the delivery of siRNA (about 15 kDa). In addition, DNA needs to enter the nucleus to initiate transcription of the encoded genes, whereas siRNA can function in the cytosol. Thus, achieving effective delivery may be relatively easier with siRNA than with DNA [34].

Many studies have investigated siRNA delivery *in vivo*. For example, in a mouse model, reporter gene silencing and downregulation of TNF-α expression was achieved after intraperitoneal adminstration of an siRNA/lipoplex complex [75]. In another study, a family of highly branched histidine-lysine (HK) polymeric peptides was found to act as an effective carrier of siRNAs [76]. The Arg-Gly-Asp (RGD) peptide ligand is specific to the integrin receptor, a marker of the activated endothelial and tumour cells. Nanoparticles containing this ligand have achieved efficient delivery to tumour xenografts or viral infected eye models through systemic administration [77, 78].

14.5.3 Airway siRNA Delivery

Generally, the negatively charged siRNAs are not thought to be able to bind to the negatively charged cell surface or cross the cell membrane. A delivery carrier is needed to aid cell interaction and uptake of siRNAs. However, siRNAs without transfection reagent or delivery carriers (unformulated) have been successfully delivered into lungs and suppressed the expression of target genes (Table 14.1). In most cases, these siRNAs were also not chemically modified to protect them from nuclease degradation. For example, one of the earliest experiments by Zhang *et al.* [42, 43], performed an IN delivery to introduce naked siRNA specific for haem oxyenase-1 (HO-1) into mouse lungs. The

Table 14.1 *Delivery of siRNAs to the lungs*

siRNA target	Delivery route	Delivery agent	Biological action in animal model	Ref
Haem Oxygenase-1	IN, 2 mg/kg	No transfection reagent	Detected siRNA in the airway and lung parenchyma as early as 4 h and persisting to 16 h. Suppressed mouse HO-1 protein expression in the lung and enhanced reperfusion-induced lung apoptosis in mouse.	[42]
GFP Fas caspase 8	IT, 100 μg	PBS	GFP siRNA reduced lung GFP expression in EGFP transgenic mouse. Fas-siRNA localized primarily in pulmonary epithelial cells 24 h after dosing, reduced Fas mRNA and protected the mouse from hemorrhagic shock and sepsis-induced ALI within 48 h. siRNAs themselves did not induce inflammatory response.	[45]
GFP/MIP-2/KC	IT, 75 μg	Saline	Suppressed target gene expression (~40%), reduced IL-6 and MPO activity after acute lung injury. siRNAs themselves did not induce IL-6.	[48]
SCV	IT for mouse, 30 μg; IN for monkey, 30 mg per dose for repeated dosing	D5W or InfaSurf	siRNA in D5W suppressed reporter DNA expression in mouse lung and suppressed SARS viral titres in the lungs of Rhesus macaque and protected lung from severe acute diffuse alveolar damage. InfaSurf is not as effective as D5W in mouse model.	[79]
Haem Oxygenase-1	IN, 2 mg/kg, repeated dosing	No transfection reagent	Suppressed hyperoxia-induced HO-1 expression in mouse lung; enhanced hyperoxia-induced mouse lung injury and mortality.	[43]
p38 MAPK	IT	PBS	~45% downregulation of p38 MAPK mRNA in mouse lung 6 h after dosing; siRNA conjugated with cell penetrating peptide induced non-specific effect and inflammatory response.	[46]
RSV (ALN RSV01)	IN spray or inhaled aerosol, multiple doses	Saline	In Phase II clinical trial	[49]
RSV para-influenza virus	IN, 5 nmol (~70 μg)	TransIT-TKO	Inhibited mouse lung RSV titres by 99.98% and para-influenza virus titre by 99% when siRNA was complexed with TransIT-TKO; siRNA in Opti-MEM was 70–80% as effective as siRNA complexed with TransIT-TKO. Suppression was not due to siRNA-mediated IFN response.	[39]

Table 14.1 (*continued*)

siRNA target	Delivery route	Delivery agent	Biological action in animal model	Ref
GFP siRNA	IN, 30 μg daily for 5 days	Chitosan	37–43% reduction of lung EGFP expression in bronchiole epithelial cells	[44]
Akt1	Aerosol, nose-only inhalation system	Poly(β-amino ester)	Suppressed lung tumour progression in K-ras and urethane-induced lung cancer mouse model.	[80]
Glyceraldehyde-3-phosphate dehydrogenase	IN, 10 μg	InfaSurf	Suppressed lung GAPDH protein by 50% at 24 h and 67% at 7 days	[41]
Influenza	Hydrodynamic i.v. injection (50 μg) plus IN (20 μg)	Oligofectamine for IN	Protected mice from lethal challenge of avian influenza	[40]
β-galactosidase	IN, 40 and 160 μg	Genzyme Lipid 67	Only visible in alveolar macrophages. Suppressed β-gal mRNA by 30% (insufficient to reduce protein expression) in the airway epithelium of K18-lacZ mice.	[54]
IL-13 receptor α2	IT, 100 ug	Encapsulated in HVJ-envelope vector, repeated dosing	Suppressed mouse lung IL-13Rα2 expression and decreased TGF-β1 secretion and collagen formation.	[47]
Luciferase Influenza	i.v., 60 to 120 μg	jetPEI	17-fold decrease of the expression of the Luc gene delivered IT in complex with jetPEI.; 10–100-fold reduction of virus titre in the mousc lung	[81]
Luciferase/ Influenza	i.v., 60 to 120 μg,	Deacylated PEI25 and PEI87	77–93% suppression of Luc expression in mouse lung 24 h post injection; 94% suppression of influenza virus titre in mouse lung 24 h post infection	[82]
Caveolin-1	i.v., 1.0–1.3 mg/kg	Cationic lipoplex (DDAB & cholesterol)	Decreased Caveolin-1 expression in lung endothelia by 90%, increased vascular permeability.	[83]
Mouse tissue factor	IP, 50 or 100 μg before and after i.v. injection of tumour cells	Lipofectamine 2000	Reduce the number of metastases in lung tissue by fivefold in mice.	[84]
siGLO Natriuretic peptide receptor	Transdermal, 2–5 nmol, repeated dosing	Imiquimod cream mixed with siRNA/ chitosan	siGLO fluorescence shown in the lung area; siNPRA reduced pulmonary inflammation in mouse asthma model.	[85]

siRNA could be detected in the airway and lung parenchyma as early as four hours and persisting to 16 hours post-delivery. After siRNA treatment, the target gene (HO-1) expression in the lung was suppressed and cell apoptosis was increased in lung injury mouse models [42, 43]. siRNAs in phosphate buffered saline (PBS) were delivered (IT) into pulmonary epithelial cells and reduced target gene expression [45].

This success of airway delivery of unformulated siRNAs was quickly extended from lung injury models to viral infection models; from mouse to monkey studies [79] and even to a human clinical trial [49]. Specifically, the effective inhibition of severe acute respiratory syndromes (SARS) coronavirus RNA was achieved in monkey lungs via IN delivery of siRNA in D5W (5% D-glucose in water). In clinical applications, Alnylam Pharmaceuticals is testing its anti-respiratory syncytial virus (RSV) siRNA ALN-RSV01 to treat RSV infection. Both nasal spray and PARI's eFlow nebulizer were used as methods to deliver unformulated siRNA into human lungs. The Phase I trials have shown that ALN-RSV01 is safe and well tolerated in healthy volunteers. Furthermore, a recent Phase II GEMINI study demonstrated a 38% relative reduction in RSV infection rate and a 95% increase in the number of infection-free subjects as compared with placebo.

A variety of polymers (TransIT-TKO, chitosan, poly(β-amino ester)) and lipids (oligofectamine, Genzyme Lipid 67) as well as a non-viral vector derived from hemagglutinating virus of Japan (HVJ) have been reported to deliver siRNAs to mouse lungs through the airway (Table 14.1) [39, 40, 44, 47, 54, 80]. For example, siRNAs targeting the viral P gene for both RSV and parainfluenza virus (PIV) were administered intranasally either naked or complexed with the cationic lipid TransIT-TKO into mice. The viral titres of both viruses in mouse lungs were reduced to greater than 99% and lung pathology was also decreased [39]. Furthermore, Bitko *et al.* also compared the effectiveness of siRNAs delivered with and without the carrier and demonstrated that siRNAs without the carrier were 70–80% as effective as those with the carrier. However, detailed reports or analyses on the cell types that received siRNAs and *in vivo* pharmacokinetics of delivered siRNAs are still scarce and more research in this area will facilitate the development of highly efficient airway delivery methods. More studies are also needed to understand how siRNA, without any help from a delivery carrier, could penetrate the multiple barriers in the airway, reach lung epithelial cells, enter the cytosol and remain intact to perform its function. Nevertheless, the success in monkey and human studies for the airway treatments of lung diseases seems very promising.

14.5.4 Intravenous siRNA Delivery

Compared to airway delivery, unformulated and unmodified siRNAs exhibit extremely short half-lives when administered via intravenous injection due to rapid degradation by serum nucleases. So far, a few delivery vehicles have been examined for i.v. delivery of siRNA to mouse lungs, including cationic lipid (dimethyldioctadecylammonium bromide, DDAB) and polymer (poly(ethylene imine), PEI) [81–83]. The study using DDAB showed a decrease of target gene expression in lung endothelial cells. The cationic polymer PEI used in the latter studies facilitated the *in vivo* transfection of influenza-specific siRNA achieving marked suppression of the viral replication in the mouse model. If the suppression of the virus was due to the delivery of siRNA into lung epithelial cells where the virus replicates, it is not clear how siRNA/PEI complexes were transported through the capillary endothelial layer, basal lamina and eventually entered the epithelial layer in the alveoli. Recently, these flu-specific siRNAs were found to induce type I interferon (IFN) in human peripheral blood mononuclear cells (PBMCs) and in mouse lungs [86]. This inflammatory response was found to be responsible for suppressing influenza virus growth *in vivo*. Although chemically modified siRNAs (removing their IFN-induction capability) against influenza were found to inhibit virus in cell culture, they were not effective *in vivo*. This suggests that PEI-mediated intravenous delivery either cannot drive siRNAs to lung epithelial cells or that any material delivered is insufficient for therapeutic benefit.

For pulmonary illnesses, intravenous administration of therapeutic compounds remains a valid route for patients who have obstructive airway diseases or people who are not able to follow instructions for aerosolized medicines. Therefore, additional experiments examining new formulations will be needed to improve the effectiveness of cell-type-specific airway delivery of siRNA via an intravenous route.

14.5.5 Other Delivery Methods

Intraperitoneal injection of tissue factor-specific siRNA in a complex with Lipofectamine 2000 was shown to reduce the number of tumour metastases in mouse lung [84]. This is probably due to the leakiness of tumour vasculature. Interestingly, siRNAs could be detected in mouse lungs after transdermal delivery of siRNA/chitosan nanoparticles mixed in a topical cream containing imiquimod [85]. Using this method, siRNAs against natriuretic peptide receptor had an anti-inflammatory effect in an asthma mouse model. Although it is not clear whether siRNAs exerted their function in the cells in the airway or in the blood stream, this transdermal administration remains another non-invasive option for the treatment of airway diseases.

14.6 siRNA-Induced Inflammatory Response

Various results have demonstrated the capability of shRNAs and siRNAs to activate mammalian immune responses [87–94]. Toll-like receptors (TLR3, TLR7, and TLR8), protein kinase R (PKR), the cytosolic RNA helicases retinoic acid-inducible gene (RIG)-I and melanoma differentiation-associated gene 5 (MDA-5) are all involved in the recognition and activation of innate immune responses [95]. Several features of RNA, including length, sequence and structure could account for the recognition by these receptors. The motif analysis defined specific GU-rich sequences such as 5′-UGUGU-3′, 5′-UUGU-3′ and a special 5′-UCA-3′ that can activate human TLR7/8 and induce IFN-α, and specific AU-rich sequences such as 5′-UAUU-3′ that can activate TLR8 and induce TNF-α1 [87, 88, 96]. In addition, the expression of these receptors that recognize different aspects of siRNAs are cell-type-dependent, rendering conflicting results on the immune responsiveness of siRNAs from groups examining different cell or tissue types [95, 96]. For instance, TLR7 is primarily expressed in plasmacytoid dendritic cells (pDC) and B-cells whereas TLR8 is highly expressed in monocytes, macrophages, and myeloid DC (mDC) [97, 98]. Therefore, the choice of the right cell types, such as purified human peripheral blood mononuclear cells (PBMCs), will be essential for the detection of siRNA-induced proinflammatory cytokines. Although it may be thought that PBMCs can be avoided if the airway, rather than the intravenous route, is used for siRNA delivery, undesired inflammatory responses may occur upon stimulation of AM and DC residing locally in the lungs. Additional contributing factors for the observed differential immune responses include the selection of the time used to detect the inflammatory cytokine effect, since the cytokine response to some siRNA formulations *in vivo* is transient, peaking between two and eight hours and fully resolving within 24 hours after administration [86]. Thus, a careful evaluation of proinflammatory cytokine induction *in vitro* as well as *in vivo* is very important in target validation as well as in the early phase of the development of siRNA therapeutics.

14.7 Case Studies

14.7.1 siRNA Therapeutics Against RSV and PIV Infection in Mouse

Respiratory syncytial virus (RSV) and parainfluenza virus (PIV) are two respiratory pathogens of paramount medical significance that exert high mortality. Using the RNAi approach, Barik *et al.* [39] showed that individual as well as joint infection by RSV and PIV can be specifically prevented and inhibited by siRNAs instilled intranasally in mice, with or without transfection reagents. The degree of protection matched the antiviral activity of the siRNA in cell culture, allowing an avenue for quick screening of an efficacious siRNA. The results suggest that inhaled siRNA might offer a fast, potent and easily administrable antiviral regimen against respiratory viral diseases in humans. Starting from this preliminary result, an siRNA-based therapeutic protocol for RSV treatment was developed and evaluated through a series of clinical studies [49].

A web-based siRNA design tool was used to design siRNA against RSV and new siRNAs against PIV mRNA. The siRNAs were synthesized chemically and their IC_{50} (concentration of siRNA producing 50% reduction of target) *ex vivo* was determined. They detected the antisense strand of the siRNA in the lung tissue by northern analyses and the possibility that the antiviral effect of siRNAs was the result of activation of interferons was ruled out. Paramyxoviruses in general use diverse mechanisms to counteract IFNs; RSV, in particular, is largely resistant to type I IFNs (IFN-α and IFN-β) although sensitive to type II IFN (IFN-γ). Direct measurement of IFN-α and IFN-γ in mouse lung following treatment with various siRNAs showed no activation of either type of IFN.

The mice that received specific siRNA (with or without transfection reagents) before being exposed to RSV behaved like uninfected mice upon visual inspection. The respiratory rate, leukotriene induction and pulmonary inflammation were measured. The respiratory rate of BALB/c mice is known to increase in response to RSV infection. Leukotrienes, products of the lipoxygenase pathway, bind to the leukotriene receptors present in bronchial smooth muscle and are elevated in the respiratory secretions of asthmatic patients and RSV-infected infants and mice. These compounds provoke airway mucus secretion, bronchoconstriction and infiltration by inflammatory cells – hallmarks of severe RSV disease. When siRNA#1 was administered prior to – or concomitant with – RSV, considerable reductions in all these parameters were observed. The values were essentially comparable to those in sham-infected mice at least up to 14 days post infection, showing that the siRNA prevented illness and did not simply postpone it. In fact, siRNA-treated mice showed no visible signs of respiratory distress after up to six weeks of observation Figure 14.3a–d). Negative control RNA or luciferase specific siRNA offered no relief in all experiments (data not shown). Additional data showed that intranasal-delivered siRNAs were also effective if given after infection.

14.7.2 siRNA Therapeutics for Treatment of SCV Infection in Monkey

The emergence of severe acute respiratory syndromes (SARS) has prompted research in diagnosis, pathogenesis and therapeutics for fast response to such emerging pathogens. For instance, the identification of SARS Corona Virus (SCV) as the causative pathogen of SARS was critical for containment and patient management. This was achieved mainly by the demonstration that exposure of cynomolgus macaques to SCV resulted in similar symptoms to those of SARS patients, while also creating the first animal disease model critical not only for understanding SARS pathogenesis but also for evaluation of potential vaccines and novel therapeutics [79, 99, 100]. Demonstration of protection by PEGylated interferon-α from SCV infection in cynomolgus macaques, as seen in SARS patients, gives further proof of effectiveness of such a model [101]. The establishment of a rhesus macaque model for SARS with intranasal instillation of SCV strain PUMC01, also showed many elements of pathology similar to those of SARS patients and the cynomologus macaque model. The pathogeneses include elevated body temperature, low appetite and acute diffuse alveolar damage (DAD) clearly visible at seven days post infection (d.p.i.), but with somewhat worse severity. Thus, it offers an ideal model for evaluation of therapeutic candidates specific for SCV as potential treatments for SARS.

Due to their targeted sequences showing a 100% homology to more than 100 strains of published SCV and a synergistic anti-SCV activity *in vitro*, two siRNA duplexes, siSC2 and siSC5, targeting the SCV genome at Spike protein – coding and ORF lb (NSP12) regions, respectively – were used for *in vivo* studies [79]. siRNA were intranasally instilled into the rhesus macaques and no delivery carriers or transfection reagents were used except D5W, a commonly used solvent in the clinic. All the macaques from both viral infection and non-specific siRNA control groups suffered from severe acute DAD with a damage score of '+++', whereas lung tissues from all three siSC2-5-treated groups (prophylactic-PL, co-delivery-CD and post exposure-PE groups) showed relatively mild severity of acute DAD while none showed lung damage beyond a score of '++'.

The body temperature of infected macaques is an important indicator not only for the severity of SARS-like symptoms but also the effectiveness of the treatment. The body temperature of each animal was recorded daily until they were sacrificed for necropsy at either seven or 20 d.p.i. The mean body temperatures of the five groups were compared (Figure 14.4a–d), showing that the viral infection (IC) and non-specific siRNA control (NS) groups both exhibited higher temperatures (above 39.1 °C), indicating a fever status. In contrast, all siSC2-5 treated groups exhibited significantly lower mean body temperatures of below 38.9 °C. Among them, the mean body temperature of the PL group was the lowest (38.7 °C), close to the normal body temperature of rhesus monkey (38.5 °C). Further analysis of average group body temperatures with trend regressions at each time point revealed that the IC and NS groups had elevated temperatures consistently above 39 °C. On the other hand, macaques in the PL and CD groups exhibited a lower average body temperature around 38.7 °C and most exhibited only transient fevers between four

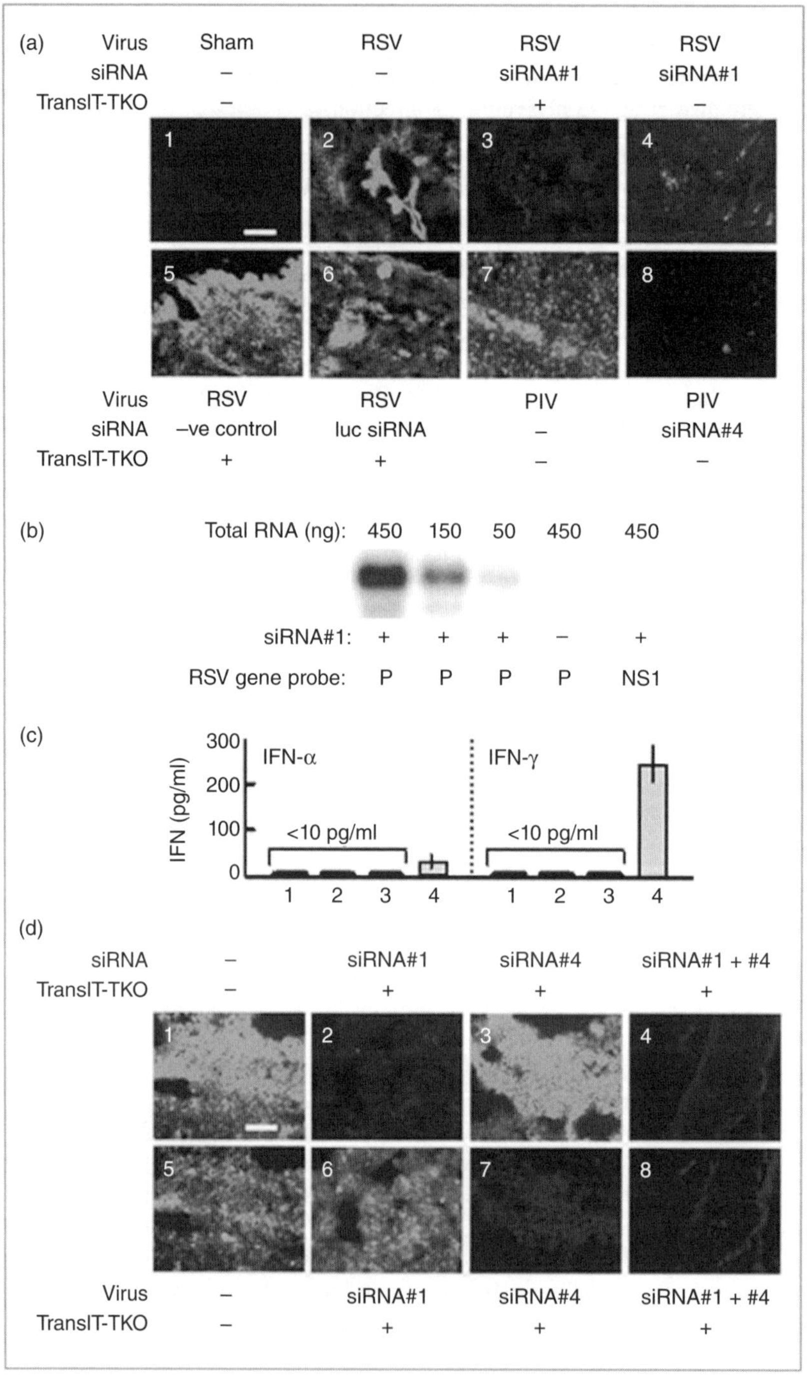

(a)
Virus
siRNA
TransIT-TKO
Sham
–
–
RSV
–
–
RSV
siRNA#1
+
RSV
siRNA#1
–
1
2
3
4
5
6
7
8
Virus
siRNA
TransIT-TKO
RSV
–ve control
+
RSV
luc siRNA
+
PIV
–
–
PIV
siRNA#4
–
(b)
Total RNA (ng): 450 150 50 450 450
siRNA#1: + + + – +
RSV gene probe: P P P P NS1
(c)
IFN (pg/ml)
300
200
100
0
IFN-α
IFN-γ
<10 pg/ml
<10 pg/ml
1 2 3 4
1 2 3 4
(d)
siRNA
TransIT-TKO
–
–
siRNA#1
+
siRNA#4
+
siRNA#1 + #4
+
1
2
3
4
5
6
7
8
Virus
TransIT-TKO
–
–
siRNA#1
+
siRNA#4
+
siRNA#1 + #4
+

Figure 14.3 *siRNA therapeutics against RSV-PIV infection in mouse lung. (a) Virus was administered four hours after siRNA, and viral antigens were detected by indirect immunohistology of lungs at four days after infection (green, RSV; red, PIV). Bar, 400 μm. (b) Antisense strand of siRNA#1 was detected by northern analysis of varying amounts of total lung RNA two days after siRNA administration using labelled RSV P DNA as probe. (c) Intranasal siRNA (10 nmol or 140 jig per mouse) did not activate pulmonary IFN-a or IFN-y above the threshold of detection (–10 pg/ml), whereas in control lungs, RSV infection activated type II and low levels of type I. Lanes: 1, siRNA#1; 2, siRNA#4; 3, Luc siRNA; 4, no siRNA but RSV-infected; error bars indicate s.e.m. (n = 4 for each graph). (d) siRNA-mediated inhibition of dual infection by RSV and PIV determined by indirect immunohistology. Virus was administered four hours after siRNA, and lung tissues were examined at days after infection*

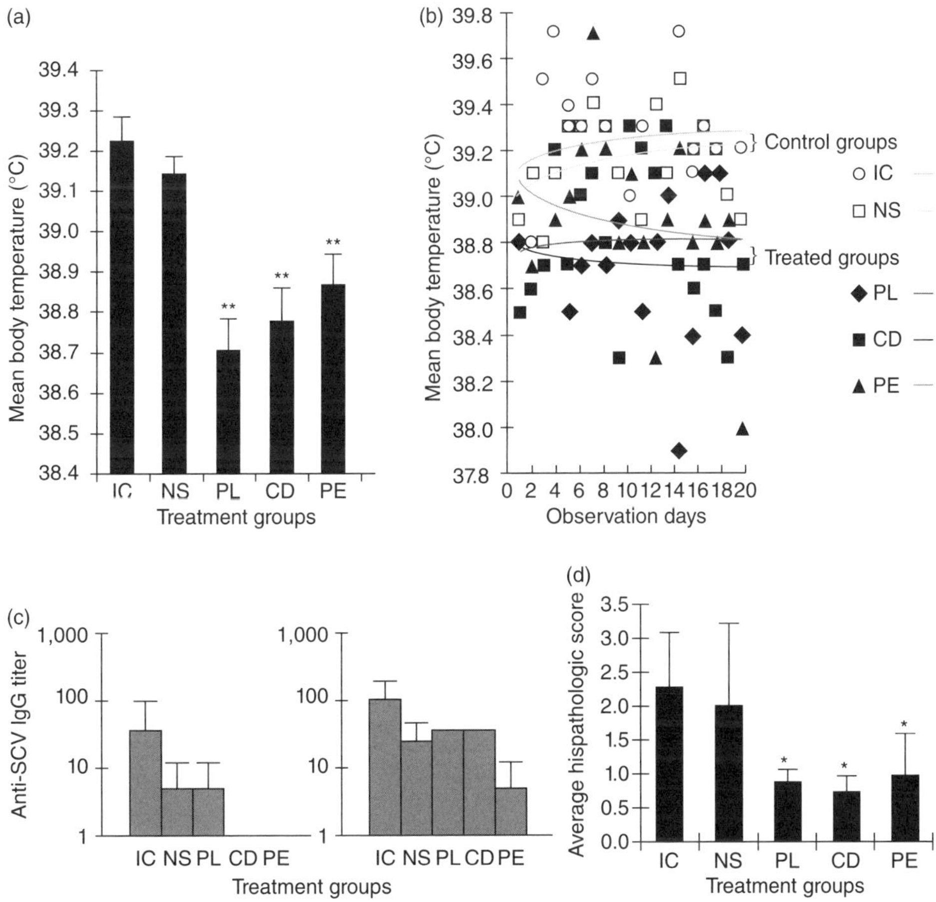

Figure 14.4 *siRNA therapeutics relieved SARS symptoms in monkey model. (a) Mean body temperature. The treated groups showed the significant lower body temperatures. (b) Distribution of the average body temperatures throughout the 20 day period. The treated groups showed the lower body temperatures and the post exposure group showed a lowing trend throughout the 20 days. (c) IgG titres of SCV-specific antibody in serum samples. (d) The average histopathological scores of each group, including lung samples collected at both seven and 20 d.p.i, were compared against the viral infection control group.*
$^*P < 0.05$, n = 4. Reprinted with permission from Macmillan Publishers, copyright 2005.

and 10 d.p.i. The average body temperature of the PE group was around 39 °C in at early time points but quickly lowered after 10 d.p.i.

The oropharyngeal swab samples from all animals at four d.p.i. were analysed repeatedly with quantitative RT-PCR. The results revealed that SCV viral RNA was detected from all samples of the IC and NS groups, but failed to show positive results in 75% of samples from all three siSC2-5 treatment groups, PL, CD and PE [79]. This indicated a possible inhibition of SCV infection and replication via siSC2-5-mediated viral RNA degradation in upper airway epithelial cells. To evaluate the SCV infection in deeper lung tissues, immunohistochemical (IHC) was used and each lung section was stained with SCV antigen specific monoclonal antibody. The SCV-infected cells include the type I pneumocytes, type II pneumocytes and macrophages, consistent with the observation in the previous studies with macaque models and SARS patients. The infected cells in lung sections were further counted and compared. The samples from IC and NS groups exhibited much higher counts of infected cells than those from siSC2-5 treated groups. Collectively, the intranasal administration of siSC2-5 in a more clinically acceptable form, such as a D5W solution, achieved a strong anti-SARS activity without any visible adverse effects, suggesting that the clinical testing of siSC2-5 duplexes as a SARS treatment is warranted.

14.8 Faster Drug for Unknown Bugs

After the success of applying siRNA therapeutic approaches to fight SCV, infection in the non-human primate model was published [79]. Alex Flemming from *Nature Review, Drug Discovery* [102] made an interesting commentary that 'there is a pressing need for *new* effective treatments that can be rapidly developed in response to previously unknown viruses. Reporting in *Nature Medicine* [79], Li *et al.* took a step towards this goal by demonstrating that intranasal delivery of SCV-specific siRNA was effective against SCV in a rhesus macaque model – providing hope that siRNA could provide the means to *move* rapidly from gene sequence to targeted therapeutics for many previously intractable diseases.' Specifically, this approach is very powerful in combating respiratory viral infection, such as SARS, influenza (avian flu H5N1 in particular) and other emerging viral pathogens. As shown in Figure 14.5, siRNA therapeutics allow us to quickly identify, select and validate viral specific inhibitors in cell culture and animal model studies in a highly specific and time saving manner. The readiness of the large scale synthesis of siRNA oligos, optimized formulation and clinically viable administration will facilitate this siRNA-based antiviral drug to be used in both prophylactic and therapeutic treatments of emerging viral infections, including those that have pandemic concerns. With the delivery hurdle overcome, a rapid response using siRNA inhibitors could be widely expanded and applied to many other emerging infectious diseases beyond the pulmonary system.

14.9 Conclusions

RNAi has become one of the essential approaches in the mechanistic study and therapeutic development against respiratory illnesses. The advantages of siRNA application are

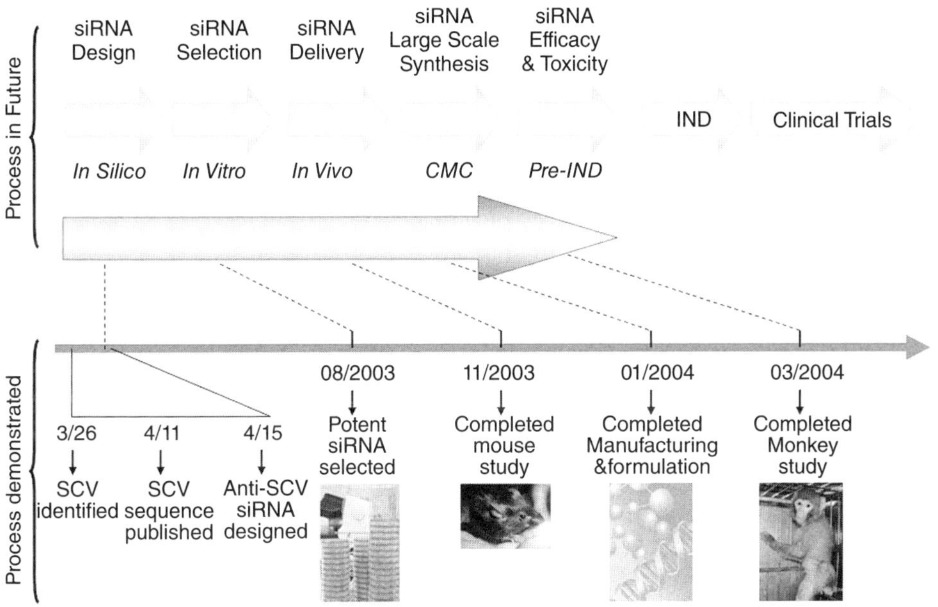

Figure 14.5 *Rapid response to potential emerging viral infectious disease. A rapid response to the SARS pandemic with siRNA drug development demonstrated a novel approach to combat the emerging viral infectious diseases. The twelve month period starting from the viral pathogen identification to completion of the non-human primate study of the new drug entity (03/2003–03/2004) represents an unprecedented speed in development of any type of anti-infectious drug. With further improvement of* in silico *drug design,* in vitro *cell culture-based siRNA drug selection,* in vivo *delivery methods, large scale manufacturing and clinical viable formulation, an even faster response will be expected.*

reflected through its highly specific inhibitory effect and its rapidity in response to previously unknown disease causing genes of emerging pathogens. This significantly shortens the time required for drug target validation and therapeutic development. The knowledge gained from the delivery of DNA and antisense oligonucleotides in the past and the recent progress made in the delivery of siRNAs into the lung will accelerate the novel drug discovery process using siRNA as an active pharmaceutical ingredient (API). Although challenges remain, the effort of developing and optimizing siRNA delivery will ultimately be translated into clinically viable administration of siRNA-based therapeutics to treat respiratory diseases.

References

1. Novina, C.D. and Sharp, P.A. (2004) The RNAi revolution. *Nature*, **430**, 161–164.
2. Filipowicz, W. (2005) RNAi: the nuts and bolts of the RISC machine. *Cell*, **122**, 17–20.

3. Hammond, S.M., Bernstein, E., Beach, D. and Hannon, G.J. (2000) An RNA-directed nuclease mediates post-transcriptional gene silencing in Drosophila cells. *Nature*, **404**, 293–296.
4. Zamore, P.D., Tuschl, T., Sharp, P.A. and Bartel, D.P. (2000) RNAi: double-stranded RNA directs the ATP-dependent cleavage of mRNA at 21 to 23 nucleotide intervals. *Cell*, **101**, 25–33.
5. Bernstein, E., Caudy, A.A., Hammond, S.M. and Hannon, G.J. (2001) Role for a bidentate ribonuclease in the initiation step of RNA interference. *Nature*, **409**, 363–366.
6. Elbashir, S.M., Lendeckel, W. and Tuschl, T. (2001) RNA interference is mediated by 21- and 22-nucleotide RNAs. *Genes Dev.*, **15**, 188–200.
7. Lee, Y.S., Nakahara, K., Pham, J.W. *et al.* (2004) Distinct roles for Drosophila Dicer-1 and Dicer-2 in the siRNA/miRNA silencing pathways. *Cell*, **117**, 69–81.
8. Pham, J.W., Pellino, J.L., Lee, Y.S. *et al.* (2004) Dicer-2-dependent 80s complex cleaves targeted mRNAs during RNAi in Drosophila. *Cell*, **117**, 83–94.
9. Tomari, Y., Matranga, C., Haley, B. *et al.* (2004) A protein sensor for siRNA asymmetry. *Science*, **306**, 1377–1380.
10. Tomari, Y. *et al.* (2004) RISC assembly defects in the Drosophila RNAi mutant armitage. *Cell*, **116**, 831–841.
11. Schwarz, D.S., Tomari, Y. and Zamore, P.D. (2004) The RNA-induced silencing complex is a Mg^{2+}-dependent endonuclease. *Curr. Biol.*, **14**, 787–791.
12. Khvorova, A., Reynolds, A. and Jayasena, S.D. (2003) Functional siRNAs and miRNAs exhibit strand bias. *Cell*, **115**, 209–216.
13. Dorsett, Y. and Tuschl, T. (2004) siRNAs: applications in functional genomics and potential as therapeutics. *Nat. Rev. Drug Discov.*, **3**, 318–329.
14. Kim, D. and Rossi, J. (2008) RNAi mechanisms and applications. *Biotechniques*, **44**, 613–616.
15. Nykanen, A., Haley, B. and Zamore, P.D. (2001) ATP requirements and small interfering RNA structure in the RNA interference pathway. *Cell*, **107**, 309–321.
16. Martinez, J., Patkaniowska, A., Urlaub, H. *et al.* (2002) Single-stranded antisense siRNAs guide target RNA cleavage in RNAi. *Cell*, **110**, 563–574.
17. Hammond, S.M., Boettcher, S., Caudy, A.A. *et al.* (2001) Argonaute2, a link between genetic and biochemical analyses of RNAi. *Science*, **293**, 1146–1150.
18. Okamura, K., Ishizuka, A., Siomi, H. and Siomi, M.C. (2004) Distinct roles for Argonaute proteins in small RNA-directed RNA cleavage pathways. *Genes Dev.*, **18**, 1655–1666.
19. Meister, G. *et al.* (2004) Human Argonaute2 mediates RNA cleavage targeted by miRNAs and siRNAs. *Mol. Cell.*, **15**, 185–197.
20. Rand, T.A., Ginalski, K., Grishin, N.V. and Wang, X. (2004) Biochemical identification of Argonaute 2 as the sole protein required for RNA-induced silencing complex activity. *Proc. Natl. Acad. Sci. USA*, **101**, 14385–14389.
21. Liu, J. Carmell, M.A., Rivas, F.V. *et al.* (2004) Argonaute2 is the catalytic engine of mammalian RNAi. *Science*, **305**, 1437–1441.
22. Song, J.J., Smith, S.K., Hannon, G.J. and Joshua-Tor, L. (2004) Crystal structure of Argonaute and its implications for RISC slicer activity. *Science*, **305**, 1434–1437.
23. Rivas, F.V. *et al.* (2005) Purified Argonaute2 and an siRNA form recombinant human RISC. *Nat. Struct. Mol. Biol.*, **12**, 340–349.
24. Natt, F. (2007) siRNAs in drug discovery: target validation and beyond. *Curr. Opin. Mol. Ther.*, **9**, 242–247.
25. Iorns, E., Lord, C.J., Turner, N. and Ashworth, A. (2007) Utilizing RNA interference to enhance cancer drug discovery. *Nat. Rev. Drug Discov.*, **6**, 556–568.
26. Smola, M., Vandamme, T. and Sokolowski, A. (2008) Nanocarriers as pulmonary drug delivery systems to treat and to diagnose respiratory and non respiratory diseases. *Int. J. Nanomedicine*, **3**, 1–19.
27. Weibel, E.R. (1973) Morphological basis of alveolar-capillary gas exchange. *Physiol. Rev.*, **53**, 419–495.
28. Sung, J.C., Pulliam, B.L. and Edwards, D.A. (2007) Nanoparticles for drug delivery to the lungs. *Trends Biotechnol.*, **25**, 563–570.

29. Nicod, L.P. (1999) Pulmonary defence mechanisms. *Respiration*, **66**, 2–11.
30. de Heer, H.J., Hammad, H., Kool, M. and Lambrecht, B.N. (2005) Dendritic cell subsets and immune regulation in the lung. *Semin. Immunol.*, **17**, 295–303.
31. Fels, A.O. and Cohn, Z.A. (1986) The alveolar macrophage. *J. Appl. Physiol.*, **60**, 353–369.
32. Muhlfeld, C. *et al.* (2008) Interactions of nanoparticles with pulmonary structures and cellular responses. *Am. J. Physiol. Lung Cell. Mol. Physiol.*, **294**, L817-L829.
33. Azarmi, S., Roa, W.H. and Lobenberg, R. (2008) Targeted delivery of nanoparticles for the treatment of lung diseases. *Adv. Drug Deliv. Rev.*, **60**, 863–875.
34. Thomas, M., Lu, J.J., Chen, J. and Klibanov, A.M. (2007) Non-viral siRNA delivery to the lung. *Adv. Drug Deliv. Rev.*, **59**, 124–133.
35. Quattrin, T. (2004) Inhaled insulin: recent advances in the therapy of Type 1 and 2 diabetes. *Expert Opin. Pharmacother.*, **5**, 2597–2604.
36. Cryan, S.A., Sivadas, N. and Garcia-Contreras, L. (2007) *In vivo* animal models for drug delivery across the lung mucosal barrier. *Adv. Drug Deliv. Rev.*, **59**, 1133–1151.
37. Sakagami, M. (2006) *In vivo, in vitro* and *ex vivo* models to assess pulmonary absorption and disposition of inhaled therapeutics for systemic delivery. *Adv. Drug Deliv. Rev.*, **58**, 1030–1060.
38. Durcan, N., Murphy, C. and Cryan, S.A. (2008) Inhalable siRNA: potential as a therapeutic agent in the lungs. *Mol. Pharm.*, **5**, 559–566.
39. Bitko, V., Musiyenko, A., Shulyayeva, O. and Barik, S. (2005) Inhibition of respiratory viruses by nasally administered siRNA. *Nat. Med.*, **11**, 50–55.
40. Tompkins, S.M., Lo, C.Y., Tumpey, T.M. and Epstein, S.L. (2004) Protection against lethal influenza virus challenge by RNA interference *in vivo*. *Proc. Natl. Acad Sci. USA*, **101**, 8682–8686.
41. Massaro, D., Massaro, G.D. and Clerch, L.B. (2004) Noninvasive delivery of small inhibitory RNA and other reagents to pulmonary alveoli in mice. *Am. J. Physiol. Lung Cell. Mol. Physiol.*, **287**, L1066–L1070.
42. Zhang, X. *et al.* (2004) Small interfering RNA targeting heme oxygenase-1 enhances ischemia-reperfusion-induced lung apoptosis. *J. Biol. Chem.*, **279**, 10677–10684.
43. Zhang, X. *et al.* (2006) Endothelial STAT3 is essential for the protective effects of HO-1 in oxidant-induced lung injury. *FASEB J.*, **20**, 2156–2158.
44. Howard, K.A. *et al.* (2006) RNA interference *in vitro* and *in vivo* using a novel chitosan/siRNA nanoparticle system. *Mol. Ther.*, **14**, 476–484.
45. Perl, M. *et al.* (2005) Silencing of Fas, but not caspase-8, in lung epithelial cells ameliorates pulmonary apoptosis, inflammation, and neutrophil influx after hemorrhagic shock and sepsis. *Am. J. Pathol.*, **167**, 1545–1559.
46. Moschos, S.A., Williams, A.E. and Lindsay, M.A. (2007) Cell-penetrating-peptide-mediated siRNA lung delivery. *Biochem. Soc. Trans.*, **35**, 807–810.
47. Fichtner-Feigl, S., Strober, W., Kawakami, K. *et al.* (2006) IL-13 signaling through the IL-13alpha2 receptor is involved in induction of TGF-beta1 production and fibrosis. *Nat. Med.*, **12**, 99–106.
48. Lomas-Neira, J.L., Chung, C.S., Wesche, D.E. *et al.* (2005) *In vivo* gene silencing (with siRNA) of pulmonary expression of MIP-2 versus KC results in divergent effects on hemorrhage-induced, neutrophil-mediated septic acute lung injury. *J. Leukoc. Biol.*, **77**, 846–853.
49. http://www.alnylam.com/Programs-and-Pipeline/Programs/index.php, accessed 2009-07-24
50. Lu, P.Y., Xie, F. and Woodle, M.C. (2005) *In vivo* application of RNA interference: from functional genomics to therapeutics. *Adv. Genet.*, **54**, 117–142.
51. Weiss, D.J. (2004) Delivery of DNA to lung airway epithelium. *Methods Mol. Biol.*, **246**, 53–68.
52. Rosenecker, J. *et al.* (2003) Interaction of bronchoalveolar lavage fluid with polyplexes and lipoplexes: analysing the role of proteins and glycoproteins. *J. Gene. Med.*, **5**, 49–60.
53. Bragonzi, A. *et al.* (1999) Comparison between cationic polymers and lipids in mediating systemic gene delivery to the lungs. *Gene Ther.*, **6**, 1995–2004.
54. Griesenbach, U. *et al.* (2006) Inefficient cationic lipid-mediated siRNA and antisense oligonucleotide transfer to airway epithelial cells *in vivo*. *Respir. Res.*, **7**, 26.

55. Behlke, M.A. (2008) Chemical modification of siRNAs for *in vivo* use. *Oligonucleotides*, **18**(4), 305–319.
56. Corey, D.R. (2007) Chemical modification: the key to clinical application of RNA interference? *J. Clin. Invest.*, **117**, 3615–3622.
57. Watts, J.K., Deleavey, G.F. and Damha, M.J. (2008) Chemically modified siRNA: tools and applications. *Drug Discov. Today*, **13**, 842–855.
58. Soutschek, J. *et al.* (2004) Therapeutic silencing of an endogenous gene by systemic administration of modified siRNAs. *Nature*, **432**, 173–178.
59. Zentilin, L. and Giacca, M. (2004) *In vivo* transfer and expression of genes coding for short interfering RNAs. *Curr. Pharm. Biotechnol.*, **5**, 341–347.
60. Li, M. and Rossi, J.J. (2008) Lentiviral vector delivery of siRNA and shRNA encoding genes into cultured and primary hematopoietic cells. *Methods Mol. Biol.*, **433**, 287–299.
61. Shen, C., Buck, A.K., Liu, X. *et al.* (2003) Gene silencing by adenovirus-delivered siRNA. *FEBS Lett.*, **539**, 111–114.
62. Narvaiza, I. *et al.* (2006) Effect of adenovirus-mediated RNA interference on endogenous microRNAs in a mouse model of multidrug resistance protein 2 gene silencing. *J. Virol.*, **80**, 12236–12247.
63. Teichler Zallen, D. (2000) US gene therapy in crisis. *Trends Genet.*, **16**, 272–275.
64. Musiyenko, A., Bitko, V. and Barik, S. (2007) RNAi-dependent and -independent antiviral phenotypes of chromosomally integrated shRNA clones: role of VASP in respiratory syncytial virus growth. *J. Mol. Med.*, **85**, 745–752.
65. Rossi, J.J. (2008) Expression strategies for short hairpin RNA interference triggers. *Hum. Gene Ther.*, **19**, 313–317.
66. Grimm, D. *et al.* (2006) Fatality in mice due to oversaturation of cellular microRNA/short hairpin RNA pathways. *Nature*, **441**, 537–541.
67. Lu, P.Y. and Woodle, M.C. (2008) Delivering small interfering RNA for novel therapeutics. *Methods Mol. Biol.*, **437**, 93–107.
68. Hayes, M.E. *et al.* (2006) Assembly of nucleic acid-lipid nanoparticles from aqueous-organic monophases. *Biochim. Biophys. Acta*, **1758**, 429–442.
69. Khoury, M. *et al.* (2006) Efficient new cationic liposome formulation for systemic delivery of small interfering RNA silencing tumor necrosis factor alpha in experimental arthritis. *Arthritis Rheum.*, **54**, 1867–1877.
70. Sonawane, N.D., Szoka, F.C. Jr. and Verkman, A.S. (2003) Chloride accumulation and swelling in endosomes enhances DNA transfer by polyamine-DNA polyplexes. *J. Biol. Chem.*, **278**, 44826–44831.
71. Heyes, J., Palmer, L., Bremner, K. and MacLachlan, I. (2005) Cationic lipid saturation influences intracellular delivery of encapsulated nucleic acids. *J. Control. Release*, **107**, 276–287.
72. Midoux, P., Breuzard, G., Gomez, J.P. and Pichon, C. (2008) Polymer-based gene delivery: a current review on the uptake and intracellular trafficking of polyplexes. *Curr. Gene Ther.*, **8**, 335–352.
73. Patil, Y. and Panyam, J. (2008) Polymeric nanoparticles for siRNA delivery and gene silencing. *Int J Pharm*, **367**(1–2), 195–203.
74. Aigner, A. (2007) Nonviral *in vivo* delivery of therapeutic small interfering RNAs. *Curr. Opin. Mol. Ther.*, **9**, 345–352.
75. Sorensen, D.R., Leirdal, M. and Sioud, M. (2003) Gene silencing by systemic delivery of synthetic siRNAs in adult mice. *J. Mol. Biol.*, **327**, 761–766.
76. Leng, Q. *et al.* (2005) Highly branched HK peptides are effective carriers of siRNA. *J. Gene Med.*, **7**, 977–986.
77. Kim, B. *et al.* (2004) Inhibition of ocular angiogenesis by siRNA targeting vascular endothelial growth factor pathway genes: therapeutic strategy for herpetic stromal keratitis. *Am. J. Pathol.*, **165**, 2177–2185.
78. Schiffelers, R.M. *et al.* (2004) Cancer siRNA therapy by tumor selective delivery with ligand-targeted sterically stabilized nanoparticle. *Nucleic Acids Res.*, **32**, e149.
79. Li, B.J. *et al.* (2005) Using siRNA in prophylactic and therapeutic regimens against SARS coronavirus in Rhesus macaque. *Nat. Med.*, **11**, 944–951.

80. Xu, C.X. *et al.* (2008) Poly(ester amine)-mediated, aerosol-delivered Akt1 small interfering RNA suppresses lung tumorigenesis. *Am. J. Respir. Crit. Care Med.*, **178**, 60–73.
81. Ge, Q. *et al.* (2004) Inhibition of influenza virus production in virus-infected mice by RNA interference. *Proc. Natl. Acad Sci. USA*, **101**, 8676–8681.
82. Thomas, M. *et al.* (2005) Full deacylation of polyethylenimine dramatically boosts its gene delivery efficiency and specificity to mouse lung. *Proc. Natl. Acad Sci. USA*, **102**, 5679–5684.
83. Miyawaki-Shimizu, K. *et al.* (2006) siRNA-induced caveolin-1 knockdown in mice increases lung vascular permeability via the junctional pathway. *Am. J. Physiol. Lung Cell. Mol. Physiol.*, **290**, L405–L413.
84. Amarzguioui, M. *et al.* (2006) Ex vivo and *in vivo* delivery of anti-tissue factor short interfering RNA inhibits mouse pulmonary metastasis of B16 melanoma cells. *Clin. Cancer Res.*, **12**, 4055–4061.
85. Wang, X. *et al.* (2008) Prevention of airway inflammation with topical cream containing imiquimod and small interfering RNA for natriuretic peptide receptor. *Genet. Vaccines Ther.*, **6**, 7.
86. Robbins, M. *et al.* (2008) Misinterpreting the therapeutic effects of siRNA caused by immune stimulation. *Hum. Gene Ther.*, **19** (10), 991–999.
87. Hornung, V. *et al.* (2005) Sequence-specific potent induction of IFN-alpha by short interfering RNA in plasmacytoid dendritic cells through TLR7. *Nat. Med.*, **11**, 263–270.
88. Judge, A.D. *et al.* (2005) Sequence-dependent stimulation of the mammalian innate immune response by synthetic siRNA. *Nat. Biotechnol.*, **23**, 457–462.
89. Sioud, M. (2005) Induction of inflammatory cytokines and interferon responses by double-stranded and single-stranded siRNAs is sequence-dependent and requires endosomal localization. *J. Mol. Biol.*, **348**, 1079–1090.
90. Kariko, K., Bhuyan, P., Capodici, J. and Weissman, D. (2004) Small interfering RNAs mediate sequence-independent gene suppression and induce immune activation by signaling through toll-like receptor 3. *J. Immunol.*, **172**, 6545–6549.
91. Schlee, M., Hornung, V. and Hartmann, G. (2006) siRNA and isRNA: two edges of one sword. *Mol. Ther.*, **14**, 463–470.
92. Kim, D.H. *et al.* (2004) Interferon induction by siRNAs and ssRNAs synthesized by phage polymerase. *Nat. Biotechnol.*, **22**, 321–325.
93. Marques, J.T. *et al.* (2006) A structural basis for discriminating between self and nonself double-stranded RNAs in mammalian cells. *Nat. Biotechnol.*, **24**, 559–565.
94. Kleinman, M.E. *et al.* (2008) Sequence- and target-independent angiogenesis suppression by siRNA via TLR3. *Nature*, **452**, 591–597.
95. Judge, A. and MacLachlan, I. (2008) Overcoming the innate immune response to small interfering RNA. *Hum. Gene Ther.*, **19**, 111–124.
96. Forsbach, A. *et al.* (2008) Identification of RNA sequence motifs stimulating sequence-specific TLR8-dependent immune responses. *J. Immunol.*, **180**, 3729–3738.
97. Ito, T. *et al.* (2002) Interferon-alpha and interleukin-12 are induced differentially by Toll-like receptor 7 ligands in human blood dendritic cell subsets. *J. Exp. Med.*, **195**, 1507–1512.
98. Zarember, K.A. and Godowski, P.J. (2002) Tissue expression of human Toll-like receptors and differential regulation of Toll-like receptor mRNAs in leukocytes in response to microbes, their products, and cytokines. *J. Immunol.*, **168**, 554–561.
99. Lien, S.P. *et al.* (2007) Identification of synthetic vaccine candidates against SARS CoV infection. *Biochem. Biophys. Res. Commun.*, **358**, 716–721.
100. Okada, M. *et al.* (2007) Development of vaccines and passive immunotherapy against SARS corona virus using SCID-PBL/hu mouse models. *Vaccine*, **25**, 3038–3040.
101. Haagmans, B.L. *et al.* (2004) Pegylated interferon-alpha protects type 1 pneumocytes against SARS coronavirus infection in macaques. *Nat. Med.*, **10**, 290–293.
102. Flemming, A. (2005) Research highlights. *Nat. Rev. Drug Discov*, **4**, 812.

15

Lipid-Based Formulations for siRNA Delivery

Camilla Foged,[1] *Pieter Vader*[2] *and Raymond M. Schiffelers*[2]
[1] University of Copenhagen, Department of Pharmaceutics and Analytical Chemistry, Faculty of Pharmaceutical Sciences, Copenhagen, Denmark
[2] Utrecht University, Department of Pharmaceutical Sciences, Faculty of Science, Utrecht, The Netherlands

15.1 Introduction

RNA interference (RNAi) is an evolutionary conserved process for post-transcriptional silencing in many species, including mammalians [1, 2]. The physiological functions of RNAi are viral defence, transposon silencing and regulation of developmental pathways. RNAi is mediated by small interfering RNA (siRNA) molecules [3], which are able to interact with several proteins to form a multienzyme complex including Argonaute 2 (AGO2) and the RNA-Induced Silencing Complex (RISC) [4–6]. Upon unwinding of the siRNA, the antisense strand guides the AGO2-RISC complex to complementary mRNA, which is subsequently cleaved [3, 7].

Using this machinery, introduction of synthetic siRNA molecules of 21–23 nucleotides into cells has been shown to result in sequence-specific silencing of genes at low concentrations compared to other antisense drugs [2]. In addition to its use in functional genomics, siRNA holds great promise as a therapeutic agent with wide applications for treatment of severe and chronic diseases [8–11]. The huge potential of RNAi for human therapy is demonstrated by the fact that several siRNA-based drugs are already in clinical trials (some in phase III). However, despite these high expectations, a number of delivery

Delivery Technologies for Biopharmaceuticals: Peptides, Proteins, Nucleic Acids and Vaccines Edited by Lene Jorgensen and Hanne Mørck Nielsen

hurdles has to be overcome before the therapeutic potential of siRNA can be realized. Firstly, naked siRNA merely has a half-life of several minutes *in vivo* due to degradation by nucleases and excretion in the kidneys, which is an insufficient length of time to induce silencing [12]. Furthermore, siRNA needs to reach the specific cells that are actively translating the targeted mRNA. Finally, cell membrane passage of a relatively large and highly charged polyanion like siRNA represents a major obstacle. The major challenge of bringing RNAi from a laboratory tool to a therapeutic strategy is therefore to develop effective delivery systems that can increase stability, cellular targeting and cell membrane passage of siRNA, and thereby reduce the amount and costs of siRNA required for therapeutic efficacy.

To date, several delivery systems for siRNA have been described, including viral vectors and non-viral vectors. For therapeutic use, non-viral vectors are preferred by virtue of their ease of synthesis, low immune response and benefits in terms of safety. Among the wide variety of non-viral vectors, lipid-based carriers have been the most extensively described in both *in vitro* and *in vivo* studies.

The majorities of lipids applied for siRNA delivery are cationic and interact readily with negatively charged siRNA molecules to form so-called lipoplexes, which are heterogeneous and metastable aggregates. However, also neutral and even negatively charged lipids have been used. Delivery systems based on these lipids rely on entrapment of siRNA in the aqueous compartment of liposomes to form lipid-based nanoparticles [13]. All the systems are able to protect siRNA from enzymatic degradation [12, 14]. Moreover, the hydrophobization of the siRNA by the lipids promotes passage across cellular membranes.

Despite these attractive features of lipid carriers for nucleic acid delivery in general, no impressive results have been obtained in clinical trials to date due to limited delivery efficiency *in vivo*. Additionally, the occurrence of non-specific immune-stimulating effects of lipids as well as siRNAs has retarded the development of more efficient carrier systems.

In this chapter an overview is provided of currently employed lipid-based delivery systems for siRNA, and of recent research into improvements of cellular delivery by enhancing target cell uptake and redirecting intracellular trafficking routes. In addition, the role of lipids in immune activation is discussed, and possible ways to overcome the response are highlighted.

15.2 Cationic Lipid-Based Delivery Systems for Nucleic Acid-Based Drugs

The multivalency of cationic liposomes provides for electrostatic interactions with the negatively charged phosphate backbone of nucleic acids, resulting in condensation of the nucleic acid into particles, also called lipoplexes [15–17]. Preformed cationic liposomes are among the most efficient *in vitro* transfection reagents for nucleic acids/siRNA. Their high efficiency is partly caused by the large amounts of siRNA that can be complexed by cationic lipids, compared to passive loading in charge neutral liposomes [12, 14]. Besides condensation of nucleic acids, cationic lipids serve to protect nucleic acids against degradation by nucleases and provide the complexes with a net positive charge, enabling efficient binding of the complexes to negatively charged heparan sulfate proteoglycans present on cell surfaces [18–20]. Transfection-competent lipoplexes possess an ordered structure with irregular morphology, which for cationic liposome-DNA complexes consists of DNA coated by cationic lipid monolayers and arranged on a so-called two-dimensional hexagonal lattice [21]. Lamellar liposome-DNA complexes are less efficient for transfection. Surprisingly, a study with siRNA showed that lamellar cationic lipid-siRNA complexes were superior for gene silencing compared to inverted hexagonal complexes, despite the general similarity between the structure of lipid-DNA and lipid-siRNA complexes [22]. This indicates that important aspects of lipoplex-mediated siRNA differ significantly from lipoplex-mediated DNA delivery.

To date a variety of cationic lipids have been synthesized for *in vitro* transfection purposes but most, if not all, of them have failed in clinical trials due to poor *in vivo* performance and serious drawbacks, such as low *in vivo* transfection efficiency, toxicity problems associated with immune activation and unfavourable biodistribution. Lipoplexes are rapidly cleared upon systemic administration due to accumulation in the capillary bed of first-pass organs such as the lungs [23, 24]. First-pass clearance is followed by hepatic elimination of the majority of the lipoplexes 60 minutes after injection due to redistribution from the lungs to the phagocytic Kupffer cells of the liver [25]. Strong interactions between the cationic surface of lipoplexes and plasma proteins prime the lipoplexes for this rapid clearance by the reticuloendothelial system (RES) [26, 27].

To circumvent the shortcomings of cationic lipids, titratable, ionizable lipids have been developed [28, 29]. The lipids are strongly charged at reduced pH, offering the possibility of efficient complex formation, but are more neutral at physiological pH. This approach still benefits from the charge interaction at reduced pH but avoids the unfavourable biodistribution of charged systems by changing the pharmacokinetics, enabling passive targeting of neutral, appropriately sized (100 nm) liposomes to pathological tissue via the enhanced permeability and retention effect (EPR) upon systemic administration [30]. Another option is to remove residual cationic surface charge upon complex formation [31] or to shield the charge of lipoplexes by covering them with an additional, neutral membrane bilayer, so-called wrapped liposomes [32, 33].

15.2.1 Helper Lipids

Some cationic lipids possess intrinsic fusogenic properties, enhancing the membrane fusion and intracellular delivery of lipoplexes [29, 34, 35]. Above the phase transition temperature, some lipids can adopt a reverse hexagonal H_{II} phase that is highly fusogenic [21, 29, 36]. For example, for titratable cationic lipids it has been shown that the fusogenicity of the liposomal system increases as the lipid becomes less saturated [29]. This suggests that a more flexible bilayer structure is important for the propensity to destabilize cell membranes. However, in addition to cationic lipids, so-called helper lipids are often included in liposomal formulations for efficient delivery [34, 35, 37, 38]. Non-bilayer forming lipids like dioleyl phosphatidylethanolamin (DOPE) can form a reverse hexagonal H_{II} phase that promotes destabilization of the lipid bilayer and fusion with the target cell membrane [21]. *In vivo*, the increased transfection competency of fusogenic systems might, though, be counterbalanced by their increased propensity to interact with components in the circulation, resulting in stability problems. However, for siRNA, a recent study indicates that inclusion of DOPE in cationic liposomes does not increase the transfection efficiency but rather causes increased toxicity and lower target-specific gene silencing [22], indicating that DOPE is not beneficial for siRNA delivery but only for plasmid DNA delivery.

15.2.2 Lipopolymers

Another way to circumvent the fast clearance of lipoplexes and delay the uptake of lipid-based nanoparticles into the RES is to shield and stabilize the cationic lipid bilayer with polymers like poly(ethylene glycol) (PEG) [39–41], which allows for passive targeting to vasculature at sites where the endothelium is leaky (EPR effect) due to increased circulation time upon systemic administration. However, PEG prevents the cell association, uptake and membrane destabilization necessary for endosomal escape, resulting in low transfection efficiency [41–43]. This possesses a contradictory dilemma of particle stabilization and charge shielding versus target cell uptake, drug release and membrane destabilization.

Therefore, coatings that may be shed have been developed; these are dissociated from the carrier upon arrival at the target site, converting the delivery system from a stable particle to a transfection-competent complex, thereby facilitating drug release and/or target cell interaction processes [44]. One such coating is diffusible PEG-lipids that can dissociate from the carrier, with dissociation constants depending on the length of the lipid anchor [45]. Diffusible PEGs have been incorporated into siRNA-containing liposomal systems (Table 15.1) called stabilized siRNA lipid particles (SNALPs) [12, 14, 46]. Zimmermann *et al.* observed dose-dependent silencing of apolipoprotein B (ApoB) in the liver of cynomolgus monkeys after a single intravenous (i.v.) injection of SNALPs encapsulating siRNA directed towards ApoB [14]. The SNALP liposomes contained distearoyl phosphatidylcholine (DSPC), cholesterol, the cationic and fusogenic lipid DLinDMA and the diffusible lipopolymer PEG-C-DMA, prepared by a carefully controlled step-wise ethanol dilution method [47]. The plasma half-life of SNALP-encapsulated siRNA was 38 minutes in mice [14], compared to less than two minutes for naked siRNA [12]. The liposomal siRNA showed 50% silencing of apoB messenger at 0.1 mg/kg SNALP-formulated ApoB-siRNA, whereas a 100 mg/kg dose was required to obtain comparable levels of efficacy with a cholesterol-conjugated ApoB siRNA, which corresponds to a 1000-fold increase in potency [48]. No immune stimulation or toxic effects were observed.

An alternative to diffusible PEGs involves the use of stimuli-responsive lipid-polymer conjugates with a chemically labile linker between the lipid anchor and the polymer. An example is incorporation of an acid labile PEG-diorthoester lipid-conjugate into the bilayer that is destabilized at the low pH found in the endosomes, facilitating the release of entrapped drug [49, 50].

15.2.3 Active Targeting

Active targeting strategies have been applied to provide sterically stabilized, long-circulating systems with increased affinity for target cells. The cellular binding and uptake of liposomes can be increased by attaching ligands for internalizing receptors to the distal ends of PEG-chains. Examples of commonly used targeting moieties are peptide and protein ligands like antibodies and smaller peptides [51, 52], vitamins-like folate and

Table 15.1 *Examples of liposomal siRNA delivery systems*

Lipid composition	Target gene	Formulation method	Size (nm)	Encapsulation (%)	Reference
DSPC:Chol:PEG-C-DMA:various cationic lipids	Luciferase (*in vitro*)	Ethanol dilution – SNALP	132–182	67–85	[29]
DSPC:Chol:PEG-C-DMA:DLinDMA or DODMA	B-galactosidase (*in vitro*)	Ethanol dilution – SNALP	100–130	90–95	[73]
DSPC:Chol:PEG-C-DMA:DLinDMA	Hepatitis B virus (*in vivo*)	Ethanol dilution – SNALP	140 ± 12	93 ± 3	[12]
DSPC:Chol:PEG-C-DMA:DLinDMA	Apolipoprotein B (*in vivo*)	Ethanol dilution – SNALP	73–83	92–97	[14]
DSPC:Chol:PEG-C-DMA:DLinDMA	Polymerase (L) gene of Ebola virus (*in vivo*)	Ethanol dilution – SNALP	71–84	90–95	[46]
DSPC:Chol:PEG-C-DMA:DLinDMA	Apolipoprotein B (*in vivo*)	Ethanol dilution – SNALP	100–130	90–95	[75]
DOTAP:DOPE:TfRscFV	None	Ethanol injection	211 ± 18	ND	[51]
DOTAP:DOPE:TfRscFV	HER-2 (*in vivo*)	Ethanol injection	100	ND	[52]
Gal-C4-Chol/DOPE	Ubc13 gene (*in vivo*)	Passive encapsulation	50 ± 4	ND	[54]
(Diethylaminoethyl) carbamoyldioleoyl glycerol:PC:Lactosylated PE	Hepatitis C virus (*in vivo*)	Passive encapsulation	ND	ND	[53]
DOTAP:Chol:DNA:protamine:DSPE-PEG-anisaminde	None	Passive encapsulation	120–150	>95	[58]
DOTAP:Chol:DNA:protamine:DSPE-PEG-anisaminde	Epidermal growth factor receptor (*in vivo*)	Passive encapsulation	120–150	>95	[57]
EPC:Chol:PEG-PE:DOTAP	Human double minute gene 2 (HDM2) (*in vitro*)	Passive encapsulation	50–200	ND	[61]
PC:DMPA:Chol	None	Passive encapsulation	880	ND	[85]
DOPC:Tween 20	EphA2 (*in vivo*)	Passive encapsulation	ND	65	[63]
DOPC:Tween 20	Focal adhesion kinase (*in vivo*)	Passive encapsulation	ND	ND	[64]
DPPG:DPPC:DPPE-PEG	Enhanced green fluorescent protein (*in vitro*)	Reverse-phase evaporation	60–90	7–14	[62]
DOTAP:DOPE:Transferrin	c-Jun/luciferase (*in vitro*)	Passive encapsulation	328–926	90	[86]

TfRscFV, anti-transferrin receptor single-chain antibody fragment.

glycolipids [53, 54]. Active targeting serves to increase the therapeutic index of drugs by increasing the fraction of drug reaching the target site, and by reducing the fraction of drug reaching non-target sites and toxicological targets. Disadvantages of active targeting approaches are though increased costs, manufacturing complexity and increased risk of immunogenicity. Development of a post-insertion technique has allowed for insertion of ligand-PEG-lipid conjugates into preformed liposomes [55].

A few studies have been published where siRNAs have been successfully delivered *in vivo* with actively targeted liposomal formulations. Pirollo *et al.* formulated siRNA into immunoliposomes targeted to upregulated transferrin receptors present on tumour cells via an anti-transferrin receptor single-chain antibody fragment [51, 52, 56]. A histidine-lysine peptide was included in the complex to enhance endosomal release. High levels of siRNA were delivered specifically to tumours upon systemic administration [51] and were able to silence the target gene HER2 and specifically inhibit tumour growth [52].

Huang and co-workers developed self-assembling siRNA nanoparticles prepared by mixing carrier DNA, siRNA, protamine and lipids (DOTAP and cholesterol), followed by post-modification with the ligand anisamide coupled to PEG-lipid [57–59]. These particles are also called LPD nanoparticles. Anisamide is a ligand for the sigma receptor present on certain tumour cells. As much as 70–80% of the injected siRNA accumulated in the tumour in a mouse xenograft model, and 40% tumour growth inhibition could be achieved.

Other ligands include cell-penetrating peptides (CPPs) [60] that can increase the transfection efficiency upon association to siRNA-liposome complexes [61].

15.3 Neutral and Anionic Lipid-Based Drug Delivery Systems

Liposomes composed of neutral or zwitterionic lipids have much longer circulation times, lower toxicities and very different clearance profiles compared to cationic particles. Neutral liposomes have been successfully applied as carriers for anti-cancer drugs and antibiotics, to achieve lower toxicities, altered distributions and higher drug efficacies than unencapsulated drugs [30]. Neutral and anionic liposomes have, though, the disadvantage compared to lipoplexes that siRNA cannot be entrapped efficiently via electrostatic interactions, and their transfection efficiencies are much reduced compared to lipoplexes [62].

Landen *et al.* used a liposomal formulation consisting of DOPC and Tween 20 (Table 15.1) to encapsulate an siRNA targeting the oncoprotein EphA2 [63]. The formulation was able to silence EphA2 gene expression 48 hours after administration and could inhibit tumour growth in an orthotopic mouse model of ovarian cancer. The same formulation was successfully applied to silence the focal adhesion kinase expressed by ovarian cancer cells, resulting in significantly reduced tumour growth in mice [64].

Another option is to pre-condensate siRNA with a polycation, followed by coating with negatively charged or neutral lipids [32, 33, 58, 65]. The polycation could be a cationic liposome or a polymer like poly(ethylene imine), protamine sulfate or polylysine.

15.4 Mechanisms of Internalization

Most studies regarding internalization routes of lipid-nucleic acid complexes have been performed with plasmid DNA. Although it seems likely that siRNA-lipoplexes essentially follow the same fate as their plasmid DNA counterparts, formal studies in this field are still lacking.

For cationic lipid-delivered siRNA, five internalization routes seem most probable: (i) Clathrin-mediated endocytosis, (ii) caveolae-mediated endocytosis, (iii) macropinocytosis, (iv) phagocytosis and (v) receptor-mediated endocytosis. Although fusion of lipoplexes at the cell surface could, in principle, occur, the majority of studies indicate that a certain kind of endocytosis contributes most to intracellular delivery. Size as well as composition of the siRNA-lipid complex can influence the route of internalization.

Large lipoplexes (in the range 0.2–0.5 μm) seem to be preferentially taken up through receptor- and clathrin-independent endocytosis. The exact nature of this uptake route is debated. Matsui *et al.* suggested phagocytosis of large lipoplexes occurred in lung epithelial cells (not considered to be specialized phagocytes), where the uptake could be inhibited by cytochalasin B and potassium depletion [66]. Rejman *et al.* suggested that larger particles are primarily taken up by caveolae [67]. This would allow for targeting of particles through a non-degradative pathway, as caveolae-mediated endocytosis is a pathway that is not coupled to acidification of the compartment and concomitant lysosomal fusion.

In contrast, smaller complexes (<0.2 μm) are mainly internalized via coated pits through a non-specific, clathrin-dependent process [68]. This route of endocytosis involves an intracellular pathway, in which lysosomes ultimately fuse with internalized vesicles, resulting in degradation of their contents.

For the remaining two internalization routes, modification of the lipoplex surface seems required. Octa-arginine modified lipoplexes appeared to be taken up by macropinocytosis [69]. Receptor-mediated endocytosis can be enforced by coupling the matching targeting ligand to the lipoplex surface. This usually requires shielding of the net cationic charge of the complex by polymers such as PEG to optimally benefit from the receptor-ligand interaction. Interestingly, cholesterol-conjugated stabilized siRNA present on the outside of low density lipoproteins was internalized via the siRNA-specific SID-1 receptor in HepG2 cells, which led to functional silencing [70]. This indicates that, for certain cell types, delivery to the cellular membrane may be sufficient, after which dissociation of siRNA and lipid may occur, followed by translocation of siRNA over the cytolemma.

After endocytic uptake, a membrane barrier still remains that separates the siRNA from the cytoplasm, where the RNAi machinery is located. Caveolar uptake provides limited environmental triggers for crossing the caveolar membrane but constitutes a non-degrading environment. In contrast, for other routes acidifying conditions in the late endosome can provide a cue for fusion of the lipoplex with the endosomal membrane. This may be achieved by inclusion of peptides or lipids that change conformation upon protonation.

15.5 Immune Activation

Although viral vectors have been shown to have superior abilities to transfect cells, they are often not the best choice for siRNA delivery for therapeutic purposes. The presence of specific immune reactions against viral particles has forced researchers to look for alternative agents for delivery. For a long time, non-viral vectors were considered to be inert, thereby avoiding these inflammatory responses. Although they have now become the first option as delivery systems, more recent studies on immune activation of lipidic particles indicate that this might not be the case at all.

For instance, Mui *et al.* found that immune stimulation by a CpG-containing oligodeoxynucleotide (ODN) is enhanced when delivered in lipid particles [71]. The therapeutic benefit from ODN-containing immune stimulatory sequences like CpG in animal models of cancer and infection had already been demonstrated. Mui *et al.* showed that after administration of CpG ODN, encapsulated in lipid particles, plasma concentrations of cytokines like interleukin (IL)-12, interferon (IFN)-γ, IL-6, monocyte chemoattractrant protein-1 and tumour necrosis factor (TNF)-α in mice were greatly increased compared with free CpG ODN. They concluded that the increased immunogenicity was largely a result of enhanced delivery of CpG ODN to macrophages and other antigen-presenting cells by the lipid particles. Indeed, liposomes are removed from the circulation by macrophages and monocytes. An increased immunostimulatory activity of CpG ODN as a result of formulation in liposomal nanoparticles was also found by De Jong *et al.* [72]. They investigated the immunopotency and also anti-tumour activity of subcutaneously administered CpG ODN in animal models of cancer.

A lipid-mediated immune reaction has also been shown for delivery of siRNA. Although not generally accepted, it seems that transfection of siRNAs results in activation of the interferon system [73]. Interferons are cytokines that are part of the innate immune response that can be triggered by double-stranded RNA molecules through the binding to Toll-like receptors (TLR). Ma *et al.* investigated this response in mice and found that although intravenous administration of a specific siRNA in this case was essentially inert, injection of cationic lipid-siRNA complexes induced a strong interferon response [74]. It was also suggested that the mechanism behind this increased reaction is enhanced uptake of siRNA by immune cells.

In addition to the immune activation induced by the lipid vector, certain siRNA sequences have also been shown to induce activation of the innate immune system, triggered primarily by the recognition of dsRNA by TLR7 [73, 75–77]. Some TLRs, including TLR7, are localized in endosomes, and immune responses elicited by siRNA are greatly potentiated by delivery systems like liposomes that facilitate intracellular delivery to endosomes. It seems possible, though, to design non-inflammatory siRNAs by avoiding GU-sequence motifs and by incorporation of 2′-O-methyl uridine or guanosine nucleosides into one strand of the siRNA duplex [75].

Lipid carriers are not only able to increase immune reactions caused by other agents, as mentioned above, but can also be immunogenic by them selves. Therefore, liposomes are used as immunological adjuvants in vaccine formulations [78–80]. The first indications for undesired immune reactions of lipid-based carriers for nucleic acid delivery arose from microarray studies that examined the effects on gene expression profiles of target cells. In one study, Omidi *et al.* investigated the effects of two lipid-based carrier systems, Lipofectin and Oligofectamine, on gene expression profiles of human A431 epithelial cells [81]. They hypothesized that incubation of cells with cationic lipid formulations had an effect on their global gene expression, based on a previous study performed by Petch *et al.* [82], showing that changes in non-target gene expression following treatment of cells with oligonucleotide-lipid complexes was not the result of the oligonucleotides alone. Using cDNA microarray expression profiling, they showed that treatment with

Lipofectin or Oligofectamine resulted in a more than twofold altered expression of 6% or 17% of the investigated genes, respectively. These changes led to downstream functional consequences, such as increased apoptosis [81].

A possible explanation for the demonstrated immunostimulation of antigen-presenting cells by cationic lipids alone was published by Vangasseri *et al.* [83]. In their study on anti-tumour immune responses through delivery of a human papillomavirus-E7 epitope antigen to dendritic cells, they found that not only the plasmid DNA but also the cationic liposomes that were used to deliver the antigen led to expression of immunostimulatory molecules [83]. Interestingly, stimulation of dendritic cells occurred via a different mechanism in both situations. Furthermore, the ability of lipids to activate an immune response appeared to be structure-specific. Both head group and hydrocarbon chain structure were found to have an influence on the immunological outcome. According to the authors, these findings point towards the possible existence of a cationic lipid binding protein, which can activate cellular signalling cascade(s) resulting in the production of (co-) stimulatory molecules [83].

Despite the fact that the pathway by which lipids stimulates the immune system is still largely unknown, the first studies on how to decrease the inflammatory response to these vectors have already been published. Liu *et al.* tried to decrease the inflammatory response to DOTAP liposomes containing plasmid DNA by incorporating two small molecules, known for their ability to specifically inhibit the NF-κB or MAPK inflammatory pathway, into the formulation [84]. These small molecules, including chenodeoxycholic acid, capsaicin and the MAP kinase inhibitor, PD98059, were efficiently encapsulated into the lipoplexes and were able to significantly decrease the TNF-α response to the liposome formulation. The gene expression levels were found to be similar for all kinds of lipoplexes. The results also suggest the possibility to reduce the immune response by an even greater extent by incorporation of several small molecules that inhibit different immune activation pathways [84].

Although the long-term effects of lipid-induced gene expression are still unknown, the results from an increasing number of studies suggest that lipid-based vectors do induce an, often undesired, rapid immune reaction. This finding might have a great influence on the development of non-viral vectors, because their inertness towards antigen-presenting cells has long been their greatest advantage in comparison to viral delivery systems.

15.6 Conclusions

Multiple *in vivo* experiments have consolidated the promising therapeutic efficacy of siRNA, but there is an urgent need for the development of novel carrier systems for siRNA. Liposomes are versatile delivery systems that can be designed and provided with properties needed for successful delivery of siRNA for a variety of pathological conditions. These properties include siRNA protection, decreased elimination, specific cellular targeting, increased membrane passage and siRNA release. Such systems serve to improve the therapeutic outcome by reducing the dose and, thereby, the costs of an expensive drug like siRNA. Multicomponent and/or multilayered systems seem to be required to equip the delivery system with the often contradictory requirements for *in vivo* administration, although simple formulations are highly desired for clinical applications. The complex composition of many lipid-based carrier systems demands a careful characterization of the formulations for a predictive and reproducible therapeutic outcome.

In addition, the development of novel, optimal and non-cationic lipid-based carrier systems that are safe and well tolerated, possess favourable pharmacokinetics, intracellular/cytoplasmic delivery and are stable upon manufacture is highly requested. To optimally tailor the lipidic delivery systems for siRNA delivery better insight is needed into the mechanisms of cell delivery. More specifically, pharmacokinetics, the nature of cell surface interaction of siRNA-lipid complexes, routes of internalization, passage of intracellular membranes and mechanisms of immune activation need further clarification. For siRNA in particular, increased knowledge on the off-target effects of the delivery systems is of special interest, since siRNA is currently being investigated in clinical trials. More research is required to elucidate the mechanism(s) behind lipid-based immune reactions and how to avoid them. Increased knowledge will help the design of better and safe lipid-based delivery systems for siRNA in the future.

Acknowledgements

We gratefully acknowledge the Danish Research Council for Technology and Production Sciences for financial support.

References

1. Fire, A., Xu, S., Montgomery, M.K. *et al.* (1998) Potent and specific genetic interference by double-stranded RNA in *Caenorhabditis elegans*. *Nature*, **391**, 806–811.
2. Elbashir, S.M., Harborth, J., Lendeckel, W. *et al.* (2001) Duplexes of 21-nucleotide RNAs mediate RNA interference in cultured mammalian cells. *Nature*, **411**, 494–498.
3. Elbashir, S.M., Lendeckel, W. and Tuschl, T. (2001) RNA interference is mediated by 21- and 22-nucleotide RNAs. *Genes Dev.*, **15**, 188–200.
4. Matranga, C., Tomari, Y., Shin, C. *et al.* (2005) Passenger-strand cleavage facilitates assembly of siRNA into Ago2-containing RNAi enzyme complexes. *Cell*, **123**, 607–620.
5. Martinez, J., Patkaniowska, A., Urlaub, H. *et al.* (2002) Single-stranded antisense siRNAs guide target RNA cleavage in RNAi. *Cell*, **110**, 563–574.
6. Liu, J.D., Carmell, M.A., Rivas, F.V. *et al.* (2004) Joshua-Tor and G. J. Hannon, Argonaute2 is the catalytic engine of mammalian RNAi. *Science*, **305**, 1437–1441.
7. Orban, T.I. and Izaurralde, E. (2005) Decay of mRNAs targeted by RISC requires XRN1, the Ski complex, and the exosome. *RNA*, **11**, 459–469.
8. Dykxhoorn, D.M., Palliser, D. and Lieberman, J. (2006) The silent treatment: siRNAs as small molecule drugs. *Gene Ther.*, **13**, 541–552.
9. de Fougerolles, A., Vornlocher, H.P., Maraganore, J. and Lieberman, J. (2007) Interfering with disease: a progress report on siRNA-based therapeutics. *Nat. Rev. Drug Discov.*, **6**, 443–453.
10. Behlke, M.A. (2006) Progress towards *in vivo* use of siRNAs. *Mol. Ther.*, **13**, 644–670.
11. Kim, D.H. and Rossi, J.J. (2007) Strategies for silencing human disease using RNA interference. *Nat. Rev. Genet.*, **8**, 173–184.
12. Morrissey, D.V., Lockridge, J.A., Shaw, L. *et al.* (2005) Potent and persistent *in vivo* anti-HBV activity of chemically modified siRNAs. *Nat. Biotechnol.*, **23**, 1002–1007.
13. Li, W. and Szoka, F.C. Jr. (2007) Lipid-based nanoparticles for nucleic acid delivery. *Pharm. Res.*, **24**, 438–449.
14. Zimmermann, T.S., Lee, A.C.H., Akinc, A. *et al.* (2006) Ai-mediated gene silencing in non-human primates. *Nature*, **441**, 111–114.

15. Felgner, P.L., Gadek, T.R., Holm, M. *et al.* (1987) Lipofection: a highly efficient, lipid-mediated DNA-transfection procedure. *Proc. Natl. Acad. Sci. USA*, **84**, 7413–7417.
16. Bennett, C.F., Chiang, M.Y., Chan, H. *et al.* (1992) Cationic lipids enhance cellular uptake and activity of phosphorothioate antisense oligonucleotides. *Mol. Pharmacol.*, **41**, 1023–1033.
17. Yeoman, L.C., Danels, Y.J. and Lynch, M.J. (1992) Lipofectin enhances cellular uptake of antisense DNA while inhibiting tumor cell growth. *Antisense Res. Dev.*, **2**, 51–59.
18. Zabner, J., Fasbender, A.J., Moninger, T. *et al.* (1995) Cellular and molecular barriers to gene transfer by a cationic lipid. *J. Biol. Chem.*, **270**, 18 997–19 007.
19. Mislick, K.A. and Baldeschwieler, J.D. (1996) Evidence for the role of proteoglycans in cation-mediated gene transfer. *Proc. Natl. Acad. Sci. USA*, **93**, 12 349–12 354.
20. Mounkes, L.C., Zhong, W., Cipres-Palacin, G. *et al.* (1998) Proteoglycans mediate cationic liposome-DNA complex-based gene delivery *in vitro* and *in vivo*. *J. Biol. Chem.*, **273**, 26 164–26 170.
21. Koltover, I., Salditt, T., Radler, J.O. and Safinya, C.R. (1998) An inverted hexagonal phase of cationic liposome-DNA complexes related to DNA release and delivery. *Science*, **281**, 78–81.
22. Bouxsein, N.F., McAllister, C.S., Ewert, K.K. *et al.* (2007) Structure and gene silencing activities of monovalent and pentavalent cationic lipid vectors complexed with siRNA. *Biochemistry*, **46**, 4785–4792.
23. Brigham, K.L., Meyrick, B., Christman, B. *et al.* (1989) *In vivo* transfection of murine lungs with a functioning prokaryotic gene using a liposome vehicle. *Am. J. Med. Sci.*, **298**, 278–281.
24. Barron, L.G., Gagne, L. and Szoka, F.C. Jr. (1999) Lipoplex-mediated gene delivery to the lung occurs within 60 minutes of intravenous administration. *Hum. Gene Ther.*, **10**, 1683–1694.
25. Barron, L.G. and Szoka, F.C. (1999) The perplexing delivery mechanism of lipoplexes, in *Nonviral Vectors for Gene Therapy* (eds L. Huang, M.C. Hung and E. Wagner), Academic Press, San Diego, pp. 229–266.
26. Tros de Ilarduya, C., Arangoa, M.A. and Duzgunes, N. (2003) Transferrin-lipoplexes with protamine-condensed DNA for serum-resistant gene delivery. *Methods Enzymol.*, **373**, 342–356.
27. Opanasopit, P., Nishikawa, M. and Hashida, M. (2002) Factors affecting drug and gene delivery: effects of interaction with blood components. *Crit. Rev. Ther. Drug Carrier Syst.*, **19**, 191 233.
28. Semple, S.C., Klimuk, S.K., Harasym, T.O. *et al.* (2001) Efficient encapsulation of antisense oligonucleotides in lipid vesicles using ionizable aminolipids: formation of novel small multilamellar vesicle structures. *Biochim. Biophys. Acta*, **1510**, 152–166.
29. Heyes, J., Palmer, L., Bremner, K. and MacLachlan, I. (2005) Cationic lipid saturation influences intracellular delivery of encapsulated nucleic acids. *J. Control. Release*, **107**, 276–287.
30. Mayer, L.D., Bally, M.B., Cullis, P.R. *et al.* (1990) Comparison of free and liposome encapsulated doxorubicin tumor drug uptake and antitumor efficacy in the SC115 murine mammary tumor. *Cancer Lett.*, **53**, 183–190.
31. Huang, Z.H., Li, W.J., Mackay, J.A. and Szoka, F.C. (2005) Thiocholesterol-based lipids for ordered assembly of bioresponsive gene carriers. *Mol. Ther.*, **11**, 409–417.
32. Yamauchi, M., Kusano, H., Saito, E. *et al.* (2006) Improved formulations of antisense oligodeoxynucleotides using wrapped liposomes. *J. Control. Release*, **114**, 268–275.
33. Yamauchi, M., Kusano, H., Saito, E. *et al.* (2006) Development of wrapped liposomes: novel liposomes comprised of polyanion drug and cationic lipid complexes wrapped with neutral lipids. *Biochim. Biophys. Acta*, **1758**, 90–97.
34. Felgner, J.H., Kumar, R., Sridhar, C.N. *et al.* (1994) Enhanced gene delivery and mechanism studies with a novel series of cationic lipid formulations. *J. Biol. Chem.*, **269**, 2550–2561.
35. Gao, X. and Huang, L. (1995) Cationic liposome-mediated gene transfer. *Gene Ther.*, **2**, 710–722.
36. Hafez, I.M., Maurer, N. and Cullis, P.R. (2001) On the mechanism whereby cationic lipids promote intracellular delivery of polynucleic acids. *Gene Ther.*, **8**, 1188–1196.

37. Farhood, H., Serbina, N. and Huang, L. (1995) The role of dioleoyl phosphatidylethanolamine in cationic liposome mediated gene transfer. *Biochim. Biophys. Acta*, **1235**, 289–295.
38. Hui, S.W., Langner, M., Zhao, Y.L. *et al.* (1996) The role of helper lipids in cationic liposome-mediated gene transfer. *Biophys. J.*, **71**, 590–599.
39. Hong, K., Zheng, W., Baker, A. and Papahadjopoulos, D. (1997) Stabilization of cationic liposome-plasmid DNA complexes by polyamines and poly(ethylene glycol)-phospholipid conjugates for efficient *in vivo* gene delivery. *FEBS Lett.*, **400**, 233–237.
40. Meyer, O., Kirpotin, D., Hong, K. *et al.* (1998) Cationic liposomes coated with polyethylene glycol as carriers for oligonucleotides. *J. Biol. Chem.*, **273**, 15621–15627.
41. Wheeler, J.J., Palmer, L., Ossanlou, M. *et al.* (1999) Stabilized plasmid-lipid particles: construction and characterization. *Gene Ther.*, **6**, 271–281.
42. Holland, J.W., Hui, C., Cullis, P.R. and Madden, T.D. (1996) Poly(ethylene glycol)-lipid conjugates regulate the calcium-induced fusion of liposomes composed of phosphatidylethanolamine and phosphatidylserine. *Biochemistry*, **35**, 2618–2624.
43. Song, L.Y., Ahkong, Q.F., Rong, Q. *et al.* (2002) Characterization of the inhibitory effect of PEG-lipid conjugates on the intracellular delivery of plasmid and antisense DNA mediated by cationic lipid liposomes. *Biochim. Biophys. Acta*, **1558**, 1–13.
44. Romberg, B., Hennink, W.E. and Storm, G. (2008) Sheddable coatings for long-circulating nanoparticles. *Pharm. Res.*, **25**, 55–71.
45. Mok, K.W., Lam, A.M. and Cullis, P.R. (1999) Stabilized plasmid-lipid particles: factors influencing plasmid entrapment and transfection properties. *Biochim. Biophys. Acta*, **1419**, 137–150.
46. Geisbert, T.W., Hensley, L.E., Kagan, E. *et al.* (2006) Postexposure protection of guinea pigs against a lethal Ebola virus challenge is conferred by RNA interference. *J. Infect. Dis.*, **193**, 1650–1657.
47. Jeffs, L.B., Palmer, L.R., Ambegia, E.G. *et al.* (2005) A scalable, extrusion-free method for efficient liposomal encapsulation of plasmid DNA. *Pharm. Res.*, **22**, 362–372.
48. MacLachlan, I. (2008) Liposomal formulations for nucleic acid delivery, in *Antisense Drug Technology. Principles, Strategies, and Applications* (ed. S.T. Crooke), CRC Press Taylor & Francis Group, Boca Raton, pp. 237–270.
49. Guo, X. and Szoka, F.C. Jr. (2001) Steric stabilization of fusogenic liposomes by a low-pH sensitive PEG-diortho ester-lipid conjugate. *Bioconjug. Chem.*, **12**, 291–300.
50. Li, W., Huang, Z., Mackay, J.A. *et al.* (2005) Low-pH-sensitive poly(ethylene glycol) (PEG)-stabilized plasmid nanolipoparticles: effects of PEG chain length, lipid composition and assembly conditions on gene delivery. *J. Gene Med.*, 67–79.
51. Pirollo, K.F., Zon, G., Rait, A. *et al.* (2006) Tumor-targeting nanoimmunoliposome complex for short interfering RNA delivery. *Hum. Gene Ther.*, **17**, 117–124.
52. Pirollo, K.F., Rait, A., Zhou, Q. *et al.* (2007) Materializing the potential of small interfering RNA via a tumor-targeting nanodelivery system. *Cancer Res.*, **67**, 2938–2943.
53. Watanabe, T., Umehara, T., Yasui, F. *et al.* (2007) Liver target delivery of small interfering RNA to the HCV gene by lactosylated cationic liposome. *J. Hepatol.*, **47**, 744–750.
54. Sato, A., Takagi, M., Shimamoto, A. *et al.* (2007) Small interfering RNA delivery to the liver by intravenous administration of galactosylated cationic liposomes in mice. *Biomaterials*, **28**, 1434–1442.
55. Iden, D.L. and Allen, T.M. (2001) *In vitro* and *in vivo* comparison of immunoliposomes made by conventional coupling techniques with those made by a new post-insertion approach. *Biochim. Biophys. Acta*, **1513**, 207–216.
56. Yu, W., Pirollo, K.F., Yu, B. *et al.* (2004) Enhanced transfection efficiency of a systemically delivered tumor-targeting immunolipoplex by inclusion of a pH-sensitive histidylated oligolysine peptide. *Nucleic Acids Res.*, **32**, e48.
57. Li, S.D., Chen, Y.C., Hackett, M.J. and Huang, L. (2008) Tumor-targeted delivery of siRNA by self-assembled nanoparticles. *Mol. Ther.*, **16**, 163–169.
58. Li, S.D. and Huang, L. (2006) Surface-modified LPD nanoparticles for tumor targeting. *Oligonucl. Ther.*, **1082**, 1–8.
59. Li, S.D. and Huang, L. (2006) Gene therapy progress and prospects: non-viral gene therapy by systemic delivery. *Gene Ther.*, **13**, 1313–1319.

60. Foged, C. and Nielsen, H.M. (2008) Cell-penetrating peptides for drug delivery across membrane barriers. *Exp. Opin. Drug Deliv.*, **5**, 105–117.
61. Zhang, C.L., Tang, N., Liu, X.J. *et al.* (2006) Torchilin, siRNA-containing liposomes modified with polyarginine effectively silence the targeted gene. *J. Control. Release*, **112**, 229–239.
62. Foged, C., Nielsen, H.M. and Frokjaer, S. (2007) Liposomes for phospholipase A2 triggered siRNA release: preparation and *in vitro* test. *Int. J. Pharm.*, **331**, 160–166.
63. Landen, C.N. Jr., Chavez-Reyes, A., Bucana, C. *et al.* (2005) Lopez-Berestein and A. K. Sood, Therapeutic EphA2 gene targeting *in vivo* using neutral liposomal small interfering RNA delivery. *Cancer Res.*, **65**, 6910–6918.
64. Halder, J., Kamat, A.A., Landen, C.N. *et al.* (2006) Lopez-Berestein and A. K. Sood, Focal adhesion kinase targeting using *in vivo* short interfering RNA delivery in neutral liposomes for ovarian carcinoma therapy. *Clin. Cancer Res.*, **12**, 4916–4924.
65. Mastrobattista, E., Kapel, R.H.G., Eggenhuisen, M.H. *et al.* (2001) Lipid-coated polyplexes for targeted gene delivery to ovarian carcinoma cells. *Cancer Gene Ther.*, **8**, 405–413.
66. Matsui, H., Johnson, L.G., Randell, S.H. and Boucher, R.C. (1997) Loss of binding and entry of liposome-DNA complexes decreases transfection efficiency in differentiated airway epithelial cells. *J. Biol. Chem.*, **272**, 1117–1126.
67. Rejman, J., Bragonzi, A. and Conese, M. (2005) Role of clathrin- and caveolae-mediated endocytosis in gene transfer mediated by lipo- and polyplexes. *Mol. Ther.*, **12**, 468–474.
68. Simoes, S., Pires, P., Duzgunes, N. and Pedrosa de Lima, M.C. (1999) Cationic liposomes as gene transfer vectors: barriers to successful application in gene therapy. *Curr. Opin. Mol. Ther.*, **1**, 147–157.
69. Khalil, I.A., Kogure, K., Futaki, S. and Harashima, H. (2006) High density of octaarginine stimulates macropinocytosis leading to efficient intracellular trafficking for gene expression. *J. Biol. Chem.*, **281**, 3544–3551.
70. Wolfrum, C., Shi, S., Jayaprakash, K.N. *et al.* (2007) Mechanisms and optimization of *in vivo* delivery of lipophilic siRNAs. *Nat. Biotechnol.*, **25**, 1149–1157.
71. Mui, B., Raney, S.G., Semple, S.C. and Hope, M.J. (2001) Immune stimulation by a CpG-containing oligodeoxynucleotide is enhanced when encapsulated and delivered in lipid particles. *J. Pharmacol. Exp. Ther.*, **298**, 1185–1192.
72. de Jong, S., Chikh, G., Sekirov, L. *et al.* (2007) Encapsulation in liposomal nanoparticles enhances the immunostimulatory, adjuvant and anti-tumor activity of subcutaneously administered CpG ODN. *Cancer Immunol. Immunother.*, **56**, 1251–1264.
73. Judge, A.D., Sood, V., Shaw, J.R. *et al.* (2005) Sequence-dependent stimulation of the mammalian innate immune response by synthetic siRNA. *Nat. Biotechnol.*, **23**, 457–462.
74. Ma, Z., Li, J., He, F.T. *et al.* (2005) Cationic lipids enhance siRNA-mediated interferon response in mice. *Biochem. Biophys. Res. Commun.*, **330**, 755–759.
75. Judge, A.D., Bola, G., Lee, A.C.H. and MacLachlan, I. (2006) Design of noninflammatory synthetic siRNA mediating potent gene silencing *in vivo*. *Mol. Ther.*, **13**, 494–505.
76. Hornung, V., Guenthner-Biller, M., Bourquin, C. *et al.* (2005) Sequence-specific potent induction of IFN-alpha by short interfering RNA in plasmacytoid dendritic cells through TLR7. *Nat. Med.*, **11**, 263–270.
77. Schlee, M., Hornung, V. and Hartmann, G. (2006) siRNA and isRNA: two edges of one sword. *Mol. Ther.*, **14**, 463–470.
78. Allison, A.G. and Gregoriadis, G. (1974) Liposomes as immunological adjuvants. *Nature*, **252**, 252.
79. Gregoriadis, G., McCormack, B., Obrenovic, M. *et al.* (1999) Vaccine entrapment in liposomes. *Methods*, **19**, 156–162.
80. Christensen, D., Korsholm, K.S., Rosenkrands, I. *et al.* (2007) Cationic liposornes as vaccine adjuvants. *Exp. Rev. Vaccines*, **6**, 785–796.
81. Omidi, Y., Hollins, A.J., Benboubetra, M. *et al.* (2003) Toxicogenomics of non-viral vectors for gene therapy: a microarray study of lipofectin- and oligofectamine-induced gene expression changes in human epithelial cells. *J. Drug Target*, **11**, 311–323.
82. Petch, A.K., Sohail, M., Hughes, M.D. *et al.* (2003) RNA expression profiling of genes involved in epidermal growth factor receptor signalling in human cancer cells treated with scanning array-designed antisense oligonucleotides. *Biochem. Pharmacol.*, **66**, 819–830.

83. Vangasseri, D.P., Cui, Z.R., Chen, W.H. *et al.* (2006) Immunostimulation of dendritic cells by cationic liposomes. *Mol. Mem. Biol.*, **23**, 385–395.
84. Liu, F., Conwell, C.C., Yuan, X. *et al.* (2007) Novel nonviral vectors target cellular signaling pathways: regulated gene expression and reduced toxicity. *J. Pharmacol. Exp. Ther.*, **321**, 777–783.
85. Kunisawa, J., Masuda, T., Katayama, K. *et al.* (2005) Fusogenic liposome delivers encapsulated nanoparticles for cytosolic controlled gene release. *J. Control. Release*, **105**, 344–353.
86. Cardoso, A.L.C., Simoes, S., de Almeida, L. *et al.* (2007) siRNA delivery by a transferrin-associated lipid-based vector: a non-viral strategy to mediate gene silencing. *J. Gene Med.*, **9**, 170–183.

16

Cellular Bioavailability of Peptide Nucleic Acids (PNAs) Conjugated to Cell Penetrating Peptides

Takehiko Shiraishi[1,2] *and Peter E. Nielsen*[2]

[1] Utrecht University, Department of Pharmaceutical Sciences, Faculty of Science, Utrecht, The Netherlands

[2] University of Copenhagen, Department of Cellular and Molecular Medicine, The Panum Institute, Faculty of Health Sciences, Copenhagen, Denmark

16.1 Introduction

RNA interference as a drug discovery principle has existed for 30 years, starting with traditional oligonucleotide antisense in the 1980s and moving into second generation oligonucleotide chemistry in the 1990s. Although significant technical and conceptual progress was achieved, only one drug, Vitravene®, was approved and the field and the businesses were suffered a confidence crisis. Thus, there is little doubt that, for both the technology and the drug discovery opportunities, the entrance of small interfering RNAs (siRNA) and subsequently micro RNA (miRNA) on the scientific scene in this century was a saviour, one which brought novel targeting principles as well as novel potential genetic drug targets. Nonetheless, the major obstacles for the development of large, hydrophilic and most often polyanionic molecules like oligonucleotides into gene therapeutic drugs, that is efficient *in vivo* delivery and bioavailability, still remain largely unsolved, and many researchers in the field believe that surmounting this hurdle will indeed unleash the potential of genetic drug discovery and development.

Delivery Technologies for Biopharmaceuticals: Peptides, Proteins, Nucleic Acids and Vaccines Edited by Lene Jorgensen and Hanne Mørck Nielsen

16.2 Peptide Nucleic Acids (PNAs)

Peptide Nucleic Acids (PNAs) were introduced in 1991 [1] as a pseudopeptide DNA mimic (Figure 16.1). Thus PNAs have the nucleobase sequence recognition properties of nucleic acids but the chemical properties of non-natural peptides. Being charge neutral molecules of very high chemical and, in particular, biological stability, with a structure that is readily synthetically accessible and provides broad opportunities for chemical modification and conjugation, PNAs have attracted broad scientific attention in the bioorganic community as well as for drug discovery and development and for exploitation in molecular biology and genetic diagnostic tools and techniques [2]. In particular, numerous studies have demonstrated the potential of designed PNA oligomers as modulators of cellular gene expression.

16.3 Cell Penetrating Peptides (CPPs)

The first cell penetrating peptides (CPPs) described were derived from functional domains of naturally occurring proteins allowing these to migrate between cells. The two original

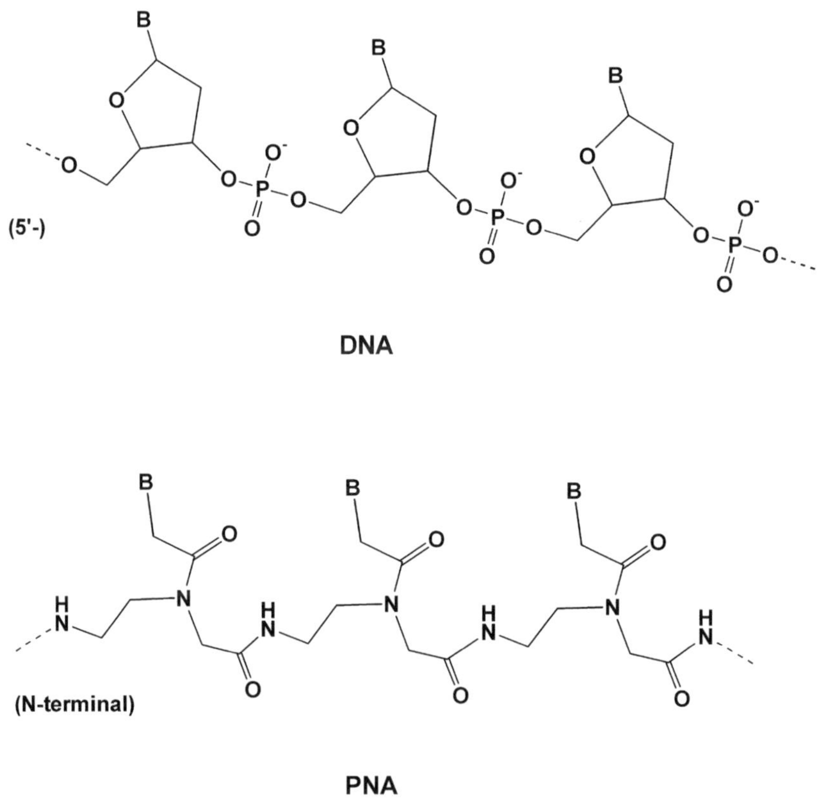

Figure 16.1 Chemical structures of DNA and PNA

peptides, penetratin [3] from the homeodomain (DNA binding domain) of Antennapedia from Drosophila, and the Tat-peptide [4] from HIV tat protein, are still being studied in terms of mechanism of cellular uptake, but a large number of novel CPPs have since been discovered [5]. Like penetratin and Tat-peptide, most of these are cationic with a significant content of arginines, and indeed oligoarginines themselves are quite effective CPPs [5, 6].

16.4 Cellular Uptake Versus Bioavailability

Cellular uptake is typically studied by fluorescence (confocal) microscopy using fluorophore-labelled compounds. Although such analyses yield information about both uptake and intracellular distribution of the compound, they are prone to artifacts due to handling of the cells (such as fixation) and detailed compartmentalization is difficult to discern [7–9]. This is especially important for gene targeting compounds, which often do not diffuse freely in the cell and for which the targets are highly compartmentalized. Therefore, cellular uptake *per se* is only a very rough guideline for target bioavailability, and a measure of genuine bioavailability is only obtained from a (molecular) biology end point of gene modulation activity. Recently, many researchers have chosen to exploit the very sensitive and easily quantifiable luciferase splicing correction pLuc705 HeLa cell system developed by Kole *et al.* [10] for structure–activity relationship (SAR) studies on CPPs conjugated to mRNA splice junction active antisense oligomers such as PNAs. This assay allows both a relative luciferase activity readout as well as an absolute splice correction readout at the mRNA level using RT-PCR.

16.5 PNA Conjugates

Thus, a very large number of different CPPs have been studied as PNA conjugates in the pLuc705 HeLa cell system as well as in other antisense contexts (Table 16.1). Unfortunately, it is not straightforward to quantitatively compare the results from different studies as slightly different conditions have been used and, for many of the studies, only luciferase activity and not mRNA splice correction readout has been performed. Nonetheless, we think that it is safe to draw some general conclusion on the basis of the compiled data (Table 16.1). Despite early reports, subsequent studies have, in general, not been able to confirm the efficient cellular delivery by the penetratin or Tat peptides. Likewise, NLS (nuclear localization signal) peptide by itself is not an efficient delivery agent. On the other hand, results from several studies have shown that the chimaeric peptide transportan and simple oligoarginines containing optimally around eight arginines are the most effective of the CPPs. Furthermore, recent results have demonstrated that attachment of a lipophilic domain, such as a fatty acid, to cationic PNA-CPP-conjugates may significantly improve the activity (two orders of magnitude in the case of Tat) [11, 12].

16.6 Efficacy Versus Efficiency

The majority of studies reporting cellular antisense efficacy of PNA-peptide conjugates report EC_{50} values at best in the low micromolar range [48] and in a few cases only just

Table 16.1 In vitro *studies performed with different CPP types*

CPP type	Author, year, reference	CPP sequence[a]	Cargo	Conjugation method for CPP/PNA
Arginine-rich peptide				
	Hu and Coey (2007) [11]	R_8F_2, R_8F_4, R_8W_2, R_8W_4, R_8H_4	PNA, PNA with lipidic domain (fatty acid)	Direct conjugation via continuous synthesis
	Koppelhus *et al.* (2008) [12]	$(RK)_3$, $(RK)_4$, $(RRA)_3$, $(RRA)_4$, R_4K_4, $K_2R_4K_2$	PNA, PNA with lipidic domain (decanoic acid)	Direct conjugation via continuous synthesis (via eg1 linker or Lys).
	Shiraishi and Nielsen (2006) [56]	$(RK)_3$, $(RK)_4$, $(RRA)_3$, $(RRA)_4$, R_4K_4, $K_2R_4K_2$	PNA	Direct conjugation via continuous synthesis (via Gly).
	Turner *et al.* (2005) [13]	R_9F_2	PNA	Chemical conjugation via disulfide bridge
Chimaeric peptide or combination of CPPs				
	Abes *et al.* (2007) [14]	R_6-Penetratin (RRRRRRR QIKIWFQN RRMKW KKGGC) and derivatives	PNA	Chemical conjugation via thioacetyl linker or disulfide bridge.
	Fabani and Gait (2008) [15]	R_6-Penetratin	PNA	Chemical conjugation via disulfide bridge
	Gros *et al.* (2006) [16]	Pep-2 (KETWFETWF TEWSQPKKKRKV)	HypNA-pPNA (trans-4-hydroxyl-L-proline/ phosphonate-PNA) chimera	Complexation
	Morris *et al.* (2007) [17]	Pep-3 (K W F E T W F T E W P K K R K) and its PEG derivatives	HypNA-pPNA	Complexation

Assay system	PNA target	Label	Detection (s)	Results
Breast cancer cells (T47D, MCF-7)	Antigene targeting of Progesterone receptor (PR) gene	None	Downregulation of PR expression	Eight or more cationic amino acids are preferred. Improved antisense potency with hydrophobic moieties.
HeLa pLuc705 cells	Cryptic splice site of the mutated luciferase gene	None	Luciferase activity (splicing correction)	Very significant improvement of the antisense potency with hydrophobic moieties.
HeLa pLuc705 cells	Cryptic splice site of the mutated luciferase gene	None	Luciferase activity (splicing correction)	All PNA conjugates showed significantly higher nuclear antisense splicing correction activity in combination with PCI.
HeLa cells	HIV-1 Tat-dependent transactivation response RNA	None	Confocal microscope, inhibition of the transactivation,	Inactive in the absence of chloroquine (CQ).
HeLa pLuc705 cells	Cryptic splice site of the mutated luciferase gene	None	Luciferase activity (splicing correction)	Efficient splicing correction occurs at 1 μM.
Huh7 cells	miR-122 (liver specific miRNA linked with lipid metabolism)	None	Inhibition of the miR122 activity	Successful inhibition of the miR-122 at 1 μM.
Mammalian cultured cell (HS-68 cells)	Cyclin B1	None	Downregulation of Cyclin B1	Downregulation of Cyclin B1 at submicromolar concentrations.
HeLa, primary human umbilical vein endothelial cells (HUVEC), Jurkat T, PC3 and MCF-7 cells. PC3cells xenografted to nude mice (i.v.)	Cyclin B1	Fluorescein	Downregulation of Cyclin B1, fluorescence microscopy.	Downregulation of Cyclin B1 at submicromolar concentrations. PEGylation of Pep-3 significantly improved complex stability *in vivo* and was used when antisense cyclin B1 administered intravenously (90% tumour growth inhibition at 10 mg/kg at 25 days).

(*continued overleaf*)

Table 16.1 (*continued*)

CPP type	Author, year, reference	CPP sequence[a]	Cargo	Conjugation method for CPP/PNA
	Turner *et al.* (2005) [13]	Combination of NLS- and Tat-, R6-penetratin	PNA	Chemical conjugation via eg1 linker (for NLS and Tat) or via disulfide bridge (for R6-Penetratin)
Human Protein origin				
	Bendifallah *et al.* (2006) [18]	SynB3 (RRLSYSRR RF, derived from natural mammalian antimicrobial peptide)	PNA	Chemical conjugation via disulfide bridge
	El-Andaloussi *et al.* (2007) [19]	M918 (MVTVLFRRL RIRRACGPPR VRV, derived from the tumour suppressor protein p12ARF)	PNA	Chemical conjugation via disulphide bridge
MAP (model amphipathic peptide)				
	Bendifallah *et al.* (2006) [18]	$(KFF)_3K$	PNA	Direct synthesis
	Faridani *et al.* (2006) [20]	$(KFF)_3K$	PNA	Direct synthesis
	Good *et al.* (2001) [21]	$(KFF)_3K$, and its derivatives	PNA	Chemical conjugation via eg1 linker or maleimide-eg1 linker for the derivatives
	Kurupati *et al.* (2007) [22]	$(KFF)_3K$	PNA	Chemical conjugation via Leu.
	Maier *et al.* (2006) [23]	Three peptides (I: GKKAFKGAGKGF KK; II: GKKAFKK AKKKFKK; III: GKKLAKLAAKL LKK). Substitution derivatives of I, II and III. Inverso isomers and retro-inverso isomers of II and III.	PNA	Chemical conjugation via ethylene glycol linker (eg1, 8-amino-2,6-dioxaoctanoic acid)

Assay system	PNA target	Label	Detection (s)	Results
HeLa cells	HIV-1 Tat-dependent transactivation response RNA	Fluorescein (eg1 linker conjugates)	Confocal fluorescence microscopy, inhibition of transactivation	R6-Penetratin PNA conjugates with disulfide bridge inhibit Tat-dependent transactivation. NLS–PNA–Tat showed general fluorescence release and no activity in the absence of CQ.
HeLa pLuc705 cells	Cryptic splice site of the mutated luciferase gene	None	Luciferase activity (splicing correction)	No activity.
HeLa pLuc705 cells	Cryptic splice site of the mutated luciferase gene	None	Luciferase activity (splicing correction)	Better splicing correction (at 5 μM) than Penetratin, TP10. Macropinocytotic uptake (inhibitor study with cytochalasin D and wortamannin)
HeLa pLuc705 cells	Cryptic splice site of the mutated luciferase gene	None	Luciferase activity (splicing correction)	KFF-PNA conjugate was the third most active conjugate.
E. coli	Sok-RNA	None	Inhibition of bacterial cell growth	Bacterial growth inhibition at 8 μM.
E. coli	lacZ, essential acyl carrier protein (*acpP*)	None	Inhibition of bacterial cell growth	Downregulation of lacZ. Downregulation of the acpP and inhibition of the bacterial growth at submicromolar concentrations.
Klebsiella pneumoniae	Essential genes (*gyrA*, *ompA*)	None	Inhibition of bacterial cell growth	Bacterial growth inhibition at higher than 20 μM.
BCL1 cells, Balb/c male mice (D-(RRFRRRAR RFARRG)-and D-(KKLLKAALKLA KKG)-	3′-end of exon 6 of the primary murine CD40 gene	Radiolabelling (^{3}H)	Downregulation of the CD40 expression, radioactivity measurement of the tissues	Antisense activities at few μM range. Highest concentration found in liver, kidney and spleen at 3 mg/Kg dose.

(continued overleaf)

Table 16.1 (*continued*)

CPP type	Author, year, reference	CPP sequence[a]	Cargo	Conjugation method for CPP/PNA
	Nekhotiaeva *et al.* (2004) [24]	$(KFF)_3K$, and its derivatives	PNA	Chemical conjugation via Leu
	Nikravesh *et al.* (2007) [25]	$(KFF)_3K$	PNA	Chemical conjugation via eg1 linker
	Pritz *et al.* (2007) [26]	KLAKLA LKALKA AKLA	PNA	Coupling of PNA and CPP by sortase-mediated ligation
	Shiraishi *et al.* (2005) [27]	$(KFF)_3K$	PNA	Chemical conjugation via eg1 linker
	Shiraishi and Nielsen (2006) [56]	(KFF)3K, KLALK LALKA LKAALKLA	PNA	Chemical conjugation via Gly
	Wolf *et al.* (2006) [28]	KLA (KLAL KLAL KAL KAAL KLA-) and its delivertives (KGL, KAL, ELA, RLA)	PNA	Chemical conjugation either AEEA linkers or disulfide bridge
New CPP sequences				
	Abes *et al.* (2007) [29]	Tat, $(R\text{-}Ahx\text{-}R)_3$ (Ahx, 6-aminohexanoic acid)	PNA	Chemical conjugation

Assay system	PNA target	Label	Detection (s)	Results
Staphylococcus aureus	Essential genes *fmhB*, *gyrA* and *hmrB*	None	Inhibition of bacterial cell growth	Antisense downregulation of reporter GFP and *S.aureus* genes. Bacterial cell growth inhibition at 10 μM.
E. coli	*acpP*	Cy5	Fluorescence microscopy, inhibition of the bacterial cell growth	Accumulation in the bacterial cytoplasm. Showed a bactericidal activity with a strong PAE (post-antibiotic effect) at submicromolar concentrations.
HeLa pLuc 705 cells	Cryptic splice site of the mutated luciferase gene	None	Luciferase activity (splicing correction)	Dose dependent antisense activity at μM concentrations.
HeLa pLuc705 cells	Cryptic splice site of the mutated luciferase gene	None	Luciferase activity (splicing correction)	Improvement of the CPP-PNA conjugates antisense effects by Ca^{2+} treatment.
HeLa pLuc705 cells	Cryptic splice site of the mutated luciferase gene	None	Luciferase activity (splicing correction)	Significantly higher nuclear antisense splicing correction activity in combination with PCI.
HeLa pLuc705 cells	Cryptic splice site of the mutated luciferase gene	Fluorescein	Confocal microscopy (for KLA conjugate), luciferase activity (splicing correction)	Improvement of the antisense activity by CQ and Ca^{2+} treatment. Nuclear uptake. Addition of 6 mM Ca^{2+} resulted in an enhanced accumulation of the PNA-KLA conjugate at the cell membrane, outside as well as inside the cell, as large aggregates, with a few spots being located within the nucleus. The effect of Ca^{2+} on the distribution of the PNA-Pen conjugate was less strong compared to that of CQ.
HeLa pLuc705 cells	Cryptic splice site of the mutated luciferase gene	None	Luciferase activity (splicing correction)	Efficient splicing correction with $(R\text{-}Ahx\text{-}R)_4$ at low concentration (0.5–1 μM).

(continued overleaf)

Table 16.1 (*continued*)

CPP type	Author, year, reference	CPP sequence[a]	Cargo	Conjugation method for CPP/PNA
	Bøe and Hovig (2006) [30]	KDEL (ER targeting peptide), SDYGRL- (trans-Golgi network targeting), (AKL (peroxisomes targeting), GHHHHHG-, a mitochondrial import signal (MSVLTPLLLRG LTGSARRLPVPR AKIHSL)	PNA	Chemical conjugation via AEEA linker
	Resina *et al.* (2007) [31]	$(R\text{-}Ahx\text{-}R)_4$	PNA	Chemical conjugation via thioacetyl linker
	Yin *et al.* (2008) [32]	Muscle-specific peptide (MSP, found from the phage display in mouse C2C12 cell line)), AAV6 and AAV8 (derived from the adeno associated virus 6 functional domains)	PNA	Chemical conjugation via disulfide bridge with the AEEA (acetyl-ethyleneglycol-ethyl-amine) linker
NLS				
	Bendifallah *et al.* (2006) [18]	NLS (PKKKRKV)	PNA	Direct conjugation
	Boffa *et al.* (2007) [33]	NLS	PNA	Direct conjugation
	Bøe and Hovig (2006) [30]	NLS	PNA	Chemical conjugation via AEEA linker, eg1 liker or disulfide bridge
	Rasmussen *et al.* (2006) [34]	NLS	PNA (Delivered as heteroduplex with complementary oligo DNA by poly(ethylene imine)	Chemical conjugation via eg1 linker

Assay system	PNA target	Label	Detection (s)	Results
FEMX III, HeLa, OHS cells	S100A4 gene	Fluorescein	Confocal fluorescence microscopy (GHHHHHG-, mitochondrial import signal), flow cytometry (KDEL, SDYGRL)	Uptake was CPP charge-dependent. Net charge 0 or −1, KDEL (ER targeting peptide), SDYGRL- (trans-Golgi network targeting). Net charge +1 (AKL (peroxisomes targeting). Net charge +5, MSVL TPLLLRGLTGSA RRLPVPRAKIHSL, a mitochondrial import signal), GHHHHHG-.
HeLa pLuc705 cells	Cryptic splice site of the mutated luciferase gene	Fluorescein	Luciferase activity (splicing correction)	Efficient splicing correction with $(R\text{-}Ahx\text{-}R)_4$ at low concentration (0.5–1 μM). Endosomal entrapment of the Rar-PNA.
H2k mdx cells, mdx mice (intramuscular injection)	Exon 23 of dystrophin pre-mRNA	None	Skipping of exon 23 of dystrophin and expression of functional dystrophin (restoration the reading frame)	Exon skipping by PNA and PNA conjugated (cultured cells). Restoration of the dystrophin expression by PNA (non-CPP conjugate) is superior to other CPP-PNA conjugates.
HeLa pLuc705 cells	Cryptic splice site of the mutated luciferase gene	None	Luciferase activity (splicing correction)	No activity.
Burkitt's lymphoma (BL) cells xenografted to SCID mice	Eμ-enhancer	None	Downregulation of c-Myc	c-Myc downregulation and tumour growth inhibition by i.v. injection (100 μl of 0.2 mM PNA).
FEMX III, HeLa, OHS cells	S100A4 gene	Fluorescein	Confocal fluorescence microscopy, downregulation of S100A4 gene, flow cytometry	Relocalization of NLS-PNA to the nucleus and downregulation of the S100A4 gene with PCI treatment.
HeLa pLuc705 cells	Cryptic splice site of the mutated luciferase gene	None	Luciferase activity (splicing correction)	Less antisense activities than unmodified PNA.

(continued overleaf)

Table 16.1 (*continued*)

CPP type	Author, year, reference	CPP sequence[a]	Cargo	Conjugation method for CPP/PNA
	Robaczewska *et al.* (2005) [35]	NLS	PNA	Chemical conjugation via eg1 linker
	Shiraishi *et al.* (2005) [27]	NLS	PNA	Chemical conjugation via eg1 linker
	Shiraishi and Nielsen (2006) [56]	NLS	PNA	Chemical conjugation via eg1 linker
	Turner *et al.* (2005) [13]	NLS	PNA	Chemical conjugation via eg1 linker
Oligo-arginine				
	Bendifallah *et al.* (2006) [18]	D-R_{7-9}	PNA	Chemical conjugation by direct conjugation or via ester maleimide or stable maleimide for R9-
	Bøe and Hovig (2006) [30]	R_7	PNA	Chemical conjugation via eg1 linker
	Hu and Corey (2007) [11]	R_8, R_{12}, D-R_8, D-R_{12}, R_5-R_5(branched)	PNA, PNA with lipidic domain (such as fatty acid)	Direct synthesis
	Koppelhus *et al.* (2008) [12]	R_{6-9}, D-R_{6-9}	PNA, PNA with lipidic domain (such as fatty acid)	Chemical conjugation by direct conjugation or via eg1 spacer, Gly or Lys

Assay system	PNA target	Label	Detection (s)	Results
Primary duck hepatocytes (PDH)	DHBV (Duck hepatitis B virus) encapsidation signal ε gene	None	Inhibition of duck hepatitis B virus reverse transcription	PNAs reproducibly inhibited DHBV RT in a dose-dependent manner with IC_{50} of 10 nM (*in vitro*), whereas up to 600-fold higher concentration of S-ODNs was required for similar inhibition.
HeLa pLuc705 cells	Cryptic splice site of the mutated luciferase gene	Fluorescein	Luciferase activity (splicing correction)	Improvement of the NLS-PNA conjugates antisense effects by Ca^{2+} treatment. Less active than other CPP-PNA conjugates.
HeLa pLuc705 cells	Cryptic splice site of the mutated luciferase gene	None	Luciferase activity (splicing correction)	Significantly higher nuclear antisense splicing correction activity in combination with PCI. Less active than other CPP-PNA conjugates.
HeLa cells	HIV-1 Tat-dependent transactivation response RNA	Fluorescein	Inhibition of the transactivation, confocal microscope	No activity.
HeLa pLuc705 cells	Cryptic splice site of the mutated luciferase gene	None	Luciferase activity (splicing correction)	Less active than Transportan-PNA conjugate.
FEMX III, HeLa, OHS cells	S100A4 gene	Fluorescein	Flow cytometry	No difference with NLS-PNA conjugate.
Breast cancer cells (T47D, MCF-7)	Antigene targeting of Progesterone receptor (PR) gene	None	Downregulation of PR expression	Eight or more cationic amino acids are preferred. Improved antisense potency with hydrophobic moieties.
HeLa pLuc705 cells	Cryptic splice site of the mutated luciferase gene	Fluorescein (R_8-conjugate)	Confocal microscope and flow cytometry (for R_8-conjugates), luciferase activity (splicing correction)	Improvement by the lipidic domain addition.

(*continued overleaf*)

Table 16.1 (continued)

CPP type	Author, year, reference	CPP sequence[a]	Cargo	Conjugation method for CPP/PNA
	Robaczewska *et al.* (2005) [35]	R_7	PNA	Chemical conjugation via eg1 linker
	Shiraishi *et al.* (2005) [27]	R_6, R_7, R_9	PNA	Chemical conjugation via Gly. Lys and eg1linker at the N terminal of PNA with fluorescein labelling
	Shiraishi and Nielsen (2006) [56]	R_7, D-R_{6-9}	PNA	Chemical conjugation via Gly or eg1 linker (for R_7-PNAs).
Oligo-lysine, Lysine-rich peptide				
	Fabani and Gait (2008) [15]	K_3	PNA	Direct synthesis
	Fletcher *et al.* (2006) [36]	K_4	PNA	Chemical conjugation via linker
	Hu and Corey (2007) [11]	K_8, D-K_4, D-K_8, D-K_{10}, D-K_{12}, D-$(AAKK)_4$, K_4K_4 (branched)	PNA, PNA with nonadecanoic acid (for R_8)	Chemical conjugation
	Maier *et al.* (2006) [23]	K_8	PNA	Chemical conjugation via eg1 linker
	Resina *et al.* (2007) [31]	K_4, K_8	PNA	Direct synthesis

Assay system	PNA target	Label	Detection (s)	Results
Primary duck hepatocytes (PDH)	DHBV (Duck hepatitis B virus) encapsidation signal ε gene	None	Inhibition of duck hepatitis B virus reverse transcription	PNAs reproducibly inhibited DHBV RT in a dose-dependent manner with IC_{50} of 10 nM, whereas up to 600-fold higher concentration of S-ODNs was required for similar inhibition.
HeLa pLuc705 cells	Cryptic splice site of the mutated luciferase gene	Fluorescein (R_7-conjugate)	Confocal microscope and flow cytometry (for R_7-conjugate), luciferase activity (splicing correction)	Colocalization of R_7-Fl-PNA with Tf. Improvement of the CPP-PNA conjugates antisense effects by Ca^{2+} treatment.
HeLa pLuc705 cells, p53R cells (expressing normal luciferase)	Cryptic splice site of the mutated luciferase gene (HeLa pLuc705 cells), Start codon of the luciferase gene (p53R cells)	None	Luciferase activity (antisense downregulation (p53R cells, D-R_7-PNA), splicing correction (HeLa pLuc705 cells)	Significantly higher nuclear antisense splicing correction activity in combination with PCI. R7-PNA showed antisense downregulation of the luciferase gene only in combination with PCI.
Huh7 cells	miR-122 (liver specific miRNA linked with lipid metabolism)	None	Inhibition of the miR 122 activity	Strong inhibition of the miR-122 activity at 1 μM.
Mdx mouse (intramuscular injection)	Muscular dystrophy exon 23 splice site	None	Skipping of Exon 23 (RT) and restoration of the dystrophin protein expression	No activity.
Breast cancer cells (T47D, MCF-7)	Antigene targeting of Progesterone receptor (PR) gene	None	Downregulation of PR expression	Eight or more cationic amino acids are preferred. Improved antisense potency with hydrophobic moieties.
BCL1 cells	3′-end of exon 6 of the primary murine CD40	None	Downregulation of the CD40 expression, radioactivity measurement of the tissues	Antisense activities at a few μM range.
HeLa pLuc705 cells	Cryptic splice site of the mutated luciferase gene	None	Luciferase activity (splicing correction)	Slightly better than Tat-PNA conjugate.

(continued overleaf)

Table 16.1 (*continued*)

CPP type	Author, year, reference	CPP sequence[a]	Cargo	Conjugation method for CPP/PNA
	Turner *et al.* (2005) [13]	K_8	PNA	Direct synthesis
Penetratin				
	Bendifallah *et al.* (2006) [18]	Penetratin (CRQIK IWFQNRRMKW KK)	PNA	Chemical conjugation via stable maleimide linker
	El-Andaloussi *et al.* (2006) [37]	Penetratin	PNA	Chemical conjugation via disulfide bridge
	El-Andaloussi *et al.* (2007) [19]	Penetratin	PNA	Chemical conjugation via disulfide bridge
	Ganguly *et al.* (2008) [38]	Penetratin	PNA	Chemical conjugation via disulfide bridge or eg1 linker
	Koppelhus *et al.* (2002) [39]	Penetratin	PNA, PNA with three Lysine backbone thymine residues.	Chemical conjugation via disulfide bridge or maleimide
	Turner *et al.* (2005) [13]	Penetratin	PNA	Chemical conjugation via disulfide bridge

Assay system	PNA target	Label	Detection (s)	Results
HeLa cells	HIV-1 Tat-dependent transactivation response RNA	Fluorescein	Inhibition of the transactivation, confocal microscopy	No inhibition with K_8-PNA. Confocal microscopy showed that such fluorescently labelled CPP-PNA conjugates were sequestered in endosomal or membrane-bound compartments of HeLa cells (no significant cytosolic release by CQ).
HeLa pLuc705 cells	Cryptic splice site of the mutated luciferase gene	None	Luciferase activity (splicing correction)	No activity.
HeLa pLuc705 cells	Cryptic splice site of the mutated luciferase gene	Fluorescein	Confocal microscopy, fluorescence spectroscopy, luciferase activity (splicing correction)	High correlation between the uptake and nuclear antisense activity. Inhibition of the uptake and antisense activity by low temperature.
HeLa pLuc705 cells	Cryptic splice site of the mutated luciferase gene	None	Luciferase activity (splicing correction)	M918 peptide is better than Penetratin conjugate.
Male Balb/c mice (*ip* or oral gavage)	Transactivation response (TAR) element of HIV-1 LTR	Radiolabelling (^{125}I)	Radioactivity measurement in the tissues	Significant accumulation of CPP-PNA conjugated in organs and tissues. Unconjugated naked PNA displayed a similar distribution and clearance profile in most organs studied although extent of its uptake was lower than the CPP-PNA conjugates.
SK-BR-3, IMR-90, HeLa H9 and U937	Starting codon of her2 oncogene	Fluorescein	Confocal microscopy	Vesicular cytoplasmic uptake (confocal microscopy).
HeLa cells	HIV-1 Tat-dependent transactivation response RNA	None	Inhibition of the transactivation	Inactive in the absence of CQ.

(*continued overleaf*)

Table 16.1 (*continued*)

CPP type	Author, year, reference	CPP sequence[a]	Cargo	Conjugation method for CPP/PNA
	Villa *et al.* (2000) [40]	Penetratin	PNA	Chemical conjugation via disulfide bridge.
	Wolf *et al.* (2006) [28]	Penetratin	PNA	Chemical conjugation either AEEA-linkers(x3) or disulfide bridge
PNA-Backbone modification				
	Dragulescu-Andrasi *et al.* (2006) [41]	D-arginine side chains within the PNA sequence backbone	GPNA (PNA containing internally linked D-arginine side chains)	N-(2-aminoethyl)-D-arginine backbone unit
	Nielsen and Shiraishi (2006) [42]	Arginine side chains within the PNA sequence backbone	GPNA (PNA containing internally linked D-arginine side chains)	N-(2-aminoethyl)-D-arginine backbone unit
	Koppelhus *et al.* (2002) [39]	Lys side chains within the PNA sequence backbone	PNA containing internally linked Lys side chains	Lys residues within the PNA sequence backbone
Tat				
	Abes *et al.* (2007) [29]	Tat (GRKK RRQR RRPPQ)	PNA	Chemical conjugation by direct synthesis or via eg1 linker, disulfide bridge
	Bendifallah *et al.* (2006) [18]	Tat and its derivatives	PNA	Chemical conjugation by direct synthesis or via ester maleimide or stable maleimide linking
	Bøe and Hovig (2006) [30]	Tat	PNA	Chemical conjugation via eg1 linker

Assay system	PNA target	Label	Detection (s)	Results
JR8 melanoma cells	RNA component of human telomerase	Biotin	Immunofluorescence detection of biotin-PNA, inhibition of the telomerase activity	Detection of the homogeneous distribution within the cells. Inhibition of the telomerase activity.
HeLa pLuc705 cells	Cryptic splice site of the mutated luciferase gene	Fluorescein	Confocal microscopy (AEEA-linker conjugate), luciferase activity (splicing correction)	Improvement of the antisense activity by CQ and Ca^{2+} treatment. Cleavable disulfide bond is not essential. No nuclear uptake. Addition of 6 mM Ca^{2+} resulted in an enhanced accumulation of the PNA-conjugate at the cell membrane, outside as well as inside the cell, as large aggregate.
Human carcinoma cultured cells (A549 cells)	5′-UTR of E-cadherin	None	Downregulation of E-Cadherin (immunofluorescent-staining, RT-PCR)	Downregulation of E-Cadherin at 10 μM.
HeLa pLuc705 cells	Cryptic splice site of the mutated luciferase gene	None	Luciferase activity (splicing correction)	The effect of having the arginine side chain (guanidine) in the backbone is similar to terminal arginine conjugation.
SK-BR-3, IMR-90, HeLa, H9 and U937 cells	Starting codon of her2 oncogene	Fluorescein	Confocal microscopy	Cell type-dependent uptake. Vesicular cytoplasmic (and diffuse cytoplasmic) uptake (confocal microscopy).
HeLa pLuc705 cells	Cryptic splice site of the mutated luciferase gene	None	Luciferase activity (splicing correction)	Significant enhancement by CQ-treatment. Stable linker conjugates are slightly better than disulfide linkage conjugates in the presence of CQ.
HeLa pLuc705 cells	Cryptic splice site of the mutated luciferase gene	None	Luciferase activity (splicing correction)	Tat-PNA conjugate was the third most active conjugate.
FEMX III, HeLa, OHS cells	S100A4 gene	Fluorescein	Flow cytometry	No significant difference with NLS-PNA conjugate.

(*continued overleaf*)

Table 16.1 (*continued*)

CPP type	Author, year, reference	CPP sequence[a]	Cargo	Conjugation method for CPP/PNA
	El-Andaloussi *et al.* (2006) [37]	Tat	PNA	Chemical conjugation via disulfide bridge
	Folini *et al.* (2007) [43]	Tat	PNA	Direct conjugation
	Ganguly *et al.* (2008) [38]	Tat	PNA	Chemical conjugation via disulfide bridge
	Koppelhus *et al.* (2002) [39]	Tat	PNA	Chemical conjugation via maleimide or disulfide bridge with eg1 linkers at the N terminal of the PNA
	Koppelhus *et al.* (2008) [12]	Tat	PNA, PNA with lipidic domain(s) (decanoic acid)	Chemical conjugation via eg1 linker or Gly
	Maier *et al.* (2006) [23]	Tat	PNA	Chemical conjugation via eg1 linker
	Resina *et al.* (2007) [31]	Tat	PNA	Chemical conjugation via eg1 linker

Assay system	PNA target	Label	Detection (s)	Results
HeLa pLuc705cells	Cryptic splice site of the mutated luciferase gene	Fluorescein	Confocal microscopy, fluorescence spectroscopy, luciferase activity (splicing correction)	High correlation between the uptake and nuclear antisense activity. Inhibition of the uptake and antisense activity by low temperature.
DU135 prostate cancer cells	mRNA of the catalytic component of human telomerase reverse transcriptase (hTERT)	Rhodamine B	Downregulation of telomerase, confocal microscopy	Antisense downregulation of the telomerase and cell growth inhibition in combination with PCI. Relocalication of Tat-PNA to the cytosol by PCI (data not shown).
Male Balb/c mice (i.p. or oral administration)	Transactivation response (TAR) element of HIV-1	Radiolabelling (^{125}I)	Radioactivity measurement of the tissues	Significant accumulation of CPP-PNA conjugates in organs and tissues. Unconjugated, naked PNA displayed a similar distribution and clearance profile in most organs studied although the extent of uptake was lower than for the CPP-PNA conjugates.
SK-BR-3, IMR-90, HeLa, H9 and U937 cells	Starting codon of her2 oncogene	Fluorescein	Confocal microscope, fluorescent spectroscopy	Cell type-dependent uptake. Vesicular cytoplasmic uptake (confocal microscopy). Dose and temperature dependent CPP-PNA conjugate uptake (fluorescence microscopy).
HeLa pLuc705 cells	Cryptic splice site of the mutated luciferase gene	None	Luciferase activity (splicing correction)	Improvement of the antisense potency with hydrophobic moieties.
BCL1 cells	3′-end of exon 6 of the primary murine CD40	None	Inhibition of CD40 expression	Antisense activities at few μM range.
HeLa pLuc705 cells	Cryptic splice site of the mutated luciferase gene	Fluorescein	Confocal microscopy, luciferase activity (splicing correction)	Showed slightly less antisense activity than K_8-PNA conjugate. Endosomal entrapment of the Tat-PNA.

(continued overleaf)

Table 16.1 (*continued*)

CPP type	Author, year, reference	CPP sequence[a]	Cargo	Conjugation method for CPP/PNA
	Richard *et al.* (2003) [9]	Tat	PNA	Chemical conjugation via AEEA linker
	Segura *et al.* (2007) [44]	Tat	PNA	Direct conjugation
	Shiraishi *et al.* (2005) [27]	Tat	PNA	Chemical conjugation via eg1 linker
	Shiraishi and Nielsen (2006) [56]	Tat	PNA	Chemical conjugation via eg1 linker
	Su *et al.* (2007) [45]	Tat	PNA	Chemical conjugation via AEEA (2-aminoethoxy-2-ethoxyacetic acid) linker
	Turner *et al.* (2005) [13]	Tat	PNA	Chemical conjugation via eg1 linker or disulfide bridge

Assay system	PNA target	Label	Detection (s)	Results
HeLa cells	Transactivation response (TAR) element of HIV-1	Fluorescein	Confocal microscopy, flow cytometry	Colocalization with Tf. Uptake inhibition by low temperature and ATP-depletion.
C2C12 cells, mouse (tail vein injection)	Unique murin erythropoietin (EPO) gene	Fluorescein or radiolabelling (^{123}I)	Fluorescence microscopy, planar gammagraphic image (mouse), radioactivity measurement (cells)	Detection of the uptake by radioactivity measurement. Downregulation of EPO at the few μM range. Accumulation of the ^{123}I-Tat-PNA in the transfected limb (dose of 10–12 MBq in the tail vein).
HeLa pLuc705 cells	Cryptic splice site within the luciferase gen	None	Luciferase activity (splicing correction)	Improvement of the CPP-PNA conjugates antisense effects by Ca^{2+} treatment.
HeLa pLuc705 cells	Cryptic splice site of the mutated luciferase gene	None	Luciferase activity (splicing correction)	Significantly higher nuclear antisense splicing correction activity in combination with PCI.
NIH 3T3 embryonic mouse fibroblasts	dsRed (red fluorescent protein from Discosoma coral) gene	Fluorescein and Gd-DOTA complex	Fluorescence spectroscopy, fluorescence microscopy, magnetic resonance imaging	Uptake confirmation by fluorescence microscopy and fluorescence spectroscopy at the concentration range 0.5–9.0 μM. The contrast agent was located predominantly in vesicles around the nucleus, whereas no uptake into the nucleus was observed. *In vitro* MR studies showed a statistically significant increase of the intracellular relaxation rate R (1,cell).
HeLa cells	Transactivation response (TAR) element of HIV-1	Fluorescein (eg1 linker conjugate)	Inhibition of the transactivation, confocal microscopy	Significant improvement of the activity by CQ. Showed punctuate cytosolic distribution (in the absence of the CQ) and uniform fluorescence in the cytosol and nucleus in the presence of CQ.

(*continued overleaf*)

Table 16.1 *(continued)*

CPP type	Author, year, reference	CPP sequence[a]	Cargo	Conjugation method for CPP/PNA
	Tyagi *et al.* (2006) [46]	Tat	PNA	Direct conjugation
	Yin *et al.* (2008) [32]	Tat	PNA	Chemical conjugation via disulfide bridge with AEEA linker at the N-terminal of the PNA
Transportan				
	Bendifallah *et al.* (2006) [18]	Transportan (CGWTLNSAG YLLGKINLKAL AALAKKIL), Transportan (ortho)	PNA	Chemical conjugation via ester maleimide, stable maleimide or disulfide bridge
	Bárány-Wallje *et al.* (2007) [47]	TP10 (AGYLLGKIN LKALAALAKKIL)	PNA	Chemical conjugation via disulfide bridge
	El-Andaloussi *et al.* (2006) [37]	Transportan	PNA	Chemical conjugation via disulfide bond
	El-Andaloussi *et al.* (2007) [19]	TP10	PNA	Chemical conjugation via disulfide bridge
	Turner *et al.* (2005) [13]	Transportan, TP10	PNA	Chemical conjugation via eg1 linker or a disulfide bridge (only for transportan).

[a] Peptides contain amino acid residues in the L-configuration unless otherwise noted.

Assay system	PNA target	Label	Detection (s)	Results
Rat (instillation to the bladder)	NGF (nerve growth factor) gene	Rhodamine	Confocal microscopy, downregulation of NGF	PNA Uptake (Confocal microscopy, dose at 0.2 ml of 30 μM) and downregulation of NGF detected by immuno histochemistry (dose: 0.5 ml of 100 μM).
H2k mdx cells. mdx mice (intramuscular injection)	Exon 23 of dystrophin pre-mRNA	None	Skipping of exon 23 of dystrophin and expression of functional dystrophin (restoration of the reading frame)	Exon skipping by PNA and PNA conjugates (cultured cells). Restoration of the dystrophin expression by PNA and PNA conjugates (non-CPP conjugate) is superior to others.
HeLa pLuc705 cells	Cryptic splice site of the mutated luciferase gene	None	Luciferase activity (splicing correction)	Transportan-PNA conjugates showed superior activity
Large unilamellar vesicles	Cryptic splice site of the mutated luciferase gene	None	Leakage of the entrapped fluorophore calcein	The leakage by PNA-Tp10 is compatible with Tp10 alone.
HeLa pLuc705cells	Cryptic splice site of the mutated luciferase gene	Fluorescein	Confocal microscopy, fluorescence spectroscopy, luciferase activity (splicing correction)	High correlation between the uptake and nuclear antisense activity. Inhibition of the uptake and antisense activity by low temperature.
HeLa pLuc705 cells	Cryptic splice site of the mutated luciferase gene	None	Luciferase activity (splicing correction)	M918 peptide is better than TP10 conjugate.
HeLa cells	HIV-1 Tat-dependent transactivation response RNA	Fluorescein (eg1 linker conjugates)	Confocal microscopy, inhibition of the transactivation	Transportan or TP10-PNA conjugates were inactive when delivered alone, but attained trans activation inhibition in the presence of CQ. Confocal microscopy showed that such fluorescently labelled CPP–PNA conjugates were sequestered in endosomal or membrane-bound compartments of HeLa cells (no significant cytosolic release by CQ).

below one micromolar [49]. This is almost certainly not sufficient for high efficacy in *in vivo* applications, and compared to the nanomolar range activities, which are achievable using oligonucleotide-type cationic delivery agents such as Lipofectamine™ in combination with anionic PNA/DNA complexes [50] or phosphonate-peptide conjugates [51], the peptide carriers identified so far unfortunately appear quite inefficient. However, detailed uptake and activity studies strongly suggest that the cellular uptake *per se* is indeed much more efficient than evident from the cellular activities obtained, because in most cases the PNA-peptide conjugates enter the cells via an endosomal pathway, and the major obstacle using such conjugates appears not to be the cellular entry step but rather endosomal escape [27, 52, 53, 56]. Therefore, one possible mechanism for strongly improving cellular (and possibly *in vivo*) efficiency of PNA-peptide conjugates could be employment of auxiliary endosome disruptive agents.

16.7 Auxiliary Agents

Several endosome disruptive agents have been identified. It has long been recognized that the antimalaria drug, chloroquine (Figure 16.2), accumulates very significantly in endosomes and upon their gradual transformation into lysosomes causes destabilization of these, proposedly via a 'proton sponge' effect due to the weak base properties of chloroquine (amino groups) in combination with the acidification of the lysosomes. Accordingly, it has been demonstrated that cellular co-administration of chloroquine (up to the cellular toxicity limit of around 100 μM) results in an enhancement of the antisense effect of PNA-peptide conjugates of up to two orders of magnitude, dependent on the peptide used [27, 52]. Analogously, hydrophobic photosensitizers such as $AlPcS_{2a}$ (aluminium phthalocyanine), which accumulate in (endosomal) membranes, induce endosomal destabilization upon irradiation with visible light through the production of highly reactive singlet oxygen, which causes disruptive oxidation of (unsaturated) membrane lipids. Thus photochemical internalization (PCI) has been shown to significantly increase the efficiency of functional gene transfection [54, 55], as well as of PNA-peptide antisense efficacy [30, 56], and the effect is quantitatively comparable to that seen with chloroquine [56].

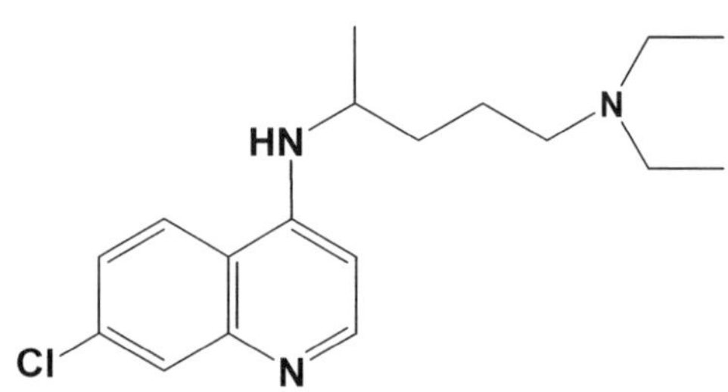

Chloroquine

Figure 16.2 Chloroquine structure

Finally, it has been discovered that low millimolar concentrations of Ca^{2+} ions also have endosome disruptive and PNA-peptide activity enhancing effects at least in HeLa cells, but the molecular/biological mechanism for this effect has not been unravelled [27].

16.8 *In Vivo* Activity of PNA Conjugates

Only a few data are so far available for *in vivo* activity of PNA oligomers and PNA-peptide conjugates (Table 16.2). It is, however, quite clear that the pharmacokinetic behaviour of unmodified PNA is not compatible with high potency as the oligomers, due to their hydrophilic character, are very quickly (t½ ~½ hour) excreted through the kidneys [23, 38]. Therefore, from a pharmacokinetic aspect, chemical modification or pharmaceutical formulation of the PNA is required. Furthermore, several of the *in vivo* studies are quite preliminary in terms of analysis (molecular biology markers), mechanistic conclusions and sufficient controls. It is also observed that when using systemic i.v. or i.p. administration, high doses (10–60 mg/kg) are needed in order to demonstrate detectable effects, indicating poor bioavailability of the PNA oligomers employed. Nonetheless, interesting *in vivo* effects have been demonstrated in several studies. In particular, the study in transgenic GFP mice by the Kole group [64] convincingly showed that systemically, i.v. administered unmodified PNA is virtually without effect, whereas a simple tetralysine conjugate was active in a variety of tissues, most pronounced in kidney, liver and intestine. From a future therapeutic point of view, it is also noteworthy that splicing redirection of dystrophin pre-mRNA was accomplished through intramuscular administration of a PNA-MSP-peptide conjugate [32].

16.9 Conclusions

Although it is difficult to make quantitative comparisons of the many different CPPs, the many studies clearly demonstrate that cellular and *in vivo* bioavailability of PNAs and similar agents may be significantly improved through conjugation to a CPP. It is also quite clear that different CPPs do not operate by fully identical mechanisms, but for the majority cellular uptake is via an endocytotic pathway. Therefore, one of the challenges to address is identifying and exploiting mechanisms for endosomal escape, as already indicated by the pronounced enhancing effects of endosomolytic agents like chloroquine and lipophilic photosensitizers. Furthermore, it is highly encouraging that simple chemical modification of CPPs such as fatty acid acylation can dramatically improve potency. However, it is also quite evident that the absolute potencies demonstrated so far yielding low or only slightly sub-millimolar activities will hardly be sufficient for effective drug development. Nonetheless, through more detailed understanding of structure–activity relationships of different (modified) peptides in terms of mechanisms of cellular uptake as well as *in vivo* pharmacokinetic behaviour and biodistribution, it should be possible to design and develop novel (modified) CPPs that assure sufficient bioavailability both at the organism level as well as on the cellular level.

Table 16.2 In vivo *studies*

Author, year, reference	CPPs	Cargo	Conjugation method for CPP/PNA	Animal	Administration method (dose)
Boffa *et al.* (2007) [33]	NLS (PKKK RKV)	PNA	Direct conjugation	SCID mice xenografted with Burkitt's lymphoma cells (BRG (Eμ responsive c-myc) and Ramos cells (Eμ irresponsive c-myc)	i.v. injection every other day (20 nmol) or pretreatment of BL cells before inoculum (10 μM for 8 h)
Boules *et al.* (2004) [57]	No CPP used	PNA	None	Rat	i.p. injection (60 mg/kg (sense PNA) or 40 mg/kg (antisense PNA))
Efimov *et al.* (2006) [58]	No CPP used	pHypNA[(a)], HypNA-pPNA[(b)]	None	Xenopus embryo	Microinjection once (2–6 nl of 0.25 mM)
Fraser *et al.* (2000) [59]	No CPP used	PNA	None	Rat	Intracerebroventricular (i.c.v.) injection twice daily for 5 days (0.45 nmol)
Ganguly *et al.* (2008) [38]	Penetratin (RQIK IWFQN RRMK WKK)	PNA	Chemical conjugation via disulfide bridge or ethylene glycol linker (eg1, 8-amino-2,6-dioxaoctanoic acid)	Mice	i.p. injection or oral administration (both with 36 nmol of labelled PNA)
Maier *et al.* (2006) [23]	(D-(RRFR RRARR FARRG), and D-(KKLL KAALK LAKKG)	PNA	Chemical conjugation via eg1 linker	Mice	i.v. injection (3 mg/kg)

PNA target	Label	Detection (s)	Results
Eμ enhancer	None	c-myc expression (Northern blot, Western blot), tumour size	Increased latency of tumour appearance and decreased final tumour size only with the BRG cells (with Em-enhancer responsive cMyc). Downregulation of c-myc expression.
Amyloid precursor protein (APP)	None	APP expression (Protein (Western Blot), mRNA (RT-PCR))	Reduction of APP protein expression by antisense PNA and antigene PNA. Insufficient controls.
Ras-like GTPase (Ras-dva) mRNA	None	Structure of Xenopus tadpole heads	Severe abnormalities of the head structure with the injected side.
s-opioid receptor mRNA	None	Pain and locomotor activity, [35S]GTPγS-binding assay	Inhibit the antinociceptive response and locomotor response to selective sigma-opioid receptor agonist. Decrease in agonist efficacy in homogenates prepared from antisense-treated rats. No direct molecular markers.
Transactivation response (TAR) element of HIV-1 LTR	Radiolabel (^{125}I)	Radioactivity measurement in the tissues	Significant accumulation of CPP-PNA conjugated in organs and tissues. Unconjugated naked PNA displayed a similar distribution and clearance profile in most organs studied although extent of its uptake was lower than the CPP-PNA conjugates.
3′-end of exon 6 of the primary murine CD40 gene pre-mRNA	Radiolabel (^{3}H)	Downregulation of the CD40 expression, radioactivity measurement in the tissues	Highest concentration found in liver, kidney and spleen at 3 mg/kg dose.

(continued overleaf)

Table 16.2 (*continued*)

Author, year, reference	CPPs	Cargo	Conjugation method for CPP/PNA	Animal	Administration method (dose)
McMahon *et al.* (2002) [60]	No CPP used	PNA	None	Spontaneously hypertensive rat	i.p. injection once or daily for 4 days (20 mg/kg, 10 mg/kg on day 2)
Morris *et al.* (2007) [17]	Pep-3 (K W F E T W F T E W P K K R K) and its PEG derivatives	HypNA-Ppna	Complexation	PC3 cells xenografted to nude mice	Intratumoral administration every 3 days by PEP-3 (1 or 5 μg) or i.v. injection every 3 days by PEP-3 and PEG-Pep3 (10 μg)
Pooga *et al.* (1998) [61]	Penetratin	PNA	Chemical conjugation via disulfide bridge	Rat	Intrathecal injection 3 times with 12 h interval (1.5 nmol)
Sazani *et al.* (2002) [62]	K, K_4	PNA	Direct conjugation	EGFP-654 transgenic mouse	i.p. injection once daily for 1 or 4 days (50 mg/kg)
Segura *et al.* (2007) [44]	Tat (GRKK RRQR RPP)	PNA	Direct conjugation	Mouse	i.v. injection (10–12 MBq)
Tyagi *et al.* (2006) [46]	Tat	PNA	Direct conjugation	Rat	Instillation to the bladder (0.2 ml of 30 μM)
Tyler *et al.* (1998) [63]	No CPP used	PNA	None	Rat	Injection to periaqueductal grey (PAG) region of the brain 3 times with 2 days interval (1.3 nmol)
Yin *et al.* (2008) [32]	Tat, Muscle-specific peptide (MSP), AAV6 and AAV8	PNA	Chemical conjugation via disulfide bridge with the AEEA (acetyl-ethyleneglycol-ethyl-amine) linker	mdx mice	Intramuscular injection (5–20 μg)

[a] chiral analogues of peptide nucleic acids with a constrained trans-4-hydroxy-N-acetylpyrrolidine-2-phosphonate backbone.
[b] hetero-oligomers consisting of alternating phosphono-PNA monomersand PNA-like monomers based on trans-4-hydroxy-L-proline.

PNA target	Label	Detection (s)	Results
Antigene targeting of angiotensinogen gene	None	Angiotensinogen expression mRNA (RT-PCR) and protein (raidoimmunoassay). Blood pressure	Decrease in plasma angiotensin I, systolic blood pressure, and liver and brain angiotensiongen mRNA levels.
Cyclin B1 mRNA	None	Tumour growth, downregulation of Cyclin B1 (Western blot)	Downregulation of tumour Cyclin B1 protein by both intratumoral injection and i.v. injection. Inhibit tumour growth with i.v. injection at 25 days.
Human galanin receptor type1 mRNA	Biotin	Autoradiographic quantification of galanin receptors in spinal cord	Inhibit galanin receptor expression in the spinal cord and modify the pain response.
Cryptic splice site of the mutated GFP pre-mRNA	None	EGFP expression in the tissues (fluorescent microscopy), splicing correction (RT-PCR)	Splicing correction and upregulation of the EGFP expression by K_4-PNA. High activity in kidney, liver and small intestine. Lower activity in the lung, heart and skeletal muscle). One Lys PNA was completely inactive.
Unique murin erythropoietin (EPO) mRNA	Radiolabel (^{123}I)	Planar gammagraphic image	Accumulation of the ^{123}I-Tat-PNA in the transfected limb.
NGF (nerve growth factor) mRNA	Rhodamine	Confocal microscopy, downregulation of NGF	PNA Uptake (confocal microscopy) and downregulation of NGF (immunohistochemistry).
Neurotensin (NT) receptor gene or opioid receptor gene mRNA	None	Behavioural responses to neurotensin and morphine. Competitional binding assay (PAG and hypothalamus)	Reduction in receptor sites. Lost of behavioural responses to nurotensin and morphine in the specific manner. No molecular markers.
Exon 23 of dystrophin pre-mRNA	None	Dystrophin expression (histology, immunohistochemistry, Western blotting)	Restoration of the dystrophin expression by PNA and PNA-CPP conjugates in mice of all ages (three stages (3 week, 2 month, 6 month).

References

1. Nielsen, P.E., Egholm, M., Berg, R.H. and Buchardt, O. (1991) Sequence-selective recognition of DNA by strand displacement with a thymine-substituted polyamide. *Science*, **254**, 1497–1500.
2. Lundin, K.E., Good, L., Stromberg, R. *et al.* (2006) Biological activity and biotechnological aspects of peptide nucleic acid. *Adv. Genet.*, **56**, 1–51.
3. Derossi, D., Joliot, A.H., Chassaing, G. and Prochiantz, A. (1994) The third helix of the Antennapedia homeodomain translocates through biological membranes. *J. Biol. Chem.*, **269**, 10 444–10 450.
4. Fawell, S., Seery, J., Daikh, Y. *et al.* (1994) Tat-mediated delivery of heterologous proteins into cells. *Proc. Natl. Acad. Sci. USA*, **91**, 664–668.
5. Morris, M.C., Deshayes, S., Heitz, F. and Divita, G. (2008) Cell-penetrating peptides: from molecular mechanisms to therapeutics. *Biol. Cell*, **100**, 201–217.
6. Wender, P.A., Mitchell, D.J., Pattabiraman, K. *et al.* (2000) The design, synthesis, and evaluation of molecules that enable or enhance cellular uptake: peptoid molecular transporter. *Proc. Natl. Acad. Sci USA*, **97**, 13 003–13 008.
7. Fotin-Mleczek, M., Fischer, R. and Brock, R. (2005) Endocytosis and cationic cell-penetrating peptides – a merger of concepts and methods. *Curr. Pharm. Des.*, **11**, 3613–3628.
8. Fuchs, S.M. and Raines, R.T. (2004) Pathway for polyarginine entry into mammalian cells. *Biochemistry*, **43**, 2438–2444.
9. Richard, J.P., Melikov, K., Vives, E. *et al.* (2003) A reevaluation of the mechanism of cellular uptake. *J. Biol. Chem.*, **278**, 585–590.
10. Kang, S.H., Cho, M.J. and Kole, R. (1998) Up-regulation of luciferase gene expression with antisense oligonucleotides: implications and applications in functional assay development. *Biochemistry*, **37**, 6235–6239.
11. Hu, J. and Corey, D.R. (2007) Inhibiting gene expression with Peptide Nucleic Acid (PNA)-peptide conjugates that target chromosomal DNA. *Biochemistry*, **46**, 7581–7589.
12. Koppelhus, U., Shiraishi, T., Zachar, V. *et al.* (2008) Improved cellular activity of antisense peptide nucleic acids by conjugation to a cationic peptide-lipid (CatLip) domain. *Bioconjug. Chem.*, **19**, 1526–1534.
13. Turner, J.J., Ivanova, G.D., Verbeure, B. *et al.* (2005) Cell-penetrating peptide conjugates of peptide nucleic acids (PNA) as inhibitors of HIV-1 Tat-dependent trans-activation in cells. *Nucleic Acids Res.*, **33**, 6837–6849.
14. Abes, S., Moulton, H., Turner, J. *et al.* (2007) Peptide-based delivery of nucleic acids: design, mechanism of uptake and applications to splice-correcting oligonucleotides. *Biochem. Soc. Trans.*, **35**, 53–55.
15. Fabani, M.M. and Gait, M.J. (2008) miR-122 targeting with LNA/2′-O-methyl oligonucleotide mixmers, peptide nucleic acids (PNA), and PNA-peptide conjugates. *RNA*, **14**, 336–346.
16. Gros, E., Deshayes, S., Morris, M.C. *et al.* (2006) A non-covalent peptide-based strategy for protein and peptide nucleic acid transduction. *Biochim. Biophys. Acta*, **1758**, 384–393.
17. Morris, M.C., Gros, E., drian-Herrada, G. *et al.* (2007) A non-covalent peptide-based carrier for *in vivo* delivery of DNA mimics. *Nucleic Acids Res.*, **35**, e49.
18. Bendifallah, N., Rasmussen, F.W., Zachar, V. *et al.* (2006) Peptides (CPPs) as vehicles for intracellular delivery of antisense Peptide Nucleic Acid (PNA). *Bioconj. Chem.*, **17**, 750–758.
19. El-Andaloussi, S., Johansson, H.J., Holm, T. and Langel, L. (2007) A novel cell-penetrating peptide, M918, for efficient delivery of proteins and peptide nucleic acids. *Mol. Ther.*, **15**, 1820–1826.
20. Faridani, O.R., Nikravesh, A., Pandey, D.P. *et al.* (2006) Competitive inhibition of natural antisense Sok-RNA interactions activates Hok-mediated cell killing in *Escherichia coli*. *Nucleic Acids Res.*, **34**, 5915–5922.
21. Good, L., Awasthi, S.K., Dryselius, R. *et al.* (2001) Bactericidal antisense effects of peptide-PNA conjugates. *Nat. Biotechnol.*, **19**, 360–364.

22. Kurupati, P., Tan, K.S., Kumarasinghe, G. and Poh, C.L. (2007) Inhibition of gene expression and growth by Antisense Peptide Nucleic Acids in a multiresistant {beta}-lactamase-producing klebsiella pneumoniae strain. *Antimicrob. Agents Chemother.*, **51**, 805–811.
23. Maier, M.A., Esau, C.C., Siwkowski, A.M. *et al.* (2006) Evaluation of basic amphipathic peptides for cellular delivery of antisense peptide nucleic acids. *J. Med. Chem.*, **49**, 2534–2542.
24. Nekhotiaeva, N., Awasthi, S.K., Nielsen, P.E. and Good, L. (2004) Inhibition of *Staphylococcus aureus* gene expression and growth using antisense peptide nucleic acids. *Mol. Ther.*, **10**, 652–659.
25. Nikravesh, A., Dryselius, R., Faridani, O.R. *et al.* (2007) Antisense PNA accumulates in *Escherichia coli* and mediates a long post-antibiotic effect. *Mol. Ther.*, **15**, 1537–1542.
26. Pritz, S., Wolf, Y., Kraetke, O. *et al.* (2007) Synthesis of biologically active peptide nucleic acid-peptide conjugates by sortase-mediated ligation. *J. Org. Chem.*, **72**, 3909–3912.
27. Shiraishi, T., Pankratova, S. and Nielsen, P.E. (2005) Calcium ions effectively enhance the effect of antisense peptide nucleic acids conjugated to cationic tat and oligoarginine peptides. *Chem. Biol.*, **12**, 923–929.
28. Wolf, Y., Pritz, S., Abes, S. *et al.* (2006) Structural remequirents for cellular uptake and antisense activity of peptide nucleic acids conjugated with various peptides. *Biochemistry*, **45**, 14944–14954.
29. Abes, S., Turner, J.J., Ivanova, G.D. *et al.* (2007) Efficient splicing correction by PNA conjugation to an R6-Penetratin delivery peptide. *Nucleic Acids Res.*, **35**, 4495–4502.
30. Bøe, S. and Hovig, E. (2006) Photochemically induced gene silencing using PNA-peptide conjugates. *Oligonucleotides*, **16**, 145–157.
31. Resina, S., Abes, S., Turner, J.J. *et al.* (2007) Lipoplex and peptide-based strategies for the delivery of steric-block oligonucleotides. *Int. J. Pharm.*, **344**, 96–102.
32. Yin, H., Lu, Q. and Wood, M. (2008) Effective exon skipping and restoration of dystrophin expression by peptide nucleic acid antisense oligonucleotides in mdx mice. *Mol. Ther.*, **16**, 38–45.
33. Boffa, L.C., Cutrona, G., Cilli, M. *et al.* (2007) Inhibition of Burkitt's lymphoma cells growth in SCID mice by a PNA specific for a regulatory sequence of the translocated c-myc. *Cancer Gene Ther.*, **14**, 220–226.
34. Rasmussen, F.W., Bendifallah, N., Zachar, V. *et al.* (2006) Evaluation of transfection protocols for unmodified and modified Peptide Nucleic Acid (PNA) oligomers. *Oligonucleotides*, **16**, 43–57.
35. Robaczewska, M., Narayan, R., Seigneres, B. *et al.* (2005) Sequence-specific inhibition of duck hepatitis B virus reverse transcription by peptide nucleic acids (PNA). *J. Hepatol.*, **42**, 180–187.
36. Fletcher, S., Honeyman, K., Fall, A.M. *et al.* (2006) Dystrophin expression in the mdx mouse after localised and systemic administration of a morpholino antisense oligonucleotide. *J. Gene Med.*, **8**, 207–216.
37. El-Andaloussi, S., Johansson, H.J., Lundberg, P. and Langel, L. (2006) Induction of splice correction by cell-penetrating peptide nucleic acids. *J. Gene Med.*, **8**, 1262–1273.
38. Ganguly, S., Chaubey, B., Tripathi, S. *et al.* (2008) Pharmacokinetic analysis of polyamide nucleic-acid-cell penetrating peptide conjugates targeted against HIV-1 transactivation response element. *Oligonucleotides*, **18**, 277–286.
39. Koppelhus, U., Awasthi, S.K., Zachar, V. *et al.* (2002) Cell-dependent differential cellular uptake of PNA, peptides, and PNA-peptide conjugates. *Antisense Nucleic Acid Drug Dev.*, **12**, 51–63.
40. Villa, R., Folini, M., Lualdi, S. *et al.* (2000) Inhibition of telomerase activity by a cell-penetrating peptide nucleic acid construct in human melanoma cells. *FEBS Lett.*, **473**, 241–248.
41. Dragulescu-Andrasi, A., Rapireddy, S., He, G.B. *et al.* (2006) Cell-permeable peptide nucleic acid designed to bind to the 5′-untranslated region of E-cadherin transcript induces potent and sequence-specific antisense effects. *J. Am. Chem. Soc.*, **128**, 16104–16112.
42. Nielsen, P.E. and Shiraishi, T. (2006) Improving the bioavailability of PNA-peptide conjugates. *Pept. Sci.*, **43**, 219.

43. Folini, M., Bandiera, R., Millo, E. *et al.* (2007) Photochemically enhanced delivery of a cell-penetrating peptide nucleic acid conjugate targeting human telomerase reverse transcriptase: effects on telomere status and proliferative potential of human prostate cancer cells. *Cell Prolif.*, **40**, 905–920.
44. Segura, J., Fillat, C., Andreu, D. *et al.* (2007) Monitoring gene therapy by external imaging of mRNA: pilot study on murine erythropoietin. *Ther. Drug Monit.*, **29**, 612–618.
45. Su, W., Mishra, R., Pfeuffer, J. *et al.* (2007) Synthesis and cellular uptake of a MR contrast agent coupled to an antisense peptide nucleic acid-cell-penetrating peptide conjugate. *Contrast Media Mol. Imaging*, **2**, 42–49.
46. Tyagi, P., Banerjee, R., Basu, S. *et al.* (2006) Intravesical antisense therapy for cystitis using TAT-peptide nucleic acid conjugates. *Mol. Pharm.*, **3**, 398–406.
47. Bárány-Wallje, E., Gaur, J., Lundberg, P. *et al.* (2007) Differential membrane perturbation caused by the cell penetrating peptide Tp10 depending on attached cargo. *FEBS Lett.*, **581**, 2389–2393.
48. Kaihatsu, K., Huffman, K.E. and Corey, D.R. (2004) Intracellular uptake and inhibition of gene expression by PNAs and PNA-peptide conjugates. *Biochemistry*, **43**, 14 340–14 347.
49. Kilk, K., Elmquist, A., Saar, K. *et al.* (2004) Targeting of antisense PNA oligomers to human galanin receptor type 1 mRNA. *Neuropeptides*, **38**, 316–324.
50. Doyle, D.F., Braasch, D.A., Simmons, C.G. *et al.* (2001) Inhibition of gene expression inside cells by peptide nucleic acids: effect of mRNA target sequence, mismatched bases, and PNA length. *Biochemistry*, **40**, 53–64.
51. Shiraishi, T., Hamzavi, R. and Nielsen, P.E. (2008) Subnanomolar antisense activity of phosphonate-peptide nucleic acid (PNA) conjugates delivered by cationic lipids to HeLa cells. *Nucleic Acids Res.*, **36**, 4424–4432.
52. Abes, S., Williams, D., Prevot, P. *et al.* (2006) Endosome trapping limits the efficiency of splicing correction by PNA-oligolysine conjugates. *J. Control. Release*, **110**, 595–604.
53. Shiraishi, T. and Nielsen, P.E. (2006) Enhanced delivery of cell-pepetrating peptide-peptide nucleic acid conjugates by endosomal disruption. *Nat. Protoc.*, **1**, 633–636.
54. Hogset, A., Prasmickaite, L., Hellum, M. *et al.* (2002) Photochemical transfection: a technology for efficient light-directed gene delivery. *Somat. Cell Mol. Genet.*, **27**, 97–113.
55. Hogset, A., Prasmickaite, L., Engesaeter, B.O. *et al.* (2003) Light directed gene transfer by photochemical internalisation. *Curr. Gene Ther.*, **3**, 89–112.
56. Shiraishi, T. and Nielsen, P.E. (2006) Photochemically enhanced cellular delivery of cell penetrating peptide-PNA conjugates. *FEBS Lett.*, **580**, 1451–1456.
57. Boules, M., Williams, K., Gollatz, E. *et al.* (2004) Down-regulation of amyloid precursor protein by peptide nucleic acid *in vivo*. *J. Mol. Neurosci.*, **24**, 123–128.
58. Efimov, V.A., Birikh, K.R., Staroverov, D.B. *et al.* (2006) Hydroxyproline-based DNA mimics provide an efficient gene silencing *in vitro* and *in vivo*. *Nucleic Acids Res.*, **34**, 2247–2257.
59. Fraser, G.L., Holmgren, J., Clarke, P.B.S. and Wahlestedt, C. (2000) Antisense inhibition of .delta.-opioid receptor gene function *in vivo* by peptide nucleic acids. *Mol. Pharmacol.*, **57**, 725–731.
60. McMahon, B.M., Stewart, J.A., Bitner, M.D. *et al.* (2002) Peptide nucleic acids specifically cause antigene effects *in vivo* by systemic injection. *Life Sci.*, **71**, 325–337.
61. Pooga, M., Soomets, U., Hällbrink, M.H. *et al.* (1998) Cell penetrating PNA constructs regulate galanin receptor levels and modify pain transmission *in vivo*. *Nat. Biotechnol.*, **16**, 857–861.
62. Sazani, P., Gemignani, F., Kang, S.H. *et al.* (2002) Systemically delivered antisense oligomers upregulate gene expression in mouse tissues. *Nat. Biotechnol.*, **20**, 1228–1233.
63. Tyler, B.M., McCormick, D.J., Hoshall, C.V. *et al.* (1998) Specific gene blockade shows that peptide nucleic acids readily enter neuronal cells *in vivo*. *FEBS Lett.*, **421**, 280–284.
64. Sazani, P., Gemignani, F., Kang, S.H. *et al.* (2002) Systemically delivered antisense oligomers upregulate gene expression in mouse tissues. *Nat. Biotechnol.*, **20**, 1228–1233.

17

DEAE-dextran-MMA Graft Copolymer for Non-Viral Delivery of DNA

Yasuhiko Onishi,[1] *Yuki Eshita*[2] *and Masaaki Mizuno*[3]
[1]Ryuju Science Co. Ltd., Seto, Japan
[2]Oita University, Department of Infectious Diseases, Faculty of Medicine, Oita, Japan
[3]Nagoya University, Graduate School of Medicine Program in Cell Information Medicine, Nagoya, Japan

17.1 Introduction

Recently, *in vivo* or *in vitro* gene delivery has allowed the study of gene expression via insertion of foreign genes or alteration of existing genes. However, some harmful adverse effects remain associated with the viral vectors. Non-viral gene delivery vectors should be a key technology in circumventing immunogenicities inherent in viral-mediated gene transfer. 2-Diethylaminoethyl (DEAE)-dextran has been used for a non-viral gene delivery vector [1, 2] by the reasons of its fast and easy procedure and of safety owing to autoclave sterilization different from other lipofection vectors. It may also be preferable for gene therapy that DEAE-dextran offers only transient transfection. However, DEAE-dextran may not be superior to viral vectors and lipofection vectors in cytotoxic and a transfection efficiency. Many efforts have been made in the field of non-viral gene delivery vector for a safety and high transfection efficiency, especially with DEAE-dextran [3–5]. DEAE-dextran has been investigated to increase its transfection efficiency, and several good conditions for a human macrophage were found [6]. DEAE-dextran, which has strong adsorbing properties with DNA or RNA due to its cationic properties, is found out to change its adsorbing power for nucleic acids by pH and ionic strength [7, 8].

Delivery Technologies for Biopharmaceuticals: Peptides, Proteins, Nucleic Acids and Vaccines Edited by Lene Jorgensen and Hanne Mørck Nielsen

It is also well known that Dextran-methyl methacrylate (MMA) graft copolymer, having a hydrophilic–hydrophobic micro-separated domain, has a good affinity to a cell membrane [9]. This case is related to a novel graft copolymer that has some possibilities for a non-viral gene delivery carrier and that is composed of a DEAE-dextran and a vinyl monomer. DEAE-dextran-methyl methacrylate graft copolymer (DDMC) is obtained by graft polymerization of MMA onto DEAE-dextran in water using ceric ammonium nitrate (CAN) to form a stable latex [10], which is very effective as a non-viral gene delivery vector [11–16].

17.2 Theoretical

A polyion complex [ES] formed by a polycation and DNA should be followed by a Michaelis Menten type reaction:

$$[E]+[S] \overset{Km}{\rightleftarrows} [ES] \tag{17.1}$$

where [E] is polycation and [S] is DNA.

Then, at a steady state:

$$[E][S]/[ES] = k2/k1 = Km \tag{17.2}$$

where Km is an equilibrium constant.

When the initial concentration of polycation is $[E_0]$:

$$[E] = [E_0] - [ES] \tag{17.3}$$

With (17.2):

$$[ES] = [E_0][S]/(Km + [S]) \tag{17.4}$$

For the core-shell particle of DDMC, at a low concentration of [E] and [S], the kinetics of k1[E][S] may be small because a coulomb force between the phosphoric acid of nucleic acids and the DEAE group of DEAE-dextran should not be enough in comparison with the hydrogen bond and the hydrophobic force of graft poly(MMA) for its conformation. That is why Km is much larger than [S].

As Km>>[S]:

$$[ES] = [E_0][S]/Km \tag{17.5}$$

For DEAE-dextran, a linear water-soluble polymer, at a low concentration of [E] and [S], the kinetics of k1[E][S] is large because it is accomplished by mainly a coulomb force between the phosphoric acid of nucleic acids and the DEAE group of DEAE-dextran. That is why [S] is much larger than Km.

As Km ≪ [S], the concentration of the complex is nearly the initial concentration of $[E_0]$

$$[ES] = [E_0] \tag{17.6}$$

17.3 Materials and Methods

17.3.1 Preparation of DDMC

Samples DDMC1, DDMC2 and DDMC3 in Table 17.1 were prepared by the following procedure. 2-Diethylaminoethyl-dextran hydrochloride (nitrogen content 3%) derived from dextran having a molecular weight of 500 000 was dissolved in a water, and then methyl methacrylate (MMA) was added. With stirring, the air in the reaction vessel was fully replaced with nitrogen gas. To the solution were added CAN and 0.1 N nitric acid, and the mixture was reacted with stirring for one hour at 30 °C. Then, hydroquinone was added to stop the reaction and the resulting latex of DDMC was purified by water dialysis using a cellophane tube in order to remove the un-reacted MMA, ceric salts and nitric acid.

17.3.2 Characterization of DDMC

DDMC is stable and soap-less in water. However, DDMC precipitated by methanol becomes insoluble in water and acetone at 25 °C. As DEAE-dextran hydrochloride is soluble in water and poly(methyl methacrylate) (PMMA) is soluble in acetone, it is evident that DDMC is not a mixture but copolymer of DEAE-dextran and PMMA.

The infrared absorption spectrum of DDMC (Figure 17.1) has some characteristic absorption bands at 1730 cm^{-1} and at 1000–1150 cm^{-1}, which are attributed to the carbonyl group of PMMA and the pyranose ring of DEAE-dextran, respectively. Thus, DDMC exhibits different solubility from DEAE-dextran and PMMA and shows the characteristic absorption in infrared absorption spectrum described above. From this fact, it is judged that the DDMC is a compound graft polymer. These graft copolymers have an amphiphilic domain forming a polymer micelle through their hairy nanoparticles and should become a stable latex of core-shell particles with a hydrophilic–hydrophobic micro-separated domain to form a spherical structure [17].

17.3.3 Reaction Between DDMC and DNA

The *in vivo* interaction between DNA and basic proteins such as histones, known by the appearance of a partially unfolded part on chromatin, plays a major role in the regulation of the gene transfer system [18].

For reaction between DDMC and DNA, 2 ml of solution (10 mg/ml) of the resulting latex of DDMC was added drop-wise to 1 ml of DNA (ex salmon sperm) solution

Table 17.1 *Properties of DEAE-dextran-MMA graft copolymers. Reprinted with permission from Elsevier. Copyright 2007.*

Sample	Weight-increase(%)[1]	Precipitation time(h) by DNA
DDMC1	150	2.0
DDMC2	100	1.0
DDMC3	130	1.5
DEAE-dextran	0	96.0

[1] Weight increase (%) = (weight of MMA used/weight of DEAE-dextran hydrochloride used) × 100

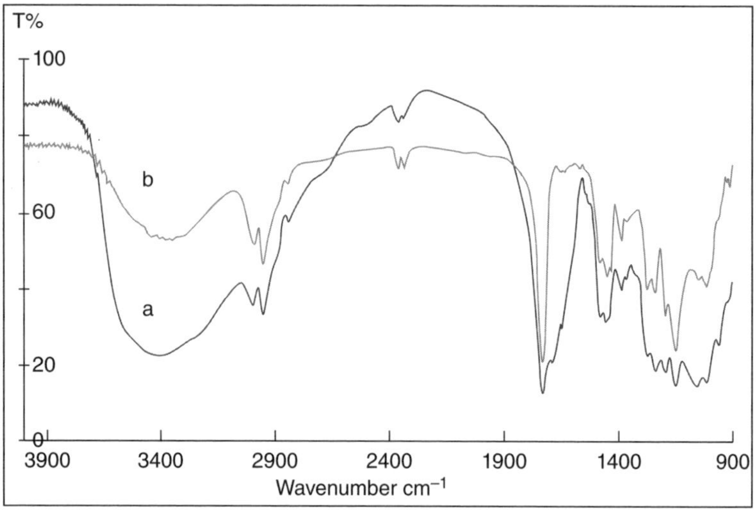

Figure 17.1 *IR absorption spectra of DEAE-dextran-MMA graft copolymer and the complexes between DNA and DEAE-dextran-MMA graft copolymer: (a) complex of DDMC2/DNA; (b) DDMC2. Reprinted with permission from Elsevier. Copyright 2007.*

(20 mg/ml) in order to obtain the complex of DDMC/DNA. The complex obtained was insoluble in water, which is a good solvent for nucleic acids. Thus, the complex between DNA and DDMC must consist of a polyion complex (PIC) formed by a coulomb force on a polymer micelle [28]. In the case of samples in Table 1, a complex between DNA and DDMC2 that had a 100% weight increase needed one hour to precipitate, and the complex between DNA and DDMC1 that had a 150% weight increases needed two hours to precipitate. However, a complex between DNA and DEAE-dextran hydrochloride needed 96 hours to precipitate at these conditions.

From the thermodynamic consideration for the complex reaction between DNA and DDMC, the Gibbs free energy change at complex reaction $\Delta G[Jg^{-1}]$ can be written as follows:

$$\Delta G = \Delta H - T\Delta S \qquad (17.7)$$

where $\Delta H[Jg^{-1}]$ is the enthalpy change and $\Delta S[Jg^{-1}K^{-1}]$ is the entropy change at the formation of the complex DNA/DDMC.

The Gibbs free energy change $\Delta G[Jg^{-1}]$ at the complex formation should be minus owing to its large plus entropy change (ΔS) by a hydrophobic force from hydrophobic domains of PMMA in DDMC, because the enthalpy change ΔH $[Jg^{-1}]$ is small in comparison with the entropy change (ΔS). As shown by its precipitation time in Table 1, this means that the complex reaction between DNA and DDMC can easily proceed and the molecule of the complex in a crystal structure may be more tightly packed compared with the complex DEAE-dextran/DNA.

This may support that the complex DDMC/DNA is compactly formed not only by a coulomb force between the phosphoric acid of DNA and the DEAE group of the DEAE-dextran copolymer but also by a force from the multi-inter-molecule hydrogen bond and a hydrophobic force from the hydrophobic domains of graft PMMA in DDMC and that it has a high protection facility for DNase.

17.3.4 The Structure of the Complex Between DNA and DDMC

The structural transition of DNA, namely a coil–globule transition, induces discrete on/off switching of transcriptional activity [19]. This collapse transition in single giant DNA chains has been reviewed as DNA condensation [20–21]. The *in vitro* collapse of DNA can be induced by various cationic compound vectors such as cationic lipids [22–24], peptides [25] or cationic polymers [26, 27]. In the case of cationic lipid vectors, the complex of dioctadecylamidoglycylspermine (DOGS)/DNA, which has a nucleosome-like structure in which DNA wraps around a micellar aggregate of DOGS so that they have an association with each other to form a network structure, is very effective for gene transfection [19]. The complex of cationic polymers/DNA in cytoplasm can be generally protected from restriction enzymes for the collapse of DNA [20, 21].

In the case of cationic dextran, the complex of DEAE-dextran/DNA in the cytoplasm can be also protected from DNase [4]. Formation of a complex between nucleic acids (DNA or RNA) and cationic graft copolymers, such as the DEAE-dextran-MMA copolymer, is accomplished by a coulomb force between the phosphoric acid of nucleic acids and the DEAE group of the DEAE-dextran-MMA copolymer. The complex so obtained with DDMC/DNA was insoluble in water, which is a good solvent for nucleic acids. The complex of DDMC/DNA should be a polyion complex formed by a polymer micelle; the infrared absorption spectrum of νOH vibration and νNH vibration at around 3100–3800 m^{-1} is shown in Figure 17.2. These results show that the complex between DNA and DDMC must form a polyion complex by coulomb force. The complex between nucleic acids (DNA or RNA) and the cationic graft copolymer of this research must typically form a polyion complex [28]. DDMC relatively formed a polyion complex between DNA and DDMC by the hydrophobic force of graft PMMA depending on its large positive entropy change (ΔS) compared with enthalpy change (ΔH).

The complex of DDMC/DNA should be thought to become a more compact structure by the shift of νNH vibration at 3450 cm^{-1} compared with that of DDMC as shown in Figure 17.1. Thus, DDMC has been confirmed as having a high protection facility for DNase degradation.

As shown in Figure 17.3, the resulting DDMC of hairy nanoparticles [29, 30], which has an amphiphilic domain in order to form a polymer micelle of core-shell particles [30], should become a stable latex with a hydrophilic–hydrophobic micro-separated domain [31, 32]. The complex of DDMC and DNA may be initially formed on the spherical structure of the amphiphilic micro-separated domain of DDMC having a good affinity to the cell membrane. When DNA (Figure 17.4e)is released in cytosol, the complex structure can be protected from degradation by DNase and dextransucrase, thereby making it easier for nuclear entry to occur as shown in Figure 17.4f.

17.3.4.1 Measurement of the Infrared (IR) Absorption Spectra for DDMC

IR measurement on samples was carried out by the potassium bromide method using Jasco FT/IR-300. The infrared absorption spectra of the resulting complex between

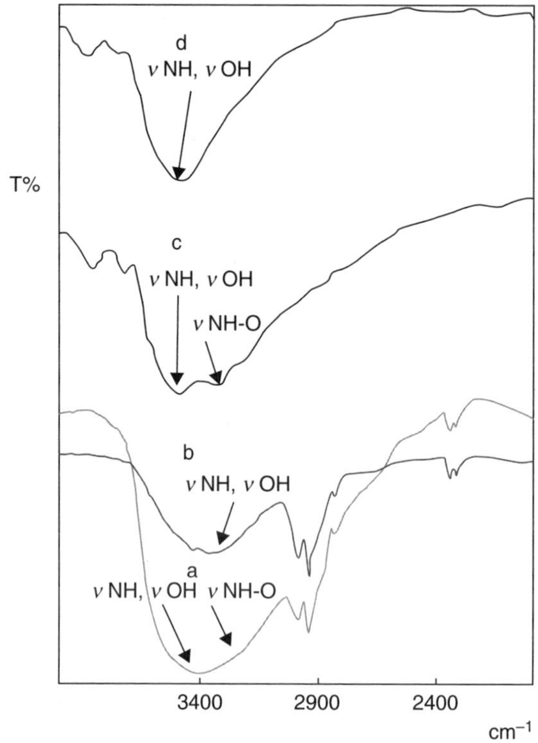

Figure 17.2 *IR absorption spectra: (a) complex of DDMC2/DNA; (b) DDMC2; (c) complex of DEAE-dextran/DNA; (d) DEAE-dextran. Reprinted with permission from Elsevier. Copyright 2007.*

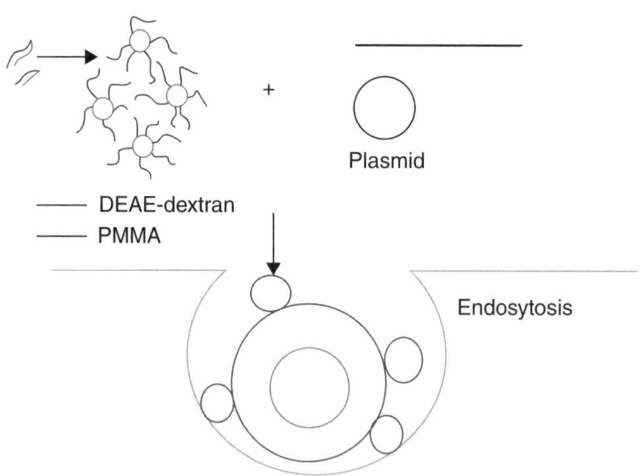

Figure 17.3 *Schematic representation of endosytosis by the complex between DDMC and plasmid DNA. The complex of DDMC/plasmid DNA may be initially formed on the spherical structure of the amphiphilic micro-separated domain of DDMC by a core-shell-type polymer micelle. Reprinted with permission from Elsevier. Copyright 2007.*

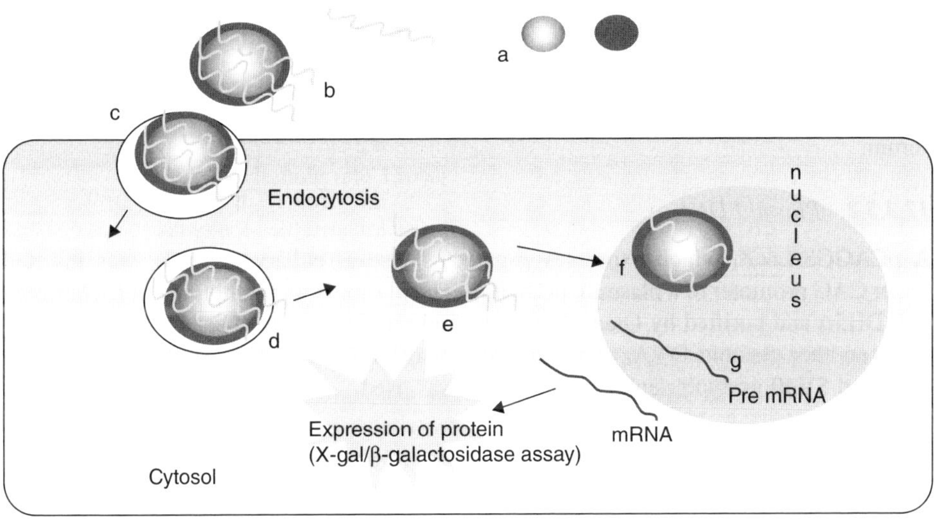

Figure 17.4 *Schematic drawing of DNA delivery pathways by DDMC*

DDMC2 and DNA are shown in Figure 17.1. The spectrum of the complex has some characteristic absorption bands at 1730 cm^{-1}, 1220 cm^{-1} and at 1000–1150 cm^{-1}, which are attributed to the carbonyl group of PMMA, P–O stretching vibration of DNA and the pyranose ring of DEAE-dextran, respectively.

Figure 17.2 also shows the infrared spectrum of both the complex by DDMC2/DNA and the complex by DEAE-dextran/DNA compared with DDMC2 and DEAE-dextran, respectively. The spectrum of the complex has some characteristic absorption bands owing to hydrogen bonds by –OH and –NH at around 3100–3800 cm^{-1}.

There also was a structural difference between the complex of DDMC/DNA and the complex of DEAE-dextran/DNA by the decrease of νNH–O vibration at around 3350 cm^{-1} and the shift of νNH vibration from 3450 cm^{-1} to 3550 cm^{-1} (Figure 17.2). This shows the complex of DDMC/DNA should form a more stable and tight PIC by multi-inter-molecule hydrogen bond compared with the complex of DEAE-dextran/DNA to reduce collapse transition in DNA chains. This may not only be due to a coulomb force between the phosphoric acid of DNA and the DEAE group of the DEAE-dextran copolymer but also a force from multi-inter-molecule hydrogen bond.

17.3.5 Transfection

17.3.5.1 Cell Line and Cell Culture

HEK293cell line, which is a permanent line of primary human embryonal kidney transformed by sheared human adenovirus type 5 DNA, was used [33].

The cell line was grown in Dulbecco's Modified Eagle's Medium (DMEM) supplemented with 10% foetal calf serum, 0.1 mM non-essential amino acids, 5 mM L-glutamine and antibiotics (streptomycin, 10 μg/ml; penicillin, 100 U/ml).

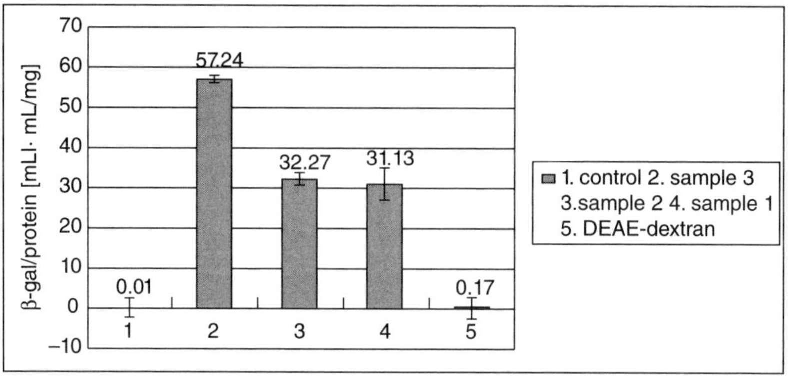

Figure 17.5 *β-Galactosidase Assay: DDMC3 (sample1 10.0 mg/ml, sample2 20.0 mg/ml, sample3 28.6 mg/ml), weight increase 130%; DEAE-dextran (10 mg/ml), weight increase 0%. Each condition represents the mean of three transfections per condition. Standard deviation was calculated and plotted for each condition.*

DEAE-dextran hydrochloride. In Figure 17.5, quantities of β-Galactosidase enzyme are shown as 50 times and more β-gal/protein [mU · ml/mg] for the case of DDMC3 (sample1, sample2, sample3) compared with DEAE-dextran. From the results, the transfection efficiency and the reaction rate of formation of the complex should increase when using DDMC hydrochloride instead of DEAE-dextran hydrochloride.

17.3.5.7 Cytotoxicity for the Transfection

Figure 17.5 shows quantities of β-Galactosidase enzyme and transfection of HEK293 cells by DDMC3 having a grafting rate of 130% for sample 1 of 10 mg/ml, sample 2 of 20 mg/ml and sample 3 of 28.5 mg/ml. The quantities of β-Galactosidase enzyme increase with the concentration of DDMC3 were 31.13, 32.27 and 57.24 [mU ml/mg] for sample 1, sample 2, and sample 3, respectively.

Figure 17.6 shows the transfection of COS7 cells by sample of DEAE-dextran and DDMC3 having grafting rate of 130% for sample 1 of 10 mg/ml, sample 2 of 20 mg/ml and sample 3 of 28.6 mg/ml. The transfection for COS7 cells was carried out by pGL3DNA/DDMC. Cells are ready to harvest 72 hours post-transfection and assayed for luciferase activity. Their luciferase activity increases with the concentrations of DDMC3, such as sample 1 < sample2 < sample 3. These show a low cytotoxicity for DDMC3.

Figure 17.7 shows the transfection activity of HEK293 cells using the X-gal staining method for sample 1 of 10 mg/ml, sample 2 of 20 mg/ml and sample 3 of 28.6 mg/ml of DDMC3 compared with DEAE-dextran the same as Figure 17.5. Cell death by apoptosis on HEK293 cells for DEAE-dextran was also detected. However, a low cytotoxicity for DDMC3 compared with DEAE-dextran was seen.

Figure 17.7 shows the change of transfection efficiency when using twice the quantity, for example using 20 μg DNA, of both DNA and DDMC as much as the protocol. The transfection of HEK293 cells by sample DDMC1 and DDMC2, carried out using twice

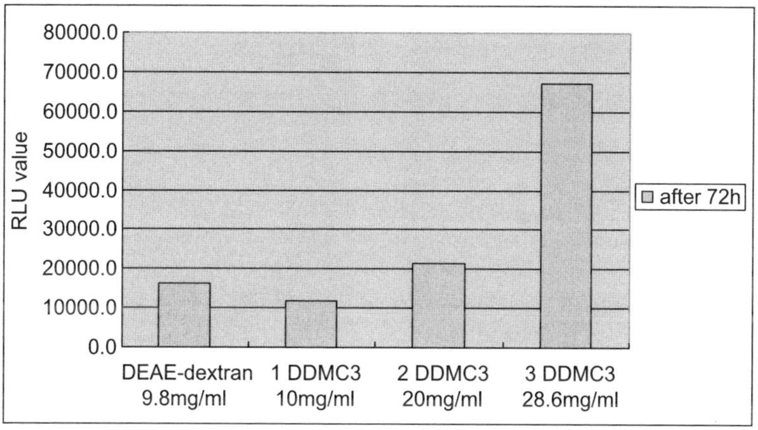

Figure 17.6 *Transfection of COS7 cells by sample of DEAE-dextran and DEAE-dextran-MMA graft copolymer having grafting rate of 130% for sample 1 of 10 mg/ml, sample 2 of 20 mg/ml, and sample 3 of 28.6 mg/ml. The transfection for COS7 cells was carried out by pGL3DNA/DDMC. Cells are ready to harvest 72 hours post-transfection and assayed for luciferase activity*

the quantity of both DNA and DDMC as much as the protocol, has shown two times higher efficiency than the original one using a transfection activity determined using the X-gal staining method. From these results, the cytotoxicity for the transfection reagent can be decreased and cell viability improved when using DDMC hydrochloride instead of DEAE-dextran hydrochloride.

17.3.5.8 Kinetics of the Transfection

The complex between DDMC and DNA should be followed by (17.8) of a Michaelis Menten type equation such as (17.5), because it is accomplished not only by a coulomb force between the phosphoric acid of nucleic acids and the DEAE group of DEAE-dextran but also by a hydrogen bond in their PIC structure and the hydrophobic force of graft PMMA.

$$\text{complex} = \text{Km1}\,(\text{DNA})\quad(\text{DDMC}) \tag{17.8}$$

The complex between DEAE-dextran and DNA should be followed by (17.9) of the first order of the concentration of DEAE-dextran such as (17.6), because it is accomplished by mainly a coulomb force between the phosphoric acid of nucleic acids and the DEAE group of DEAE-dextran.

$$\text{complex} = \text{Km2}\,(\text{DEAE-dextran}) \tag{17.9}$$

Figure 17.8 shows the transfection activity of COS7 cells by samples of DEAE-dextran and DDMC3 having a grafting rate of 130% for DNA 0.075 μg when maximum relative luciferase expression (RLU) within each experiment was set to 100%. The RLU values were calculated by (17.8) and (17.9) in Figure 17.8 using Km1 = 1.055×10^{-7} (μg/well)^2

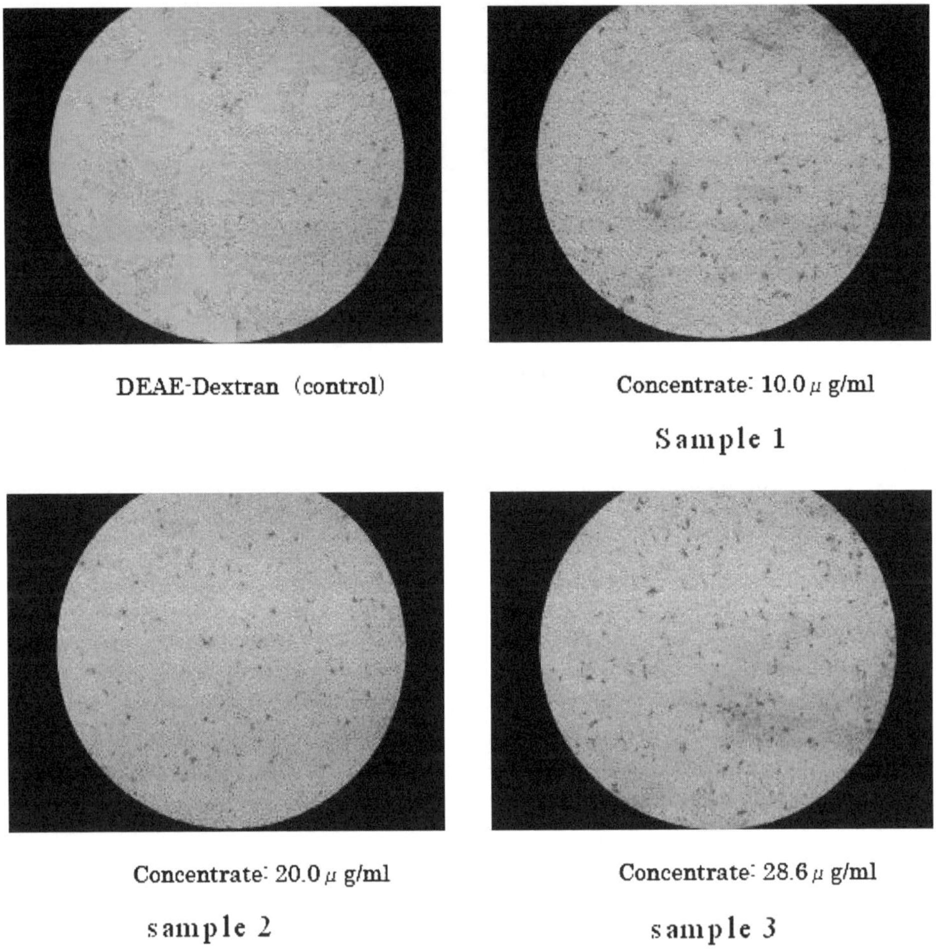

Figure 17.7 *Transfection of a monolayer of HEK 293 cells by DEAE-dextran-MMA graft copolymer (DDMC3) of weight increase 130% for sample 1 of 10 mg/ml, sample 2 of 20 mg/ml and sample 3 of 28.6 mg/ml.; DEAE-dextran, weight increase 0%, for a sample of 10 mg/ml. Reprinted with permission from Elsevier. Copyright 2007.*

and Km2 = $1.626 \times 10^{-5}\,\mu$g at maximum RLU value, respectively. As a quantum of the complex is obviously in proportion to RLU value, (17.8) and (17.9) can be correctly estimated.

17.3.6 Protection from DNase Degradation

The complex of DDMC/DNA is thought to become a more compact structure by its shift of νNH vibration at 3450 cm^{-1} as shown in Figure 17.1. DDMC has also been confirmed as having a high protection facility for DNase degradation (Figure 17.9). DNase I degrades

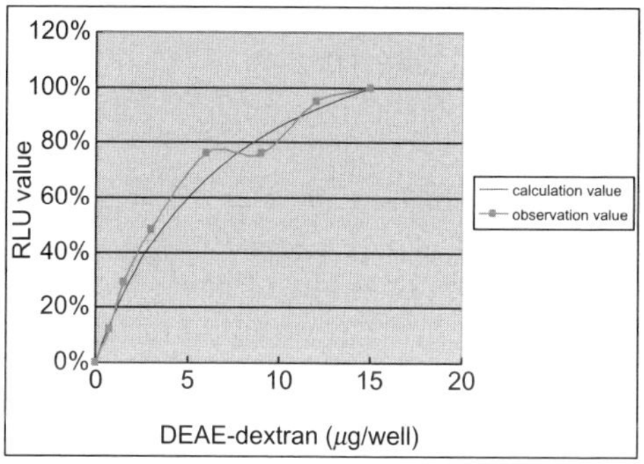

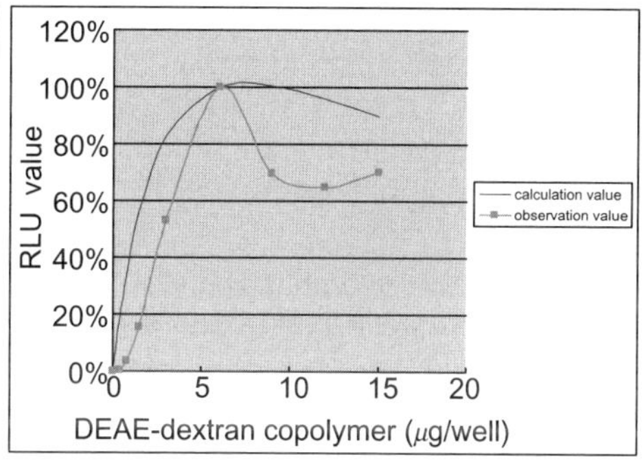

Figure 17.8 *Transfection of COS7 cells by samples of DEAE-dextran and DEAE-dextran-MMA graft copolymer (DDMC3) having grafting rate of 130% for DNA 0.075 μg. Maximum luciferase expression within each experiment was set to 100%*

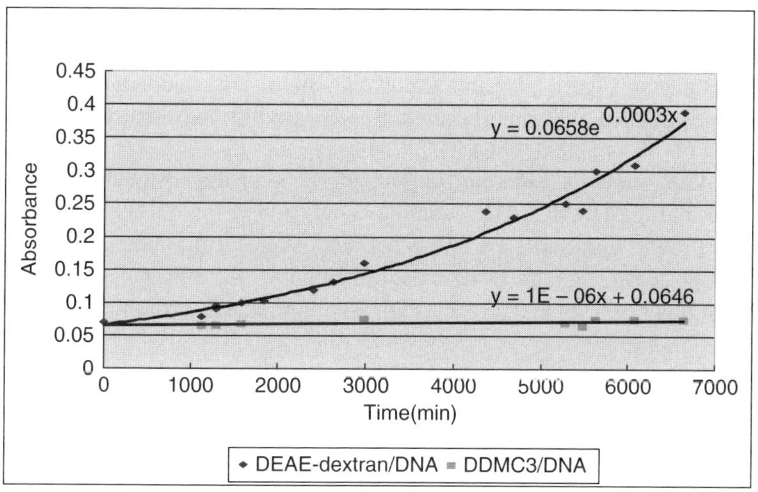

Figure 17.9 *DNase degradation: The samples were added to 4 ml of a distilled water, then 10 μl of DNase (RQ1 RNase-Free DNase, Promega) and 0.1 ml of 10× Reaction Buffer (400 mM Tris-HCl, 100 mM $MgSO_4$, 10 mM $CaCl_2$, pH8) and incubated at 37 °C. The wavelength used for this experiment was 633 nm for Toluidine Blue isolated from DNA. Reprinted with permission from Elsevier. Copyright 2007.*

both double-stranded and single-stranded DNA endonucleolytically, producing 3′-OH oligonucleotides.

17.3.6.1 Measurement of DNase Degradation

Toluidine blue (TB) and DNA form a complex by stain reaction [32]. The stain reaction between TB and DNA was carried out by adding 1 ml of 0.005% TB solution (pH 7.0) to 1 ml of DNA solution (10 mg/ml, ex salmon sperm, Wako, Osaka, Japan). For the reaction of DDMC/DNA, to 2 ml of DNA solution mixed with TB solution, the solution (10 mg/ml as DEAE-dextran) of DDMC was added to obtain the complex of DDMC/DNA stained by TB.

The resulting precipitation of the complex by DDMC/DNA stained by TB was separated by filter (ADVANTEC 5A, Toyo Roshi, Tokyo, Japan), then added to 4 ml of distilled water, mixed with 0.01 ml (10 units) of DNase (RQ1 RNase-free DNase, Promega, Madison, WI) and 0.1 ml of 10× reaction buffer (400 mM Tris-HCl, 100 mM magnesium sulfate, 10 mM calcium chloride, pH 8.0), and was incubated at 37 °C for 110 hours.

DNase degradation was determined by measuring the absorbance for TB isolated from DNA in the water with a spectrophotometer. TB in water absorbs broadly in the red range, with a maximum absorption at 633 nm and a shoulder near 600 nm.

The wavelength at which the absorbance was followed was chosen as 633 nm to give a reasonably large initial absorbance.

17.3.6.2 DNase Degradation

TB is isolated in water from DNA at the degradation, when DNA is stained by TB. Figure 17.9 shows the absorbance for TB isolated from the DNA of samples in the water with a spectrophotometer. Figure 17.9 also shows the rate of DNase degradation with a slope of both the absorbance and time, because Beer's law plot between the absorbance and concentration of TB can be applied [37].

In Figure 17.9, some degradation by DNase can be observed for the DEAE-dextran/DNA sample, but the DDMC3/DNA sample shows a very low DNase degradation.

It was also found that the DNA at the colloidal stage of the interaction with DEAE-dextran was protected against DNase degradation [4, 38, 39]. As a result, obviously DDMC has a higher protective effect on DNase than DEAE-dextran.

17.3.7 DNA Delivery Pathways by DDMC

As in Figure 17.4, DNA delivery pathways by DDMC for transfection of cells should be carried out using the steps below: (a) formation of a complex between DNA and DDMC; (b) uptake; (c) endosytosis (endosome); (d) escape from endocytotic vesicle; (e) DNA release in cytosol; (f) nuclear entry; (g) DNA release and transcription in nucleus.

For transfection efficiency, it is very important to examine factors such as uptake in step (b), resistance of nuclease in step (c), escape from endocytotic vesicle in step (d), nuclear targeting in step (f) and DNA release in step (g). The positively charged DEAE-dextran copolymer interacts with the negatively charged phosphate backbone of DNA. The resulting complex in step (a) is absorbed into cells by endocytosis. The specifically designed molecular structure of DDMC having a positive charge and a hydrophilic–hydrophobic micro-separated domain ensures easy entry of DNA into cells for steps (b), (c), (d), (f) and (g), especially DNA penetration of the plasma membrane, and intracellular release of DNA in these steps because of its affinity for a cell membrane by its hydrophilic–hydrophobic micro-separated domain [9]. DDMC, has the thixotropy property of a non-Newtonian fluid. The transfection solution can flow and wet the cells by simple sheer stress to overcome the barrier of the transfection, such as steps (c).

Regarding the transfection efficiency for all steps, it was also considered to be important to examine DNA protection from degradation by DNase I attack.

However, DNA is tightly packed in native genomes by formation of the complex in step (a) and the manner of this packaging is involved in the mechanism of gene expression. The formation of the complex should induce the structural transition of DNA, which is known as DNA condensation, by a coil-globule transition [19, 21, 40, 41]. This may induce discrete on/off switching in transcriptional activity [21] and may promote a protection from the degradation of entrapped DNA and associated complexes.

17.4 Conclusions

In the first step (a) of delivery pathways, the formation of a complex between nucleic acids (DNA or RNA) and DDMC is, of course, accomplished by a coulomb force between the phosphoric acid of nucleic acids and the DEAE group of DEAE-dextran. Figure 17.1 shows the infrared absorption spectra of the resulting complex between DDMC (sample DDMC2) and DNA. The spectrum of the complex has some characteristic absorption bands at 1730 cm^{-1}, 1220 cm^{-1}, 1000–1150 cm^{-1} and 3450 cm^{-1}, which are attributed to the carbonyl group of PMMA, P–O stretching vibration of DNA, the pyranose ring of DEAE-dextran in DDMC and the DEAE group of DEAE-dextran in DDMC, respectively. The absorption spectrum shift at around 3450 cm^{-1} of the complexes should mean the formation of more compact structures by a coulomb force between the phosphoric acid of DNA and the DEAE group of DEAE-dextran in DDMC to conclude to DNA condensation theory [21]. Moreover, there were also many differences between the complex of DDMC/DNA and the complex of DEAE-dextran/DNA with the hydrogen bond in their PIC structure by the decrease of νNH–O vibration at around 3350 cm^{-1} and the shift of νNH vibration from 3450 cm^{-1} to 3550 cm^{-1} in Figure 17.2.

As shown in Figure 17.3, the resulting DDMC of hairy nanoparticles having amphiphilic domain, forming a polymer micelle of core shell particles, should become a stable latex with a hydrophilic–hydrophobic micro-separated domain [28]. The complex of DDMC and plasmid DNA may be formed initially on the spherical structure of the amphiphilic micro-separated-domain of DDMC to be stable and have a good affinity to cell membrane. When (e) DNA is released in cytosol, this structure should be protected from degradation by DNase and dextransucrase so as to be easy for (f) nuclear entry such as Figure 17.4.

From these viewpoints, DDMC must have a lower cytotoxicity and the higher transfection efficiency compared with DEAE-dextran.

The higher transfection efficiency of this graft copolymer autoclaved can make it a valuable tool for gene delivery *in vivo* or *in vitro*.

Specifications of DDMC Vector from DEAE-Dextran MMA copolymer were as below:

1. Stable for autoclaving sterilization (FDA 121 °C, 15 min)
2. Fast and easy procedure
3. Applicable in high-throughput-screening (HTS)
4. No serum inhibition
5. High efficiency by use of small DNA amounts
6. An excellent reproducibility
7. A very low toxicity in comparison with DEAE-dextran

Acknowledgement

We would like to express our sincere gratitude to Dr Jun Yoshida, Professor Emeritus of Nagoya Universitity, for his stimulating advice.

References

1. Murata, J., Ohya, Y. and Ouchi, T. (1966) Possibility of application of quaternary chitosan having pendant galactose residues as gene delivery tool. *Carbohydr. Polym.*, **29**, 69–74.
2. Sato, T. (2002) Carbohydrate polymer for gene delivery. *Kobunshi*, **51**, 837–840.
3. McCutchan, J.H. and Pagano, J.S. (1968) Enhancement of the infectivity of simian virus 40 deoxyribonucleic acid with diethylaminoethyl-dextran. *J. Nat. Cancer Inst.*, **41**, 351–358.
4. Warden, D. and Thorne, H.V. (1969) Influence of diethylaminoethyl-dextran on uptake and degradation of polyoma virus deoxyribonucleic acid by mouse embryo cells. *J. Virol.*, **4**, 380–387.
5. Constantin, T. and Vendrely, C. (1969) Effect of DEAE-dextran on the incorporation of tritiated DNA by cultured rat cells. *C. R. Soc. Biol.*, **163**, 300–305.
6. Mack, K.D., Wei, R., Elbagarri, A. *et al.* (1998) A novel method for DEAE-dextran mediated transfection of adherent primary cultured human macrophages. *J. Immunol. Methods*, **211**, 79–86.
7. Onisi, Y. and Kikuchi, Y. (2003) Study of the complex between DNA and DEAE-Dextran. *Kobunshi Ronbunshu*, **60**, 359–364.
8. Onisi, Y. and Kikuchi, Y. (2004) Study of the complex between RNA and DEAE-Dextran. *Kobunshi Ronbunshu*, **61**, 139–143.
9. Onishi, Y., Maruno, S., Kamiya, S. *et al.* (1978) Preparation and characteristics of dextran-methyl methacrylate graft copolymer. *Polymer*, **19**, 1325–1328.
10. Onishi, Y. (1987) Cationic graft-copolymer.US Patent 4816540.
11. Onishi, Y., Eshita, Y., Murashita, A. *et al.* (2005) Synthesis and characterization of 2-diethylaminoethyl(DEAE)-dextran-MMA graft copolymer for non-viral gene delivery vector. *J. Appl. Polym. Sci.*, **98**, 9–14.
12. Higashihara, J., Onishi, Y., Mizuno, M. *et al.* (2005) Transfection of foreign genes into culture cells using novel DEAE-dextran copolymer as a non-viral gene carrier. The 55th Annual Meeting of Southern Region, Japan Society of Medical Entomology and Zoology, Miyazaki Prefecture, Japan, 23 October 2005, Abstract15.
13. Onishi, Y., Eshita, Y., Murashita, A. *et al.* (2006) 2- diethylaminoethyl(DEAE)dextran-MMA graft copolymer for non-viral gene delivery. *Bull. Res. Center Environ. Sci. Technol., Nippon Bunri University*, **5**, 8–13.
14. Onishi, Y., Eshita, Y., Murashita, A. *et al.* (2007) Characteristics of 2-diethylaminoethyl(DEAE)-dextran-MMA graft copolymer as a non-viral gene carrier. *Nanomed. Nanotechnol. Biol. Med.*, **3**, 184–191.
15. Onishi, Y., Eshita, Y., Murashita, A. *et al.* (2008) A novel vector of 2- diethylaminoethyl(DEAE) dextran-MMA graft copolymer for non-viral gene delivery. *J. Gene Med.*, **10**, 430.
16. Eshita, Y., Higashihara, J., Onishi, M. *et al.* (2009) Mechanism of introducing exogenous genes into cultured cells using DEAE-Dextran-MMA graft copolymer as non-viral gene carrier. *Molecules*, **14**, 2669–2683.
17. Price, C. and Woods, D. (1973) A method for studying micellar aggregates in block and graft copolymers. *Eur. Polymer Sci.*, **9**, 827–830.
18. Minagawa, K., Matsuzawa, Y., Yoshikawa, K. *et al.* (2002) Direct observation of the biphasic conformational change of DNA induced by cationic polymers. *FEBS Lett.*, **295**, 67–69.
19. Yoshikawa, Y., Emi, N., Kanbe, T. *et al.* (1996) Folding and aggregation of DNA chains induced by complexation with lipospermine: formation of a nucleosome-like structure and network assembly. *FEBS Lett*, **396**, 71–76.

20. Yamasaki, Y. and Yoshikawa, K. (1999) Collapsing transition of single DNA chains and their morphorogical variations. *Kobunshi Ronbunshu*, **56**, 772–785.
21. Yoshikawa, Y., Tsumoto, K. and Yoshikawa, K. (2002) Switching of higher-order structure of DNA and gene expression. *Seibutsu Butsuri*, **42**, 179–184.
22. Behr, J.P., Demeneix, B., Loeffler, J.P. and Perez-Mutul, J. (1989) Efficient gene transfer into mammalian primary endocrine cells with lipopolyamine-coated DNA. *Proc. Natl. Acad. Sci. USA*, **86**, 6982–6986.
23. Gao, X. and Huang, L. (1991) A novel cationic liposome reagent for efficient transfection of mammalian cells. *Biochem. Biophys. Res. Commun.*, **179**, 280–285.
24. Kojima, H., Ohishi, N., Takamori, M. and Yagi, K. (1995) Cationic multilamellar liposome-mediated gene transfer into primary myoblasts. *Biochem. Biophys. Res. Commun.*, **207**, 8–12.
25. Farber, F.E., Melnick, J.L. and Butel, J.S. (1975) Optimal conditions for uptake of exogenous DNA by Chinese hamster lung cells deficient in hypoxanthine-guanine phosphoribosyltransferase. *Biochim. Biophys. Acta*, **390**, 298–311.
26. Holter, W., Fordis, C.M. and Howard, B.H. (1989) Efficient gene transfer by sequential treatment of mammalian cells with DEAE-dextran and deoxyribonucleic acid. *Exp. Cell Res.*, **184**, 546–551.
27. Haensler, J. and Szoka, F.C. (1993) Polyamidoamine cascade polymers mediate efficient transfection of cells in culture. *Bioconj. Chem.*, **4**, 372–379.
28. Michaels, A.S. (1965) Polyelectrolyte complexes. *Ind. Eng. Chem.*, **57**, 32–40.
29. Tanaka, S., Shibayama, M., Sasabe, R. and Kawaguchi, H. (2003) Preparation and structure characterization of hairy nanoparticles consisting of hydrophobic core and thermosensitive hairs. *Polymer*, **44**, 495–501.
30. Horie, K., Yamada, S., Machida, S. *et al.* (2003) Dansyl fluorescence and local structure of dansyl-labeled core-shell and core-hair type microspheres in solution. *Macromol. Chem. Phys.*, **204**, 131–138.
31. Onishi, Y., Maruno, S., Kamiya, S. *et al.* (1978) Preparation and characteristics of dextran-methyl methacrylate graft copolymer. *Polymer*, **19**, 1325–1328.
32. Onisi, Y., Maruno, S. and Hokkoku, S. (2003) Graft copolymerization of methyl methacrylate onto dextran and some properties of copolymer. *Kobunshi Ronbunshu*, **60**, 359–364.
33. Graham, F.L. and van der Eb, A.J. (1973) A new technique for the assay of infectivity of human adenovirus 5 DNA. *Virology*, **52**, 456–467.
34. McCormik, C.L. and Park, L.S. (1981) Water-soluble copolymers. III. Dextran-g-poly(acrylamides) control of grafting sites and molecular weight by Ce(IV)-induced initiation in homogeneous solutions. *J. Polym. Sci. Polym. Chem. Ed.*, **19**, 2229–2241.
35. Onishi, Y., Butler, G.B. and Hogen-Esch, T.E. (2004) 1,2-propanediol-cellulose-acrylamide graft copolymers. *J. Appl. Polym. Sci.*, **92**, 3022–3029.
36. Schreier, J.B. (1969) Modification of deoxyribonuclease test medium for rapid identification of Serratia marcescens. *Am. J. Clin. Pathol.*, **51**, 711–716.
37. Ogren, P.J., Henry, I., Fletcher, S.E.S. and Kelly, I. (2003) Chemical applications of a programmable image acquisition system. *J. Chem. Educ.*, **80**, 699–703.
38. Maes, R., Sedwick, W. and Vaheri, A. (1967) Interaction between DEAE-dextran andnucleic acids. *Biochim. Biophys. Acta*, **134**, 269–276.
39. Pagano, J.S., McCutchan, J.H. and Vaheri, A. (1967) Factors influencing the enhancement of the infectivity of poliovirus ribonucleic acid by diethylaminoethyl-dextran. *J. Virol.*, **1**, 891–897.
40. Bloomfield, V.A. (1997) DNA condensation by multivalent cations. *Biopolymers*, **44**, 269–282.
41. Chaszczewska-Markowska, M., Ugorski, M. and Langner, M. (2004) Plasmid condensation induced by cationic compounds: hydrophilic polylysine and amphiphilic cationic lipid. *Cell. Mol. Biol. Lett.*, **9**, 3–13.

18

Liposomes in Adjuvant Systems for Parenteral Delivery of Vaccines

Dennis Christensen,[1] *Karen Korsholm,*[1] *Grith Krøyer Wood,*[1] *Afzal Mohammed,*[2] *Vincent Bramwell,*[2] *Peter Andersen,*[1] *Else Marie Agger*[1] *and Yvonne Perrie*[2]

[1] Statens Serum Institute, Adjuvant Research, Department of Infectious Disease Immunology, Copenhagen, Denmark

[2] Aston University, School of Life and Health Sciences, Birmingham, United Kingdom

18.1 The Structural Basis of Liposomes

Liposomes were first described by Bangham *et al.* in 1965 [1] when he discovered that phospholipids in water form closed vesicles. Today the term liposome is a broad definition for vesicles composed of lipid bilayers enclosing aqueous compartments. The membrane-forming lipids are amphiphilic and, accordingly, contain a polar and an apolar region. The polar region typically consists of a phosphate group, an acidic group and/or tertiary or quaternary ammonium salts and can either have a net negative (anionic), neutral or positive (cationic) surface charge, depending on the composition of the lipid head groups. The apolar region typically consists of one or more fatty acid chains with eight or more carbon atoms and/or cholesterol.

The chain length, degree of saturation and head-group composition of the lipids has great influence on the properties of the formed liposomes. Hence, liposomes based on longer fatty acid chains have a higher gel–fluid phase transition temperature than those of shorter chains [2–4] and the gel–fluid phase transition temperature is increased with the degree of saturation [4, 5]. The size of the head group has great influence on the stability of the liposomes, since lipids generally should have a cylindrical shape to form stable liposomes.

Delivery Technologies for Biopharmaceuticals: Peptides, Proteins, Nucleic Acids and Vaccines Edited by Lene Jorgensen and Hanne Mørck Nielsen

An important characteristic to take into consideration is the overall charge of the bilayer surface, which is determined by the charge of the polar head groups. Cationic liposomes tend to interact and form complexes with the, predominantly, negatively charged surfaces of biomolecules such as proteins and nucleic acids. Cationic liposomes are, for the same reason, very potent transfection agents for, for example, nucleic acids [6, 7]. In contrast, the surface of neutral and anionic liposomes will only interact modestly with negatively charged molecules. This makes them suitable for encapsulation of proteins, peptides and nucleic acids to obtain, for example, systemic or organ-specific drug delivery and sustained release after intravenous (i.v.) injection. However, in order to avoid rapid clearance, additional constituents such as PEG-ylated lipids have to be added. This approach has been used in licensed products like Doxil®/Caelyx® [8, 9].

18.2 Key Characteristics Which Make Liposomes Advantagous as Vaccine Delivery Systems

Ever since Paul Ehrlich launched his idea about magic bullets – compounds with specific structures for combating specific diseases that would leave everything else undisturbed – scientists have had these bullets as a goal for their medical developments. With the observation that phospholipids in excess water form lamellar structures [1] (Figure 18.1), later to be known as liposomes, and the subsequent development of the drug delivery concept [10, 11], the development of magic bullets has taken a big leap forward. Currently, there are several liposome-based clinical products that improve the delivery of drugs to target sites and, in turn, reduce systemic toxicity (e.g. doxorubicin formulated as Doxil®/Caeylx®).

The use of liposomes as delivery systems is also of special interest in the field of vaccinology [12–14]. The function of liposomes as delivery systems of vaccine antigens is dual. Firstly, due to the fact that most antigens developed today are recombinant proteins or synthetic peptides which, without a carrier, will be cleared from the body before they are recognized by the immune system, the liposomes must act as a depot for the vaccine antigens; protecting and avoiding clearance from the body. Secondly, the liposomes must present/deliver antigens and immunomodulators to antigen-presenting cells (APC) like dendritic cells (DC) and macrophages (Mφ). Furthermore, some liposomes themselves have immunostimulating properties, for example liposomes based on quaternary ammonium lipids such as N,N-dimethyl-N,N-dioctadecylammonium (DDA, Figure 18.2). Mixed with hapten-carrier complexes, DDA produces hapten-specific delayed-type hypersensitivity (DTH) in guinea pigs [15] and mice [16, 17], suggesting that DDA can generate a cell-mediated immune (CMI) response. In addition, cationic liposomes based on DDA are able to enhance the cellular uptake of ovalbumin (OVA), both *in vitro* and *in vivo* [18]. This enhanced antigen-uptake results in efficient presentation of the antigen to T-cells, leading to their subsequent proliferation and release of IFN-γ.

In addition, liposomes based on other cationic lipids, such as 1,2-dioleoyl-*sn*-glycero-3-ethylphosphocholine (DOTAP) and 3β-[N-(N′,N′-dimethylaminoethane)-carbamoyl] cholesterol (DC-Chol), also exhibit immunostimulating properties [18]. However, alone none of these liposomes have yet been sufficient to generate an appropriate immune response against vaccine antigens in animal trials, and the combination with

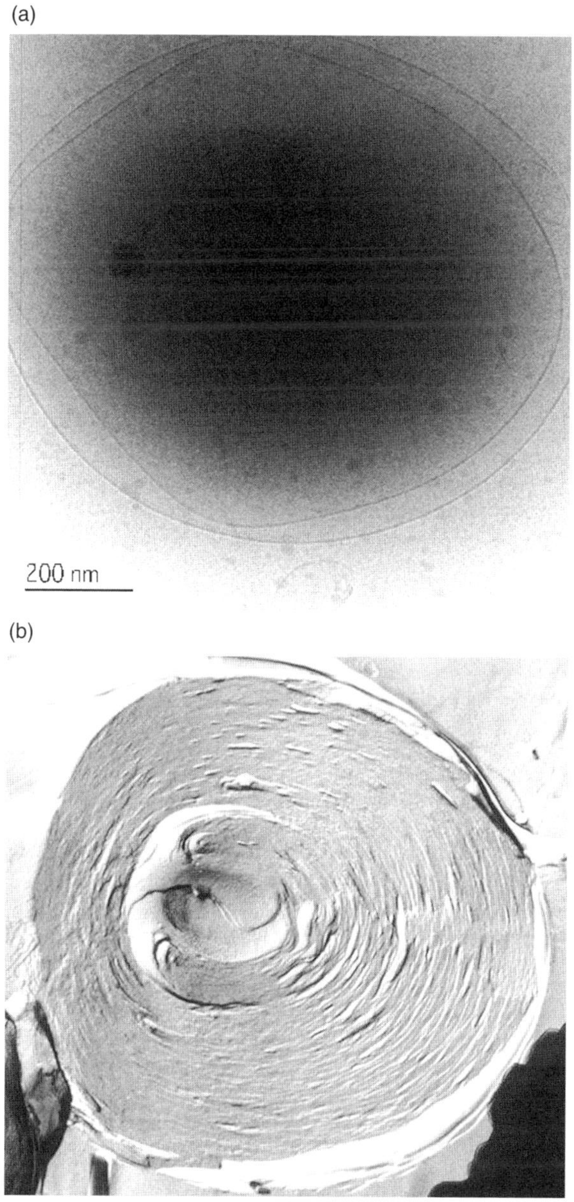

Figure 18.1 *(a) Cryo-TEM micrograph of dimethyldioctadecylammonium liposomes prepared by the lipid film hydration method and (b) freeze–fracture electron micrograph of liposomes prepared by the dehydration–rehydration vesicle method, incorporating phosphatidyl choline with cholesterol (image kindly supplied by Brigette Sternberg-Papahadjopoulos)*

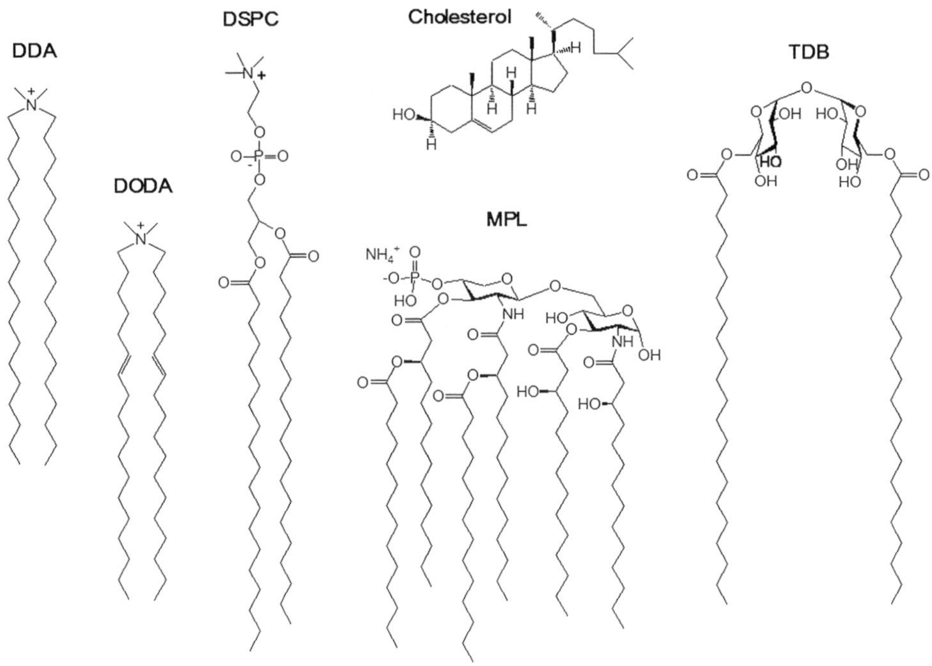

Figure 18.2 Chemical structures of the most important lipids mentioned in this chapter: N,N-dimethyl-N,N-dioctadecylammonium (DDA), N,N-dioleoyl-N,N-dimethylammonium (DODA), 1,2-distearoyl-sn-glycero-3-phosphocholine (DSPC), cholesterol, monophosphoryl lipid A (MPL) and trehalose 6,6′-dibehenate (TDB)

immunomodulators significantly improves their immunogenicity [18]. Furthermore, the immunostimulating properties, or adjuvanticity, of liposomes appear not only to be dependent on their surface charge but also their structural characteristics, such as vesicle size and fluidity of the bilayers [12, 19]. This is discussed in detail in the next chapter.

The ability of adjuvants to deliver and present antigens to APC can be highly dependent on the interaction between the adjuvant and the antigens [20]. This can, therefore, be an important aspect to take into consideration when developing liposomal delivery systems for vaccines. Negatively charged proteins are adsorbed to the cationic liposomal surface because of electrostatic attraction, whereas positively charged proteins are not adsorbed due to electrostatic repulsion. This is exemplified when formulating antigens of different pI values in either the cationic liposomal adjuvant CAF01, consisting of DDA with 11 mole% of the immunomodulating lipid trehalose 6,6′-dibehenate (TDB, Figure 18.2), or in anionic liposomes consisting of DSPG with the same amount of TDB (unpublished results).

In the case of CAF01, adsorption is, however, not a prerequisite for an immune response against these antigens. Studies have investigated the immune response in mice, either immunized with CAF01 mixed before injection with a *Mycobacterium tuberculosis* vaccine candidate which was adsorbed (Ag85B-ESAT-6, pI = 4.9) or a *Chlamydia*

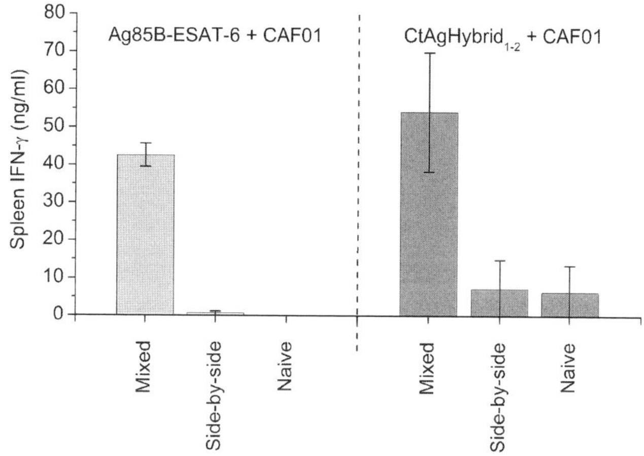

Figure 18.3 *Release of IFN-γ from spleen cells isolated from C57BL/6J mice. The mice were immunized subcutaneously (s.c.) at the base of the tail three times with a two-week interval with 2 μg of either Ag85B–ESAT-6 adsorbed to CAF01 or CtAgHybrid$_{1-2}$ mixed with CAF01 (not adsorbed) or the two antigens administered side-by-side with CAF01, thus draining to the same lymph node. Spleen cells were isolated eight weeks after the first immunization and restimulated in vitro with the corresponding antigen (5 μg/ml) (method described in [23, 33, 89]). All values shown are the average of three independent samples ± standard deviation. The results show that both antigens elicit high IFN-γ production by spleen cells after vaccination together with CAF01, whereas IFN-γ production after side-by-side vaccination was minimal. This indicates that for CAF01 to function optimally, it has to be administered together with the antigen which, however, does not have to be adsorbed to the adjuvant*

trachomatis vaccine candidate which was not adsorbed (CtAgHybrid$_{1-2}$, pI = 9.0), or immunized by injecting the adjuvant and antigen side-by-side, thus draining to the same lymph node. High levels of antigen-specific IFN-γ were detected for both the adsorbed and the non-adsorbed antigen when mixed and injected together, whereas the cytokine levels were comparable to those of naïve mice when injected side-by-side (Figure 18.3). This indicates that antigens do not have to be adsorbed to CAF01, as long as they are mixed and injected together.

18.3 Formulation Approaches to Improve Adjuvanticity and Enhance Immunological Targeting

Liposomes composed of phosphatidylcholine-based lipids are inherently non-immunogenic and can be used as drug delivery systems, targeting sites of enhanced permeability such as tumours and sites of inflammation. Alternatively, liposomes composed of cationic lipids, such as DDA, can act as functional drug delivery systems of vaccine antigens targeting APC to elicit immune responses. However, their ability to generate a powerful and directed immune response suitable for a given disease target can be greatly improved by combination with immunomodulators such as monophosphoryl lipid A (MPL, Figure 18.2) [21, 22], TDB [21, 23] or other lipids from, for example, the mycobacterial cell wall [23]. Thus the unspecific immune response generated by DDA can potentially be modulated by immunomodulators to fit the immunological profile required to prevent a given disease.

18.3.1 Immunomodulators

18.3.1.1 Monophosphoryl Lipid A

Lipopolysaccharides (LPS) of the cell wall of gram-negative bacteria are potent endotoxins. These toxic properties can be beneficial for immunopotentiation and the induction of endotoxin tolerance [24]. However, LPS are not appropriate for human use due to the fact that their toxic effects are too severe. This has led to the development of detoxified monophosphoryl lipid A (MPL®, Figure 18.2). This minimal active moiety of LPS is extracted from cell cultures of either *Salmonella typhimurium* or *Salmonella minnesota* and the cellular mechanism by which MPL has been demonstrated to induce activation is predominantly through the TLR4 receptor inducing TNF-α, IL-10 and IL-12 [25, 26].

Being a membrane-derived glycolipid, MPL is well suited for incorporation into liposomal membranes and several studies have investigated the adjuvant effect of liposomes containing MPL [27–31]. The GSK adjuvant AS01 comprises MPL incorporated into liposomes together with QS21 [32]. The combination of MPL with DDA-based liposomes has a synergistic effect, enhancing the immunogenicity significantly compared to the two components alone, and generating a very strong T_H1 response [21, 33].

18.3.1.2 Trehalose Dibehenate

Several mycolic acids found in the wax-like membrane of mycobacterium spp are immunogenic. The best characterized is trehalose 6,6′-dimycolate (TDM), also known as Cord Factor, which elicits both humoral [34, 35] and cellular immune responses [35–37]. In addition, TDM can enhance the protection against bacteria [35, 38, 39], viruses [40–42] and parasites [43, 44]. However, the long branched side chains of TDM seem to be an important co-factor for the toxicity of mycobacteria [45–47], rendering TDM unacceptable for clinical use today.

Trehalose 6,6′-dibehenate (TDB) (Figure 18.2) is a synthetic analogue of TDM with two saturated fatty acid chains of 22 carbons each replacing the branched mycobacterial mycolic acids of >70 carbons. TDB alone does not lead to consistent protection against schistosomiasis [48] and tumour suppression in mice [49], but application of TDB in an emulsion results in tumour suppression in rats [50]. The incorporation of TDB into DDA liposomes has, like MPL, a synergistic effect on the two adjuvant components [21, 23]. In tuberculosis (TB) challenge studies in mice and guinea pigs, CAF01 (DDA/TDB) is as effective as DDA/MPL [21, 22] and CAF01 elicits a strong T_H1 response characterized by high levels of IFN-γ, low levels of IL-5 and the production of IgG2b, but also high levels of IgG1 comparable to those of $Al(OH)_3$ [23]. This combined T_H1/antibody response makes CAF01 a promising adjuvant for many diseases including TB [21, 22], hepatitis B [51] and malaria [52].

18.3.2 Physicochemical Properties

As mentioned previously, the function of liposomes and their interaction with cells can be highly dependent on their physicochemical properties, such as particle size, surface charge and fluidity.

18.3.2.1 Particle Size

The uptake of nanoparticles by dendritic cells [53] and the resulting immune response is much dependent on the size of the particles [54]. The theory behind these observations is that small particles (20–200 nm) mimic viruses and are taken up by endocytosis by dendritic cells, whereas larger particles (500–5000 nm) mimic bacteria and are taken up by macropinocytosis and even larger particles are taken up by phagocytosis by macrophages. Thus, small particles may generate virus-like responses, whereas larger particles induce bacteria-like responses [54]. This has, to our knowledge, never been fully exploited for liposomes in general. However, the CAF01 adjuvant made by the film method is comprised of multivesicular vesicles that are highly polydisperse in particle size [23]. Attempts to make a more homogenous particle size distribution of CAF01 formulations have not led to enhanced or more specific immune response (data not shown).

18.3.2.2 Surface Charge

The surface charge has a major influence on the vaccine delivery effect of liposomes and most *in vivo* studies demonstrate that cationic liposomes are superior compared to both anionic and neutral liposomes [55–58]. This was investigated with CAF01, in which DDA was partly or completely replaced by the neutral 1,2-distearoyl-*sn*-glycero-3-phosphocholine (DSPC, Figure 18.2). Replacement of up to 40% of the DDA with DSPC did not affect the zeta potential or immune response significantly (Figure 18.4). However, the zeta potential was reduced from ~65 mV to ~0 mV at higher proportions of DSPC and the amount of antigen-specific IFN-γ produced by blood lymphocytes one week after the last immunization decreased proportionally (Figure 18.4), indicating that the ability of the liposomes to generate a CMI response can be diminished by neutralizing the liposomal surface charge.

18.3.2.3 Fluidity

The gel–fluid phase transition of liposomes is another parameter affecting immunogenicity that can be modulated. By replacing some of the DDA lipids in the CAF01 liposomes with the unsaturated analogue N,N-dioleoyl-N,N-dimethylammonium (DODA, Figure 18.2), the phase transition enthalpy is reduced. In addition, the ability of the liposomes to generate a CMI response is decreasing with the gradual replacement of DDA with DODA, indicating that the solid particulate form of the DDA liposomes is a stronger inducer of CMI than the fluid form of the DODA liposomes [59]. This has also been observed with neutral phosphatidyl choline (PC) liposomes [60]. Furthermore, the immunogenicity of the liposomes was found to be enhanced with longer acyl chain length, and hence higher gel–fluid phase transition temperature. This was the case for both PC liposomes [60, 61] and liposomes based on quaternary ammonium lipids [62]. This supports the hypothesis that liposome formulations in their gel state are able to generate stronger

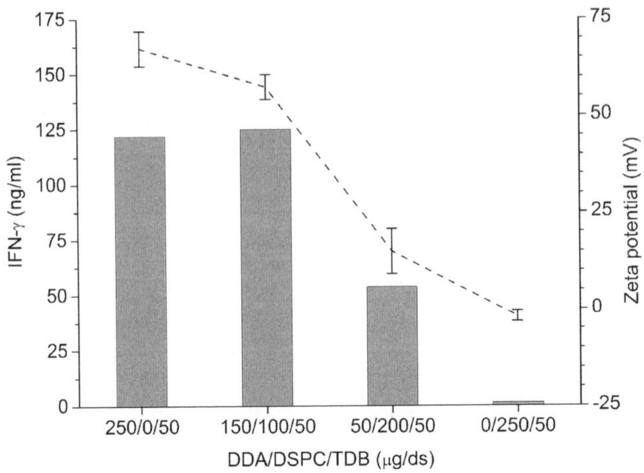

Figure 18.4 *Release of IFN-γ from blood lymphocytes isolated from C57BL/6J mice (grey bars, left y-axis). The mice were immunized subcutaneously (s.c.) at the base of the tail three times with a two-week interval with 2 μg of Ag85B–ESAT-6 administered with CAF01 liposome formulations where DDA was gradually replaced by DSPC. Blood lymphocytes were isolated eight weeks after the first immunization and restimulated in vitro with Ag85B–ESAT-6 (5 μg/ml) (method described in [23, 33, 89]). The zeta potential was determined for the adjuvant formulations using a Malvern Zetasizer nanoZS (dashed line, right y-axis). All values shown are the average values of three independent samples ± standard deviation. The results show that neither IFN-γ release nor zeta potential was affected significantly with replacement of up to 40% of DDA with DSPC, whereas replacement of more DDA reduced both the cytokine levels and zeta potential. This indicates that there is a direct proportional relationship between the zeta potential and immunogenicity*

CMI responses than liposomes in their fluid state. Liposomes based on DDA thus appear to have the optimum characteristics for a CMI-inducing adjuvant, having a high positive surface charge and being particulate at physiological temperature.

18.4 Instability Considerations of Liposomes

Since DDA is a quaternary ammonium compound with no moieties that can bind water, the surface of DDA-based liposomes cannot be hydrated, making the liposomes unstable. On the other hand, DDA liposomes cannot, unlike phospholipids, undergo chemical hydrolysis.

Incorporation of TDB or other glycolipids enhances the stability of DDA liposomes [63], probably by enhancing the hydration of the lipid surface [23, 64]. However, incorporation of TDB into DDA liposomes disrupts the ordering of the lipids and thus affects the gel–fluid phase transition of the membrane. This has been shown by DSC, where incorporation of different ratios of TDB into DDA liposomes shifts the main phase

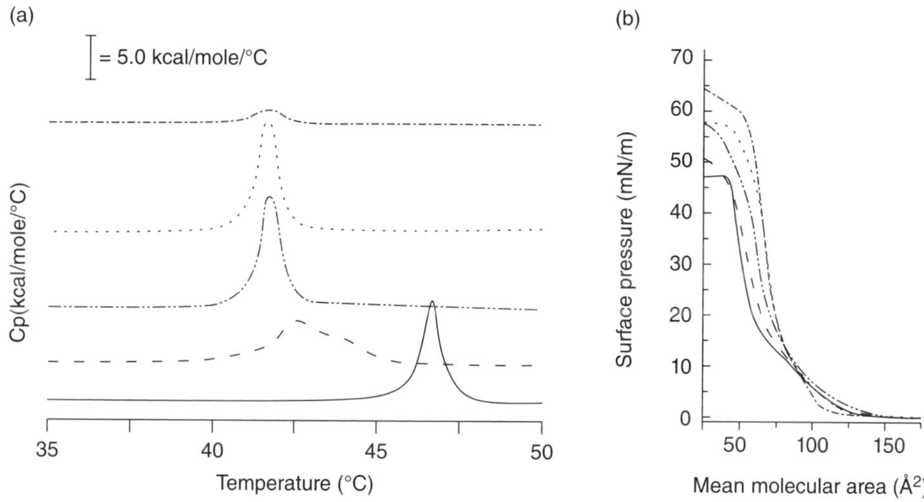

Figure 18.5 *(a) Differential scanning calorimetry (DSC) heat capacity curves for DDA liposomes incorporating TDB in w/w ratios of 5 : 0 (solid), 5 : 1 (dash), 5 : 3 (dash-dot-dot), 5 : 5 (dot) and 1 : 5 (dash-dot). The curves have been normalized to molar content. Notice that the scans have been displaced on the heat capacity axis for clarity. The curves are averages of three experiments. Introduction of TDB results in a shift in T_m towards lower temperatures but with no effect on the phase transition enthalpy. This demonstrates that the acyl chains of the TDB glycolipids are incorporated into the hydrophobic core of the DDA lipid bilayers causing a change in the lipid chain packing properties (these data have previously been published [64]) (Reprinted with permission from Elsevier. Copyright 2008). (b) Monolayers were formed by spreading 6.3 mmol lipid mixtures in chloroform : methanol (9 : 1) on a KSV minithrough 1 (KSV instruments Ltd, Helsinki, Finland). Pressure/area isotherm of Langmuir monolayers containing DDA/TDB in w/w ratios of 5 : 0 (solid), 5 : 1 (dash), 5 : 3 (dash-dot-dot), 5 : 5 (dot) and 1 : 5 (dash-dot) on 10 mM tris-buffer subphases was generated as described in [64]. The total molar lipid concentration of the different monolayers is identical for all experiments. The curves are averages of three experiments. The results show that TDB enhances the surface pressure of the compressed monolayer, indicating that it enhances the interaction with the subphase, probably through hydrogen bonding between the water molecules and the trehalose head group of TDB (these data have previously been published in [64]) (Reprinted with permission from Elsevier. Copyright 2009.)*

transition temperature from ~47 °C for DDA alone down to ~41.5 °C for DDA containing 20 mole% TDB and above (Figure 18.5a) [23, 64]. Hydration of the DDA liposome membrane mediated by incorporated TDB has been investigated with the Langmuir–Blodgett technique, which measures the lipid–water surface pressure, where the hydration resulted in enhanced interaction with water and, hence, an increased surface pressure (Figure 18.5b) [64].

A traditional way to stabilize liposomes in suspensions is to incorporate cholesterol (Figure 18.2). The beneficial role of cholesterol in liposomal drug carriers is well recognized. Early developmental studies of liposomal drug delivery demonstrated that inclusion of 50 mole% cholesterol within a liposome formulation increases the stability and reduces the permeability of liposomal bilayers [65]. The inclusion of cholesterol within liposomal bilayers results in increased packing densities of phospholipid molecules [66], which is thought to result from the accommodation of cholesterol in the molecular cavities formed by the surfactant monomers assembled into liposomes [67]. This is documented by surface pressure measurements that show a decrease in the effective area per molecule as the cholesterol content of the monolayer is increased [68]. Addition of cholesterol reduces the phase transition enthalpy of liposomes [69] and, yet, small amounts of cholesterol can enhance their immunogenicity [60]. Accordingly, incorporation of cholesterol into CAF01 liposomes reduced the enthalpy of the gel–fluid phase transition of the membrane, but the enhanced fluidity also appeared to decrease the ability of the liposomes to generate a CMI response [59]. As mentioned earlier, exchanging DDA with DODA had the same effect on fluidity as cholesterol and the ability to generate a CMI response was consistently reduced. Therefore, incorporation of membrane stabilizers that will bring the liposomes from the gel phase to the fluid phase at room temperature appears to reduce their ability to generate a CMI response. Consequently, cholesterol is not a suitable stabilizer for CAF01 liposomes.

18.5 Techniques to Formulate Long-Term Stable Liposome Formulations

Most of the major infectious diseases for which CAF01 is a suitable adjuvant are mainly represented in the developing world. This requires a vaccine formulation that is thermostable and/or can stand the often harsh environment to which it is exposed during distribution. Even though CAF01 liposomes have a long shelf-life at 4 °C [23], it is likely that changing temperatures and exposure to, for example, sunlight will affect their stability. To achieve long-term stability of liposomes under such conditions, lyophilization (freeze-drying) is commonly employed. Lyophilization offers the opportunity to remove water from the system, thus producing a solid phase product with low molecular mobility which will suppress both chemical and physical instabilities. Indeed, freeze-drying is the main process used to produce stable proteins and polypeptides that are physically and/or chemically unstable in aqueous solution [70].

The process of freeze-drying involves three stages: pre-freezing (initial freezing of the liposomal suspension), primary drying (water loss from the formulation by sublimation without the intermediate liquid stage) and secondary drying (removal of adsorbed surface-bound water and reduction in residual water content to obtain a free-flowing dry powder). However, liposomes may be damaged during lyophilization and/or subsequent rehydration if appropriate stabilizers are not employed. Such instabilities generally lead to aggregation and fusion of liposomes into significantly larger constructs. For example, freeze-drying and rehydration of small unilamellar vesicles (SUVs) in the absence of additional stabilizing agents results in over a 30-fold increase in vesicle diameter independently of the freeze-drying protocol used [71]. Therefore, to promote vesicle stability during the freeze-drying process cryoprotectants, including saccharides (e.g. sucrose or trehalose) and their derivatives, are employed.

18.5.1 Protection During Freeze-Drying

A detailed review by Crowe and Crowe [72] outlines the mechanism of stabilization offered by saccharides. Briefly, the increased vesicle size and leakage of encapsulated material from un-protected liposomes during freeze-drying is attributed to two parameters, phase separation of lipids around their gel–fluid transition temperature and fusion of the liposome membranes. Thus, materials that could promote the formation of stable boundaries between the liposomes could also prevent aggregation of the vesicles. This mechanism of protection offered by cryo-/lyoprotectants has been ascribed to different functions: water replacement, kosmotropic effect and vitrification. According to the water replacement theory, the water bound to the membrane surface, via hydrogen bonding to the phospholipid head group, is replaced by the cryo-/lyoprotectant, preventing fusion between adjacent bilayers and the formation of larger aggregates. Furthermore, this binding of cryo-/lyoprotectant to the membrane surface depresses the gel–liquid phase transition of the membrane in the dry state, thereby preventing leakage during rehydration [73]. The kosmotropic effect theory states that kosmotropes, such as sucrose and trehalose, in addition to the functions stated in the water replacement theory, also displace water in the bulk phase and thereby preserve the three-dimensional structure of the liposomes [74]. The vitrification theory states that the cryo-/lyoprotectants form a glassy matrix around the liposomes, preventing membrane-bound water from evaporating and thereby maintaining the liposome structure and preventing fusion of the liposomes [75]. Since these theories are not mutually exclusive, they may all be, at least partly, true.

Cryo-/lyoprotectants require careful optimization of concentration for effective stabilization of liposomes and the protectant/lipid ratio required to stabilize a given system depends on both the liposome formulation and the protectant used. However, as a general guideline, disaccharides such as sucrose and trehalose are effective protectants of phospholipid-based liposomes and optimal protection is achieved when amorphous sugar is present at ratios above, for example, 2 g sugar/g phospholipid. To ensure the formation of an effective cake after freeze-drying (which will support easy re-dispersion of the liposomes etc.), the total solute concentration should be at least 2.5% (w/v) [76].

18.5.1.1 Cryo-/Lyoprotection of CAF01 Using Trehalose

Trehalose, and to a certain degree sucrose, is able to protect CAF01 during freeze-drying with at least 8% (w/v) trehalose or more required to enable rapid rehydration and obtain the pre-freeze-drying average particle size of the liposomes [77]. This concentration is above the one suggested for phospholipid-based liposomes and it is ascribed to the fact that trehalose cannot interact directly with the quaternary head group of DDA but only the trehalose head group of TDB. Consistently, DDA liposomes alone cannot be protected by trehalose [77]. The direct interaction between bulk trehalose and TDB has been investigated by the Langmuir–Blodgett technique [64]: the addition of trehalose to a DDA monolayer containing small amounts of TDB results in an increased surface pressure, which is not observed for DDA monolayers without TDB, suggesting that trehalose head groups of TDB, present at the lipid–water interface, attract the free trehalose to the liposome surface, probably by hydrogen bonding. Similar protective effects have also been reported for other glycolipids, such as galactocyldiacylglycerols [78–81], malto- and galactosyltriethoxycholesterol [82], trisaccharide-dioleoylphosphatidylethanolamine conjugates [83] and galactosyl-stearoylphosphatidylcholine conjugates [84], during freeze-drying of phospholipids with trehalose, and were similarly ascribed to direct interaction between the glycolipids and trehalose.

18.5.1.2 Freeze-Drying of DDA/TDB Using Alternative Cryo-/Lyoprotectants

The use of saccharides as cryo-/lyoprotectants is associated with some disadvantages, including the requirement for high concentrations (in some cases up to 30%), which may compromise the viscosity of the liposome formulation post hydration and trigger unacceptable adverse reactions [85]. As an alternative to saccharides, amino acids have been investigated as stabilizing moieties during the freeze-drying of liposomes [71, 73, 86]. Besides being much cheaper, amino acids hold the advantages of not only being able to offer hydrogen bonds, but also to provide electrostatic interactions for effective lyophilization.

Investigations into the cryo-/lyoprotectant efficacy of a series of amino acids suggested that lysine offers the most effective retention of liposome vesicle size compared to the other amino acids tested [71]. In contrast to trehalose, the cryo-/lyoprotection profile of lysine is biphasic in nature (Figure 18.6) with maximal protection against vesicle aggregation occurring at a lysine/lipid molar ratio of 4/1. Histidine and arginine exhibit similar biphasic properties with neutral SUV liposomes [71]. The biphasic nature of cryoprotectants has previously been demonstrated in alternative systems [87, 88].

Using lysine at the required molar ratio, CAF01 liposomes with adsorbed Ag85B-ESAT-6 were freeze-dried and subsequently reconstituted to obtain vesicles not signifi-

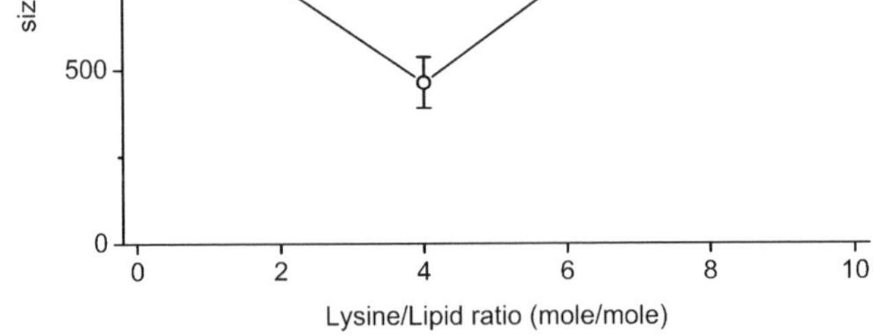

Figure 18.6 *CAF01 liposomes were freeze-dried in the presence of increasing molar ratios of lysine. After re-constitution the mean volume diameter of vesicles were measured. All values shown are the average values of at least three independent samples ± standard deviation. The results show that the cryo-/lyoprotection profile of lysine is biphasic, with maximal protection against vesicle aggregation occurring at a lysine/lipid molar ratio of 4/1*

Table 18.1 *Properties of CAF01 before and after freeze-drying*

	Before freeze-drying	After freeze-drying
Vesicle size (nm)	435 ± 84	463 ± 74
Zeta Potential (mV)	40 ± 4.6	35 ± 36
Viscosity (mPas)	0.900 ± 0.002	0.918 ± 0.005
pH	7.43 ± 0.02	7.57 ± 0.02
IFN-γ (pg/ml)	1128 ± 773	1795 ± 481

CAF01 with adsorbed antigen was subjected to freeze-drying and re-constitution with distilled water and their size, surface charge, suspension dynamic viscosity and pH measured. Results represent mean ± standard deviation, $n = 4$. IFN-γ from restimulated mouse splenocytes. IFN-γ was detected in cell culture supernatants after 48 hours of incubation of viable splenocytes (approximately 1×10^7 cells/ml) recovered from animals vaccinated with 2 μg/ml Ag85B-ESAT-6 fusion protein. The cell medium was separated by centrifugation, collected and analysed using DuoSet® capture ELISA kits (mouse IFN-γ) purchased from R&D systems, Abingdon, UK, according to the manufacturers instructions. Results represent mean ± standard deviation, $n = 5$.

cantly different in size, surface charge or viscosity compared to the liposomes prior to freeze-drying (Table 18.1). The inclusion of lysine resulted in a minor increase in the pH of the suspension but this was not to a biologically relevant extent (Table 18.1).

In terms of preservation of immunological characteristics, the freeze-dried formulations with both trehalose and lysine had similar immunological activity to freshly prepared CAF01 formulations. Comparing antigen-specific IFN-γ production in mice immunized with vaccines formulated with either freshly-prepared CAF01 or freeze-dried CAF01 demonstrated that biological efficacy was maintained when trehalose or lysine was used as a cryoprotectant for CAF01 [71, 77].

18.6 Conclusion

Cationic liposomes based on DDA are highly efficient antigen delivery systems that enhance the uptake and presentation of antigens by APC. In addition, these liposomes are suitable for presenting immunomodulators like TDB and MPL to the APC, thereby directing the immune system towards a specific immune response.

However, liposomes based on DDA are unstable during long-term storage. Stabilizing the liposomes with cholesterol, which is the method usually used, decreases the ability of the liposomes to generate an appropriate CMI response, making it unsuitable for the stabilization of this adjuvant system. Glycolipids also have a stabilizing effect and incorporation of the immunomodulator TDB not only enhances the immune response but also stabilizes the DDA membrane. The CAF01 liposomes, based on DDA and TDB, are therefore highly immunogenic and have a long shelf-life.

The fight against infectious diseases has intensified during the last 20 years and, with that, so too has the focus on developing new vaccine strategies. Production and transportation aspects are major focus points in the development of new vaccines and cost, stability and dependence on a cold chain are key issues, since the majority of the most serious infectious diseases are spreading from rural areas in Africa, Asia and South America. It is, therefore, encouraging that CAF01 liposomes are readily rehydrated after freeze-drying without loss in immunogenicity, using either trehalose or lysine as cryo-/lyoprotectors.

Hence CAF01 liposomes represent a simple and stable adjuvant system for vaccines against a vast range of infectious diseases, including tuberculosis, hepatitis, malaria and chlamydia. Furthermore, combining this adjuvant system with other immunomodulators makes it possible to generate immune responses against other infectious diseases, such as HIV, autoimmune diseases and cancer.

References

1. Bangham, A.D., Standish, M.M. and Watkins, J.C. (1965) Diffusion of univalent ions across the lamellae of swollen phospholipids. *J. Mol. Biol.*, **13**, 238–252.
2. Cevc, G. and Marsh, D. (1985) Hydration of noncharged lipid bilayer membranes. Theory and experiments with phosphatidylethanolamines. *Biophys. J.*, **47**, 21–31.
3. Feitosa, E., Jansson, J. and Lindman, B. (2006) The effect of chain length on the melting temperature and size of dialkyldimethylammonium bromide vesicles, *Chem. Phys. Lipids*, **142**, 128–132.
4. Koynova, R. and Caffrey, M. (1998) Phases and phase transitions of the phosphatidylcholines. *Biochim. Biophys. Acta*, **1376**, 91–145.
5. Feitosa, E., Alves, F.R., Niemiec, A. *et al.* (2006) Cationic liposomes in mixed didodecyldimethylammonium bromide and dioctadecyldimethylammonium bromide aqueous dispersions studied by differential scanning calorimetry, Nile Red fluorescence, and turbidity. *Langmuir*, **22**, 3579–3585.
6. Zuhorn, I.S., Engberts, J.B. and Hoekstra, D. (2007) Gene delivery by cationic lipid vectors: overcoming cellular barriers. *Eur. Biophys. J.*, **36**, 349–362.
7. Zuhorn, I.S. and Hoekstra, D. (2002) On the mechanism of cationic amphiphile-mediated transfection. To fuse or not to fuse: is that the question? *J. Membr. Biol.*, **189**, 167–179.
8. Gabizon, A.A., Shmeeda, H. and Zalipsky, S. (2006) Pros and cons of the liposome platform in cancer drug targeting. *J. Liposome Res.*, **16**, 175–183.
9. Immordino, M.L., Dosio, F. and Cattel, L. (2006) Stealth liposomes: review of the basic science, rationale, and clinical applications, existing and potential. *Int. J. Nanomed.*, **1**, 297–315.
10. Gregoriadis, G., Leathwood, P.D. and Ryman, B.E. (1971) Enzyme entrapment in liposomes. *FEBS Lett.*, **14**, 95–99.
11. Gregoriadis, G. and Ryman, B.E. (1972) Fate of protein-containing liposomes injected into rats. An approach to the treatment of storage diseases. *Eur. J. Biochem.*, **24**, 485–491.
12. Gregoriadis, G. (1990) Immunological adjuvants: a role for liposomes. *Immunol. Today*, **11**, 89–97.
13. Gregoriadis, G. (1994) Liposomes as immunoadjuvants and vaccine carriers: antigen entrapment. *ImmunoMethods*, **4**, 210–216.
14. Gregoriadis, G., McCormack, B., Obrenovic, M. *et al.* (1999) Vaccine entrapment in liposomes. *Methods*, **19**, 156–162.
15. Dailey, M.O. and Hunter, R.L. (1974) The role of lipid in the induction of hapten-specific delayed hypersensitivity and contact sensitivity. *J. Immunol.*, **112**, 1526–1534.
16. Snippe, H., Belder, M. and Willers, J.M. (1977) Dimethyl diotadecyl ammonium bromide as adjuvant for delayed hypersensitivity in mice. *Immunology*, **33**, 931–936.
17. Snippe, H., Johannesen, L., Inman, J.K. and Merchant, B. (1978) Specificity of murine delayed-type hypersensitivity to conjugates of large or small haptens on protein carriers bearing lipid groups. *Immunology*, **34**, 947–954.
18. Christensen, D., Korsholm, K.S., Rosenkrands, I. *et al.* (2007) Cationic liposomes as vaccine adjuvants. *Expert Rev. Vaccines*, **6**, 785–796.
19. Gupta, R.K., Relyveld, E.H., Lindblad, E.B. *et al.* (1993) Adjuvants – a balance between toxicity and adjuvanticity. *Vaccine*, **11**, 293–306.
20. Hem, S.L. and Hogenesch, H. (2007) Relationship between physical and chemical properties of aluminum-containing adjuvants and immunopotentiation. *Expert Rev. Vaccines*, **6**, 685–698.

21. Holten-Andersen, L., Doherty, T.M., Korsholm, K.S. and Andersen, P. (2004) Combination of the cationic surfactant dimethyl dioctadecyl ammonium bromide and synthetic mycobacterial cord factor as an efficient adjuvant for tuberculosis subunit vaccines. *Infect. Immun.*, **72**, 1608–1617.
22. Olsen, A.W., Williams, A., Okkels, L.M. *et al.* (2004) Protective effect of a tuberculosis subunit vaccine based on a fusion of antigen 85B and ESAT-6 in the aerosol guinea pig model. *Infect. Immun.*, **72**, 6148–6150.
23. Davidsen, J., Rosenkrands, I., Christensen, D. *et al.* (2005) Characterization of cationic liposomes based on dimethyldioctadecylammonium and synthetic cord factor from M. tuberculosis (trehalose 6,6′-dibehenate) – a novel adjuvant inducing both strong CMI and antibody responses. *Biochim. Biophys. Acta*, **1718**, 22–31.
24. Madonna, G.S., Peterson, J.E., Ribi, E.E. and Vogel, S.N. (1986) Early-phase endotoxin tolerance: induction by a detoxified lipid A derivative, monophosphoryl lipid A. *Infect. Immun.*, **52**, 6–11.
25. Evans, J.T., Cluff, C.W., Johnson, D.A. *et al.* (2003) Enhancement of Antigen-Specific Immunity via the TLR4 Ligands MPL™ Adjuvant and Ribi.529, *Expert Rev. Vaccines*, **2**,(2), 219–229.
26. Martin, M., Michalek, S.M. and Katz, J. (2003) Role of innate immune factors in the adjuvant activity of monophosphoryl lipid A. *Infect. Immun.*, **71**, 2498–2507.
27. Alving, C.R. (1993) Lipopolysaccharide, lipid A, and liposomes containing lipid A as immunologic adjuvants. *Immunobiology*, **187**, 430–446.
28. Childers, N.K., Miller, K.L., Tong, G. *et al.* (2000) Adjuvant activity of monophosphoryl lipid A for nasal and oral immunization with soluble or liposome-associated antigen. *Infect. Immun.*, **68**, 5509–5516.
29. Hall, M.A., Stroop, S.D., Hu, M.C. *et al.* (2004) Intranasal immunization with multivalent group A streptococcal vaccines protects mice against intranasal challenge infections. *Infect. Immun.*, **72**, 2507–2512.
30. Richards, R.L., Rao, M., Wassef, N.M. *et al.* (1998) Liposomes containing lipid A serve as an adjuvant for induction of antibody and cytotoxic T-cell responses against RTS, S malaria antigen. *Infect. Immun.*, **66**, 2859–2865.
31. Zhou, F. and Huang, L. (1993) Monophosphoryl lipid A enhances specific CTL induction by a soluble protein antigen entrapped in liposomes. *Vaccine*, **11**, 1139–1144.
32. Stewart, V.A., Walsh, D.S., McGrath, S.M. *et al.* (2006) Cutaneous delayed-type hypersensitivity (DTH) in a multi-formulation comparator trial of the anti-falciparum malaria vaccine candidate RTS,S in rhesus macaques. *Vaccine*, **24**, 6493–6502.
33. Brandt, L., Elhay, M., Rosenkrands, I. *et al.* (2000) ESAT-6 Subunit vaccination against *Mycobacterium tuberculosis*. *Infect. Immun.*, **68**, 791–795.
34. Bekierkunst, A., Yarkoni, E., Flechner, I. *et al.* (1971) Immune response to sheep red blood cells in mice pretreated with mycobacterial fractions. *Infect. Immun.*, **4**, 256–263.
35. Lima, V.M., Bonato, V.L., Lima, K.M. *et al.* (2001) Role of trehalose dimycolate in recruitment of cells and modulation of production of cytokines and NO in tuberculosis. *Infect. Immun.*, **69**, 5305–5312.
36. Ryll, R., Watanabe, K., Fujiwara, N. *et al.* (2001) Mycobacterial cord factor, but not sulfolipid, causes depletion of NKT cells and upregulation of CD1d1 on murine macrophages. *Microbes Infect.*, **3**, 611–619.
37. Silva, C.L., Ekizlerian, S.M. and Fazioli, R.A. (1985) Role of cord factor in the modulation of infection caused by mycobacteria. *Am. J. Pathol.*, **118**, 238–247.
38. McBride, B.W., Mogg, A., Telfer, J.L. *et al.* (1998) Protective efficacy of a recombinant protective antigen against *Bacillus anthracis* challenge and assessment of immunological markers. *Vaccine*, **16**, 810–817.
39. Woodard, L.F., Toone, N.M. and McLaughlin, C.A. (1980) Comparison of muramyl dipeptide, trehalose dimycolate, and dimethyl dioctadecyl ammonium bromide as adjuvants in *Brucella abortus* 45/20 vaccines. *Infect. Immun.*, **30**, 409–412.
40. Masihi, K.N., Lange, W., Brehmer, W. and Ribi, E. (1986) Immunobiological activities of nontoxic lipid A: enhancement of nonspecific resistance in combination with trehalose dimycolate against viral infection and adjuvant effects. *Int. J. Immunopharmacol.*, **8**, 339–345.

41. Masihi, K.N., Brehmer, W., Azuma, I. *et al.* (1984) Stimulation of chemiluminescence and resistance against aerogenic influenza virus infection by synthetic muramyl dipeptide combined with trehalose dimycolate. *Infect. Immun.*, **43**, 233–237.
42. Masihi, K.N., Brehmer, W., Lange, W. *et al.* (1984) Protective effect of muramyl dipeptide analogs in combination with trehalose dimycolate against aerogenic influenza virus and *Mycobacterium tuberculosis* infections in mice. *J. Biol. Response. Mod.*, **3**, 663–671.
43. Coelho, E.A., Tavares, C.A., Kde, M. Lima *et al.* (2006) Fernandes, mycobacterium hsp65 DNA entrapped into TDM-loaded PLGA microspheres induces protection in mice against Leishmania (Leishmania) major infection. *Parasitol. Res.*, **98**, 568–575.
44. Masihi, K.N., Bhaduri, C.R., Werner, H. *et al.* (1986) Effects of muramyl dipeptide and trehalose dimycolate on resistance of mice to *Toxoplasma gondii* and *Acanthamoeba culbertsoni* infections. *Int. Arch. Allergy Appl. Immunol.*, **81**, 112–117.
45. Hunter, R.L., Olsen, M., Jagannath, C. and Actor, J.K. (2006) Trehalose 6,6′-dimycolate and lipid in the pathogenesis of caseating granulomas of tuberculosis in mice. *Am. J. Pathol.*, **168**, 1249–1261.
46. Hunter, R.L., Venkataprasad, N. and Olsen, M.R. (2006) The role of trehalose dimycolate (cord factor) on morphology of virulent *M. tuberculosis in vitro*. *Tuberculosis*, **86**, 349–356.
47. Lee, R.E., Brennan, P.J. and Besra, G.S. (1996) *Mycobacterium tuberculosis* cell envelope. *Curr. Top. Microbiol. Immunol.*, **215**, 1–27.
48. Olds, G.R., Chedid, L., Lederer, E. and Mahmoud, A.A. (1980) Induction of resistance to *Schistosoma mansoni* by natural cord factor and synthetic lower homologues. *J. Infect. Dis.*, **141**, 473–478.
49. Yarkoni, E., Rapp, H.J., Polonsky, J. and Lederer, E. (1978) Immunotherapy with an intralesionally administered synthetic cord factor analogue. *Int. J. Cancer*, **22**, 564–569.
50. Pimm, M.V., Baldwin, R.W., Polonsky, J. and Lederer, E. (1979) Immunotherapy of an ascitic rat hepatoma with cord factor (trehalose-6, 6′-dimycolate) and synthetic analogues. *Int. J. Cancer*, **24**, 780–785.
51. Vangala, A., Bramwell, V.W., McNeil, S. *et al.* (2007) Comparison of vesicle based antigen delivery systems for delivery of hepatitis B surface antigen. *J. Control. Release*, **119**, 102–110.
52. Vangala, A., Kirby, D., Rosenkrands, I. *et al.* (2006) A comparative study of cationic liposome and niosome-based adjuvant systems for protein subunit vaccines: characterisation, environmental scanning electron microscopy and immunisation studies in mice. *J. Pharm. Pharmacol.*, **58**, 787–799.
53. Foged, C., Brodin, B., Frokjaer, S. and Sundblad, A. (2005) Particle size and surface charge affect particle uptake by human dendritic cells in an *in vitro* model. *Int. J. Pharm.*, **298**, 315–322.
54. Xiang, S.D., Scholzen, A., Minigo, G. *et al.* (2006) Pathogen recognition and development of particulate vaccines: does size matter? *Methods*, **40**, 1–9.
55. Joseph, A., Itskovitz-Cooper, N., Samira, S. *et al.* (2006) A new intranasal influenza vaccine based on a novel polycationic lipid–ceramide carbamoyl-spermine (CCS) I. Immunogenicity and efficacy studies in mice. *Vaccine*, **24**, 3990–4006.
56. Latif, N. and Bachhawat, B.K. (1984) The effect of surface sugars on liposomes in immunopotentiation. *Immunol. Lett.*, **8**, 75–78.
57. Nakanishi, T., Kunisawa, J., Hayashi, A. *et al.* (1997) Positively charged liposome functions as an efficient immunoadjuvant in inducing immune responses to soluble proteins. *Biochem. Biophys. Res. Commun.*, **240**, 793–797.
58. Perrie, Y., Frederik, P.M. and Gregoriadis, G. (2001) Liposome-mediated DNA vaccination: the effect of vesicle composition. *Vaccine*, **19**, 3301–3310.
59. Christensen, D., Smith Korsholm, K., Lindenstrøm, T. *et al.* Altered membrane fluidity changes the adjuvant effect of DDA/TDB liposomes (Submitted).
60. Bakouche, O. and Gerlier, D. (1986) Enhancement of immunogenicity of tumour virus antigen by liposomes: the effect of lipid composition. *Immunology*, **58**, 507–513.
61. Yasuda, T., Dancey, G.F. and Kinsky, S.C. (1977) Immunogenicity of liposomal model membranes in mice: dependence on phospholipid composition. *PNAS*, **74**, 1234–1236.

62. Gall, D. (1966) The adjuvant activity of aliphatic nitrogenous bases. *Immunology*, **11**, 369–386.
63. Davidsen, J., Rosenkrands, I. and Andersen, P. (2006) Compositions and Methods for Stabilizing Lipid based Adjuvant formulations using Glycolipids, WO 2006002642.
64. Christensen, D., Kirby, D.J., Foged, C. *et al.* (2008) α,α'-trehalose 6,6′-dibehenate in non-phospholipid based liposomes enables direct interaction with trehalose, offering stability during freeze-drying. *Biochim. Biophys. Acta*, **1178**, 1365–1373.
65. Gregoriadis, G. and Davis, C. (1979) Stability of liposomes *in vivo* and *in vitro* is promoted by their cholesterol content and the presence of blood cells. *Biochem. Biophys. Res. Commun.*, **89**, 1287–1293.
66. Semple, S.C., Chonn, A. and Cullis, P.R. (1996) Influence of cholesterol on the association of plasma proteins with liposomes. *Biochemistry*, **35**, 2521–2525.
67. Devaraj, G.N., Parakh, S.R., Devraj, R. *et al.* (2002) Release studies on niosomes containing fatty alcohols as bilayer stabilizers instead of cholesterol. *J. Colloid Interface Sci.*, **251**, 360–365.
68. Rogerson, A., Cummings, J. and Florence, A.T. (1987) Adriamycin-loaded niosomes: drug entrapment, stability and release. *J. Microencapsul.*, **4**, 321–328.
69. Papahadjopoulos, D., Jacobson, K., Nir, S. and Isac, T. (1973) Phase transitions in phospholipid vesicles. Fluorescence polarization and permeability measurements concerning the effect of temperature and cholesterol. *Biochim. Biophys. Acta*, **311**, 330–348.
70. Manning, M.C., Patel, K. and Borchardt, R.T. (1989) Stability of protein pharmaceuticals. *Pharm. Res.* **6**, 903–918.
71. Mohammed, A.R., Coombes, A.G. and Perrie, Y. (2007) Amino acids as cryoprotectants for liposomal delivery systems. *Eur J. Pharm. Sci.*, **30**, 406–413.
72. Crowe, J.H. and Crowe, L.M. (1988) Factors affecting the stability of dry liposomes. *Biochim. Biophys. Acta*, **939**, 327–334.
73. Crowe, J.H., Leslie, S.B. and Crowe, L.M. (1994) Is vitrification sufficient to preserve liposomes during freeze-drying? *Cryobiology*, **31**, 355–366.
74. Koynova, R.D., Tenchov, B.G. and Quinn, P.J. (1989) Sugars favour formation of hexagonal (H_{II}) phase at the expense of lamellar liquid-crystalline phase in hydrated phosphatidylethanolamines. *Biochim. Biophys. Acta*, **980**, 377–380.
75. Wolfe, J. and Bryant, G. (1999) Freezing, drying, and/or vitrification of membrane-solute-water systems. *Cryobiology*, **39**, 103–129.
76. Zuidam, N.J., van Winden, E., de Vrueh, R. and Crommelin, D.J. (2003) Stability, storage and sterilization of liposomes, in *Liposomes – A Practical Approach* (eds V.P. Torchilin and V. Weissig), Oxford University Press, Oxford, pp. 149–164.
77. Christensen, D., Foged, C., Rosenkrands, I. *et al.* (2007) Trehalose preserves DDA/TDB liposomes and their adjuvant effect during freeze-drying. *Biochim. Biophys. Acta*, **1768**, 2120–2129.
78. Bakaltcheva, I., Williams, W.P., Schmitt, J.M. and Hincha, D.K. (1994) The solute permeability of thylakoid membranes is reduced by low concentrations of trehalose as a co-solute. *Biochim. Biophys. Acta*, **1189**, 38–44.
79. Hincha, D.K. (1989) Low concentrations of trehalose protect isolated thylakoids against mechanical freeze-thaw damage. *Biochim. Biophys. Acta*, **987**, 231–234.
80. Popova, A.V. and Hincha, D.K. (2003) Intermolecular interactions in dry and rehydrated pure and mixed bilayers of phosphatidylcholine and digalactosyldiacylglycerol: a Fourier transform infrared spectroscopy study. *Biophys. J.*, **85**, 1682–1690.
81. Popova, A.V. and Hincha, D.K. (2005) Effects of the sugar headgroup of a glycoglycerolipid on the phase behavior of phospholipid model membranes in the dry state. *Glycobiology*, **15**, 1150–1155.
82. Goodrich, R.P. and Baldeschwieler, J.D. (1991) The cryoprotective action of synthetic glycolipids. *Cryobiology*, **28**, 327–334.
83. Park, Y.S. and Huang, L. (1992) Cryoprotective activity of synthetic glycophospholipids and their interactions with trehalose. *Biochim. Biophys. Acta*, **1124**, 241–248.
84. Bendas, G., Wilhelm, F., Richter, W. and Nuhn, P. (1996) Synthetic glycolipids as membrane-bound cryoprotectants in the freeze-drying process of liposomes. *Eur. J. Pharm. Sci.*, **4**, 211–222.

85. Peer, D., Florentin, A. and Margalit, R. (2003) Hyaluronan is a key component in cryoprotection and formulation of targeted unilamellar liposomes. *Biochim. Biophys. Acta*, **1612**, 76–82.
86. Osterberg, T. and Wadsten, T. (1999) Physical state of L-histidine after freeze-drying and long-term storage. *Eur. J. Pharm. Sci.*, **8**, 301–308.
87. Kundu, C.N., Das, K. and Majumder, G.C. (2001) Effect of amino acids on goat cauda epididymal sperm cryopreservation using a chemically defined model system. *Cryobiology*, **42**, 21–27.
88. Suzuki, T., Komatsu, H. and Miyajima, K. (1996) Effects of glucose and its oligomers on the stability of freeze-dried liposomes. *Biochim. Biophys. Acta*, **1278**, 176–182.
89. Rosenkrands, I., Agger, E.M., Olsen, A.W. *et al.* (2005) Cationic liposomes containing mycobacterial lipids: a new powerful Th1 adjuvant system. *Infect. Immun.*, **73**, 5817–5826.

19

Influenza Virosomes as Delivery Systems for Antigens

Mario Amacker,[1] *Stefan Moese,*[2] *Andreas R. Kammer,*[1] *Ari Helenius*[3] *and Rinaldo Zurbriggen*[1]

[1] Pevion Biotech Ltd., Ittigen/Bern, Switzerland
[2] 3-V Biosciences, Schlieren, Switzerland
[3] Institute of Biochemistry, ETH Hoenggerberg, Zurich, Switzerland

19.1 Introduction

Influenza viruses are the cause of outbreaks of acute respiratory disease, known as influenza or 'flu', which have afflicted man and animals since ancient times [1, 2]. Influenza viruses are classified in the Orthomyxoviridae family. They are divided into three genera, influenza A, B and C viruses, which can be distinguished on the basis of antigenic differences between their nucleoproteins (NPs) and matrix (M) proteins. Influenza A viruses are primarily avian viruses, but also infect a wide variety of animals, including pigs, horses, and seals. Type A viruses are further divided into subtypes based on the antigenic nature of their surface glycoproteins haemagglutinin (HA) and neuraminidase (NA).

19.1.1 Immune Stimulating Reconstituted Influenza Virosomes (IRIVs) as Adjuvant and Delivery System

Immunostimulating reconstituted influenza virosomes (IRIVs) are influenza virus envelopes lacking viral genetic material (Figure 19.1) [3–7]. IRIVs are spherical, unilamellar vesicles with a mean diameter of less than 200 nm. Their base is a liposome comprised of phosphatidylcholine (PC), phosphatidylethanolamine (PE) and lipids derived from the

Delivery Technologies for Biopharmaceuticals: Peptides, Proteins, Nucleic Acids and Vaccines Edited by Lene Jorgensen and Hanne Mørck Nielsen

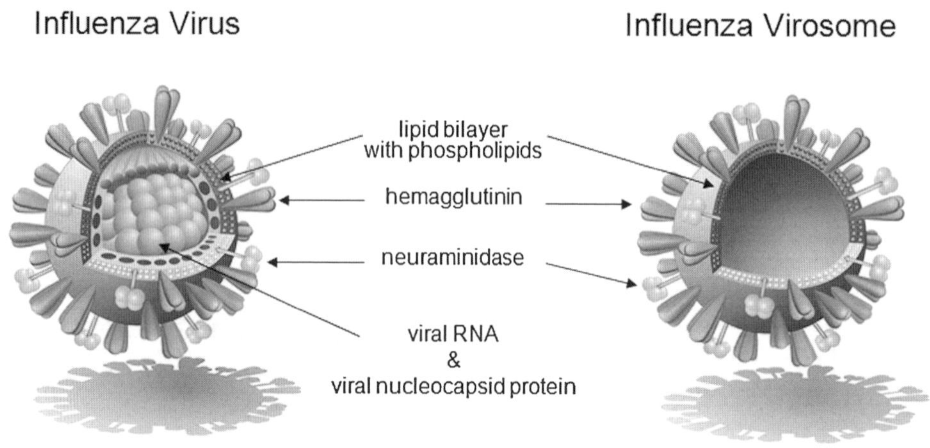

Figure 19.1 *The structure of influenza virosomes (IRIVs). Graphic representations of the influenza virus (left) and of an influenza virosome (IRIV) (right) are shown. The virus and the virosome are composed of a lipid bilayer with phospholipids and two surface membrane proteins, haemagglutinin and neuraminidase. The viral RNA and the viral nucleocapsid protein are only present in the virus, but not in the virosome*

influenza virus. In contrast to liposomes, virosomes contain functional viral envelope glycoproteins: influenza virus HA and NA intercalated in the phospholipid bilayer. Thus, reconstituted influenza virosomes retain the receptor-binding and membrane fusion activity of influenza virus [8, 9]. Virosomes have been demonstrated to be a versatile and efficient carrier system for a variety of antigens, for example proteins, peptides, nucleic acids and carbohydrates [8, 10–23]. IRIVs are one of the few adjuvant systems approved by regulatory authorities for human use and the only one that has carrier capabilities. Furthermore, they have an excellent tolerability, as shown by the two products already on the market [24, 25].

When using IRIVs as a carrier, it is preferred to physically associate the antigen of interest with the virosomal particle in order to exploit the full adjuvant effect [26, 27]. The biochemical properties and the intended effect of the payload define the optimal location and the most suitable method for association. A set of formulation procedures have been developed, yielding distinct IRIV types outlined in Table 19.1, each optimized for a specific application. Ideally, but rarely, antigens adhere spontaneously to the IRIV surface via non-covalent association with proteins or lipid molecules. These antigens can be formulated by simply adding them to IRIVs, as applied in the licensed virosomal hepatitis A vaccine. Antigens featuring sufficiently lipophilic domains, for example a membrane anchor, are readily integrated into the virosomal structure by simply adding the antigen to the formulation mixture before detergent removal (IIRIV). Antigens unable to associate directly with the virosomal membrane can be covalently linked to a phospholipid, which acts as synthetic anchor. Efficient encapsulation of antigen inside the virosomes can be achieved using chimaeric IRIVs (CIRIVs) [8, 13, 28, 29] or, alternatively, by reconstituting lyophilized IRIVs (TIRIVs) with antigen solution [9]. TIRIVs represent the latest formulation method, which is suitable for both association and

Table 19.1 *Description of different types of IRIVs. (Adapted from Expert Rev. Vaccines,* **6**, *711–721 (2007) [7] with permission of Expert Reviews Ltd.)*

Name	Description	Purpose
IRIV	Influenza virosomes with supplementary lipids	Induction of Influenza specific B-cell and CD4+ T-cell responses
IIRIV	IRIV formulated with antigens containing a membrane anchor and thus integrated in the virosomal membrane	Induction of antigen-specific B-cell and CD4+ T-cell responses
CIRIV	Chimaeric IRIV with two types of influenza HA and with antigens inside non-leaky IRIVs	Induction of antigen-specific CTL responses
TIRIV	Lyophilisable IRIV where the antigens are either integrated in the virosomal membrane or located inside of IRIVs	Induction of antigen-specific B-cell and CD4+ T-cell responses or induction of antigen-specific CTL responses or a combination of both

encapsulation of antigens. In contrast to conventional IRIVs, TIRIVs retain their physicochemical and adjuvant properties during the lyophilisation/reconstitution process [9]. Antigens can be added to TIRIVs during the formulation process or upon reconstitution. Antigen encapsulation in TIRIVs is technically straightforward and fast, but encapsulation efficiency is lower than in CIRIVs.

19.1.2 Mechanisms of Action

Influenza virosomes are thought to enhance the immune response against the co-delivered antigen at multiple levels (Figure 19.2). To initiate an immune response, the virosome particles need to interact with antigen presenting cells (APCs) and in some cases with B-lymphocytes, which in turn present the antigen to T-lymphocytes. These antigen delivery functions of virosomes are well documented. Additional immunostimulating activities of influenza virosomes or their components are plausible yet supported only by fragmentary experimental data. Influenza virosomes bind to APCs and are taken up via receptor-mediated endocytosis [30–32]. HA mediates membrane fusion upon acidification of the endosome [8, 33]. At this point the positioning of the antigen in the context of the virosome determines its processing and presentation pathway [5–7, 34].

For the induction of a humoral immune response the antigen of interest (protein, peptide, carbohydrate etc.) is linked to the surface of the virosomes (Figure 19.2, left). The virosomes with the antigen (including HA and NA) are then degraded in endosomes of the antigen-presenting cell (APC). The resulting peptides, originating from the antigen of interest or from influenza virus, associate with MHC class II molecules and are thereby presented to CD4+ helper T-cells. Presentation of the MHC class II/peptide complexes on the surface of the APC to CD4+ T-cells will induce a strong T-helper response that is essential for the support of antibody-forming B-cells. Membrane-associated immunoglobulin receptor molecules can recognize the specific antigen of choice on the surface of the virosomes on B-lymphocytes. The repetitive arrangement of the antigens on the virosome surface presumably enables cross-linking of these immunoglobulin receptors on the

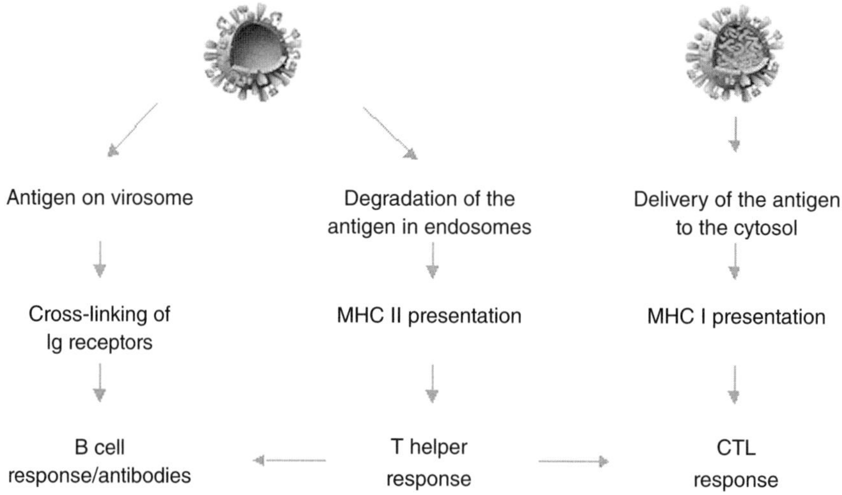

Figure 19.2 *Virosome-based antigen processing for generation of humoral and cellular immune response. Vaccines from Pevion Biotech are composed of virosome-based formulated antigens of choice on the surface of IRIVs or encapsulated in the IRIVs. Depending on the location of the antigen, different immune responses are generated. Left: When the antigen of interest (protein, peptide, carbohydrate etc.) is linked to the surface of the virosomes a strong humoral immune response is induced. Right: When the antigen of interest (protein, peptide, DNA etc.) is encapsulated into virosomes a strong cellular immune response is induced. Further details are given in the text*

B-cells, which is known to be an exceptionally strong activation signal. Together this results in a specific and high-quality immune response.

For the induction of a cellular immune response the antigen of interest (protein, peptide, DNA etc.) is encapsulated into virosomes (Figure 19.2, right). Inside the endosome, the virosome starts fusion with the membrane of the endosome, thereby releasing the antigen into the cytosol of the antigen-presenting cell. The antigen recombines with a MHC I molecule and is presented to CD8+ T-cells, resulting in a cytotoxic T cell response. Although virosomes fuse efficiently with the endosomal membrane after uptake by APCs, a fraction of the particles will inevitably be degraded within the endosomal/lysosomal compartment, resulting in the generation of peptides derived from the viral spike proteins as well as from virosome-encapsulated antigen. The resulting peptides eventually associate with MHC class II molecules. Presentation of the MHC class II/peptide complexes on the surface of the antigen-presenting cell to CD4+ T-cells will induce a strong T-helper response that is essential for the stimulation of CTLs [8, 13, 30–33]. Together this results in a specific and high-quality cellular immune response. A combined cellular and humoral immune response is obtained by combining the two types of virosomes.

Virus-like particles are known as powerful immune stimulants due to their size and structure [35–37]. In the context of influenza virosomes, HA and NA as well as any other antigen displayed on the particle surface in a repetitive array represent a strong activation signal for antigen-specific B-cells via cross-linking of immunoglobulin (Ig) receptors [36, 37]. After uptake of the virosome into the B-cell via Ig receptor-mediated endocytosis, influenza-derived as well as heterologous antigens are degraded in the endosome and the resulting peptides are presented on MHC class II molecules at the surface of the B-cells. Recognition of these peptides by APC-primed helper T-cells provides the activated B-cells with the co-stimulation required for full differentiation and antibody production. The induction of high antibody titres against a malaria-derived peptide devoid of CD4 epitopes illustrates that the co-stimulated influenza-specific CD4 help is indeed functional and sufficient, both in mice [15, 38] and in humans (unpublished data).

Registered virosomal vaccines against hepatitis A and influenza have validated the excellent characteristics of virosomes as an adjuvant and carrier system. The virosomal hepatitis A vaccine has shown an excellent local tolerability and has been proven to induce a very rapid immune response with 100% seroconversion with a good long term protection [25, 39–42]. The virosomal influenza vaccine has been shown to be highly immunogenic and well tolerated, also in children, young adults and the elderly [24, 43–45]. The concept of IRIV as a vaccine carrier and adjuvant system has been tested most extensively in the context of a prophylactic malaria vaccine with strong and highly specific immune responses that significantly reduced parasite growth rates (clinical phase I and phase I/IIa) [18, 19, 22]. Clinical phase I studies for virosomal vaccine candidates against breast cancer and hepatitis C are ongoing.

In order to infect, viruses must first bind to the cell surface. The surface structures that they bind to are of two general types depending on the functional consequences of the interaction. Attachment factors serve to bind the particles and thus help to concentrate viruses on the cell surface. Such interactions can be relatively non-specific. Unlike attachment factors, virus receptors actively promote entry. They can do so by initiating conformational changes in the virus particle, by activating signalling pathways, and by promoting endocytic internalization. Often the receptors accompany the virus into the cell during endocytic uptake and may then play a role intracellularly in the penetration reaction. Of the endocytic pathways taken by viruses, the most commonly used is the clathrin-mediated endocytic route. It transports incoming viruses together with their receptors into early and late endosomes. Clathrin-mediated endocytosis is a continuous process, and, for virus entry, it is usually rapid and efficient [46]. The incoming viruses are often exposed to the acidic milieu of endosomes within minutes after internalization, and many respond to the pH drop by undergoing changes that lead to penetration. Depending on the pH threshold, the site of penetration is either the early (pH 6.5–6.0) or the late endosome (pH 6.0–5.5). Single-particle tracking of influenza virus in cultured simian kidney epithelial cells shows that, although 60% of the particles enter via clathrin-coated pits, 40% use a clathrin-independent pathway [47]. Surprisingly, of those that use the clathrin-independent pathway, only 5% associate with pre-existing clathrin-coated domains in the membrane; the majority enter via coated pits that assemble underneath the surface-bound viral particles. This implies induction of transmembrane signalling by the receptor-bound virus particles. Endocytic clathrin-coated vesicles deliver their contents to early endosomes. That some viruses use clathrin-independent pathways for endocytosis and infection is now widely accepted.

19.2 A Case Study – Antigen-Virosome Drug Delivery System

19.2.1 Particle Characterization

Virosomes are spherical and unilamellar vesicles. Photon correlation spectroscopy (PCS), or dynamic light scattering, is a powerful technique used extensively to determine the size distribution profile of various types of nanoparticles in the size range of virosomes. This fast, non-invasive and non-destructive method is used to determine the average hydrodynamic diameter and the polydispersity index (PI) as an indicator of the size homogeneity of samples in solution under physiological conditions without any pretreatment. The following four samples were measured with PCS: inactivated influenza virus (150 nm, PI = 0.1), IRIV (110 nm, PI = 0.05), TIRIV (145 nm, PI = 0.08) and liposomes (100 nm, PI = 0.01). All samples showed a very good homogeneity. These results were confirmed by freeze–fracture electron microscopy analysis (Figure 19.3). Several freeze–fracture preparations of each sample were analysed and mostly spherical structures were detected. All these structures displayed their shadows in front and behind the structures (concave and convex fracture planes), characteristic for the fracture behaviour of bilayer vesicles as virosomes and liposomes, but obviously also for lipid bilayer enveloped influenza virus. The average diameters of the particles were measured to be around or below 100 nm with a quite narrow size distribution profile. Spikes could be detected on the outer leaflet of the envelope bilayer or on cross fractions of the influenza virus particles, but were not detectable on the virosomes because of their lower concentration of influenza membrane proteins. Fusion activity of virus and virosomes was demonstrated by an *in vitro* fluorescence resonance energy transfer assay [8, 9].

The first steps of virosomes entry into cells were then analysed by transmission electron microscopy (TEM). Influenza A virus (strain X-31) and IRIV preparations were applied to HeLa cells for one hour and subjected to electron microscopy analysis (Figure 19.4). Complete viruses could be found tightly bound to the plasma membrane (Figure 19.4a). The viral envelope can be distinguished as electron dense ring and spikes around the envelope corresponding to HA and NA on the surface of the virus. Furthermore, the glycoprotein spikes on the viral envelope can be recognized in a very regular pattern. Most virus particles were filled with electron dense material corresponding to the viral ribonuclear proteins (vRNPs). Diameters of influenza particles were determined to be between 80 and 120 nm. In contrast, virosomes hardly showed any electron dense staining inside the particles and its envelope didn't show a regular spike like pattern on the surface of the particles (Figure 19.4b). Accumulation of electron dense staining was visible at the interface between virosomes and host cells, which could correspond to the glycoproteins on the virosome envelope. Diameters of virosomal particles were determined to be between 125 and 200 nm.

19.2.2 Virosome Uptake into HeLa Cells

In order to characterize virosomes functionally the binding kinetics to HeLa cells were analysed first. Therefore, fluorescence-labelled IRIVs were bound to HeLa cells, which were seeded on live cell imaging dishes. These were imaged with excitation at 633 nm for 10 frames with 10 ms exposition time at indicated time points. Intensities of 10 particles were averaged and normalized to the starting intensities. The maximum number of

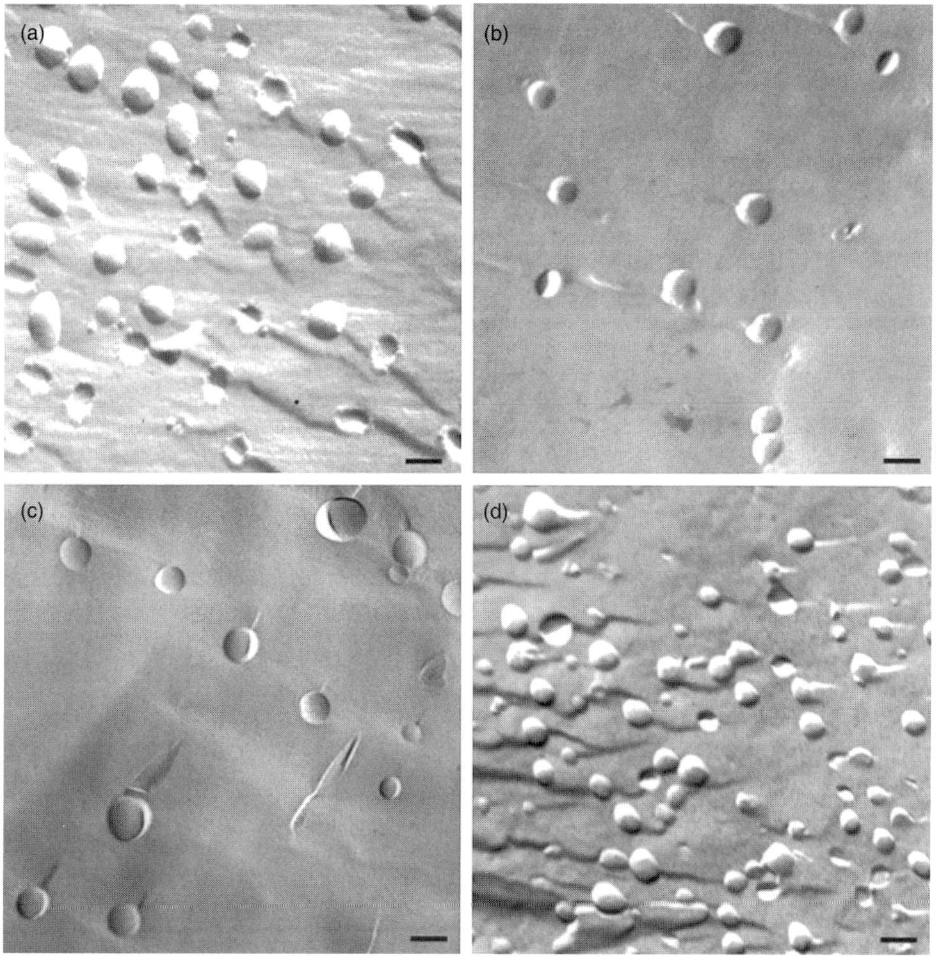

Figure 19.3 *Selected pictures of influenza virus, influenza virosomes and liposomes. Freeze–fracture electron micrograph of inactivated influenza virus (a), influenza virosomes, IRIV (b), lyophilized and reconstituted influenza virosomes, TIRIV (c) and liposomes (d) are shown. Black bar represents 100 nm*

input virosomes was determined by the amount of virosomes bound after four hours at 4 °C and set to 100%. Virosomes binding appeared to be nearly linear until 30 minutes and then asymptotically approached 70% between 40 and 120 minutes (Figure 19.5a). Further analysis of the endocytosis of virosomes was performed making use of the lysosomotropic agent ammonium chloride (NH_4Cl). First, the fusion kinetic was detected microscopically. Labelled virosomes were bound to HeLa cells at 4 °C for one hour, then non-bound virosomes were removed by sequential washing and the supernatant was exchanged to imaging medium. Widefield fluorescence microscopy was conducted at each time point or condition for five minutes. The total number of non-stationary virosome

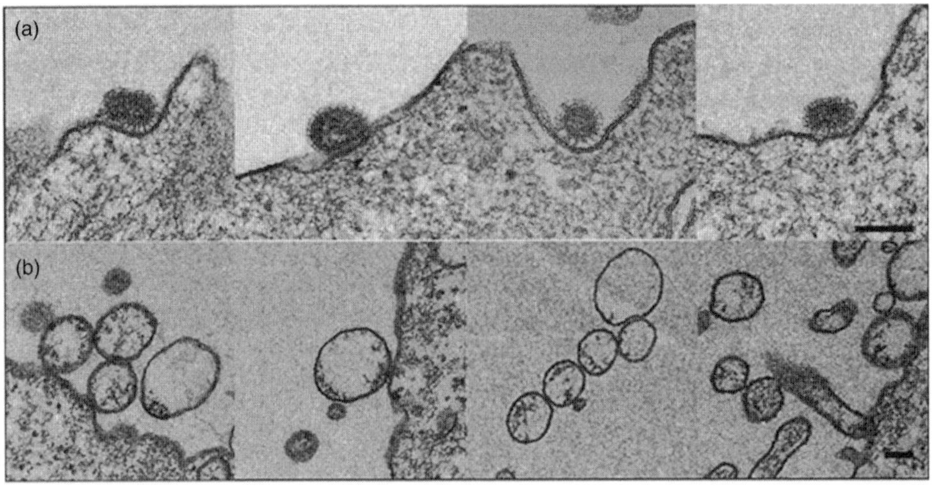

Figure 19.4 *Selected pictures of influenza virus and influenza virosomes. TEM of HeLa cells incubated with influenza virus strain X-31 (a) or influenza virosomes (b) are shown. Black bar represents 100 nm. Influenza virus and virosomes were incubated with HeLa cells at 4 °C for 1 h, followed by supernatant removal, sample fixation with 4% glutaraldehyde and processing for electron microscopy*

particles was determined by time lapse analysis (Figure 19.5b). With increasing amounts of ammonium chloride (0.1–50 mM) inhibition of virosome fusion could be determined. Total inhibition of fusion could be observed at 20 mM ammonium chloride.

In order to further characterize the time point of fusion, ammonium chloride was added to a final concentration of 20 mM at indicated time points and time lapse microscopy was conducted at each time point for five minutes (Figure 19.5c). Ammonium chloride dependence significantly decreased after 30 minutes, indicating that most of the virosomes had fused already. Finally, in order to determine the fusion time window ammonium chloride washout experiments were carried out. Ammonium chloride was added at time point zero and exchanged to imaging medium at indicated time points. After 120 minutes time lapse microscopy was carried out for 10 minutes and number of non-stationary particles was quantified, but no clear dependence could be observed indicating that virosomes did not enter in a synchronized fashion (Figure 19.5d).

Endocytosis of influenza virus has been described to occur via two independent endocytosis pathways, clathrin-dependent and clathrin- and caveolin-independent endocytosis pathways. HeLa cells were transfected with clathrin heavy chain-EGFP and uptake of IRIV was monitored by total internal reflectance fluorescence (TIRF)/widefield-microscopy (data not shown). Virosomes were taken up by HeLa cells with partial colocalization with the clathrin signal indicating uptake by clathrin mediated endocytosis. Additionally, clathrin-independent endocytosis of virosomes was found to a very similar extent as active influenza virus (S. Moese, personal communication). Uptake of virosomes by plasmacytoid dendritic cells (PDC) has been demonstrated recently [30]. PDC incubated with IRIV at 4 °C showed a membranous fluorescence (Figure 19.6a). Conversely, cells incubated with IRIV at 37 °C displayed a vesicular fluorescence into the cytosol, suggesting an endosomal localisation of IRIV-associated HA. By transmission electron microscopy, the

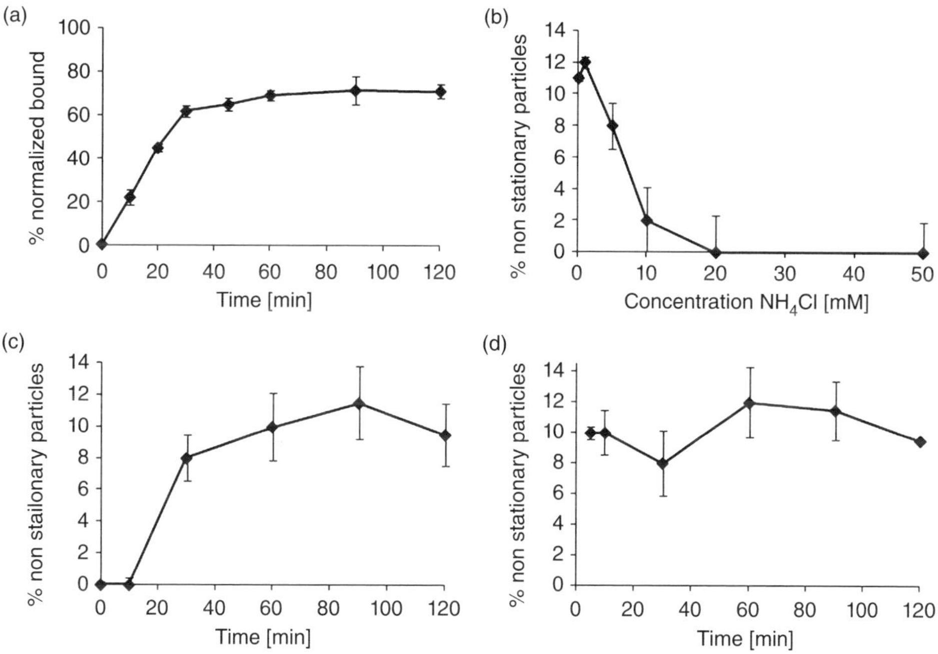

Figure 19.5 *Virosome entry into HeLa cells. (a) Binding kinetic of fluorescence-labelled virosomes. Fluorescence-labelled IRIVs were bound to HeLa cells seeded on live cell imaging dishes and exited at 633 nm for 10 frames with 10 ms exposition time at indicated time points. Intensities of 10 particles were averaged and normalized to the starting intensities. Maximum number of input virosomes was determined by amount of virosomes bound after four hours incubation at 4 °C and set to 100%. (b) Ammonium chloride inhibition kinetics: Labelled IRIVs were bound to HeLa cells at 4 °C for 1 h, followed by exchange of supernatant to imaging medium and widefield fluorescence microscopy was conducted at each time point or condition for 5 min. The total number of non-stationary virosome particles was determined by time lapse analysis. Ammonium chloride titration: Ammonium chloride was added at time point zero and the number of non-stationary particles was scored after 30 min. (c) Ammonium chloride addition: Ammonium chloride was added to a final concentration of 20 mM at indicated time points and time lapse microscopy was conducted at each time point for 5 min. (d) Ammonium chloride washout: Ammonium chloride was added at time point zero and exchanged to imaging medium at indicated time points. After 120 min time lapse microscopy was carried out for 10 min and number of non-stationary particles was quantified*

localisation of IRIV within vesicles in the cytoplasm of PDC was observed (Figure 19.6b), and one endosome usually contained more than one virosome.

19.2.3 Humoral Immune Response to Antigen Virosomes

For the induction of a strong humoral immune response, the presence of the antigen on the outer surface of the virosome is considered to be optimal. Therefore, the antigen of choice should contain a lipid anchor or a transmembrane domain to assure an intercalation in the lipid bilayer membrane. To demonstrate this, chicken avidin was chosen as the

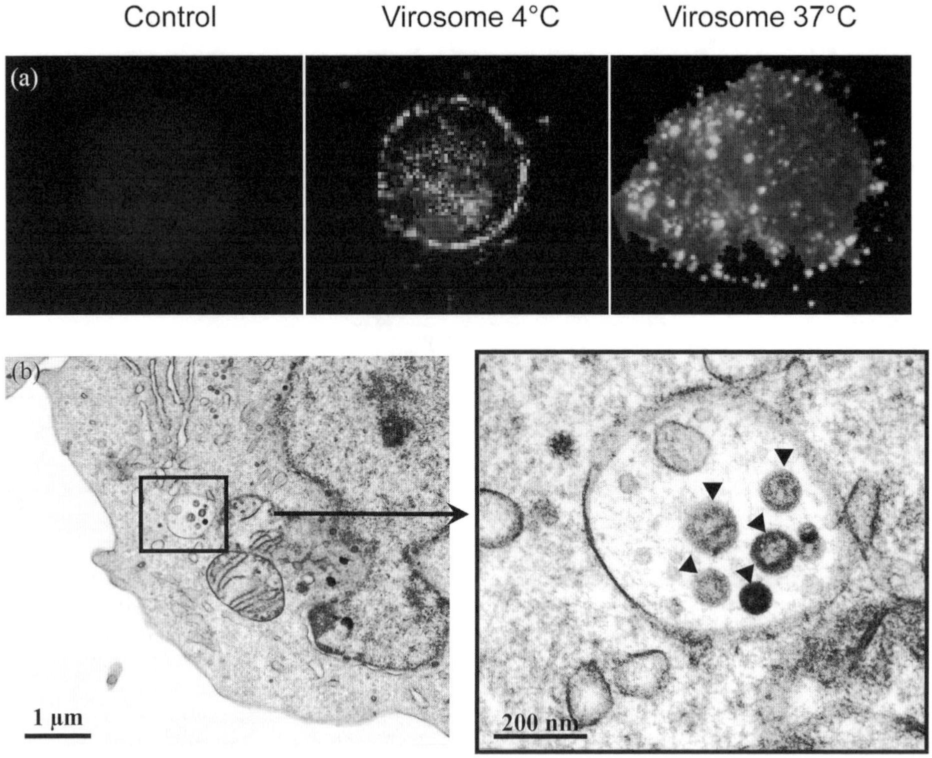

Figure 19.6 Virosome entry into plasmacytoid dendritic cells. (a) Optical microscopy of IRIV endocytosis by a PDC cell line (GEN2.2). Cells were incubated with medium (left), with IRIV for 1 h at 4 °C (middle) or with IRIV at 37 °C (right) and prepared for microscopy. Cells were labelled with a rabbit anti-HA followed by a goat anti-rabbit-FITC. (b) Electron microscopy of IRIV endocytosis by a PDC cell line. Black arrowheads show virosomes into a vesicle. (Adapted from Vaccine, **25**, 3913–3921 (2007) [30] with permission of Elsevier. Copyright 2007.)

specific antigen in a mouse model. Balb/c mice were immunized intramuscularly with different avidin formulations and the specific antibody response was analysed by ELISA (Figure 19.7). Soluble avidin alone gave a weak immune response, indicating the need of an adjuvant or a structured antigen presentation for the induction of antibodies. A mixture of soluble avidin with IRIVs increased the specific antibody response slightly, similar to a formulation where avidin was coupled to the surface of liposomes. Only the combination of both resulted in a strong humoral immune response: when avidin was coupled to the surface of IRIVs, a significant increase in the specific antibody titre was evident.

19.2.4 Cellular Immune Response to Antigen Virosomes

Protein antigens are often poor immunogens when administered alone. Therefore, an adjuvant and/or carrier is required to potentiate the immune response to the co-administered antigen. IRIVs are a safe adjuvant registered for human use. Figure 19.8 depicts the

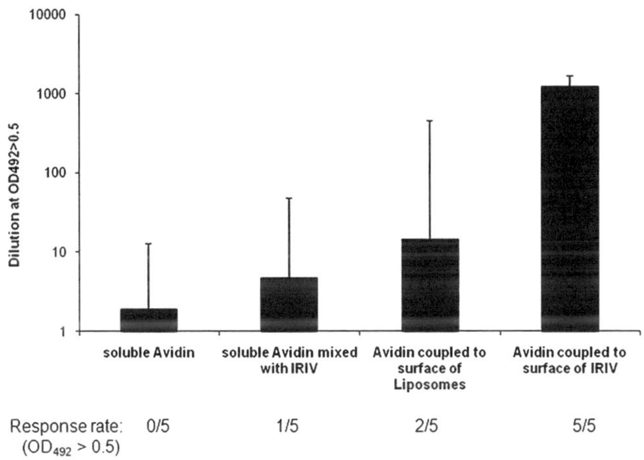

Figure 19.7 *Antibody titres induced by avidin formulations. Balb/c mice were immunized with four different avidin vaccine formulations. Antibody titres were measured after two immunizations. Results are expressed as the geometrical mean (GMT) of reciprocal serum dilutions. Groups of five mice were either immunized with soluble avidin, with soluble avidin mixed with IRIVs, with avidin covalently attached to liposomes or with avidin covalently attached to virosomes. The response rates to the different formulations are indicated below the graph*

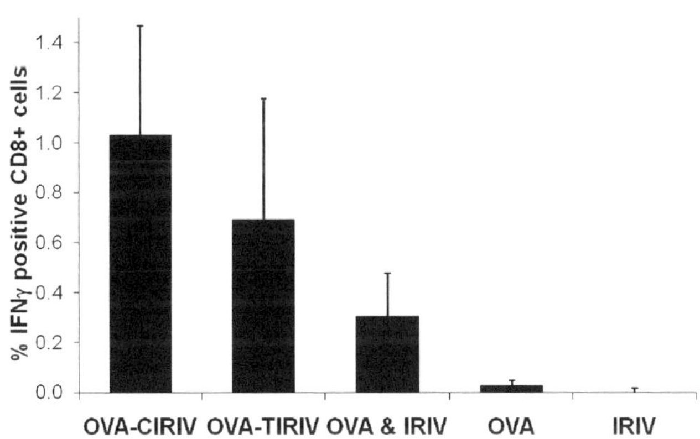

Figure 19.8 *CD8+ T-cell response in mice immunized with ovalbumin protein (OVA) in different formulations. C57Bl/6 mice (3–6 mice/group) were immunized twice subcutaneously at a three week interval with OVA in CIRIVs, in TIRIVs, mixed with IRIVs or protein alone. Spleen cells were isolated two weeks after the second immunization and the frequency of OVA 257–264 (SIINFEKL) peptide specific IFNγ-producing CD8+ T-cells was analysed by intracellular cytokine staining. Shown are mean values and standard deviations*

adjuvant and carrier effect of IRIVs in a mouse model. C57Bl/6 mice were immunized with different formulations all containing ovalbumin protein (OVA) as antigen. CTL responses were measured by quantifying the IFNγ producing CD8+ T-cells. While OVA alone was not immunogenic, mixing the protein with virosomes already induces an OVA-specific CTL response, demonstrating the adjuvant properties of virosomes. Encapsulation of the antigen inside the virosome, as in CIRIVs and TIRIVs, once more improved the immune response. This additional effect demonstrates the carrier properties of the virosomes, as antigens encapsulated inside influenza virosomes are thought to be released into the cytoplasm upon membrane fusion and then presented to CTL via the MHC class I pathway.

19.2.5 Efficiency of the Antigen Virosome Drug Delivery System

Conventional adjuvants, such as alum, are thought to enhance the antigen-specific immune response via the depot effect and danger signals resulting from tissue damage [48], both of which are likely to cause unspecific inflammation at the injection site. By contrast, influenza virosomes achieve efficient immunostimulation via mechanisms independent of inflammation, facilitated antigen processing and specific co-stimulatory signals [4, 5, 7, 51]. This fact is the key to the superior tolerability of virosomal vaccines. With more than 40 million people immunized so far with the licensed virosomal vaccines, the excellent safety profile of IRIVs has been confirmed [7]. At the same time, the robustness of the GMP compliant production process at the industrial scale and the economic viability of IRIV-based products have been validated, thereby accelerating the development of novel products [24, 25].

IRIVs can also be described as liposomes with intercalated influenza membrane proteins. This semisynthetic carrier system mimics the parenteral influenza virus in its capability to enter cells and deliver genetic material. In analogy, influenza virosomes can transport encapsulated molecules to the cytoplasm of cells [8, 9]. Numerous reports demonstrate that IRIVs represent an attractive delivery vehicle that is not restricted to protein antigens but also suitable for carbohydrates, DNA and siRNA drugs [10–12, 14, 49–56]. The availability of an efficient and safe delivery system might facilitate the clinical breakthrough of nucleic acid-based therapeutics.

Influenza viruses bind to sialic acid residues present on cell surface glycoproteins or glycolipids through the receptor-binding site in the distal tip of the HA molecule [57]. Although the interaction of HA with sialic acid is of fairly low affinity, a high avidity of the virus for cell surfaces is probably achieved by multiple low-affinity interactions. We have demonstrated that IRIV preparations are very homogenous and look similar to virus preparations except that IRIVs lack the inner core structure. Furthermore, IRIVs can bind to HeLa cells at similar efficiency to influenza virus despite of the lower concentration of membrane proteins in the virosomes. However, HA and NA may most likely freely diffuse in the lipid bilayer of virosomes as they are not anchored to the inner core as is the case in the virus. This would allow diffusion of HA in IRIVs to the local site where receptor-binding takes place, with a similar HA concentration as in the virus. This hypothesis is supported by the amount of bound virosomes to cells after a defined time interval, which is in a very similar range as for the virus (S. Moese, manuscript in preparation).

Influenza virus enter cells by receptor-mediated endocytosis, the high-capacity endocytic activity by which components of the medium are internalized in clathrin-coated membrane-bound vesicles, formed by invagination of specialized coated-pit domains of the plasma membrane [46, 47]. After internalization, the clathrin coat is removed and vesicles fuse with endosomes, beginning with mildly acidic primary endosomes and progressing to late endosomes. The uncoating of influenza virions in endosomes is dependent on the acidic pH of this compartment, and agents such as acidotropic weak bases (e.g. ammonium chloride) and carboxylic ionophores raise the pH of endosomes and block influenza virus uncoating [58]. The inhibitory effect of ammonium chloride was also observed with IRIVs. Furthermore, we have observed that the uptake of IRIVs is also clathrin-dependent, although some virosomes may enter cells with the clathrin-independent pathway as it is the case also for the virus (S. Moese, manuscript in preparation). Taken together, influenza virosomes have very similar properties to influenza viruses and enter cells in a very similar manner, making them an efficient but safe carrier system.

IRIVs are capable of inducing a highly specific humoral and cellular immune response, depending on the localization of the specific antigen in the virosomes. A frequently observed drawback of synthetic peptides and recombinant proteins is their limited *in vitro* and *in vivo* stability. Lyophilization is a process particularly useful to enhance the stability of biological materials, such as proteins and vaccines. However, conventional influenza virosomes cannot be lyophilized without loss of immunogenicity, particle size homogeneity and functional activity of the influenza HA. TIRIVs, the lyophilizable IRIVs [9], provide a flexible antigen delivery vehicle with long-term storage capability that supports the induction of strong cellular and/or humoral immune responses. In addition, virosomal formulations also protect the antigens from *in vivo* degradation. Therefore, IRIVs represent a modular and flexible delivery platform, which can be used to induce potent immune responses, and already have an impressive clinical record with regards to safety and tolerability.

19.3 Methods

19.3.1 Virosome Formulation

Preparation of virosomal formulation IRIVs, CIRIVs, TIRIVs, liposomes and the antigen-attachment procedures has been described in detail elsewhere [8, 9, 13, 16].

19.3.2 Size Determination

Size determination by dynamic light scattering using a Zetasizer 1000HS instrument (Malvern Instruments) has been described in detail elsewhere [8, 9].

19.3.3 Freeze–Fracture Electron Microscopy

Freeze–fracture electron microscopy was done by Dr B. Papahadjopoulos-Sternberg at the Nano Analytical Laboratory (San Francisco, CA). The samples were quenched using the sandwich technique and liquid nitrogen-cooled propane [59, 60]. Using this technique a cooling rate of 20 000 Kelvin/s is reached, avoiding ice crystal formation and artefacts possibly caused by the cryofixation process. The cryo-fixed sample was stored in liquid nitrogen for less than two hours before processing. The fracturing process was carried out in JEOL JED-9000 freeze-etching equipment and the exposed fracture planes are shadowed with platinum for 30 seconds in an angle of 25–35 °C and with carbon for 35 seconds (2 kV/ 60–70 mA, 1×10^{-5} Torr). The replicas produced this way were cleaned with concentrated, fuming nitric acid for 24 hours followed by repeating agitation with fresh chloroform/methanol (1 : 1 v/v) at least five times. The replicas cleaned this way were examined at a JEOL 100 CX electron microscope.

19.3.4 Transmission Electron Microscopy

Concentrated influenza virus strain X-31 (10^8 pfu/ml) or IRIVs were bound to cells for one hour on ice, unbound virus was removed, and DMEM with 20 mM HEPES pH 6.8 was added. After incubation at 37 °C for various time points, samples were fixed with 2% glutaraldehyde/2% osmium tetroxide. Sample preparation was performed according to standard EM protocols.

19.3.5 TIRF Microscopy

Microscopy was performed on a custom modified Olympus IX71 inverted microscope. Modifications included a heated incubation chamber that surrounded the microscope stage set to 37 °C, an objective-type total internal reflection fluorescence microscopy setup from TILL Photonics (Grafeling, Germany), and a monochromator for epi-fluorescence excitation with a controller allowing hardware-controlled fast switching between total internal reflection fluorescence and epi-fluorescence excitation and acquisition (TILL Photonics). The 488 and 568 nm line of a Spectra Physics laser as well as monochromator light were coupled into the microscope condenser (TILL Photonics) through optical fibres. Images were acquired using a TILL Image QE charge-coupled device camera and TILLVISION software (both from TILL Photonics). The total internal reflection angle was manually adjusted for every experiment. HeLa cells on 18 mm coverslips were mounted in custom-made chambers. To avoid changes in membrane or cytoskeleton, the medium was not exchanged when mounting cells. Movies were recorded at a rate of 20 frames per second for 1000 or 2000 frames in TIRF mode. After each experiment, the cells that had been recorded were imaged by differential interference contrast microscopy to check for viability.

19.3.6 Mice and Immunizations

Immunization experiments were performed in C57Bl/6 and Balb/c (Charles River). Mice were housed in appropriate animal care facilities and handled according to international guidelines. Mice were immunized subcutaneously for cellular response or intramuscularly for humoral response with 100 μl of the appropriate formulation. Mice received two injections at a three-week interval and the response was analysed two weeks after the last injection.

19.3.7 Intracellular IFNγ Staining

The preparation of spleen cells for IFNγ staining has been described in detail elsewhere [8, 9].

19.3.8 ELISA

ELISA has been described in detail [9, 16, 19].

Acknowledgements

We thank Anne Wieilly, Dr Nicole Westerfeld and Dr Christian Moser for their technical and administrative support.

References

1. Fodor, E. and Brownlee, G.G. (2002) Influenza virus replication, in *Influenza* (ed. C.W. Potter), Elsevier, Amsterdam, The Netherlands, pp. 1–29.
2. Lamb, R.A. and Krug, R.M. (2001) Orthomyxoviridae: the viruses and their replication, in *Fields Virology* (eds D.M. Knipe and P.M. Howley), Lippincott Williams & Wilkins, Philadelphia, pp. 1487–1531.
3. Gluck, R. and Metcalfe, I.C. (2003) Novel approaches in the development of immunopotentiating reconstituted influenza virosomes as efficient antigen carrier systems. *Vaccine*, **21**, 611–615.
4. Huckriede, A., Bungener, L., Stegmann, T. *et al.* (2005) The virosome concept for influenza vaccines. *Vaccine*, **23** (Suppl. 1), S26–S38.
5. Moser, C., Metcalfe, I.C. and Viret, J.F. (2003) Virosomal adjuvanted antigen delivery systems. *Expert. Rev. Vaccines*, **2**, 189–196.
6. Zurbriggen, R. (2003) Immunostimulating reconstituted influenza virosomes. *Vaccine*, **21**, 921–924.
7. Moser, C., Amacker, M., Kammer, A.R. *et al.* (2007) Influenza virosomes as a combined vaccine carrier and adjuvant system for prophylactic and therapeutic immunizations. *Expert. Rev. Vaccines.*, **6**, 711–721.
8. Amacker, M., Engler, O., Kammer, A.R. *et al.* (2005) Peptide-loaded chimeric influenza virosomes for efficient *in vivo* induction of cytotoxic T cells. *Int. Immunol.*, **17**, 695–704.
9. Kammer, A.R., Amacker, M., Rasi, S. *et al.* (2007) A new and versatile virosomal antigen delivery system to induce cellular and humoral immune responses. *Vaccine*, **25**, 7065–7074.
10. Cusi, M.G., Terrosi, C., Savellini, G.G. *et al.* (2004) Efficient delivery of DNA to dendritic cells mediated by influenza virosomes. *Vaccine*, **22**, 735–739.
11. Felnerova, D., Viret, J.F., Gluck, R. and Moser, C. (2004) Liposomes and virosomes as delivery systems for antigens, nucleic acids and drugs. *Curr. Opin. Biotechnol.*, **15**, 518–529.
12. Liu, X., Siegrist, S., Amacker, M. *et al.* (2006) Enhancement of the immunogenicity of synthetic carbohydrates by conjugation to virosomes: a leishmaniasis vaccine candidate. *ACS Chem. Biol.*, **1**, 161–164.
13. Schumacher, R., Amacker, M., Neuhaus, D. *et al.* (2005) Efficient induction of tumoricidal cytotoxic T lymphocytes by HLA-A0201 restricted, melanoma associated, L(27)Melan-A/MART-1(26–35) peptide encapsulated into virosomes *in vitro*. *Vaccine*, **23**, 5572–5582.
14. Waelti, E.R. and Gluck, R. (1998) Delivery to cancer cells of antisense L-myc oligonucleotides incorporated in fusogenic, cationic-lipid-reconstituted influenza-virus envelopes (cationic virosomes). *Int. J. Cancer*, **77**, 728–733.
15. Westerfeld, N. and Zurbriggen, R. (2005) Peptides delivered by immunostimulating reconstituted influenza virosomes. *J. Pept. Sci.*, **11**, 707–712.
16. Zurbriggen, R., Amacker, M., Kammer, A.R. *et al.* (2005) Virosome-based active immunization targets soluble amyloid species rather than plaques in a transgenic mouse model of Alzheimer's disease. *J. Mol. Neurosci.*, **27**, 157–166.

17. Okitsu, S.L., Mueller, M.S., Amacker, M. *et al.* (2008) Preclinical profiling of the immunogenicity of a two-component subunit malaria vaccine candidate based on virosome technology. *Hum. Vaccin.*, **4**, 106–114.
18. Genton, B., Pluschke, G., Degen, L. *et al.* (2007) A randomized placebo-controlled phase Ia malaria vaccine trial of two virosome-formulated synthetic peptides in healthy adult volunteers. *PLoS ONE*, **2**, e1018.
19. Okitsu, S.L., Silvie, O., Westerfeld, N. *et al.* (2007) A virosomal malaria peptide vaccine elicits a long-lasting sporozoite-inhibitory antibody response in a phase 1a clinical trial. *PLoS ONE*, **2**, e1278.
20. Okitsu, S.L., Boato, F., Mueller, M.S. *et al.* (2007) Antibodies elicited by a virosomally formulated Plasmodium falciparum serine repeat antigen-5 derived peptide detect the processed 47 kDa fragment both in sporozoites and merozoites. *Peptides*, **28**, 2051–2060.
21. Okitsu, S.L., Kienzl, U., Moehle, K. *et al.* (2007) Structure-activity-based design of a synthetic malaria peptide eliciting sporozoite inhibitory antibodies in a virosomal formulation. *Chem. Biol.*, **14**, 577–587.
22. Thompson, F.M., Porter, D.W., Okitsu, S.L. *et al.* (2008) Evidence of blood stage efficacy with a virosomal malaria vaccine in a phase IIa clinical trial. *PLoS ONE*, **3**, e1493.
23. Correale, P., Vecchio, D., Renieri, M.T. *et al.* (2008) Anti-angiogenetic effects of immune-reconstituted influenza virosomes assembled with parathyroid hormone-related protein derived peptide vaccine. *Cancer Lett.*, **263**, 291–301.
24. Gluck, R. (1997) Influenza immunization. *Biologicals*, **25**, 221–225.
25. Loutan, L., Bovier, P., Althaus, B. and Gluck, R. (1994) Inactivated virosome hepatitis A vaccine. *Lancet*, **343**, 322–324.
26. Zurbriggen, R. and Gluck, R. (1999) Immunogenicity of IRIV-versus alum-adjuvanted diphtheria and tetanus toxoid vaccines in influenza primed mice. *Vaccine*, **17**, 1301–1305.
27. Zurbriggen, R., Novak-Hofer, I., Seelig, A. and Gluck, R. (2000) IRIV-adjuvanted hepatitis A vaccine: *in vivo* absorption and biophysical characterization. *Prog. Lipid. Res.*, **39**, 3–18.
28. Adamina, M., Guller, U., Bracci, L. *et al.* (2006) Clinical applications of virosomes in cancer immunotherapy. *Expert Opin. Biol. Ther.*, **6**, 1113–1121.
29. Adamina, M., Schumacher, R., Zajac, P. *et al.* (2006) Advanced liposomal vectors as cancer vaccines in melanoma immunotherapy. *J. Liposome Res.*, **16**, 195–204.
30. Angel, J., Chaperot, L., Molens, J.P. *et al.* (2007) Virosome-mediated delivery of tumor antigen to plasmacytoid dendritic cells. *Vaccine*, **25**, 3913–3921.
31. Bungener, L., Serre, K., Bijl, L. *et al.* (2002) Virosome-mediated delivery of protein antigens to dendritic cells. *Vaccine*, **20**, 2287–2295.
32. Schumacher, R., Adamina, M., Zurbriggen, R. *et al.* (2004) Influenza virosomes enhance class I restricted CTL induction Through CD4+ T cell activation. *Vaccine*, **22**, 714–723.
33. Bungener, L., Huckriede, A., de Mare, A. *et al.* (2005) Virosome-mediated delivery of protein antigens *in vivo*: efficient induction of class I MHC-restricted cytotoxic T lymphocyte activity. *Vaccine*, **23**, 1232–1241.
34. Huckriede, A., Bungener, L., ter Veer, W. *et al.* (2003) Influenza virosomes: combining optimal presentation of hemagglutinin with immunopotentiating activity. *Vaccine*, **21**, 925–931.
35. Grgacic, E.V. and Anderson, D.A. (2006) Virus-like particles: passport to immune recognition. *Methods*, **40**, 60–65.
36. Bachmann, M.F., Kalinke, U., Althage, A. *et al.* (1997) The role of antibody concentration and avidity in antiviral protection. *Science*, **276**, 2024–2027.
37. Justewicz, D.M., Doherty, P.C. and Webster, R.G. (1995) The B-cell response in lymphoid tissue of mice immunized with various antigenic forms of the influenza virus hemagglutinin. *J. Virol.*, **69**, 5414–5421.
38. Pfeiffer, B., Peduzzi, E., Moehle, K. *et al.* (2003) A virosome-mimotope approach to synthetic vaccine design and optimization: synthesis, conformation, and immune recognition of a potential malaria-vaccine candidate. *Angew. Chem. Int. Ed.*, **42**, 2368–2371.
39. Bovier, P.A., Bock, J., Loutan, L. *et al.* (2002) Long-term immunogenicity of an inactivated virosome hepatitis A vaccine. *J. Med. Virol.*, **68**, 489–493.

40. Clarke, P.D., Adams, P., Ibanez, R. and Herzog, C. (2006) Rate, intensity, and duration of local reactions to a virosome-adjuvanted vs. an aluminium-adsorbed hepatitis A vaccine in UK travellers. *Travel Med. Infect. Dis.*, **4**, 313–318.
41. D'Acremont, V., Herzog, C. and Genton, B. (2006) Immunogenicity and safety of a virosomal hepatitis A vaccine (Epaxal) in the elderly. *J. Travel Med.*, **13**, 78–83.
42. Holzer, B.R., Hatz, C., Schmidt-Sissolak, D. *et al.* (1996) Immunogenicity and adverse effects of inactivated virosome versus alum-adsorbed hepatitis A vaccine: a randomized controlled trial. *Vaccine*, **14**, 982–986.
43. Gaeta, G.B., Stornaiuolo, G., Precone, D.F. *et al.* (2002) Immunogenicity and safety of an adjuvanted influenza vaccine in patients with decompensated cirrhosis. *Vaccine*, **20** (Suppl. 5), B33–B35.
44. Kanra, G., Marchisio, P., Feiterna-Sperling, C. *et al.* (2004) Comparison of immunogenicity and tolerability of a virosome-adjuvanted and a split influenza vaccine in children. *Pediatr. Infect. Dis. J.*, **23**, 300–306.
45. Schaad, U.B., Buhlmann, U., Burger, R. *et al.* (2000) Comparison of immunogenicity and safety of a virosome influenza vaccine with those of a subunit influenza vaccine in pediatric patients with cystic fibrosis. *Antimicrob. Agents Chemother.*, **44**, 1163–1167.
46. Marsh, M. and Helenius, A. (1989) Virus entry into animal cells. *Adv. Virus Res.*, **36**, 107–151.
47. Rust, M.J., Lakadamyali, M., Zhang, F. and Zhuang, X. (2004) Assembly of endocytic machinery around individual influenza viruses during viral entry. *Nat. Struct. Mol. Biol.*, **11**, 567–573.
48. Schijns, V.E. (2000) Immunological concepts of vaccine adjuvant activity. *Curr. Opin. Immunol.*, **12**, 456–463.
49. Correale, P., Cusi, M.G., Sabatino, M. *et al.* (2001) Tumour-associated antigen (TAA)-specific cytotoxic T cell (CTL) response *in vitro* and in a mouse model, induced by TAA-plasmids delivered by influenza virosomes. *Eur. J. Cancer*, **37**, 2097–2103.
50. Cusi, M.G. and Gluck, R. (2000) Potential of DNA vaccines delivered by influenza virosomes. *Vaccine*, **18**, 1435.
51. Daemen, T., de Mare, A., Bungener, L. *et al.* (2005) Virosomes for antigen and DNA delivery. *Adv. Drug Deliv. Rev.*, **57**, 451–463.
52. de Jonge, J., Holtrop, M., Wilschut, J. and Huckriede, A. (2006) Reconstituted influenza virus envelopes as an efficient carrier system for cellular delivery of small-interfering RNAs. *Gene Ther.*, **13**, 400–411.
53. Schoen, P., Bron, J. and Wilschut, J. (1993) Delivery of foreign substances to cells mediated by fusion-active reconstituted influenza virus envelopes (virosomes). *J. Liposome Res.*, **3**, 767–792.
54. Schoen, P., Chonn, A., Cullis, P.R. *et al.* (1999) Gene transfer mediated by fusion protein hemagglutinin reconstituted in cationic lipid vesicles. *Gene Ther.*, **6**, 823–832.
55. Tuffin, G., Waelti, E., Huwyler, J. *et al.* (2005) Immunoliposome targeting to mesangial cells: a promising strategy for specific drug delivery to the kidney. *J. Am. Soc. Nephrol.*, **16**, 3295–3305.
56. Waelti, E., Wegmann, N., Schwaninger, R. *et al.* (2002) Targeting her-2/neu with antirat Neu virosomes for cancer therapy. *Cancer Res*, **62**, 437–444.
57. Weis, W., Brown, J.H., Cusack, S. *et al.* (1988) Structure of the influenza virus haemagglutinin complexed with its receptor, sialic acid. *Nature*, **333**, 426–431.
58. Matlin, K.S., Reggio, H., Helenius, A. and Simons, K. (1982) The entry of enveloped viruses into an epithelial cell line. *Prog. Clin. Biol. Res.*, **91**, 599–611.
59. Sternberg, B. (1993) Freeze-fracture electron microscopy of liposomes, in *Liposome Technology*, vol. 1 (ed. G. Gregoriadis), CRC Press, London, pp. 363–383.
60. Sternberg, B. (1996) Liposomes as a model for membrane structures and structural transformations: a liposome album, in *Handbook of Nonmedical Applications of Liposomes*, vol. IV (eds D.D. Lasic and Y. Barenholz), CRC Press, London, pp. 271–297.

20

GRAZAX®: An Oromucosal Vaccine for Treating Grass Pollen Allergy with Immunotherapy

Annette Römmelmayer Lundegaard, Lise Lund and Jørgen Nedergaard Larsen
ALK-Abelló A/S, Hørsholm, Denmark

20.1 Introduction

The prevalence of allergic diseases is increasing in developed countries where allergy is becoming the most widespread disease among adolescents and young adults [1]. Allergy is a systemic manifestation with symptoms most often presenting from the nose and eyes (i.e. rhino-conjunctivitis or hay fever) or from the lungs (i.e. asthma). Hay fever and asthma are both caused by common immunological mechanisms expressed both systemically and in the airway mucosa [2].

Specific immunotherapy (SIT) for allergic diseases, that is specific allergy vaccination, has been performed for almost a century, since it was originally described in 1911 [3]. Recommended indications include mild to moderate allergic asthma and moderate to severe hay fever. Traditional SIT is performed by repeated subcutaneous injections over a three- to five-year period using a standardized allergen vaccine, such as Alutard SQ® from ALK-Abelló [4].

This chapter describes a new convenient concept for SIT, where patients can treat themselves at home. The new treatment is based on GRAZAX®, a fast dissolving tablet containing a standardized allergen vaccine. The tablet is placed under the tongue once daily and the treatment continues for three years.

Delivery Technologies for Biopharmaceuticals: Peptides, Proteins, Nucleic Acids and Vaccines Edited by Lene Jorgensen and Hanne Mørck Nielsen

20.2 Specific Immunotherapy

Management of allergic diseases is a combination of four measures: allergen avoidance, symptomatic treatment, specific immunotherapy and patient education [2]. Having diagnosed the allergen source(s), to which the individual is sensitized, allergen avoidance is always recommended. However, in many cases effective avoidance is not possible. Effective drugs for the control of allergic symptoms are available but, in spite of the fact that they are inexpensive and easy to use, patient compliance is not sufficiently satisfactory [5]. Often a combination of drugs (e.g. antihistamines in combination with steroids) is prescribed to control all symptoms. Specific immunotherapy is a causal treatment that interferes specifically with the underlying immunological mechanisms; SIT therefore has simultaneous effect on all symptoms as well as persistent effect on the immunological status of the patient [4].

Vaccines for SIT are biological products, due to the fact that the source material is of biological origin like pollens, house dust mites or animal hair. Each manufacturer uses different raw materials, different processes and different procedures for quality control. It is not possible to relate the potency of allergen vaccines of different composition, and therefore each manufacturer uses individual units of measure. Alutard SQ® is the best documented product for SIT and in general considered the 'gold standard' of allergen specific injection immunotherapy [6, 7].

A simple vaccination strategy as used for infectious diseases, with only one or a few high doses administered to the patient, is impossible in allergy treatment, as the patient already has a pathological immune response against the vaccine. Instead, a cautious approach is taken with gradual dose increase in weekly injections over an 11–16-week up-dosing period. Having reached the recommended maintenance dose this is administered every six weeks for three years. The effect of SIT is dose dependent; the higher the dose, the better and faster the onset of protection. The optimal dose, however, is limited by a small but significant risk of inducing allergic side reactions. For this reason, subcutaneous injection therapy needs to be conducted by specialists in clinics where relevant emergency procedures are maintained [4].

In order to improve the convenience of SIT, ALK-Abelló has developed a fast dissolving tablet for sublingual administration, GRAZAX®, which was introduced to the market in late 2006. GRAZAX® represents the first clinical development programme within allergen SIT designed to fulfil the regulatory requirements for a central procedure resulting in simultaneous market authorisation in 27 European countries [8, 9]. A total of 10 trials have so far been completed, among which is the world's largest immunotherapy trial including 855 patients [10]. More than 3000 patients have participated in finished and ongoing trials, so the concept is thoroughly tested.

GRAZAX® is well tolerated. Each tablet contains grass pollen extract comparable to three quarters of one dose of Alutard SQ®, and there is no up-dosing. Patients take one tablet each day for three years, and thus the monthly dose administered is roughly 30 times higher when compared to treatment with Alutard SQ®. The mechanisms of delivery to the immunological centres in the oral mucosa are not known, but immunological changes corresponding to those seen with injection treatment have been observed. Although many patients experience local itching, which tend to cease after short duration, adverse reactions are mild, and the treatment is safe for self administration at home. The safety profile is thus the major difference between the two treatments (subcutaneous and sublingual administration), and patients may choose the treatment best in line with their preferences.

20.3 Description of Grass Pollen Extract

Allergen extracts and vaccines based on allergen extracts are used for both specific diagnosis and treatment of allergic diseases [11]. Allergen extracts are aqueous solutions of extractable material from allergenic source materials, such as grass pollens. No structural feature defining an allergen has hitherto been described in literature or in the Pharmacopoeia, and the definition of an allergen is based upon the functional criterion of being able to elicit an allergic IgE response in susceptible individuals. Airborne allergens are proteins carried by particles in the micrometre range abundant in the air that we breathe. The particle carrying the allergen is inhaled and the allergen is deposited on the mucosal surface of the airways, thereby stimulating the immune system. Thus, an allergen is defined by the immune system of the individual patient.

By this definition, any immunogenic protein (antigen) has allergenic potential, even though most allergic patients have IgE specific for a relatively limited number of 'major' allergens. An important objective in the manufacture of allergen extracts, therefore, is to secure an adequate complexity reflecting the composition of water-soluble components of the allergenic source material. Another important aspect of allergen extract manufacturing is controlling the major allergen content, which has distinct importance for the activity of allergen extracts/vaccines in diagnosis as well as treatment. The third important matter of batch-to-batch control is standardization of potency, that is the overall IgE binding capacity, which is a reflection of the anaphylactic potential of the preparation, and hence an important safety parameter. Raw materials of biological origin are inherently variable in composition, and without standardization this variability would be reflected in the final product. All aspects of the manufacturing procedure from culturing and collection of raw materials, extract preparation and storage, to validation of assays and reagents, have an impact on extract quality and should be considered part of the standardization procedure [11].

A central aspect of the standardization procedure is the in-house reference preparation (IHRP). Since allergen extracts of different composition cannot be compared in a meaningful way, there is no international standard in effect, and each manufacturer maintains individual IHRPs and uses individual units of measure. According to guidelines, the IHRP needs to be thoroughly characterized and tested for biological potency by skin prick testing in a panel of allergic patients [12]. A reflection of the unique standardization quality (SQ) of ALK-Abelló products is the fact that the IHRP for grass pollen extract has been tested in clinical trials of specific immunotherapy. This enables ALK-Abelló to label the potency in SQ units reflecting the treatment dose in specific immunotherapy.

Grass pollens for the production of GRAZAX® are collected by vacuuming cultivated fields of Timothy grass, that is *Phleum pratense*. Different harvests are combined to ensure consistency in composition. Acceptance limits define the maximal content of irrelevant pollens and plant debris. The grass pollen extract is produced by aqueous extraction followed by filtering and stabilization yielding the drug substance (mixture of allergens). Batch-to-batch standardization is performed on the drug substance by comparing every new batch to the IHRP using validated assays addressing (i) complexity, (ii) major allergen content and (iii) overall potency. The new batch has to match the IHRP in every aspect analysed [11].

The methods used for assessment of antigen complexity are standard biochemical separation techniques, such as crossed immunoelectrophoresis, polyacrylamide gel electrophoresis and isoelectric focusing. Crossed immunoelectrophoresis is particularly useful, as it represents a semi-quantitative technique and, furthermore, can be probed with patient serum to detect allergens. However, the technique is dependent on an animal serum as reagent. Major allergen determination can be performed by in-gel-precipitation techniques, such as single radial immunodiffusion or rocket immunoelectrophoresis, or by ELISA. All techniques are dependent on mono-specific animal anti-sera or monoclonal antibodies. Assessment of overall IgE-binding potency is performed by solid phase immunoinhibition assays. Every assay is run with the IHRP in parallel and readouts must match within predefined acceptance limits [11].

20.4 Technologies of Fast Dissolving Tablets

The dose form of choice for GRAZAX® is a fast dissolving tablet.

Fast dissolving tablets have many benefits for the patient compared to traditional oral tablets. The most obvious benefit is a standardized delivery of a concentrated dose of the pharmaceutically active ingredient to the sublingual mucosa. Also, the patient does not need to swallow a tablet and there is no need for drinking water. In addition, the fast dissolving tablets will disintegrate or dissolve rapidly in the saliva while maintaining a pleasant taste and mouth feeling. Most of the fast dissolving tablet technologies use sugar-based excipients. Sugars taste pleasant, are highly water soluble and dissolve quickly in saliva. Fast dissolution of the tablets benefits fast release of the drug substance in the saliva. This makes the drug substance potentially ready for absorption if sufficient solubility in the saliva is available.

Different technologies for fast dissolving tablets have been introduced to the market as different companies have focused on a specific drug delivery system. Before the development work of GRAZAX® was initiated, different technologies were carefully evaluated. Selection of a contract manufacturing organization (CMO) has to be based on different perspectives and evaluation of both 'soft skills' and 'hard skills'. 'Hard skills' of a CMO could be defined as: drug delivery system, technical fit, supply of clinical batches during development, commercial lines and batch sizes, size of CMO, location, price, intellectual property rights, history and experience with authorities, history and experience with technology. 'Soft skills' of a CMO could be defined as: reputation, way of cooperation, company policy and behaviour, personnel, building of partnership and project team, and so on. 'Soft skills' are very difficult to measure but are at least as important as 'hard skills'.

The advantages and disadvantages of various technologies are described in Table 20.1.

The ZYDIS® freeze-dried, fast dissolving tablet was selected due to the technical suitability for an allergen/protein-based product, experience with authorities worldwide and proven track record of commercialization.

Table 20.1 *Comparison of advantages and disadvantages of different fast dissolving tablet technologies [13]*

Process technology	Advantages	Disadvantages	Brand/technology
Freeze-drying	Immediate dissolution (2–10 s) Protein stability often good in freeze-dried formulations	Poor physical resistance High cost of production Sensitive to humidity Low dose of water-soluble drugs	ZYDIS®, Quicksolv®, Lyoc®, Nanocrystals™, Pharmafreeze
Moulding	Very rapid dissolution (5–15 s) High dose	High cost of production Weak mechanical strength	Wowtab® Fast melt (Flashtab®)
Tableting (standard)	Low cost of production Use of standard equipment/materials High dose Good physical resistance	Significant effects of the size and hardness of the tablets on disintegration property	AdvaTabs®, DuraSolv, Pharmburst™, Frosta®, RapiTab™
Tableting (effervescent)	Use of standard equipment High dose Good physical resistance	Operation in controlled low humidity Need for totally impermeable blister	OraSolv®
Tableting (humidity treatment)	Good physical resistance Pleasant mouth feeling	No comments due to lack of information of technology	Wowtab®

20.5 Formulation of GRAZAX®

The choice of a formulation principle (and thereby CMO) is based on many different factors, which may potentially conflict, such as: stability of drug substance and excipients, cost of excipients, available manufacturing equipment and manufacturing capacity, route of administration, patient-friendly, intellectual property right, know-how and experience, and so on.

GRAZAX® is a fast dissolving tablet produced as a lyophilisate designed to rapidly disintegrate in the oral cavity. GRAZAX® contains grass pollen extract as the active pharmaceutical ingredient (API). The tablet is intended for sublingual administration, where it disintegrates on the oral mucosa within seconds to release the allergens. Hereby, contact between the relevant allergens and the immune system is achieved to ensure clinical efficacy. As a consequence, the allergen content of the formulation must be released as quickly as possible in the oral cavity to ensure maximum exposure to these areas before swallowing. Therefore, it is recommended to avoid swallowing for one minute after intake.

Besides, as allergens are proteins, in general they are sensitive to heat and mechanical stress during processing. As a consequence, the manufacturing process must be gentle in order to avoid degradation. Based upon these considerations the ZYDIS® technology provides a good match between formulation properties and manufacturing conditions.

GRAZAX® is intended for allergic subjects and no unnecessary substances are added. The tablets do not contain colour excipients or masking of taste to avoid potential hypersensitivity reactions in the allergic patient. Due to the presence of mannitol, the taste is neutral sweet. The other excipients are chosen for technical reasons to ensure optimal physical specifications of the tablets. All excipients used in the tablet formulation are well known and meet the requirements of the European Pharmacopoeia (EP) [12]. They have been used by the drug product manufacturer over many years in the manufacture of the ZYDIS® dosage form. The main components conveying the ZYDIS® dosage form with its fast dispersion properties while not compromising adequate strength are gelatine and mannitol. Gelatine forms a glassy amorphous structure, which imparts strength and resilience to the structure of the tablet during handling and ensures that the tablet retains some flexibility. Mannitol is included in the formulation as it crystallizes during the freezing process and gives the tablets their rigidity. It also has the function of improving appearance, texture and taste. It is used as a pharmaceutical agent in, for example, chewable tablets for its non-hygroscopic properties. Mannitol resists moisture sorption even at high relative humidity. Purified water is used during production as a vehicle and as solvent for the grass pollen extract and excipients. During the manufacture of GRAZAX®, the pH of the drug solution is adjusted to maintain the stability of the grass pollen extract. Each unit is a freeze-dried debossed solid preparation, referred to as a tablet.

20.6 Production of GRAZAX®

The development and commercial cultivation of grass pollen/source material and development and commercial manufacture of grass pollen extract is performed by ALK-Abelló. The pharmaceutical development and commercial manufacture of the drug product part of GRAZAX® was performed by Catalent Pharma Solutions, LLC. The manufacturing process of GRAZAX® includes five key stages: mixing, dosing, freezing, freeze-drying and primary packaging in blister packs. The secondary packaging is performed at ALK-Abelló. During the mixing stage, the carrier materials gelatine and mannitol are dissolved in purified water. The pH of the matrix solution is adjusted to ensure stability of the allergens and the grass pollen extract is ad-mixed. This step involves conventional mixing equipment. The mixture is then dosed by weight into pockets of preformed aluminium blister trays. The dosing process is fully automatic and the wet dose weight is set at the start of each batch and is controlled and kept within an acceptable range throughout the whole dosing period by in-process controls.

Once dosed, the solution is frozen within the blister pockets by passing the blister trays through a liquid nitrogen freeze tunnel. The freezing temperature of the tunnel and the time the cards remain in the tunnel are controlled to ensure frozen units, which have the required structural characteristics. After freezing, the trays of frozen products are stored in low temperature storage cabinets to maintain the product frozen prior to freeze-drying.

The frozen units are loaded on the shelves of the freeze-driers designed to enable rapid processing by removal of frozen water rapidly by sublimation at low pressure. The freeze-drying cycle is optimized and developed specifically for GRAZAX®. Once dried, the GRAZAX® units in the aluminium trays are sealed with aluminium lidding foil by passing through the blister sealing machine. Finally, the blisters are cut into blisters cards of 10 units. The last step of the manufacturing process is packaging of the blister cards into secondary cartons with insert labels.

GRAZAX® has an excellent stability and can be stored for at least three years without any special requirements regarding storage conditions.

The whole manufacturing process of GRAZAX® from cultivation of grass pollen to secondary packaging of the finished product consists of several steps:

1. Cultivation of grass pollen *Phleum pratense*
2. Harvesting of grass pollen
3. Processing/drying of grass pollen
4. QA/QC release of grass pollen
5. Extraction of allergens from grass pollen
6. Filtration of extracted proteins
7. Stabilization of grass pollen extract
8. Biologic standardization and QA/QC release of grass pollen extract
9. Mixing of water, excipients and grass pollen extract
10. Dosing of GRAZAX® solution
11. Freezing of GRAZAX®
12. Freeze-drying of GRAZAX® frozen units
13. Packaging of blister cards
14. Secondary packaging of blister cards into cartons
15. QA/QC release of GRAZAX®

20.7 Specifications and Analytical Procedures

The pharmaceutical dosage form oral lyophilisate is exempted from the EP monographs, *Tablets and Oromucosal preparations* [12]. Therefore, no specific requirements were defined for the GRAZAX® tablets besides those mentioned in the EP monograph, *Allergen products*. The requirements of the monographs for *Allergen products* and *Tablets* are compiled in Table 20.2.

Even though no specific requirements are stated for this dosage form in the EP, the tests specified in the monographs *Tablets and Oromucosal preparations* were evaluated during development. Appropriate tests for GRAZAX® tablets were chosen as a part of the release specification. The tests applied were those necessary to document the physical, chemical, immunochemical and microbiological attributes of the GRAZAX® tablets.

20.7.1 Physical and Chemical Tests

Besides the water content of the GRAZAX® tablets, the physical appearance is part of the release specifications. The accepted criterion for water content was based on batch data generated during development of the product. The protein profile is examined by conventional SDS-PAGE (Sodium-Dodecyl Sulfate Poly-Acrylamid Gel Electrophoresis), which separates the proteins by their molecular weight. The pattern of bands is compared to IHRP for positive identification.

Table 20.2 *Requirements from the European Pharmacopeia Monographs*

Tests	EP monographs	
	Allergen product	Oromucosal preparations/tablets
Physical	Water content (<5%)	Uniformity of mass/content uniformity Disintegration Dissolution Friability of uncoated tablets
Chemical	Protein content (80–120%) Protein profile (comparable to IHRP)	Uniformity of dosage units Uniformity of dose
Immunochemical	Antigen profile (identification by antigen-specific animal antibodies) Allergen profile (identification by antigen-specific human antibodies) Total allergenic activity (50–200% of stated amount of specific IgE reactivity) Individual allergens (50–200% of stated amount)	
Microbiological		Total viable count (≤10^3 cfu/gram) Fungal count (≤10^2 cfu/gram) Absence of Pathogens (*E. coli* in 1 g)

IHRP: In-House Reference Preparation.

According to the monograph *Uniformity of dosage units*, this can be measured by either content uniformity or mass variation. Since GRAZAX® tablets are solids that are packaged in single unit containers that have been prepared from true solutions and freeze-dried in the final containers, the test for mass variation may be considered and justified instead of test for content uniformity.

One of the characteristics of GRAZAX® is that the tablets are fast dissolving and the drug substance is highly soluble. According to the ICH guidelines, dissolution testing for immediate release of solid oral drug products made from highly water soluble drug substances may be replaced by disintegration testing. For GRAZAX®, both dissolution and disintegration time are tested *in vitro* to ensure that the tablets are dissolved within one minute, during which the recommendation is that it is retained in the mouth before swallowing. Furthermore, the dissolution and disintegration tests can be used to evaluate the batch-to-batch consistency variability.

20.7.2 Immunochemical Tests

Grass pollen extract consists of several allergens designated major, intermediate or minor allergens. This designation is related to the prevalence of IgE antibodies in a population of patients allergic to grass pollen. The allergen designated as the major allergen is therefore selected as test parameter. In grass pollen extracts this is referred to as Phl p 5. The assay platform used is an ELISA comprising both monoclonal and polyclonal antibodies against the major allergen. The actual batch is compared with an IHRP. The specifications in the EP on this analysis states a release criteria of '50–200% of the stated amount', but the in-house criteria for GRAZAX® have been narrowed to a smaller interval: 65–135%.

The total allergenic activity in the drug product is measured. The analytical principle is based on the reactivity of the allergens in the grass pollen extract with IgE antibodies in a pool of human serum from grass allergic patients. The assay platform is based on paramagnetic beads. Regarding antigen and allergen profiles, these are routinely performed on the grass pollen extracts (API) as described in Section 20.3.

20.7.3 Microbial Tests

In the European Pharmacopoeia monograph for tablets, requirements for total viable count, fungal counts as well as absence of pathogens are stated. No additional requirements to the microbial purity are defined specifically for non-sterile allergen products in the EP monograph for allergen products. All tests described in the tablet monograph are performed on GRAZAX®.

20.8 Feasibility Study: Additional Analyses

As an additional analysis, friability testing was performed on GRAZAX® tablets during the feasibility phase of the development of GRAZAX®.

According to the EP, the friability test equipment is designed to be used for tablets manufactured by compression. The pack is an integral part of the product and the blister pocket effectively acts as a mould for the dosed solution. Once freeze-dried, the dried unit is not free to move within the blister pocket so during storage and distribution the aluminium blister card provides sufficient protection from breakage of the tablet.

The friability method in the EP is based on an evaluation of the mechanical resistance of the tablets by difference in weight before and after rotation 100 times in a drum. To enhance the weight of the tablets, they were rotated directly in the blister packages as single dose units. An immunochemical quantitative test on single allergens was used instead of weighing, as this method is sufficiently sensitive to measure small amounts of the active drug substance. This analysis was performed in order to detect potential product residuals in blister package after rotation and removal of the GRAZAX® tablet from the package. During feasibility, GRAZAX® tablets remained intact and the results were lower than the limit of detection of the immunochemical test.

20.9 Conclusions

The development of GRAZAX® consisted of a comprehensive pharmaceutical development programme, including up-scaling and optimization of cultivation of the source material (grass pollen), optimization of the manufacturing process of the drug substance (grass pollen extract) and development of the drug product using the ZYDIS® technology to develop the freeze-dried fast dissolving tablet.

GRAZAX® is the first allergen-based tablet to address the underlying cause of grass pollen allergy on the market in Europe. GRAZAX® is a fast dissolving once-daily immunotherapy tablet, safe for home administration. This new treatment concept enables more allergic patients to benefit from specific immunotherapy, the only treatment persistently preventing symptoms and disease aggravation.

References

1. Linneberg, A., Jørgensen, T., Nielsen, N.H. *et al.* (2000) The prevalence of skin-test-positive allergic rhinitis in Danish adults: two cross-sectional surveys 8 years apart. The Copenhagen allergy study. *Allergy*, **55**, 767–772.
2. Bousquet, J., Van Cauwenberge, P., Khaltaev, N. and Aria Workshop Group (2001) World Health Organization, allergic rhinitis and its impact on asthma. *J. Allergy Clin. Immunol.*, **108** (5 Suppl.), S147–S334.
3. Noon, L. (1911) Prophylactic inoculation against hay fever. *Lancet*, **1**, 1572–1573.
4. Bousquet, J., Lockey, R.F. and Malling, H.J. (eds) (1998) WHO position paper – allergen immunotherapy: therapeutic vaccines for allergic diseases. *Allergy*, **53** (Suppl. 44), 1–42.
5. Nolte, H., Nepper-Christensen, S. and Backer, V. (2006) Unawareness and undertreatment of asthma and allergic rhinitis in a general population. *Respir. Med.*, **100**, 354–362.
6. Bousquet, J., Lockey, R. and Malling, H.J. (1998) Allergen immunotherapy: therapeutic vaccines for allergic diseases. A WHO position paper. *Allergy*, **53**, 1–42.
7. Calderon, M., Alves, B., Jacobson, M. *et al.* (2007) Allergen injection immunotherapy for seasonal allergic rhinitis. *Cochrane Database of Systematic Reviews* 1 (Art. No. CD001936). DOI: 10.1002/14651858.CD001936.pub2.
8. Kay, A.B. (2007) An extract of Timothy-grass pollen used as sublingual immunotherapy for summer hay fever. *Drugs Today (Barc.)*, **43**, 841–848.
9. Frew, A.J. (2008) Sublingual immunotherapy. *N. Engl. J. Med.*, **358**, 2259–2264.
10. Durham, S.R., Yang, W.H., Pedersen, M.R. *et al.* (2006) Sublingual immunotherapy with once-daily grass allergen tablets: a randomized controlled trial in seasonal allergic rhinoconjunctivitis. *J. Allergy Clin. Immunol.*, **117**, 802–809.
11. Larsen, J.N., Houghton, C.G., Vega, M.L. and Løwenstein, H. (2008) Manufacturing and standardizing allergen extracts in Europe. *Clin. Allergy. Immunol.*, **21**, 283–301.
12. *European Pharmacopoeia* (2001), vol. 2, supplement 5.8, 5th edn, Directorate for the Quality of Medicines of the Council of Europe, Strasbourg, France.
13. Fu, Y., Yang, S., Jeong, S.H. *et al.* (2004) Orally fast disintegration tablet: developments, technologies taste-masking and clinical studies. *Crit. Rev. Ther. Drug Carrier Syst.*, **26**, 433–475.

Final Comments

21

Delivery Technologies for Biopharmaceuticals: A Critical Assessment

Marco van de Weert

University of Copenhagen, Department of Pharmaceutics and Analytical Chemistry, Faculty of Pharmaceutical Sciences, Copenhagen, Denmark

21.1 Introduction

Biopharmaceuticals constitute a special class of drug products, in particular due to their delivery challenge. Their large size usually makes transport over biomembranes very difficult, and this is further complicated by the often limited physicochemical and enzymatic stability of biopharmaceuticals. As a result, biopharmaceuticals are almost invariably administered by injection or infusion. The drug delivery technologies described in this book include a number of methods to circumvent the parenteral route of administration, but also methods are discussed to reduce the number of injections and/or the amount of active compound required to achieve the desired effect.

The reasons to develop a delivery technology for a biopharmaceutical are highly dependent on the nature of the biopharmaceutical and its desired mode of action. A delivery technology can thus be used to:

1. circumvent the parenteral route, either to reduce systemic exposure and deliver the drug locally or to increase patient compliance;
2. achieve sustained plasma levels of the drug, thus reducing the number of injections. This may be used to increase patient compliance, but may also reduce side-effects and costs;

Delivery Technologies for Biopharmaceuticals: Peptides, Proteins, Nucleic Acids and Vaccines Edited by Lene Jorgensen and Hanne Mørck Nielsen

3. obtain some level of targeting. This both reduces systemic exposure to the drug (fewer side-effects) and the required dose. Moreover, much higher concentrations may be obtained at the site of action, increasing efficacy;
4. modulate the effect of the drug, specifically in the case of vaccines, where the delivery system can modify the ratio of humoral and cellular response;
5. reduce potential side effects such as immunogenicity;
6. increase patent life or circumvent patents of competitors.

The delivery technology may also result in a combination of the above.

Those new within the field of delivery technologies will, unfortunately, notice that there is an enormous amount of literature available, but very few products on the market. This discrepancy cannot solely be explained by the novelty of many of the drug delivery technologies. In this chapter several of the most important hurdles for new drug delivery technologies are discussed, including areas that are relatively poorly explored.

21.2 Large Scale Production and Reproducibility

Proof-of-concept in a laboratory setting is the most common endpoint in the scientific literature. Unfortunately, the production (but also performance) of delivery technologies does not always scale up well from laboratory to industrial scale, and significant modifications to the production process are common. For example, the method used to entrap growth hormone in PLGA microspheres in the industrial production of Nutropin Depot (Genentech/Alkermes) is significantly different from the most common production methods described in the scientific literature [1, 2]. Despite the optimization, Genentech and Alkermes decided to discontinue production of Nutropin Depot after six years, citing problems in maintaining the significant resources required to produce the product [3].

The academic challenge in scaling up production processes is enormous, but there are very few scientific publications on this topic (see [2] for a very rare example). However, since scalability is an important factor in the costs (see also below), it will be an important parameter in deciding which delivery technology(ies) make it to the market. The same applies to the ability to consistently reproduce the system. Although physicians and patients may be able to handle products that do not perform consistently, and thus may require frequent dose adjustments, both physicians and patients are likely to prefer the more consistent product, even when it performs worse.

Finally, production of the final product also requires that the system used is stable for a prolonged period, preferably several years. It may require significant reformulation to obtain such a stable system. Fortunately, this area has seen some more, yet still limited, academic interest, as shown for example by several papers on lyophilization of various delivery systems [4–6].

21.3 Safety Concerns

Probably the most important hurdle for the development of any delivery technology is the often unknown safety, either alone or in conjunction with the drug in question.

Although animal studies can give an indication of the safety of the delivery technology, the ultimate proof can only come from clinical studies in humans. This is especially the case for protein pharmaceuticals: many protein pharmaceuticals, even those with fully human amino acid sequences, are known to cause an immune response [7, 8]. This response may be benign, as is the case for, for example, human insulin. In other cases the immune response may be more severe and lead to, for example, anaphylactic shock. A complicating factor is that for some proteins the immune response may become clinically relevant only after months or years of use of the protein drug. Formulation factors are often implicated in this delayed immune response, but the knowledge within this field is still very limited. Thus, the delivery technology itself may have a profound (negative) effect on the safety, but this effect may take months of use to be observed. It is especially this long-term safety that is very poorly investigated.

Nucleic acid-based therapeutics can also induce an immune response, although there is limited research aimed at explaining the origin [9, 10]. It may be hypothesized that the delivery of oligonucleotides *should* result in some immune response, since oligonucleotides are intracellular compounds. Release of oligonucleotides into the extracellular space would indicate some major pathological event. Thus, the immune activation observed for small interfering RNA (siRNA) [9] may be due to the body recognizing siRNA as a compound that should not be present outside the cells. However, little work has been done to elucidate to which extent delivery technologies may aggravate an immune response.

The seemingly innocuous change of the route of administration can sometimes have profound effects. The development of the pulmonary insulin product Exubera provides an excellent example of this point. Exubera was developed to allow patients a needle-free administration device for their meal-time insulin. Clinical studies showed that the pulmonary delivery resulted in a higher level of antibodies than subcutaneously injected insulin [11, 12]. This could be countered by administering a higher dose. Moreover, the lung capacity of those using Exubera is slightly reduced, but within the time frame of the study this returned to normal after discontinuation of Exubera [12]. A more troubling issue was the observation of a higher incidence of lung cancer in the cohort treated with Exubera, compared to the control group [13]. The association was, reportedly, not statistically significant. A possible association does not preclude pulmonary delivery, however, since this may be due to the specific drug (insulin is a growth factor), specific additives in the formulation, the chronic and frequent nature of insulin delivery, or due to the administration of particulate matter to the lungs.

Also, other alternative routes of administration may provide some safety concerns that become apparent only after long-term use. For example, buccal and nasal delivery often include various absorption enhancers. It may well be that any adverse effects of these enhancers only become apparent after many months, whereas animal studies usually only last a few weeks. Oral delivery of intact biopharmaceuticals is still an enigma, since the gastrointestinal tract is designed to degrade foodstuff and only absorb small molecules. A proposed route of entry, among others when using polymeric carrier systems, is through Peyer's patches [14]. This should raise some alarm bells, since Peyer's patches are involved in immune surveillance. On the other hand, many proteins injected subcutaneously are absorbed through the lymphatic system, and thus are also in direct contact with the immune system [15].

Particulate delivery systems sometimes have unexpected side-effects in the absence of a drug. For example, it has been shown that PEGylated liposomes can trigger an immune response by activating the complement system in a subpopulation of patients, resulting in anaphylactic shock [16]. Moreover, many, if not all, patients develop transient antibodies against these liposomes, meaning that they can only be administered with at least a six-week interval [17, 18]. Pharmacological activity of the delivery system itself should also be considered. For example, certain block copolymers have been shown to affect various efflux systems [19]. Thus, the system will need to be tested in animals as early as possible in the development, in order to prevent major disasters at a later stage.

An additional concern for particulate systems is the complexity of the system, making it more difficult to properly characterize the product, including the stability of the drug compound and the excipients during storage.

Although immunogenicity development is the aim in the case of vaccines, there are obvious safety concerns here also. In fact, safety regulations are often much more stringent, since vaccines are prophylactic and the main targets are paediatric patients. Thus, side-effects are difficult to accept. Moreover, clinical studies are more difficult to get approved when dealing with children as the main patients. This partly explains why there are so few adjuvant systems currently approved.

21.4 Cost–Benefit Ratio

Safety concerns are frequently discussed in the literature, although usually confined to immediate toxicity. Discussions on cost–benefit ratios are, however, very limited, and most chapters in this book are not an exception to this rule. This is unfortunate, since the most effective system may simply be too expensive for patients.

Calculating a cost–benefit ratio is not easy and depends also on factors that are impossible to quantify objectively. For example, some patients may be willing to pay significant amounts to circumvent parenteral administration, whereas others may not. This partly explains why Exubera only was marketed for one year, showing poor sales despite clear benefits for a selected subpopulation of the patients. The influence of health insurance companies should also be taken into account, as they are likely to financially support the cheapest treatment, unless the novel product provides a clear and 'objective' benefit.

21.5 Conclusion

Despite the many promising proof-of-concepts of various delivery technologies, there is still a long road ahead. With few exceptions, most notably the modification of proteins, there are still numerous safety concerns, in particular with respect to the repeated and/or long-term use of the various technologies. Going from laboratory to industrial scale also is a poorly investigated area, despite its obvious importance in the ultimate goal of the drug delivery scientist: to develop a product that actually benefits the patients. Ultimately, even the most effective and safe technology will fail if it cannot be produced at an affordable cost and in the required amounts. This means there are still many challenges ahead, which, in turn, means there are still many chances for the academic and industrial scientist to make a decisive impact.

Perhaps within ten years many will be laughing at the somewhat bleak picture that this chapter may have painted. Hopefully, the safety and large-scale production at affordable costs of the delivery technologies described in this book will be found not to be a problem. However, this does require that these factors are taken into account more comprehensively, and that a significant research effort is made within these areas.

References

1. Cleland, J.L., Johnson, O.L., Putney, S. and Jones, A.J.S. (1997) Recombinant human growth hormone poly(lactic-co-glycolic acid) microsphere formulation development. *Adv. Drug Deliv. Rev.*, **28**, 71–84.
2. Tracy, M.A. (1998) Development and scale-up of a microsphere protein delivery system. *Biotechnol. Prog.*, **14**, 108–115.
3. Jorgensen, L., Moeller, E.H., van de Weert, M. *et al.* (2006) Preparing and evaluating delivery systems for proteins. *Eur. J. Pharm. Sci.*, **29**, 174–182.
4. Abdelwahed, W., Degobert, G., Stainmesse, S. and Fessi, H. (2006) Freeze-drying of nanoparticles: formulation, process and storage considerations. *Adv. Drug Deliv. Rev.*, **58**, 1688–1713.
5. Hinrichs, W.L.J., Manceñido, F.A., Sanders, N.N. *et al.* (2006) The choice of a suitable oligosaccharide to prevent aggregation of PEGylated nanoparticles during freeze thawing and freeze drying. *Int. J. Pharm.*, **311**, 237–244.
6. Mohammed, A.R., Bramwell, V.W., Coombes, A.G.A. and Perrie, Y. (2006) Lyophilisation and sterilisation of liposomal vaccines to produce stable and sterile products. *Methods*, **40**, 30–38.
7. van de Weert, M. and Moeller, E.H. (eds) (2008) *Immunogenicity of Biopharmaceuticals*, Springer/AAPS Press, New York.
8. Schellekens, H. and Casadevall, N. (2004) Immunogenicity of recombinant human proteins: causes and consequences. *J. Neurol.*, **251**, II/4–II/9.
9. Schlee, M., Hornung, V. and Hartmann, G. (2006) siRNA and isRNA: two edges of one sword. *Mol. Ther.*, **14**, 463–470.
10. Klinmann, D.M., Xie, H. and Ivins, B.E. (2006) CpG oligonucleotides improve the protective immune response induced by the licensed anthrax vaccine. *Ann. NY Acad. Sci.*, **1082**, 137–150.
11. Hollander, P.A., Blonde, L., Rowe, R. *et al.* (2004) Efficacy and safety of inhaled insulin (Exubera) compared with subcutaneous insulin therapy in patients with type 2 diabetes. *Diabetes Care*, **27**, 2356–2362.
12. Skyler, J.S., Hollander, P.A., Jovanovic, L. *et al.* (2008) Safety and efficacy of inhaled human insulin (Exubera) during discontinuation and readministration of therapy in adults with type 1 diabetes: a 3-year randomized controlled trial. *Diabetes Res. Clin. Pract.*, **82**, 238–246.
13. Pfizer Inc (2008) Pfizer statement on Exubera labeling update in the United States, http://mediaroom.pfizer.com/portal/site/pfizer/index.jsp?ndmViewId=news_view&ndmConfigId=1010794&newsId=20080409005647&newsLang=en (accessed 29 January 2009).
14. Shakweh, M., Ponchel, G. and Fattal, E. (2004) Particle uptake by Peyer's patches: a pathway for drug and vaccine delivery. *Expert Opin. Drug Deliv.*, **1**, 141–163.
15. Tang, L., Persky, A.M., Hochhaus, G. and Meibohm, B. (2004) Pharmacokinetic aspects of biotechnology products. *J. Pharm. Sci.*, **93**, 2184–2204.
16. Moghimi, S.M., Hamad, I., Andresen, T.L. *et al.* (2006) Methylation of the phosphate oxygen moiety of phospholipid-methoxy(polyethylene glycol) conjugate prevents PEGylated liposome-mediated complement activation and anaphylatoxin production. *FASEB J.*, **20**, 2591–2593.

17. Ishida, T., Ichihara, M., Wang, X.Y. *et al.* (2006) Injection of PEGylated liposomes in rats elicits PEG-specific IgM, which is responsible for rapid elimination of a second dose of PEGylated liposomes. *J. Control. Release*, **112**, 15–25.
18. Judge, A., McClintock, K., Phelps, J.R. and MacLachlan, I. (2006) Hypersensitivity and loss of disease site targeting caused by antibody responses to PEGylated liposomes. *Mol. Ther.*, **13**, 328–337.
19. Kabanov, A.V., Batrakova, E.V. and Alakhov, V.Y. (2002) Pluronic block copolymers as novel polymer therapeutics for drug and gene delivery. *J. Control. Release*, **82**, 189–212.

Index

Note: page numbers in *italic* refer to tables and figures.

absorption enhancers
  buccal 44–7
  intranasal 31–2, *38–9*
  pulmonary 36, 37–40
  transdermal 53–6
acylation, peptide 4, 16–17, 177
  see lipidization
Adagen® 13
adeno-associated viruses 94, *103*
  viral delivery examples 106–7
adenoviruses 94, 99, 101, *103*
  expression 100
  and siRNA 274
  viral delivery examples 105–6
adjuvants, immunomodulatory 114–17, 360–4
  and signalling receptors 122–3
aerosols 40, 42, 136, 231, 249
albulin 18
alginate microparticles 213, 217, 257–8
allergic diseases 395, 396–7
  oromucosal vaccine *see* Grazax®
Allovectin-7® 86
Altrigel® 21
AlutardSQ® 396
amino acid substitution 4, 17
amphotericin B 137
antigens 113–14
  and dentritic cells 117–20
  liposomal delivery 35, 142, 360
  particulate delivery 120–2, 257
  virosome delivery *see* virosomes
Apirdra® 185
AraNESP® 16
Atrigel® 199
auto-injectors 22–3
avidin–biotin trigger 82–3

bacterial viruses 107
baculoviruses 107–8
Basulin® 197
bioadhesive systems
  buccal absorption 45, *46–7*
  nasal absorption 32–4
  pulmonary delivery 41
Biochronomer® 196–7
Biopharmaceutical Classification System 131
biotin–avidin trigger 82–3
block copolymers 19, 197–8, 200–2, 260, 261–2
blood–brain barrier 30–1, 36, 138
  and rabies virus protein 75–6
buccal absorption *43*
  adsorption enhancers 44–7
  enzyme inhibitors 44
  insulin 43
  mucoadhesive systems 45, 46–7
  peptide/protein drugs 42–7, *57*
burst release 20, 250

Delivery Technologies for Biopharmaceuticals: Peptides, Proteins, Nucleic Acids and Vaccines Edited by Lene Jorgensen and Hanne Mørck Nielsen

Caeylx® 360
CAF01 adjuvant 362–3, *366*, 368, 371–2
  cryo-/lypoprotection of 369–71
calcitonin
  intranasal delivery 30
  oral delivery 216
  pulmonary delivery 42, 237, *238*
  and solid lipid nanoparticles 139
  transdermal absorption 56
carbon nanotubes 6, 20
carbopol 212
carrier systems, overviews of 4–6, 131–3, 139
cationic dendrimers 74–5, 155–62
cationic lipids 137, 140–1, 292–3, 343, 371
  and immune reaction 298
cationic polymers 73–4, 255–6, 343
cell penetrating peptides 75, 306–7
chemical enhancers
  buccal absorption 44
  intranasal absorption 31–2
  transdermal absorption 53, *55*
chemical modification
  buccal peptide/protein drugs 44–5
  oral peptide/protein drugs 52–3
chitosan 4
  nucleic acid delivery 258
  peptide/protein delivery
    buccal 44
    nasal 31–2, 33, 34–5
    oral 212–13, 216
    pulmonary 41, 237, *238*
cholesterol *362*
  DC-chol 360
  and liposome stability 367–8
Cimzia® 14
cis-acting viral elements 94, *95*, 101, *102*
coagulation factor VIII 4
colon specific drug delivery 52, 210
complexing 4, 34–5
condensation, DNA 140
conjugates
  lipid–drug 135–6
  PAMAM-PLGA microparticle 156–8
  peptide nucleic acid–cell penetrating peptide 307
  see also lipidization; PEGylation
copolymers 19, 197–8, 200–2, 260, 261–2
Cord Factor 384
Coxsackie Adenovirus Receptor 97
cryo-/lipoprotectants 369–71
cryopreparation of DNA 252
CTAB surfactant 137
cyclosporin A 48
cystic fibrosis gene 85
darbepoetin-alfa 15–16
DC-Chol 360
DDA liposome 360, *362*, 363, *366*, 371
  freeze-drying 369–70
  instability issues 366–8
DDAB lipid 278
DEAE-dextran–MMA copolymer (DDMC) 339–40
  characterization/preparation of the polymer 341
  complex formation
    general polyion–DNA kinetics 340
    in vivo DDMC–DNA interaction 141–3
  DNase degradation 350–2
  structure of DDMC–DNA complex 343, *344*
    infrared absorption spectra 343–5
  transfection
    charge ratio of DNA/DDMC 347
    cytotoxicity 348–9
    delivery pathways 352–3
    kinetics 349–50
    procedures and protocol 345–7
    transfection efficiency 347–8, 352
DEC-205 receptor 119–20
Decapeptyl® 19
delivery technologies, overview of 407–8
  carrier systems 4–6, 131–3, 139
  cost–benefit ratio 410
  large scale production/reproducibility 408
  overcoming delivery barriers 4–5
  safety issues 406–10
dendrimers 4
  defined and described 149–50
  and nucleic acid-based drugs 74–5, 256
  properties/molecular interactions 150–2
  studies 155
    gene delivery
      PAMAM-PLGA microparticles 75, 156–8
      siRNA delivery 158–62
    large molecular weight drugs 153–5
    vaccine delivery 162–5
dendriplexes 74–5
dendritic cells 360
  antigen capture 119–20
  immunobiology 117–18
Depofoam™ 5, 20
depot forming systems 21
  cross-linked systems 200–1
  precipitation systems 199
  thermal gelling systems 199–200
  thermoplastic semisolids 201–2
desmopressin 16, 30, 48
  lipidization 52

detemir insulin 16
see also acylation; lipidization
dextran
nanoparticles 198–9, 217
porous microparticles *233*
see also DEAE-dextran–MMA copolymer
2-diethylaminoethyl-dextran *see* DEAE-dextran
N,N-dimethyl-N,N-dioctadecylammonium *see* DDA
dioleoyl phosphatidylethanolamine 85, 293–4
N,N-dioleoyl-N,N-dimethylammonium *362*, *365*, 368
1,2-distearol-*sn*-glycero-3-phosphocholine *362*, 365, 366
DMRIE liposomes 85
DNA
delivery *see* DEAE-dextran–MMA copolymer; nucleic acid-based drugs
mimics 75
polyion complex formation 340
DODA liposomes *362*, 365, 368
DOPE lipid 85, 293–4
DORA receptor 120
dose dumping 20
DOTAP liposomes 85, 137, 141, 296, 299
and vaccine delivery systems 360
double emulsion solvent evaporation 251–2
drug delivery *see* delivery technologies
DSPC liposome *362*, 365, *366*
Dynamic PolyConjugates 73–4

elcatonin 33
electrochemical trigger 83
electroporation 56, 85
Eligard® 21
Eligen® technology 48
Enbrel® 17–18
encapsulation 4
endocytosis 5, 70–1
and viral infection 381
endosome disruption 330–1
Enhanced Permeability and Retention effect 70
enoxaparin 153
enzyme inhibitors
buccal absorption 44
oral absorption 48–9
pulmonary absorption 37, 40
EPOPC lipid 72
erythropoietin 6, 15–16
see also glycosylation
etanercept 17–18

excipients 5–6
bioadhesive 33

Fendrix vaccine 117
FLUAD vaccine 117
free-fracture electron microscopy 382, *383*, 389–90
freeze drying 369–71
fusion, protein 17–18

galactose 76–7
gas auto-injector 22–3
gastrointestinal tract 52, 209–10
GelSite® 201
gene therapy
overviews 69, 83, 139–41, 305, 339–40
see also nucleic acid-based drugs; siRNA; viral vectors
glycoengineering 15–16
glycosylation 15
gold nanoparticles 83
granulocyte colony-stimulating factor (G-CSF) 42, 84, 184
grass pollen extract 397–8
GRAZAX®
background 395
grass pollen extract 397–8
specific immunotherapy 396–7
tablet technologies 398–9
formulation and production 399–401
specifications/analytical procedures 401–3

haemagglutinin 96–7
HeLa cell study 382–5
heparin 153, 233
hepatitis B 142
hepatitis vaccine 378, 381
HIV 83–5, 101, 103
homogenization, high-pressure 135
Humalog® 17, 185
human growth hormone 85, *182*, 184, 194
hyaluronan/hyluronic acid 82, 259–60
hydrogels 49, 50, 199–201, 211
dextran 198–9
polyacrylamide 263
polyphosphoesters 195–6
hyperthermic delivery 81–2

idarubacin 138
immune responses 6, 113–14
adjuvants 114–17, 122–3, 360–4
dendritic cells 360
antigen capture 119–20
immunobiology 117–18
particulate antigen systems 120–2, 257

signalling receptors/role of adjuvants 122–3
siRNA-induced 279–80, 297–9
study of dendrimers in vaccine delivery 162–5
virosome induced 379–81, 385–8
cellular 386–8
humoral 385–6
immunomodulatory adjuvants 114–17, 360–4
virosomes as 377–9
implant systems 21–2
in situ depot forming systems *see* depot forming systems
infliximab 18
influenza haemagglutinin 96–7
influenza vaccine 381
influenza virosomes *see* virosomes
influenza viruses 94, 377, *378*, 382, *383*, 384, 389
inhalers 232
injection devices 22–3
insertional mutagenesis 101, 104–5
insoluble powder formulations 33–4
insulin 5
analogues 16, 17, 18, 185
buccal absorption 43
crystalline isophane insulin 22
and hydrogels 211
injection 22–3
intranasal absorption 33
lipidization 52
oral delivery 50, 213–14, 215, 216
pulmonary absorption 42, *233*, 235
subcutaneous injection 195, 197
transdermal absorption 56
within SLNs 139
interferons 15, 43, 198
interleukins 17, 43
intestinal absorption *50*
intranasal delivery
insulin 33
and lipid nanoparticles 136, 142
peptide/protein drugs 30–6, *38–9*, *57*
siRNA 274
iontophoresis 56
IRIVs *see* virosomes
isoelectrical point 185–6

Lantus® 17, 185, 197
LAR® 195
lectin 51
and bioadhesive systems 32, 51
lentiviruses 94, 274
HIV-1 101, 103
viral delivery examples 103–5
leptin, glycosylated 15, 16
leuprolide acetate 21, 22
ligand–protein coupling 52–3
lipid carrier systems *see* lipid nanoparticle; liposomes
lipid nanoparticle-based systems 131–42
background 131–3
delivery/targeting 136–8
lipid-drug conjugates 135–6
overview of applications
nucleic acids 139–42
peptides and proteins 20, 49, 51, 139
vaccines 120–2, 142
solid lipid nanoparticles
cytocompatibility 138–9
defined/described 133–4
drug incorporation approaches 135
preparation techniques 134–5
toxicity and safety 138–9, 298–9
lipid–drug conjugates 135–6
lipidization 52, 177–81
see also acylation peptide
methods 177–8
non-parental delivery 179–80
pharmakinetic improvements 178–9
prodrug concept/reversible lipidization 181
solubility and surface activity 180–1
structural and stability effects 181
lipoamino acids 52
Lipofectamine™ 81
lipoplexes 4, 71–2, 140–1
siRNA 275, 292
lipopolymers 74, 294
lipopolysaccharides 364
liposome-polycation-DNA 72
liposomes 4, 6, 49
defined 359
and freeze drying 369–71
nucleic acid-based drug delivery
clinical trials 85–6
lipoplexes 71–2
liposome-polycation-DNA formulations 72
see also siRNA delivery
peptide/protein drugs
buccal absorption 45
intranasal absorption 35–6
oral absorption 51
pulmonary absorption 41–2
transdermal absorption 54, 56
physiochemical properties and stabilization 365–6
stabilization 366–8
structures 359–60, *362*
vaccine systems *see* vaccine delivery
LPD complexes 72

lung physiology 228, *229*, 270–2
Lupron Depot® 195
lutenizing-hormone-releasing hormone (LHRH) 42

macro-albumin aggregates 260
Macugen® 14
magic bullets 360
magnetic nanoparticles 80–1
Magnetofection™ 80–1
mannose receptor 119, 120
mannosylation 162–5
Medusa® technology 197–8
melarsoprol 137
methyl methacrylate copolymer 340
Miacalcin® 30
microemulsion 36
microspheres/microparticles 49, 75, 177, 194–5
  bioadhesive microspheres 32–3
  mucoadhesive microspheres 40
  porous microparticles 232–6, *238*
  see also lipid nanoparticle; polymeric nano/microparticles
molecular engineering *see* protein engineering
monophosphoryl lipid A 364
MPL® 364
mucoadhesive systems
  biodegradable polymeric particles 213–15
  buccal absorption 45, *46–7*
  nasal absorption 32–4
  oral drug delivery 51, 210, 212–13
  pulmonary delivery 40, 41, 232, 237
murine leukaemia virus 98

nafarelin 30
naked DNA 76
  barriers to delivery 35, 70–1
  clinical trial and HIV 83–5
  with electroporation 85
  and hybrid oligonucleotides 85
  and Magnetofection® 80–1
nano/microparticles *see* lipid nanoparticles; polymeric nano/microparticles
nanocomplexes
  intranasal absorption 35
  self-assembling pectin-liposome 51
nanotoxicology 7, 138–9, 240, 256, 348–9
nasal absorption *see* intranasal
nasal-associated lymphoid tissue (NALT) 30, 35
nebulizers 42, 232
nerve growth factor 195
Neulastra® 14
non-viral DNA delivery *see* DEAE-dextran–MMA copolymer; nucleic acid-based drugs
NovoRapid® 17, 185
nucleic acid-based drugs (chemical vector delivery)
  barriers to drug delivery 70–1
  cell penetrating peptides 75–6
  delivery systems 4, 139–41, 247–8
    dendrimer 74–5, 155–62
    liposomal 71–2, 85–6
    see also DEAE-dextran-MMA copolymer; polymeric nano/microparticles
  naked DNA 76, 83–5
  recent clinical trials 83
    cationic liposomes 85–6
    naked DNA 83–5
    polymer-based carriers 86
  targeted delivery 76–7
    tumours/metastasis 77–9
  triggered release 79–80
    biotin–avidin 82–3
    electrochemical 83
    hyperthermic delivery 81–2
    magnetic delivery 80–1
    proton sponge effect 80
    reductive agents 82
  see also viral vector delivery
Nutropin Depot® 19, 195

octreotide 16, 19
ocular drug administration 136
Oncaspar® 14
oral delivery
  chitosan particles 50
  lipid-based nanoparticles 136
  polymer-based systems 207–19
  see also under peptide/protein drugs (non-parenteral)
ovalbumin antigen 162–5

Paclimer® 196
paclitaxel 196
PAMAM dendrimers *see* dendrimers
PAMAM-PLGA microparticle conjugates 156–8
paracellular absorption route 5, 47, 48, 49, *50*, 210
parainfluenza virus 280–1, *282–3*
parvovirus 94
patch systems, polymer based 212
pattern recognition receptors 122
PBAVE polymer 74
pectin 213

PEG-Intron® 14
PEGASYS® 14
PEGylation 4, 13–15
  applications
    dendrimers 152
    insulin/PEGylated chitosan 217
    RNase A microspheres 177
  methodology
    chemical procedures 14, 172–3
    PEG size and structure 173–4
    site and linker 173
  pharmaceutical effects 175
    self-association 176–7
    solution stability/aggregation 175–6
  pharmakinetic/immunogenic effects 174–5
pen-injectors 22
Pep-3 peptide 75
peptide modification *see* amino acid substitution; lipidization; glycosylation; PEGylation
peptide nucleic acids (PNAs) 305–6
  cell penetrating peptides (CPPs) 75, 306–7
  cell uptake *vs* bioavailability 307
  PNA–CPP conjugates 307
    auxiliary agents/endosome disruption 330–1
    efficacy *vs* efficiency 307, 330
    in vitro studies reviewed *308–29*
    in vivo studies reviewed 331, *332–5*
peptide restructuring *see* peptide modification; protein engineering
peptide/protein drugs (non-injectable) 29–57
  background 29–30, 207–8, 227–8
  buccal administration 42–4, *57*
    bioadesive sysems 45, *46–7*
    chemical enhancers 44
    chemical modification 44–5
    enzyme inhibitors 44
    liposomes 45
  intranasal delivery 30–1, *57*
    absorption enhancers 32–3
    increasing retention time 32–4
    liposomes 35–6
    microemulsions 36
    nanoparticle/nanocomplex carriers 34–5
  oral delivery 47–8, *57*, 208–11
    absorption enhancers 48–9
    chemical modification 52–3
    colon specific systems 52, 210
    enzyme inhibitors 48–9
    gastrointestinal challenges 209–11
    hydrogels 50, 211
    mucoadhesive systems 51, 210, 212–13
    particlulates/encapsulation 49–51, 213–16
    tablets/patches 212–13
  pulmonary delivery 36–7, *57*
    challenges/factors affecting 228–31
    liposomes 41–2
    particulates/encapsulation 40–1
      nano-particles 231–2
      porous microparticles 232–6, 237, *238*
    permeation enhancers 37–40
    protease inhibitors 37–40
    stability 240
    sustained release 41, 235, 239
    toxicity 240
  transdermal delivery 53, *57*
    chemical enhancers 53–4
    liposomes 54, 56
    micro/nanoparticles 56
    physical approaches 56
    structure modification 54, *55*
  see also peptide/protein drugs (injectable)
peptide/protein drugs (injectable)
  background and challenges 11–13
    barriers to drug transport 12–13
    degradation/instability *13*
  chemical modification
    acylation 4, 16–17
    amino acid substitution 4, 17
    glycoengineering 15–16
    PEGylation 4, 13–15
  delivery systems
    implant/depot forming 21–2, 199–202
    nanoparticle systems 18–20, 131–3
    see also polymeric nanoparticles
  injection device technology 22–3
  see also peptide/protein drugs (non-injectables)
peptides, cell penetrating 75, 306–7
permeation enhancers 37, 40, 213
  toxicity 48–9
Peyer's patches 49, *59*, 210, 216
phosphotidyl choline 363, 365
PIV infection 280–1, *282–3*
PLA polymer *see* poly(lactic acid)
plasmid DNA 69, 200, 250, 346
  barriers to delivery 70–1, 76
PLHP copolymer 260
Pluronic copolymers 261–2
P(MAA-g-PEG)-based particles 50, 215
PNA conjugates *see* peptide nucleic acids
POE polymers 19, 196–7, 201, 262
polyacrylamide 263
polyacrylates 212, 218
polyalkylcyanoacrylates 218
polyanhydrides 19, 196
polycaprolactones 19, 195, 217
polycarbophil 212

polycations 254–6
complex formation with DNA 340
poly(D,L-lactide-*co*-4-hydroxy-L-proline) 260–1
polyethylene glycol *see* PEG
poly(ethylene imine) 35, 73, 255, 260, 278
poly(lactic acid) (PLA) 214, 217, 218, 219
microsphere formulations 194–5
poly(lactic-*co*-glycolic acid) (PLGA) 49–50
nucleic acid delivery 75, 248
combining polycations with 254–6
current applications 256–7
design parameters 248–50
fabrication strategies 251–2
optimization of fomulations 253–4
PAMAM-PLGA microparticle conjugates 156–8
protein/peptide delivery 194–5
oral delivery 49–50, 214, 217–18, 219
parenteral delivery 19, 21
pulmonary delivery 41, 237
PolyMAG 81
polymer-based delivery systems 207–9
gastrointestinal challenges 209–11
hydrogels 211
patches and tablets 212–13
see also polymeric nano/microparticles
polymeric nano/microparticles
antigen delivery systems 120–2
nucleic acid delivery 73–4, 86, 120–2, 247–8
alginate 257–8
chitosan 258–9
hyaluronan 259–60
macro-albumin aggregates 260
Pluronic copolymers 261–2
polyacrylamide 263
poly(D,L-lactide-*co*-4-hydroxy-L-proline) 260–1
poly(lactic-*co*-glycolic acid) 248–57
polyorthoesters 262
peptide/protein delivery 19–20, 34, 40–1, 49–50, 213–19
cross-linked dextran 198–9, 217
lactic/glycotide polymers 194–5
Medusa copolymer 197–8
polyanhydrides 19, 196
polycaprolactones 19, 195, 217
polyorthoesters 196–7, 201
polyphophoesters 195–6
pulmonary delivery, advanced 231–2, *234*
nanoparticles 237–40
porous microparticles 232–6, *238*
polymers, biodegradable/biocompatible *46–7*, 49–50, 149, 213–15, *234*
see also polymer-based delivery; polymeric nano/microparticles
poly(methacrylic acid) 215
polyomaviruses 108
polyorthoesters 19, 196–7, 201, 262
polypex delivery 73–4
polyphosphazenes 19
polyphosphoesters 195–6
polysine 35
porous microparticles 232–6, *238*
precipitation depot systems 22, 199
Prepandrix vaccine 117
prodrugs 4
production, large scale 408
prokaryotic viruses 107
Proloxamer® 200
protease inhibitors 37, 40, 213
protein crystallization 22
protein design automation 182
protein drugs *see* peptide/protein drugs
protein engineering 181–2
isoelectrical point 185–6
pharmacokinetics 185
protein stability 183–5
rational/semi-rational design strategies 182–3; *see also* peptide modification; protein fusion; protein modification
protein fusion 17–18
protein modification *see* amino acid substitution; glycosylation; lipidization; PEGylation; protein fusion
protein precipitation 22, 199
protein stability 183–4
protein–ligand coupling 52–3
proton sponge effect 73, 80, 330
pulmonary delivery 36–7, *57*, 227–8
barriers/factors affecting absorption 228–31
bioadhesion/sustained release 41
protease inhibitors/permeation enhancers 37–40
stability issues 240
sustained/release 41, 235, 239
delivery systems 231–3
liposomes 41–2
nanoparticles 40–1, 136, 237–9
PAMAM dendrimer study 153–5
polymeric matrices *234*
porous microparticles 232–6, *238*
nanotoxicity 240
siRNA *see* siRNA
PulmoSphere 235

rabies virus glycoprotein 75–6
rational based protein engineering *see* protein engineering
receptors, targeting 76–9
rectal drug administration 136
reductive gene delivery agents 82
ReGel® 21, 199–200
Remicade® 18
respiratory syntactal virus 280–1, *282–3*
respiratory tract 228, *229*, 270–2
retention time, increasing 32–4
reticuloendothelial system 70
retroviruses 94, 98–9
  expression 100
  insertional mutagenesis 101
  viral delivery examples 103–5
RGD peptide ligand 78–9, 275
RNA *see* nucleic acids
RNA interference *see* siRNA
RSV infection 280–1, *282–3*

SABER® system 21, 199
safety *see* toxicity and safety
Sandostatin® 19
sargramostim 84
SARS infection 281, *283*, 284
severe acute respiratory syndrome 281, *283*, 284
SHOX deficiency 84–5
signalling receptors 122–3
single emulsion spray-drying 252
siRNA delivery
  background 269, 291–2, 305
    challenges 273–4, 293
    mechanism of RNA interference 270
  dendrimer carrier 158–62
  intraperitoneal injection 279
  intravenous delivery 278–9
  lipid-based delivery systems 275, 291–2, *295*
    active targeting strategies 294, 296
    cationic liposomes 292–3
      helper lipids/fusion enhancement 293–4
    immune activation 297–9
    internalization mechanisms/cytosis 296
    lipopolymers 294
    neutral and anionic liposomes 296
  pulmonary delivery
    airway delivery 275–8
      intravenous/intraperitoneal compared 278–9
    challenges 273–4
    inflammatory response/immune activation 297–9
    lung physiology 270–2
    mouse models 273
    RSV and PIV infection in mouse 280–1, *282–3*
    SCV infection in monkey 281, *283*, 284
    therapeutic developments/rapid response 284, *285*
  viral *vs* non-viral vectors 274–5, 292
skin
  structure 53, *54*
  transdermal absorption 53–6, *57*
SLNs *see* solid lipid nanoparticles
small interfering RNA *see* siRNA
solid lipid nanoparticles 132, 133–5
somatropin 85
Somatuline® 19
Somavert® 14
sonophoresis 56
specific immunotherapy 395, 396–7
spray-drying 251, 252
stabilization, summary 4–6
stealth SLNs 137
stearylamine 255
sucrose acetate isobutyrate 21
sulfhydryl delivery systems 82
supercritical fluid drying 232
Supercur® 19
superparamagnetic iron oxide 80–1
superporous hydrogels 211
Synarel® 30

T-cells 117, *118*, 119
tablets
  fast dissolving 398–9
  polymer based 212
targeted delivery 76–9
targeting ligands 6, 76
  galactose 76–7
  solid tumour/metastasis
    folate 77
    RGD sequence 78–9
    sigma receptor 78
    transferrin 77
TDB *see* trehalose 6,6′-dibehenate
thermal gelling systems 199–200
thermoplastic semisolids 201–2
thiolated polymers/thiomers 212
tobramycin 136
toxicity and safety 7, 138–9, 240, 256, 348–9, 408–10
  viral vectors 101, *102*
trans-acting viral elements 94, *95*, 101, *102*
transcellular absorption route 5, 47, 48, 49, *50*, 210

transferrin
fusion proteins 17–18
targeting tumours/metastasis 77
TransMAG^PEI 80, 81
trehalose 6,6′-dibehenate (TDB) 362, 364
and cryo-/lipoprotection 369–71
enhancing liposome stability 366–8
Trelstare® 19, 195
triggered release 79–80
trimethyl chitosan 31–2, 44, 216, 256
Trojan particles 239
trypanosomiasis 135, 137
tumour necrosis factor 17–18
tumours
and cell-penetrating peptides/DNA mimics 75
intratumoral delivery of nucleic acid-based drugs 79
and naked DNA plasmid delivery 76
receptor over-expression 76
targeted drug delivery 77–9
vasculature 70
Turner syndrome 84–5

vaccines 6, 113
adjuvants 114–17, 142, 362
liposomal delivery 359–60, 371–2
advantages/characteristics of liposomes 360–3
immunomodulators 363–4
liposome instability issues 366–8
physiochemical properties 365–6
stable formulations/freeze-drying 369–71
mucosal absorption
intranasal 35–6
pulmonary 36
nano/microparticle systems
alginate 257
dendrimer 162–3
microparticle release rate 250
polyorthoester 262
virosomal delivery *see* virosomes
see also specific immunotherapy
Vantas® 22
vascular endothelial growth factor (VEGF) 31, 84, 200, 201
vasoactive intestinal inhalation therapy 41
Viadur® 21, 22
viral infection studies 280–4
viral vector delivery (nucleic acids)
background 93–5, 140
cell entry/viral topologies 95
enveloped viruses and membrane fusion 94, 95–7
non-enveloped viruses 94, *96*, 97
delivery and targeting 98–100
expression 94, 100–1
safety issues 101, *102*
and siRNA 274–5, 292
viral technology examples 103
adeno-associated vectors 106–7
adenovirus vectors 105–6
mixed delivery systems 107–8
retroviral and lentiviral vectors 103–5
virosomes, influenza 108, *116*, 377–91
IRIVs as adjuvant and delivery system 377–9
IRIV types *379*
virus and virosome structure *378*
mechanisms of action 379–81
antigen processing *380*
study
cellular immune response 386–8
efficiency of virosome delivery 388–9
humoral immune response 385–6, *387*
methodology 389–91
particle characterization 382, *383*
virosome uptake in HeLa cells 382–5, *386*
viruses *see* viral vector; virosome
Vitravene® 305

water-soluble lipopolymers 74

Zoladex® 21
ZYDIS® tablets 398–9